ELEMENTS OF ORGANIC EVOLUTION

ENCYCLOPAEDIA OF EVOLUTIONARY BIOLOGY SERIES

ELEMENTS OF ORGANIC EVOLUTION

By

Richa Arora

ANMOL PUBLICATIONS PVT. LTD.

NEW DELHI - 110 002 (INDIA)

ANMOL PUBLICATIONS PVT. LTD.
4374/4B, Ansari Road, Daryaganj
New Delhi - 110 002
Phones: 23261597, 23278000
Visit us at: www.anmolpublications.com

Elements of Organic Evolution

First Edition, 2004

ISBN 81-261-1524-6

[The responsibility for the facts stated, conclusions reached and plagiarism, if any, in this book is entirely that of the Author. And the Publisher bears no responsibility for them, whatsoever.]

PRINTED IN INDIA

Published by J.L. Kumar for Anmol Publications Pvt. Ltd., New Delhi and Printed at Mehra Offset Press, Delhi.

Preface

All living organisms share a master plan of structural and functional organization because they arose from a common ancestor long ago by the process of evolution. The tremendous work conducted by scientists in this field has increased our knowledge of evolution many folds. In the limited space, it is not possible to go into the details. Hence, a brief and comprehensive account has been produced in the light of recent work and latest information available from various sources.

The present title is an attempt to present an account of what is known or what is postulated about the life, and the subsequent evolution of the main animal groups. Certain basic scientific facts are included, it is true, but the majority of the contents consist of hypothesis which are never likely to be experimentally verified; and the reader should treat the book as a source of facts, ideas and theories on which thought and discussion can be based.

The present title claims to be a complete, thorough and authentic exposition of the topics dealt with. Emphasis has been laid on introducing basic as well as modern concepts. The language of the book is kept simple in lucid style. The subject matter has been interwoven for keeping up interest and curiosity of readers.

Though care has been taken to verify all passages quoted, but it is hardly likely that, in so large a mass of material, all errors shall have been avoided. The author and the publishers would welcome as a favour any suggestions or corrections submitted by interested readers.

Thanks are due to all friends and colleagues whose continuous inspiration have initiated me to bring out the present work.

There can be no claim to originality except in the manners of treatment and much of the information has been obtained from the books and scientific journals available in the different libraries.

—Author

Chapter—1

The Origin of Life

The Study of Origins

Problems of origin carry a peculiar intellectual status. They are of quite outstanding interest and yet in a sense they are insoluble. It can be argued that as the past will never be repeated it will never be known what it contained. Yet this is one of those anthropocentric attitudes that seem self-evident but may be only a manifestation of our limited view of nature. We insist on our uniqueness, but if nature is full of recurrences then in time it will repeat us.

Interest in origins has increased in recent decades among a wide range of scientists, from astronomers to biochemists. We can here deal only in a very general way with the origin of galaxies and solar systems, though a view about these can hardly be avoided by anyone who attempts to understand man. The subject has expanded vastly, for, among other reasons, widely increased astronomical data and the knowledge of the nature of matter that has come from recent advances in physics. The origin of life has been increasingly closely examined since it was first discussed in its modern form by the biochemists Oparin in 1924 and Haldane in 1929. Understanding of the nature of living processes has made study of their possible origins conceptually more precise, though not necessarily more easy. Knowledge of the possible sequence of events in the early history of the earth, while still very inadequate, gives us new ideas about the conditions in which life may have arisen. At the same time a new range of experiments by chemists has explored the conditions under which organic molecules can be formed.

These studies give us no certain answer as to how life arose, but they make a serious scientific discipline of the subject. As to their practical importance we only have to think of their relevance to the question of the artificial production of life on the one hand and of the possible discovery of life in other worlds on the other. These may perhaps rank as the greatest of all scientific problems. Moreover our knowledge about origins determines our modes of thought and reasoning, and hence all other inquiries. Clearly we cannot fully explore these great matters here but can only try to give an indication of some of the ways in which they are being considered.

The Origin of the Galaxy

If we were to pursue the programme of search for understanding to its limit we should be compelled to consider the origin of the universe. This majestic subject is indeed the concern not only of astronomers but of all of us, and widely differing views are held about it. Astronomers are making important new discoveries every year. Previously unsuspected types of celestial body are being found, for example the quasars and pulsars. Views about the universe are therefore likely to change very much in the future. It may well be that sings of order that we cannot yet discern will revolutionize all knowledge, so that to our descendants it may seem that our thoughts must have been extremely primitive since they lacked appreciation of the fundamental principles of the cosmos. 'The universe is enormous, strange and untouchable; man's technical means and intellect are small and short-lived. Discussing stellar or galactic evolution is a large task made no easier by the lack of astronomical meaning in such commonly used words as history, evolution, birth, life and death of atoms and stars'.

Modern cosmologies assume that the distribution of matter in the universe is statistically homogenous for a 'sufficiently large sample', but the galaxies form local clusters. The universe has no centre and no axis. The spectral lines of distant galaxies show a red shift, which is supposed to be a Doppler effect due to the fact that they are all receding from each other. From such facts Einstein and de Sitter developed a model of the universe as expanding from an infinitely condensed state. In the version of Friedmann the expansion reaches a state of maximum distension and minimum density, and them contracts back to the original state, and so the cycle repeats. The apparent 'age of the universe' would then be the time since the last expansion began. It is interesting that the data suggest an approximately similar age for the earth, for meteorites and for star-clusters—namely $\sim 10^{10}$ years. It may be that this is the age of the current expansive phase of this stupendous cycle. But these are, of course, very complicated matters and there are other versions of steady-state cosmologies according to which new galaxies are continuously being formed by newly created matter.

Recent measurements of the intense cosmic black-body radiation discovered in 1965 provide further evidence that the universe is evolving, rather than steady. The expansion is the same in all directions, confirming that the universe is homogenous (isotropic) and that there was a singularity at its space-time origin. In other words it began with one big bang (Gamow). Physics has no 'explanation' for such a singular event. It is suggested that during the initial 100 seconds there was an extremely high temperature ($\sim 10^9$ K) in which the primordial nucleons were 'produced' and there was thermonuclear fusion to give helium.

Something can be said from modern physics and astronomy about the later processes of evolution of the heavier elements. Stars on the 'main sequence' are born, pass through a series of stages, and die. They are probably first differentiated from clouds of hydrogen in which some non-uniformity develops, producing gravitational forces leading to contraction. The consequent angular momentum and magnetic field stresses produce intense heating until nuclear reactions begin and the hydrogen fuses to form more helium (^4He), carbon (^{12}C), oxygen (^{16}O), and neon (^{20}Ne), and then still heavier elements at higher temperatures. At about $5X10^8$ K there is a catastrophic gravitational collapse and the star first implodes and then explodes as a brilliant supernova. Alternatively if there is no explosion the star burns off, becoming a red giant, which grows gradually fainter. The ultimate fate of these is not known but probably they collapse to form the white dwarf stars, which continue to cool for a very long time. In the recently discovered pulsars, emitting pulses of radiation with periods of a fraction of a second, the gravitational collapse is believed to proceed so far that the electrons have sufficient energy for the inverse beta reaction, electron+proton—neutron+neutrino, to be important. The electrons disappear and the protons are converted to neutrons. With the disappearance of the kinetic pressure of the electrons there is no force to restrain further gravitational collapse. The pulsars are believed to be neutron stars with the incredible density of 10^{14} g/cm^3.

There is also some evidence about the evolution of galaxies. They have various different forms, which can be classified in a way consistent with the hypothesis of an 'evolution from nearly amorphous irregular galaxies to the very open spirals, through progressive stages where the arms become thinner, more nearly circular and tightly wound'. However it is not clear how these changes begin or end, nor what they signify.

Besides stars and galaxies there are objects in the universe that are even less understood. Radio telescopes have identified enormously strong centres emitting 10^{44}-10^{45} erg/as in radio wavelengths alone. Photographs show that associated with some of these there are quasi-stellar images (quasars) with very peculiar spectra including much ultra-violet and emitting 100 times more light than any galaxy. They have large red-shifts and must be very distant, as much as $2x10^9$ light years away, that is approaching the limits of observability. Some of them change their light over a period of a few years or less and these must be very small objects. There is no explanation of what they are or how their enormous energies are generated.

The Origin of the Earth

Currently the most acceptable theory for the origin of planets seems to be that as the sun condensed from a nebular cloud it rotated with increasing speed. This would produce magnetic fields in the ionized gas, canalized along spiral lines as the plasma 'particles' moved outwards. This would produce small solid bodies, planetesimals, later aggregated into planets.

The age of the earth can be estimated fairly precisely from the rate of decay of thorium and uranium to lead to be about 4.5 x 10^9 years. This radioactive decay is the end phase of the process of breakdown of the heaviest elements that were formed during the sun's explosion. In this sense the earth, like the sun, is active and gives out heat, but at a rate 100 million times less than that of the sun. Moreover the earth is becoming less active; that is, it gives out only perhaps one-tenth as much heat as when it was formed.

Composition of Earth

Curiously enough there seems to be less agreement about how the earth reached its present distribution of materials than as to how they were first formed. Indeed, there is considerable uncertainty about the distribution itself at the present day. The earth has a metallic *core* of iron and nickel, mostly molten, but solid a the centre. The diameter of the core is about 6000 km and around it is a shell, the *mantle,* 3000 km thick. This is composed mainly of silicates of iron and magnesium and is 'solid' enough to transmit transverse waves, but plastic, enough to permit slow deformation.

The only part of the earth with which we are directly in contact is a thin *crust.* This consists of lighter silicates and many other elements. It is not a uniform shell, but is much thicker under the continents (20-50 km) than the oceans (7 km.).

The uneven distribution of land and ocean is one of the most conspicuous features that have dominated the history of life. There are various theories as to the origin and evolution of the continents, and knowledge of the subject is developing rapidly. It has been hard to acquire information even about the modern conditions inside our planet, let alone their history. It is also hard for the imagination to grasp the implications of geophysical changes, certainly for the layman. We have to school ourselves to think of such processes as the raising of mountains in the same way as we think of the wrinkling of the peel of a dried orange. Perhaps still harder, we must believe that the continents float and drift in the mantle below the crust, breaking up like icefloes into smaller pieces.

Leaving aside questions of the origin of the crust, its uneven distribution must presumably be a result of the strains produced by movements within. The continents are formed of lighter matter, floating in the denser mantle. It is likely that there was originally only one land mass, which then split into fragments, the continents, which have since drifted apart. Geophysicists and geologists now mostly agree that some such process of continental drift is at work. One question is the energy source for such vast movements. Presumably the heat production from decay of the radioactive elements in the mantle provide the motor. This is supported by the fact that heat transfer is uneven at different parts of the surface of the earth, being greatest along the central parts of the mid-ocean ridges. This suggests that the rocks are turning over, rising at the ridges in the ocean floors, and carried by convection currents in the mantle, passing along the ocean floors to be pressed into trenches at the edges of the continents. Magnetic anomalies and the pattern of earthquakes suggest that material spreading eastwards from the East Pacific Rise has produced oceanic crust material that is now close to the coast of Chile. This would have been generated 60×10^6 years ago and have moved at 4.6 cm/year. Spreading rates from 1.5 to 6.5 cm/year have been computed. Mountain ranges are formed where the ocean floor is being forced under continent, as around the Pacific, or where continental masses are being driven against one another.

The important point for the present discussion is the concept of change in the earth's surface. Geologists have long been aware that there have been periodic movements upwards and downwards of the continents (isostasy), and that these have been of first importance in their influence on the history of life.

The Age of the Earth

If life originated upon the earth it must have been within 4.5×10^9 years ago, which is not very long by astronomical standards. The light now reaching us from some of the more distant objects known left them at about the time the earth was formed. It is estimated that at that time they must have been a million times brighter than our sun. Perhaps they still are; or possibly they ceased to exist long before we see them. In all these questions of origin, however, it is well to remember how recently our knowledge has increased. Estimates of the age of the earth have grown continually since. Archbishop Usher's 4004 B.C. in the seventeenth century. However, 4500 million years is not likely to be far wrong as an estimate of the age of the earth; this figure is based on radioactive decay of thorium and uranium. But of course we may possibly be deceived about this process, just as Kelvin was when he said the earth could not be more than a million years old, an estimate based on what seemed then to be completely sound physical principles of cooling. And that was only in 1890. Kelvin could not anticipate the discovery of radioactive sources of heat within the earth. Incidentally, current estimates of the age of the sun are about 5×10^9 years and of the galaxy, as we have said, $\sim 1.5 \times 10^{10}$ years.

The age determined from the rates of decay of radioactive elements is of course that at which the earth was formed as a spheroidal mass, including the heaviest elements. This must have been the result of some process of condensation such as that already discussed. None of the actual rocks found are as old as this. The oldest are some 3×10^9 years and they are at the centres of the continents. The edges are much younger, some indeed only a few years old, and still growing. In the ocean basins the oldest rocks found are all quite young ($<10^8$ years) because all the present-day ocean floor has been formed in geologically recent time, by spreading from the mid-ocean ridges.

The Origin of Life

The first question on this subject should perhaps be 'Did life originate at all or has it in some sense always existed?' This unfortunately is one of those great questions, like the origin of the universe, that we cannot handle yet. Looking at the problem from the point of view of terrestrial life we may ask four questions.

1. Did life arrive on the planet by migration from some other body?
2. Was life produced on earth by some 'life fore' operating outside the laws controlling matter as defined by physics?

3. Do the laws that control the matter of the universe contain factors, besides the known laws of physics, which dictated the necessity for life to begin (and presumably also to evolve)?
4. Can we show that life may have arisen by the operation of forces known to operate in the terrestrial physical world?

Over all of these also of course lies the question whether life originated once only and on the earth, or once, or more often, elsewhere.

In the present state of knowledge none of these possibilities can be excluded. It has been alleged that micro-organisms are found in meteorites arriving from space. The possibility that these are in fact the result of terrestrial contamination has not been excluded and some hold it impossible that organisms could survive within a meteorite. This question remains to be resolved. The answer to the problem of the origin of life on earth may still be that it came from outer space, but probably most scientists regard this as unlikely.

Similarly it is not excluded that God or some vital force so manipulated the materials of the early earth that they acquired the property we have called information storage and that controlled systems thus began. Once they had begun, it is easier to see how they proceeded to evolve and there is nowadays little doubt that the processes of homeostasis have continued since then without further intervention. The laws of the physical world will adequately account for the behaviour of organisms, provided that their conduct is controlled by the pre-existing order in an information store. It is the origin of the latter that is in question, and it cannot at present be excluded that it was the result of the operation of an outside agency. The uniformation assumptions of science would of course be opposed to such an explanation, but this is not enough to disprove it. The fact that the operation of the agency would seem to be limited in time somehow suggests that some other solution is correct. It certainly alters the emotional appeal of a deity to consider him as intervening only to make the first DNA. Indeed, such an interventionist view may well be thought blasphemous by those who would wish to understand more of the Deity than is possible by thinking only of a being who was simply an initiator.

This brings us to the third possibility, perhaps the most elusive. Since physics has been revolutionized several times in the last few centuries it can hardly be held that we yet 'know its laws'. But we do know something about the operation of some of them, at least within certain limits. The qualification is important because we are no yet able to apply even the laws that we do know to complex systems such as organisms. The question whether other laws that we do know to complex systems such as organisms. The question whether other laws that are yet beyond our ken are in operation has not therefore much meaning.

The view that it is possible to understand the origin and subsequent course of life by the operation of what may be called well-tried physical laws is therefore the only one that can be fully submitted to rational examination. As we shall see, this does not by any means imply that we have sufficient knowledge to compel its complete acceptance. In these matters it is important that we should both humbly acknowledge our ignorance *and* firmly assert our knowledge. The latter is much greater today than even 10 years ago and this provides a further warning that any conclusions reached are provisional and are likely to be changed before long.

THEORIES OF THE ORIGIN OF LIFE

Spontaneous Generation

Of the many theories of the origin of life, perhaps the oldest is the theory of spontaneous generation, according to which even complicated forms of life might arise spontaneously from non-living matter. Thus Aristotle believed that mosquitos and fleas arose from putrefying matter. Tadpoles, worms, and many other small organisms have been supposed to arise from mud. Flies were supposed to be formed from putrefying

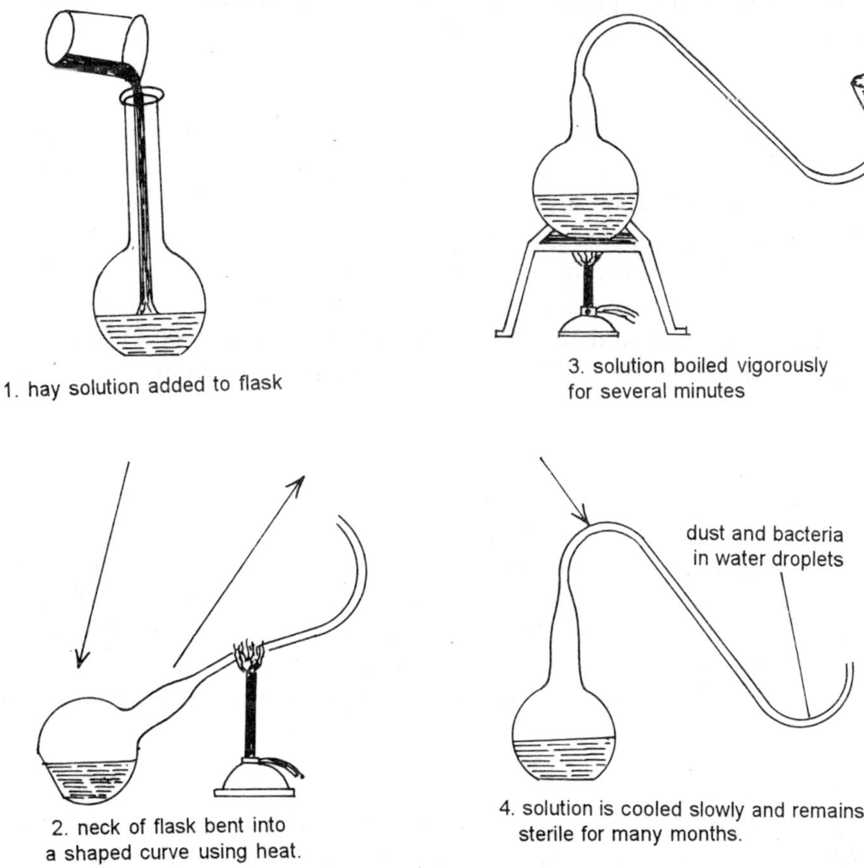

Fig.1.1: Pasteur's experiment with bend-necked flask to disproove Theory of Spontaneous Generation.

flesh. Before the motor age, every child heard that a horsehair, left in water, would transform into a horsehair worm. And meal worms have often been supposed to arise from flour spontaneously. Even such large and complex animals as rats have been supposed to arise spontaneously from nonliving matter. Redi, an Italian physician of the seventeenth century, attacked the theory of spontaneous generation experimentally, and he left it badly damaged. He exposed meat in containers which were covered over with fine mesh cloth. No maggots appeared on the putrefying meat, but flies laid their eggs on the cloth covers, and maggots developed there. It was evident, then, that the maggots which ordinarily appeared in spoiling meat were not spontaneously produced, but were developed from eggs laid in the meant by adult flies. A century later, similar experiments were performed by the Italian priest Spallanzani, who also showed that, if meat were boiled in a sealed container, no organisms developed in it, even if it had been previously infected. This fact was soon applied to the practical problem of food preservation by means of canning.

After the work of Redi and Spallanzani, the theory of spontaneous generation no longer commanded the respect of bioligists. But the discovery of bacteria changed this. Here were organisms simpler than any previously imagined. Their occurrence was practically universal, and it was very difficult to exclude them from any medium suitable for their growth. The possibility that they might be produced spontaneously within every sort of organic medium was most suggestive, and it had many adherents. The famous experiments of Pasteur disproved this completely. Boiled broth was kept in a closed container, with air entering by a capillary tube which was bent to form a trap for solid particles. Thus the broth was freely exposed to oxidation, yet no bacteria appeared in it. Hence it was evident that air-borne bacteria ordinarily infected exposed broth (or other suitable media), and that the bacteria themselves arose only from pre-existing bacteria. This dealt the death blow to the theory of spontaneous generation of complex organisms.

The Cosmozoic Theory

A second theory of the origin of life is the cosmozoic theory, that the original spores of life reached the earth accidentally from some other part of the universe. This theory is completely unsatisfactory for two reasons. First, because of the intense cold, extreme dryness, and the intense radiation of interstellar space, the probability that even the most resistant of living spores could withstand exposure to interstellar space is vanishingly small. Second, the theory does not explain the origin of life at all, but merely changes the scene of origin from the earth to some remote and undefined part of the universe.

Viruses and the Origin of Life

The discovery of the viruses has placed the problem of the origin of life in a new light. The higher forms of life cannot be descended from viruses as the latter are now known, for they are all parasitic, and parasites must always be descended from free-living ancestors. But the viruses show a unique combination of characters of living and nonliving systems. They reproduce and metabolize, as undoubted organism do. That they reproduce is shown by the fact that inoculation of a susceptible organism with a very small amount of virus containing solution gives rise to a heavy infection. That they metabolize is shown by the production of disease symptoms in the host by the products of metabolism. The chemical nature of viruses is that of the nucleoproteins, and it has been suggested that they are "escaped" genes. Like the genes, viruses are ordinarily reproduced without change, but they can mutate as genes do, that is, they can undergo an inheritable change which does not interfere with their capacity for self-reproduction. Such a mutation is detectable by a change in the disease symptoms produced by the virus or by a change in the degree of toxicity. Unlike undoubted organisms, however, the viruses do not respire. But the most striking property which viruses share with nonliving systems is the fact that they can be crystallized and stored indefinitely without loss of infective powers. This was first demonstrated in 1935 by the chemist W.M. Stanley, who succeeded in crystallizing the virus which causes tobacco mosaic disease. The crystals turned out to be a nucleoprotein. Chemical purity is indicated not only by crystallization but by a sharp sedimentation boundary when a virus suspension is ultracentrifuged. Finally, viruses can be broken down to a protein and a nucleic acid, both of which are inactive, then they can be recombined to form infective virus again. And so the viruses appear to be homogenous, or nearly so, in contrast to all undoubted organisms.

Thus viruses are on the border line between the living and the non-living, even though their parasitism cannot be primitive. The possibility cannot be excluded that there may be free-living viruses, for we know the viruses primarily by their effects (disease production), and no one knows what type of effect should be looked for from a free-living virus. The existence of these bodies which are intermediate between the living and the nonliving and which have fairly simple chemical properties suggests the possibility that something like a free-living virus may have been produced by chemical evolution under the influence of the unique conditions which prevailed when the earth was a young planet just cooling down toward a temperature range which could support life. Such a free-living, self-reproducing unit might be regarded as a single gene. Mutation could then lead to formation of gene-aggregates, with differentiation among the members of each aggregate. Such gene-aggregates could be regarded as independently existing chromosomes, and it has been suggested that some of the smallest bacterial represent such a stage in the evolution of life. Further mutations might then lead to the accumulation of metabolites around the chromosomes, and the complex so produced could then be regarded as an exposed nucleus. Bacteria in general are sometimes regarded as representing this level of organization. Next, cytoplasm might be acquired but not separated from the nuclear material, as in the blue-green algae, or in the larger bacteria. Finally, mutation could then result in the formation of typical cells with nucleus and cytoplasm separated by a membrane. Within such cells lie all of the potentialities of higher plants and animals. The possible details of such a chemical evolution of the most elementary forms of life have been particularly well worked out by Oparin, and his account will be followed for the most part below. It may be noted at the outset that this is actually a special case of spontaneous generation, though it is a plausible one because it does not involve direct origin of complicated organisms from nonliving matter.

OPARIN AND CHEMICAL EVOLUTION

The earth may have originated as a fragment from the sun. The probable chemical composition of the earth in the earliest geological periods can be surmised from spectrographic analysis of the sun and of stars in early stages of their physical evolution. All of the elements which enter into the composition of protoplasm were probably present as inorganic compounds. Free nitrogen, hydrogen, and oxygen, gases which now form so large a portion of the earth's atmosphere, probably were present at the very beginning, because they are present as free elements in the sun. But they were probably soon lost to outer space, for it is unlikely that the gravitational pull of the earth is great enough to hold such light elements at the high temperatures which prevailed during the earliest ages of the earth's history. So these elements were probably left only in compounds. A large quantity of the hydrogen and oxygen were probably united as water, but this was certainly for long ages present as superheated steam. The hot vapors would rise toward the cold outer layers of the atmosphere, condense and fall as rain, only to be again converted to steam before striking the earth. Gradually the earth cooled sufficiently to permit the rainfall to strike the earth, then to begin the formation of pools and larger bodies of water. Optimum conditions for solubility and reaction thus existed, and the entire earth was a great crucible for random compound formation and re-formation.

Origin of Organic Compounds

In such a situation, where the most important elements of organic compounds—carbon, nitrogen, oxygen, and hydrogen—are reacting at random with each other and with many other elements, forming countless compounds which in turn react in such combinations as may be possible, it is highly probable that, sooner or later, organic compounds or their precursors would appear. It is known that methane (CH_4), the simplest of organic compound, is present in the atmospheres of some of the cooler stars. And more complex hydrocarbons (compounds of carbon and hydrogen) have been found in meteorites, and so it is certain that they can originate without the intervention of living organisms. These may have been preceded by the compounds of carbon with metals, for such metallic carbides give rise to hydrocarbons when treated with steam. So it is altogether probable that such compounds were formed in abundance while the cooling earth was still much too hot to permit the existence of life. In a similar way, one would expect the formation of ammonia by the reaction of steam and metallic nitrides. Further, cyanogen (CN) and dicyanogen (NC-CN) were probably also formed.

Other known pathways for the abiotic origin of organic compounds include ultraviolet irradiation of formic acid, which yields large organic molecules, including amino acids; the action of cosmic rays upon carbon dioxide and water vapor, which yields organic acids; and the action of lightning upon an atmosphere of methane, hydrogen, ammonia, and water vapor, which yields a mixture of organic compounds, including amino acids. All of these reactions have been proven in the laboratory. Using the last-mentioned method, with electrical sparks discharged into an artificial atmosphere in a sealed flask, Miller obtained a mixture of many amino acids and other organic compounds, some in fairly high yields.

Further reactions of the compounds discussed above have great potentialities. The hydrocarbon are not particularly reactive; however, one hydrogen atom in a hydrocarbon molecule could be readily replaced by chlorine or bromine. The new compound would be highly reactive. It might, for example, be hydrolyzed to form an alcohol and an inorganic acid. The alcohol could, then be oxidized to form the corresponding aldehyde or ketone. These could, in turn, be further oxidized to form organic acids. The simplest of the aldehydes, formaldehyde, is an intermediate compound in the photosynthetic production of sugar. The reaction to produce glucose ($C_6H_{12}O_6$) from formaldehyde (CH_2O) proceeds extremely slowly in the absence of chlorophyll and sunlight. But it does proceed, however slowly, under the influence of sunlight alone, and, given time on a geologic scale, it is quite conceivable that large quantities of sugar might be accumulated. If a double alcohol, or glycol, were formed, then one alcohol group could be oxidized to form an acid, while the other alcohol group could react with ammonia (NH_3) to give water and an amine. The result would be an amino acid, one of the building blocks of the proteins. The simplest possible amino acid is glycine, $CH_2.NH_2.COOH$.

These compounds can enter into reactions which lead to compounds intimately associated with protoplasm. Sugars can be polymerized to form starches, glycogen, and cellulose. The essential precursors of fats are long-chain hydrocarbons in which an end member has been oxidized to form an acid, and glycerine. Glycerine is simply a three-carbon chain in which one hydrogen on each carbon has been replaced by an hydroxyl group ($CH_2OH.CHOH.CH_2OH$). Now if each hydroxyl group reacts with a long-chain organic acid, the resulting compound is a fat molecule. But the most significant possibility is held by the amino acids, for these can react with one another to form aggregates of great molecular weight, the proteins. The acid group of one amino acid reacts with the amino group of another in what amounts to a salt-forming reaction. This is called the *peptide linkage*. As it always leaves an acid radical exposed on one of the reacting amino acids and an amino group on the other, the reaction is subject to indefinite repetition, and thus leads to the great molecular weights of the proteins.

Colloids, Coacervates, and Individuality

By the time the earth had cooled sufficiently to permit the formation of permanent bodies of water, there was probably a large amount of organic material of great variety, and the variety was continually increasing through the natural experiment of reaction between whatever substances could react and happened to be brought together. Many of these large organic compounds would tend to form colloidal solutions in water. Wherever such colloidal particles included electrically active groups, as all proteins do in abundance, water molecules would tend to become oriented around the surface of such a particle. If colloidal droplets of opposite electrical charges should be mixed, then they would be mutually precipitated to form droplets of a complex mixture called a coacervate. Such coacervates would adsorb water on their surfaces to form a sort of membrane, which would thus establish the beginnings of individuality.

Autocatalytic Systems, Genes, and Viruses

These large, complex, colloidal aggregates, the coacervates, continue to undergo random chemical reactions in the course of which still larger aggregates may be formed, or the existing ones may be broken down. It is most probable that some of the systems thus formed would be enzymatic in character, that is, they would tend to increase, to catalyze, the rates of specific reactions. Many enzymes catalyze the reactions of substances unrelated to the enzyme itself. Thus lipase will break down fats to glycerine and long-chain acids, or it will catalyze the synthesis of fats under appropriate conditions. But there are catalysts known which, given a suitable substrate, tend to cause the production of more of the catalyst itself. The duplication of the chromosomes is an example *par excellence*. Such enzymes are said to be autocatalytic. If an autocatalytic substance were to be formed in the primordial seas, it would be expected to increase in quantity at the expense of those organic compounds which lack this property. And thus is the characteristic of self-reproduction introduced. Compounds possessing this property might well be regarded as free genes. We need only add the characteristic of mutability, that is, the capacity for undergoing changes which are reproduced without interfering with the autocatalytic properties of the molecule, and we have the most fundamental characteristics of the hereditary units, the genes. Such a structure would be comparable to a free-living virus, and it is probable that, like the viruses of today, these primitive anlagen of life were nucleoproteins. These nucleoproteins would utilize the complex organic compounds of their environment for the auto-catalytic synthesis of more nucleoprotein, that is, for reproduction. Calvin has shown that even in the pre-organismic stages, a sort of natural selection, based upon thermodynamic principles, favors the events described.

Origin of Bacteria

As larger self-reproducing units were formed, mutation might result in the different parts of each aggregate becoming differentiated. Or the same result might be obtained by the coalescence of originally different "genes." At this stage, the structure would resemble some of the smallest known bacteria, and might properly be called an organism. The organisms dealt with up to this point would have to be completely heterotrophic, that is, dependent upon complex food materials present in their environment, and "feeding" would occur only by absorption. The fact that viruses can metabolize and reproduce only within the protoplasm of higher organisms indicates that they are also heterotrophic. They can utilize proteins, starches, fats, and vitamins already preset, but they cannot synthesize these from amino acids, sugars, and other simpler organic precursors. On the other hand, all of the undoubted organisms are able to synthesize at least some of their required foods from simpler compounds. Green plants can synthesize their required foods from the elements. Animals cannot do this, but they can synthesize proteins from amino acids, complex carbohydrates from sugars, and some of the vitamins from simpler precursors. These syntheses are possible because the organisms possess enzymes which are specific for the necessary reactions. But these reactions are frequently quite complex, requiring a long series of steps with a different enzyme controlling each step. Thus it has been shown that the synthesis of arginine, an amino acid, by the mold *Neurospora* requires at least seven different genically controlled enzymes. Each of these is useless without all of the others. That such "useless" genes should be preserved by natural selection is incomprehensible. But the probability that all of them should appear simultaneously, and so become useful, is equally incomprehensible.

Horowitz has devised a very clever solution to this dilemma, namely that the enzymes might have been originally acquired in a sequence opposite to that in which they are used by existing organisms. Thus a primitive organism requires a substance *A*, which is abundantly present in its environment, as are also substances *B* and *C*, from which *A* could be synthesized in the presence of an appropriate enzyme. As long as *A* remains abundantly present in the environment, presence or absence of the enzyme can have no bearing on survival. But eventually the food requirements of the growing population of primitive organisms may be expected to outstrip the stock pile of organic compounds which was built up during the abiotic ages. Thus when *A* becomes rare in the environment, those organisms which have the enzyme for the production of *A* from *B* and *C* will have a selective advantage, and will replace the original type. Such *preadaptation* is a common phenomenon in evolution. The enzyme has now become a permanent part of the organism's biological equipment. But *B* might also be synthesizable from *D* and *E* in the presence of the necessary enzyme. Then when *B* becomes scarce, the possession of this enzyme, acquired by mutation, will also have selective advantage. There is no theoretical reason why such a process should not continue until an organism had acquired the ability to synthesize all of its requirements from the elements. This has happened in the case of the green plants. Such an organism is said to be autotrophic.

Diversification of the Bacteria

It seems probable, then, that the primitive free-living viruses and bacteria gradually used up the available supply of proteins and other complex compounds in the environment, and that they simultaneously, step by step, developed the systems of enzymes necessary for the biosynthesis of the same compounds. Or it may be that only the primitive bacteria were sufficiently complex to permit this development, the viruses remaining completely dependent upon preformed compounds. If so, this would account for their obligate parasitism, for only within the protoplasm of undoubted organisms could an environment be found to contain all necessary food substances in the completely elaborated form. The evolution of the bacteria made parasitism possible; the impoverishment of the earth's supply of complex organic compounds made it desirable for such organisms as could not synthesize their food requirements at least in part.

Respiratory Systems and Photosynthesis

The most primitive organisms, living in an environment which included an abundance of the most complex substances which the organisms might require, must have hand a metabolism which was mainly catabolic; that is, it consisted of the breaking down of complex compounds to release the energy stored in them. But, as the environmental supply of high-energy compounds was reduced, the development of a new type of metabolism became necessary. This is anabolic metabolism, whereby complex compounds are built up from simpler ones, and energy thereby stored. Expenditure of energy by the organism is required, and so the evolving bacteria explored possible sources of energy, that is, systems of respiration. Several types of respiratory reactions, and correlated nutritive systems, have been exploited by bacteria. These may be classified an heterotrophic and autotrophic systems. The heterotrophic bacteria are generally parasitic, deriving their energy by oxidation of the carbohydrates or other organic compounds of their hosts. Or they may be *saprophytic,* that is, they live by absorption of dissolved organic matter from their environment, much like the primitive ones in the scheme of Oparin. On the other hand, the autotrohic bacteria derive energy from chemical reactions involving simple inorganic compounds. They are thus independent of external sources of high-energy compounds. There are three principal groups of such *chemotrophic* bacteria, the sulfur bacteria, the nitrifying bacteria, and the iron bacteria. Within each group, there are many different species, and several different respiratory reactions. A few examples may be given:

$$2\ H_2S + O_2 \rightarrow 2\ H_2O + 2S + 81{,}600 \text{ calories per mole}$$

$$S + 2\ H_2O + 3\ O_2 \rightarrow 2\ H_2SO_4 + 214{,}000 \text{ calories}$$

$$Na_2S_2O_3 + 2\ O_2 \rightarrow Na_2SO_4 + S + 75{,}000 \text{ calories}$$

$$2\ NO_2 + O_2 \rightarrow 2\ NO_3 + 46{,}000 \text{ calories}$$

$$2NH_3 + 3\ O_2 \rightarrow 2\ NO_2 + 2\ H_2O + 156{,}000 \text{ calories}$$

Iron bacteria oxidize ferrous compounds to ferric compounds with a much lower energy yield than is indicated for the above reactions. But the best of these is vastly inferior to photosynthesis, the autotrophic mechanism of all green plants, and perhaps the real basis for the evolution of the higher plants and animals, for this is the mechanism which has made it possible for the world of life to tap the great reservoir of radiant energy from the sun. The photosynthetic reaction may be summarized as follows:

$$6\ CO_2 + 6\ H_2O + 677{,}000 \text{ calories} \rightarrow C_6H_{12}O_6 + 6\ O_2$$

This large energy reserve is subsequently released by the sugar metabolism of the organism. Photosynthesis, then, for which the chlorophyll of green plants is the catalyst not only vastly increases the energy potentially available to organisms, but it also releases oxygen from its compounds, thus making possible the oxygen respiratory systems of animals and of some bacteria. Photosynthesis is not generally known among the bacteria, but some of the sulfur bacteria contain a green pigment which is capable of absorbing sunlight, and this makes some contribution to the metabolism of these bacteria.

Oparin was of the opinion that this chemical evolution of life could have occurred only once, because, it would require a sterile environment. That is, it could not occur in a world already inhabited by organisms ready to seize upon any beginnings of organic compounds to utilize them as food. But, as Plunkett pointed out ten years earlier, this is an assumption for the proof of which no data are available, and it is entirely possible that life is in process of origin on earth at the present time. On the other hand, Plunkett also acknowledged that there is no evidence that this is true.

EVOLUTION OF THE BACTERIAL CELL

The problems of the development of the nucleus, mitotic division (cell reproduction), and the evolution of sex are by no means clear. It has already been mentioned that the most primitive bacteria may be more or less equivalent to a single chromosome. Further evolution certainly involved the accumulation of a quantity of accessory material. Some authorities have described the bacteria as cells in which cytoplasmic and nuclear materials are intermingled, while others regard the bacteria cell as an isolated nucleus without cytoplasm. Perhaps the observable facts can be adequately described in either way until much more critical data are obtained. In either case, the evolution of the bacterial must have involved an increase in the quantity of the genic material, and increasing differentiation within it. Study of the mechanics of cell division among the bacteria is hampered by the small size of most bacterial cells. It has long been generally believed that simple fission, without qualitative division of the genes, is the rule among bacteria. This requires a genetic system in which rather few equivalent genes are scattered throughout the cell, so that division need not result in genetically unlike offspring. Thus occasional mutation and the immediate action of natural selection on the mutants would be the only means of evolution.

Reports of indications of mitotic divisions, that is, qualitatively equal divisions of chromosomes, have been published for many bacteria. Some of the published photographs are quite convincing, and the possibility is open that mitotic division, so fundamental for all of the higher plants and animals, may have been first evolved among the bacteria. Genetic evidence also indicates serial arrangement of genes on bacterial chromosomes. Nothing even suggestive of sexual reproduction, the formation and fusion of gametes, has every been observed among the bacteria, but Lederburg has obtained genetic evidence that it does occur, because recombination of characters occur when different mutant strains of the same species are "crossed." And so it may be that even sexual reproduction was first developed among the bacteria. If so, this is certainly among the most important events in the whole story of evolution, for sexual reproduction makes possible the accumulation of variability in the heterozygous state (that is, the two genes of a pair are different, with the possibility that one may mask the effect of the other). Also, sexual reproduction makes possible the reshuffling of the genes already present and the testing of different combinations by natural selection. The result is a much greater probability of the achievement of good adaptation, particularly under changing conditions, than could otherwise be obtained.

Bacteria and Phylogeny

The phylogenetic relationships of the bacteria are by no means clear, primarily because of the great difficulties of taxonomic studies on the bacteria. Figure illustrates typical bacterial morphology. Two major alternative theories of the phylogenetic sequence within the bacteria have been presented. The first of these maintains that the autotrophic bacteria are the more primitive ones, and that these have given rise, through a series of intermediates, to the heterotrophic bacteria. The iron bacteria have sometimes been regarded as the most primitive of all, perhaps because of their low energy output. Adherents of this theory believe that the most primitive bacteria must have been autotrophic because there could have been no source of organic compounds for a heterotrophic organism. This theory was generally held before the publication of Oparin's work. But, as this demonstrated the plausibility of the thesis that organic material might have been accumulated by a slow process of chemical evolution long before the advent of life, the theory that the hetcrorophic bacteria maybe more primitive, with the autotrophs derived from them, gained prestige. According to this idea, the several chemosynthetic mechanisms represent a natural trial-and-terror experiment, resulting in a superior new energy source which culminated in the development of photosynthesis.

The knowledge of the relationship of the bacteria to higher organisms is also unsatisfactory, but some indications of possible relationships are at hand. As has been mentioned above, some of the sulfur bacteria contain a green pigment, *bacteriochlorophyll* or *bacteriochlorin*. This pigment is chemically quite similar to chlorophyll, and it has been shown by van Niel to have photosynthetic activity. Hence there is a reasonable probability that the blue-green algae, the most primitive of the green plants, have been derived from the sulfur bacteria. Some other possibilities have also been suggested. The spirochaetes have a type of motility which suggests that of some of the Protozoa, but they do not have a well-defined nucleus. They have been variously classified with the bacteria and with the Protozoa. Although their general biology appears to tie them in more closely with the bacteria, it is possible that they are in fact a transitional group. Finally, the slime bacteria show some similarities to the slime molds, and it has been suggested that this is based upon phylogenetic relationship. However, the cells of slime bacteria live in a secreted nonliving slime, while slime molds are multinucleate masses of protoplasm. Any or none of these hypotheses may be correct. There is no direct evidence available, and perhaps there never will be, for these events must have preceded by long ages the formation of the oldest rocks which bear fossils of any value for study. Further, the lowest plants, the algae and fungi, include many groups of quite diverse nature, their primitiveness being their principal common character. While some of these may have had a common origin from the bacteria, it is possible that each of these phyla of primitive plants may have arisen independently from a different bacterial ancestor.

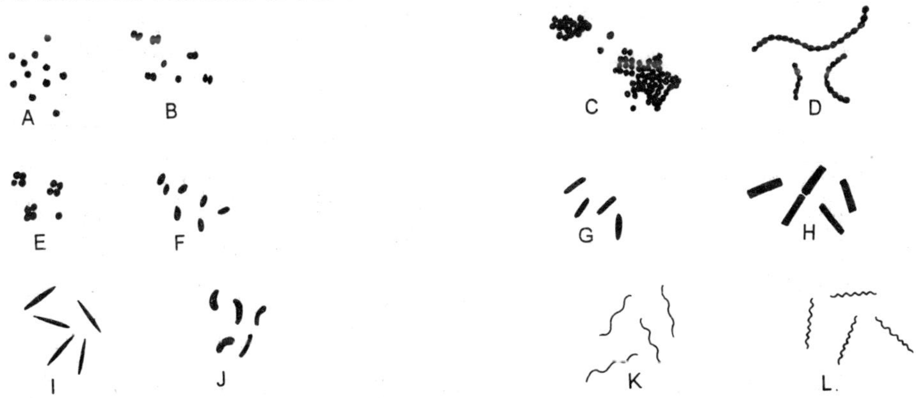

Fig. 1.2: Morphologic types of bacteria. A, micrococcus; B, diplococcus; C, staphylococcus; D, streptococcus; E, sarcina; F, coccobacilli; G and H, commonly occurring rod forms; I, fusiform bacilli; J, curved rods or vibrios; K and L, spiral forms.

SUBKINGDOM THALLOPHYTA

The plant phyla which formerly were grouped together as a single phylum Thallophyta are those which are most closely related to the bacteria. The Thallophyta are now treated as a subkingdom, including all plants which reproduce without the formation of an embryo within the ovary of the maternal plant. This subkingdom includes ten phyla ranging in complexity from the bacteria to algae nearly as complex as the simpler vascular plants. Of these, three phyla, the Schizomycophyta or bacteria; the Myxomycophyta or slime molds; and the Eumycophyta or true molds and yeasts comprise the old group Fungi. The relationships of these groups are not at all clear, but the latter two appear to be terminal groups, that is, they have not given rise to others. As was already indicated, the slime molds may have been derived from the slime bacteria. But the slime molds move by ameboid movement. They reproduce by means of flagellate, ameboid swarm cells. As these swarm spores fuse in pairs, a type of sexual reproduction is represented. Many mycologists feel that the Myxomycophyta are more closely similar to the Protozoa than to any other group, and they have even been classified as Protozoa under the name Mycetozoa. Those who support this viewpoint regard them as at least transitional between the two kingdoms. The Eumycophyta are similarly obscure as to their origin. In structure and function, they show many parallels to the green algae, and so they have been thought to be derived from this group. But their zoospores resemble those of the slime molds and show ameboid movement. Hence it is possible that the Myxomycophyta have given rise to the Eumycophyta. These flagellated zoospores closely resemble some Protozoa, and the viewpoint that the Eumycophyta may be of Protozoan origin is currently gaining much support.

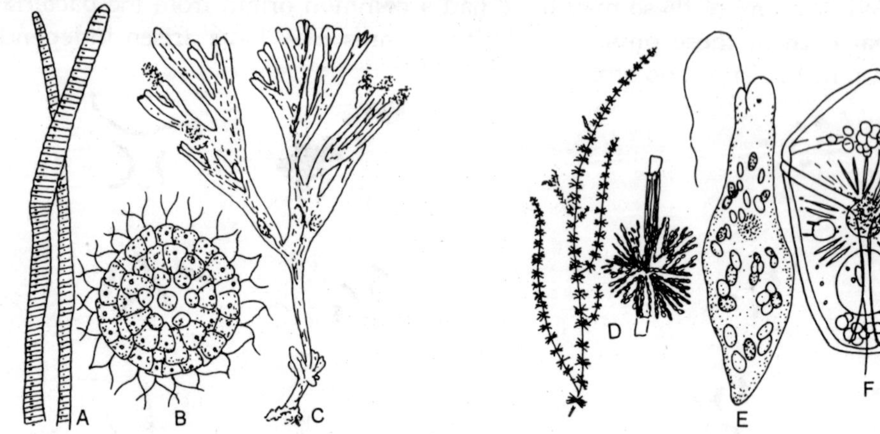

Fig. 1.3: Representative algae of several phyla. *A, Oscillatoria*, a blue-green alga; *B, Synura*, a golden-brown alga; *C, Fucus*, a brown alga; *D, Batrachospermum*, a red alga, showing growth habit and details of one whorl; *E, Euglena; F, Amphidinium*, a dinoflagellate.

The remainder of the old phylum Thalophyta comprises seven phyla, the various types of algae, ranging in complexity from the extremely simple, unicellular blue-green algae to the green algae of large size and complexity only a little less than that of the vascular plants. Most of these phyla are of uncertain origin and have given rise to no further groups, hence they need not be discussed here in detail, even though some of them have attained a considerable degree of specialization. These are the phylum Chrysophyta, including yellow-green algae, golden-brown algae, and diatoms; the phylum Pyrrophyta, including cryptomonads and dinoflagellates; the phylum Phaeophyta or brown algae which Dillon and Hutner believe to be of especial importance for the origin of animals; and the phylum Rhodophyta or red algae. Three more phyla of algae (this term means seaweeds, and is based upon the larger members of this diverse group) are of more especial interest for the present discussion: these are the Cyanophyta or blue-green algae; the Euglenophyta, including *Euglena* and its allies; and the Chlorophyta or green algae.

The Cyanophyta

The Cyanophyta or blue-green algae are undoubtedly the most primitive of all green plants. The individual plant is always unicellular, but these may form colonies of moderate size. There is, however, no differentiation within such colonies. Cytoplasm is much more abundant than in the bacterial cell, but there is still no morphological separation of nucleus and cytoplasm. No evidence of sexual processes has ever been observed in the blue-green algae, reproduction apparently occurring exclusively by simple fission. In all other green plants, the chlorophyll is confined to cell organelles, the plastids, but that of the blue-green algae is diffusely spared throughout the outer part of the cell. Their bluish color is caused by a blue pigment, phycocyanin, and a red pigment chemically similar to that of the red algae may also be present. Unlike most other algae, they lack flagella, and are in fact generally non-motile. The oldest known fossils, from rocks on the order of a billion years old, are remains of blue-green algae, and their descendants of today are regarded as little-changed survivors from that remote time. Whether other groups of algae have been derived from blue-green ancestors is unknown, but the fact that the blue-green algae share with more advanced groups so fundamental a trait as the possession of chlorophyll at least indicates that this is a possibility.

THE EUGLENOPHYTA AND THE SEPARATION OF THE KINGDOMS

The Euglenophyta, typified by the common *Euglena* of elementary biology laboratories, show many advances over the blue-green algae. They have a definitely organized nucleus which is separated from the cytoplasm by a membrane. The chlorophyll is no longer free in the cytoplasm, but rather it is concentrated in numerous ovoid bodies, the chloroplasts. Further, the color is grass-green rather than blue-green. Unlike the green algae, the Euglenophyta are not provided with a cellulose cell wall. The cells are provided with one or two flagella, and they are active swimmers. There is a gullet at the anterior end, yet it appears that *Euglena* is autotrophic. Near the gullet there is a red pigmented eyespot, which seems to be sensitive to light. Reproduction is always by simple mitotic division, although sexual reproduction has been reported for one genus (*Scytomonas*).

The Euglenophyta are also described in zoological works as the order Euglenoidina of the class Flagellata and the phylum Protozoa. This class includes representatives of most of the unicellular algal phyla as well as some indisputably animal flagellates. The group as a whole shows a curious mixture of plant and animal characteristics.

But before taking up this subject, it may be well to consider what are the typical differences between plants and animals. Generally speaking, the mode of life of animals is aggressive, while that of plants is passive. Animals are heterotrophic, eating other organisms to obtain the complex organic compounds which they require as foods. Plants, on the other hand, are generally autotrophic, being able to synthesize all of their food requirements from the elements. But there are exceptions in both kingdoms. Many of the higher plants, such as the sundew (*Drosera*), have developed mechanisms for the capture and digestion of insects, and many animals are saprozoic, that is they absorb decaying organic matter from their environment. In plants, differentiation of organs is predominantly external, while in animals it is predominantly internal. In plants, growing tissue, the meristem, is present at all stages of the life cycle, whereas most animals have a definitely limited growth. Finally, plants are generally sedentary, and the individual cells are surrounded by a rigid cellulose wall, while such rigid walls are generally lacking on animal cells, and the animal moves about freely in its environment.

As measured against such a group of criteria, *Euglena* and its allies are difficult to place. *Euglena* itself is supplied with an abundance of chlorophyll, yet it can be raised on completely inorganic media only with difficulty. Traces of amino acids or peptones facilitate culture. Nonetheless, there is no evidence that *Euglena* ever ingests other organisms; it seems more probable that its normal nutrition is predominantly holophytic (by photosynthesis) with a supplement being obtained saprozoically. Some other flagellates, however, are completely holophytic. Others lack chlorophyll, and these of course cannot be autotrophic. Some of them are entirely saprozoic, but some ingest other organisms in typical protozoan fashion and so may be said to be holozoic. Thus the whole range of nutritional possibilities occurs within a single group, and opposite extremes may occur within a single genus. Stored food in plants is ordinarily starch, while in animals it is either glycogen (similar to starch) or fat. In the euglenoids, it is paramylum, a carbohydrate different from both starch and glycogen. The embryological criteria of external or internal organ formation and presence or absence of a continuously growing meristem obviously have no applicability to unicellular organisms. As might be expected of an animal, *Euglena* moves freely in its environment, but its near relatives include sedentary forms. Finally, while Euglenoids lack a cellulose cell wall, they do have a pellicle which in some species is quite rigid.

One result of this mixture of plant and animal characteristics among the unicellular, flagellate organisms is a much confused area in taxonomy. Some biologists have treated the whole array as plans, a procedure which makes it necessary to treat as plants such organisms as the trypanosomes, blood parasites of vertebrates which do not show any plant-like characteristics. Others have tried to designate some forms as plants and others as animals on the basis of the above criteria or similar ones. This involves one in the absurd situation of assigning different members of the same genus to different kingdoms in some instances. Often the whole series of algae and Protozoa have been lumped together as a single kingdom Protista, with only the Metazoa left in the Animal Kingdom and only the vascular plants and the bryophytes left in the Plant Kingdom. This is again unsatisfactory because the higher algae are obviously much more closely related to the vascular plants than to many of the Protozoa, and conversely the animal nature of many of the Protozoa, such as the ciliates, is not open to doubt.

Copeland has made a very radical suggestion, but with much good reason. He suggests that the living world be divided into four kingdoms. Kingdom I, the Mychota, would include the bacteria, and the blue-green algae, that is, all organisms in which the nuclear-cytoplasmic differentiation is not complete. Kingdom II, the Protoctista, would include most of the algal phyla, the fungi, and Protozoa. Thus the kingdom Protoctista includes all of those primitive organisms from which higher plants and animals may have arisen, as well as their relatives which have given rise to no further groups. The kingdom Plantae is thus restricted to green algae, vascular plants, and bryophytes, while the kingdom Animalia is similarly restricted to the Metazoa. While there is much to justify this idea, it presents some serious difficulties. First, it is subject to the same criticism as is the Protista, that the extreme members of the Protoctista are jut as clear-cut plants and animals as are the vascular plants and Metazoa, respectively. This system also substitutes three areas of confusion for one in the conventional system. For it would, under Copeland's system, be necessary to decide whether an organism were a mychotan or a protoctistan; a protoctistan or a plant; or a protoctistan or an animal.

It seems altogether probable that this confusion exists because the organisms concerned are closely related to the actual common ancestors of the higher plants and animals ancestors which need not be thought of as either plant or animal, though they are usually assigned to the Plant Kingdom. The living algae and Protozoa then represent various stages in the evolution of the characteristic differences between the kingdoms, together with many specific adaptations to the unicellular (or better, acellular) grade of construction. There is, theoretically, no reason why there should be a sharp line of separation between those developing along plant-like lines and those developing along animal-like lines. Indeed, this would be in conflict with the idea of origin by evolution. This mixture of plant and animal characters among microorganisms has led Dillon to recommend that the entire world of life be included in one kingdom, Plantae. Perhaps the most important feature of *Euglena* and its allies is this intermediacy between the kingdoms, suggesting as it does the probability that the euglenoids may be primitive organisms, fairly close to the stem group from which both plants and animals have come.

THE CHLOROPHYTA

The Chlorophyta or green algae are an extra ordinarily varied group, the simplest members of which are unicellular, but there is a definite separation of nucleus and cytoplasm and the chlorophyll is contained in a single plastid. They may show considerable specialization of different cells in the multicellular species, and the higher green algae may attain large size. While the more primitive species reproduce by simple fission, sexual reproduction and alternation of generations are well developed in the phylum. The green algae appear to be on or near the main line of evolution leading to the higher plants, and hence the great interest which attaches to this phylum.

Chlamydomonas and the Origin of Sex

Sex probably originated in an unknown green alga, which may have resembled the living *Chlamydomonas* (Figure 1.4). Each plant consists of a single cell. It has a well-defined nucleus and a single large chloroplast. It swims by means of two flagella which are located at the anterior end of the cell. The cell is protected by a heavy cellulose wall. The plant may reproduce by simply dividing to form two, four, or eight *zoospores* (so called because they are actively swimming cells, like an animal). These zoospores are then released by the dissolution of the cell wall, and each swims away, an independent plant, like the parent in every respect except size. This difference is soon bridged by growth. However, sexual reproduction may also occur, for the parent plant may divide to form eight, sixteen, or thirty-two gametes, cells which resemble the zoospores and the adults, except that they are much smaller. Like the zoospores, these gametes are released into the water, where those from different parent cells unite in pairs to form zygotes. The zygote forms a thick wall about itself and remains quiescent for a time. It is in this highly resistant encysted condition that the plant survives unfavorable conditions such as the drying of ponds. In time, the zygote again becomes active. It then undergoes two divisions, the meiotic or reduction divisions, with the production of four zoospores, which are released to form adult algae.

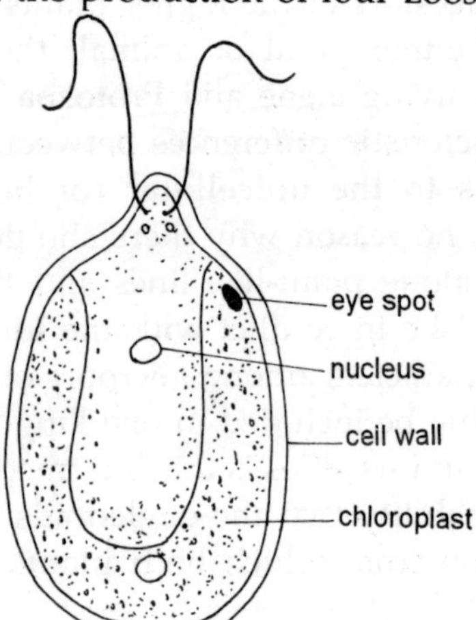

Fig. 1.4: *Chlamydomonas.*

Chlamydomonas appears to be very close to the origin of sex, and may afford some insight into that problem. It may be noted that most of the life cycle is passed with only the haploid number of chromosomes, for the reduction divisions proceed immediately when the zygote becomes active. Haploidy, that is single representation of each type of chromosome or genetic factor, was undoubtedly the normal situation for all organisms before the origin of sex, and it is still the normal thing for organisms which do not reproduce sexually. Diploidy is a necessary consequence of sexual reproduction, for the union of two gametes can have no other result. Gametes could not be reduced below the haploid condition without qualitative loss of genetic material. It appears that in these organisms, so close to exclusively haploid ancestors, meiosis serves primarily to restore the ordinary, physiological, haploid chromosome number; while diploidy is introduced as a temporary concomitant of a "new" method of reproduction. Why sexual reproduction should ever have been developed under such circumstances is not clear. A "hunger theory of sex" has been proposed, according to which the gametes are simply undersized spores, individually lacking the food and energy necessary to complete development. They therefore pool their resources by means of two-by-two fusions, and thus each zygote obtains a sufficient supply of the materials necessary for development. This theory would be more satisfying were it not for the fact that the four zoospores produced by the zygote are also very small cells, yet they complete their development satisfactorily without further fusions. Whatever the original stimulus to sexual reproduction may have been, the great selective advantages which have made it so nearly a universal property of all higher plants and animals are clear. Sexual reproduction causes a relatively rapid reshuffling of the various possible re-combinations of characters within a species, with the possibility that the most favorable combinations may be formed and tested by natural selection. Further, diploidy, which results from sexual reproduction, makes it possible to accumulate a store of genetic variability in the heterozygous state. Diploidy may also be physiologically advantageous because of the increased production of nuclear enzymes, or in other less obvious ways.

The degree of sexual differentiation varies greatly in different species of *Chlamydomonas*. The adults are morphologically identical, and cannot be sexed by inspection. In some isogamous species, the gametes are also morphologically identical. Yet they are physiologically differentiated, for fertilization is possible between some pairs of clones (asexually produced) descendants of a single cell) but not between others. The male-female alternative of higher plants and animals is not present here, for as many as eight mating types (sexes?) have been identified in some species of *Chlamydomonas*. Moevus believes that all of these are base upon varying concentrations of the isomeres of a single substance, namely the *cis*-and-*trans*-forms of dimethyl crocetin (a fairly simple hydrocarbon, which is, curiously, based upon isoprene, the unit of structure of synthetic rubber). In other species, both macrogametes (egg-like) and microgametes (sperm-like) are produced, both of these being flagellate and actively motile. In some of these species, all possible types of fertilization (large—large, large—small, and small—small) occur, while in others, one macrogamete and one microgamete always form a pair. In such cases, one speaks of heterogamy. Finally, there are species in which the macrogamete is nonflagellate and must be sought by the flagellate microgametes, thus approximating the condition of egg and sperm in the higher animals, or ovule and pollen in the higher plants (oogamy).

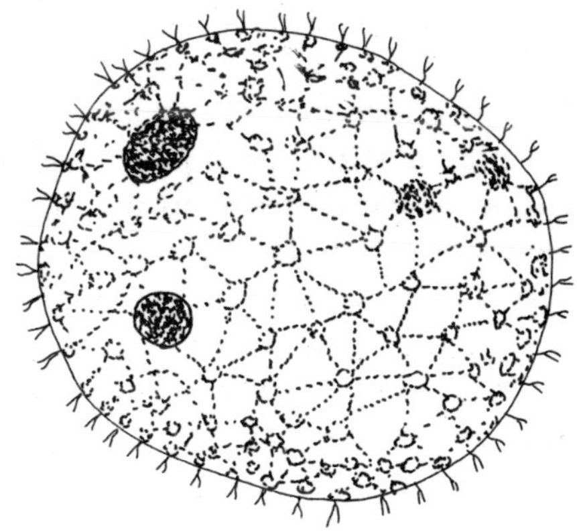

Fig. 1. 5: *Volvox*

Closely related to *Chlamydomonas* is the family Volvocidae, known to all students of elementary biology as the standard example of an evolutionary trend toward increasing complexity of colonies. The simplest members of this family (*Gonium*) consist of four to sixteen or thirty-two identical cells, any of which can form gametes. But these are advanced beyond *Chlamydomonas* in that the gametes are always morphologically different from the vegetative cells. The most specialized member of the family, *Volvox,* comprises as many as 40,000 cells arranged as a hollow sphere (Figure 1.5). Most of these cells are purely vegetative and substantially identical. The sex cells are localized in antheridia (sperm) and oogonia (eggs). The eggs of *Volvox* are always large, nonmotile cells, and the sperm are always actively swimming flagellate cells which seek out the egg. As both of these are produced by a single colony, the colony may be regarded as hermaphroditic, a condition which is very much more common in plants and in lower organisms generally than it is among higher animals. It may be, therefore, that hermaphroditism is primitive.

Multicellular Individuals

In another line of descent, which may be typified by *Ulothrix, Draparnaldia,* and *Ulva,* the green algae have developed not merely colonies of substantially independent organisms, but multicellular individuals, the various cells of which are interdependent. These are the filamentous algae, the typical "seaweeds" of laymen, and it is from this group of algae that the bryophytes and vascular plants appear to have developed. *Ulothrix* has the form of a simple, unbranched, multicellular filament. The basal cell is specialized as a holdfast to anchor the plant to a rock or other substrate. Cell divisions occur, but, whereas this results in asexual reproduction in the unicellular algae, in *Ulothrix* and its allies mitosis results in growth without reproduction. Reproduction may occur either sexually or asexually. Asexual reproduction occurs by fragmentation of the plants with subsequent regeneration by each fragment, or it occurs by the formation of zoospores from the vegetative cells of the plant. These zoospores are not unlike those of *Chlamydomonas,* except that they have four flagella instead of two. Each zoospores develops into a plant like the parent plant. But any vegetative cell of the plant may also give rise to gamete. These are smaller than the zoospores, and they have only two flagella. All of the gametes are identical morphologically, but physiologically there must be a sexual differentiation, for zygotes are always formed by the union of two gametes from different plants. After a quiescent period, the zygote undergoes the two meiotic divisions, forming four zoospores which then develop into the adult plants. Thus most of the life cycle of *Ulothrix* is passed with only the haploid chromosome complement, while the zygote alone is diploid.

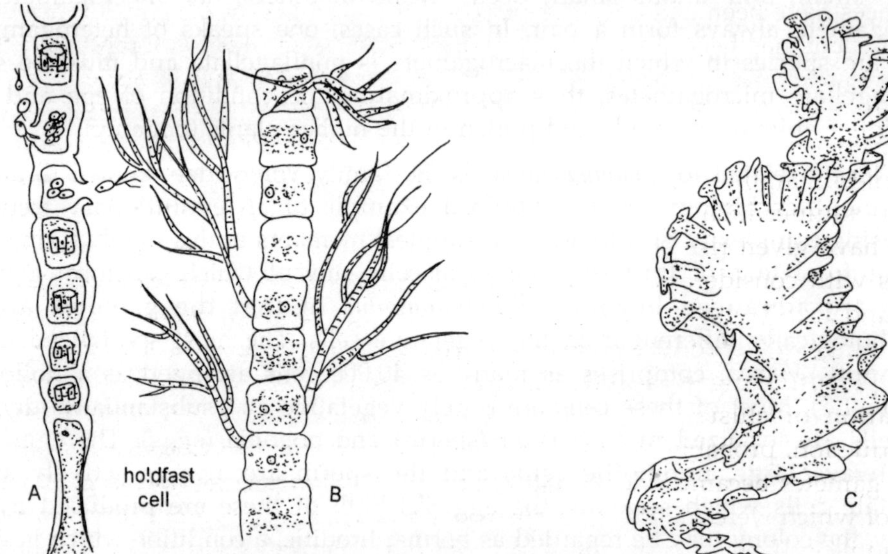

Fig. 1.6: *Ulothrix* (A), *Draparnaldia* (B), and *Ulva* (C).

Draparnaldia is advanced over *Ulothrix* principally by the more complex development of the vegetative body. While the latter is composed only of simple, unbranched filaments, the former has a major, basal filament from which many branches arise, and from these there are secondary branches. Reproduction again is accomplished either by zoospores or by isogametes.

Ulva, the common sea lettuce, is advanced over the previously discussed algae in several ways. The vegetative body forms large, leaf-like sheets which, for the first time, are more than one cell thick. Reproduction is again either by zoospores or by isogametes. But the is a significant development in the life cycle. In all of the algae discussed above, the reduction divisions occur in the zygote, with the result that in the whole life cycle only the zygote is diploid. In *Ulva,* however, the divisions of the zygote are ordinary mitotic divisions, with the result that a diploid plant is formed. Some of the cells of the adult plant then undergo the meiotic divisions, which result in the formation of haploid zoospores. This diploid plant which reproduces by the formation of haploid zoospores is called a *sporophyte.* The spores then develop into haploid plants much as do the spores of other algae. These haploid plants then reproduce by isogamy. This haploid, gamete-forming plant is called a *gametophyte.* The sporophytic and gametophytic generations of *Ulva* are morphologically indistinguishable. This alternation of a diploid, sporophytic generation with a haploid, gametophytic generation is one of the most fundamental features of plant biology. Alternation of generations is first introduced with the origin of sex, and it is present in all sexually reproducing algae; but in the lower algae, the sporophytic generation is represented only by the zygote. In *Ulva,* and in many other green algae, however, the sporophyte is as highly developed as is the gametophyte.

Thus within the algae at large, and more especially within the green algae, advances have been made which approach the condition of the simpler vascular plants. The nucleus has been delimited from the cyto-plasm by a membrane, and the mitotic mechanism has been perfected. The grass-green chlorophyll is no longer masked by the blue phycocyanin, and it is contained in chloroplasts like that of higher plants, not dissolved in the cytoplasm as it is in the blue-green algae. The most primitive algae are unicellular, but colonies of increasing complexity have been formed, leading finally to true multicellular individuals. The simplest of these are unbranched filaments, but these have given rise to branched and re-branched plants, and finally to large, fleshy plants with considerable differentiation of tissues. These may include root-like, stem-like, and leaf-like structures. Cellulose walls are present in some of the algae. Also, a great range of reproductive mechanisms has been developed among the algae. The most primitive algae reproduce only asexually, either by simple fission, or by the formation of clusters of zoospores. Isogamy, the most primitive form of sexual reproduction, probably began with the pairwise fusion of undersized zoospores. Later, these gametes became differentiated into small microgametes and large macrogametes, both of which were motile (heterogamy). Finally, the macrogamete became a very large, nonmotile cell which was sought by the microgamete. Meanwhile, the alternation of generations was developed, with the diploid generation, as first a minor incident in the life cycle, becoming increasingly prominent.

The most advanced characters of the algae are all carried over into the vascular plants. There is no single alga in which all of these characters are present, yet it is clear that the trend of development in the green algae especially is toward the type of organization characterizing the vascular plants, and it is highly probable that the most primitive vascular plants were derived from green algae. But the evolution of the vascular plants and the bryophytes will be taken up in the next chapter.

The events discussed in this chapter are undoubtedly among the most important in the whole history of evolution. In point of time, they must have occupied much the greater portion of the entire history of life. Yet all of these events must have occurred long before the earliest known useful fossils were formed. Thus it is altogether probable that decisive fossil evidence on the problems discussed in this chapter will never be obtained, and that these subjects must always remain speculative, even though some inferences may be made with a fair degree of probability on the basis of primitive or archaic organisms now living.

Chapter—2

Evidence of Evolution

Now-a-days we firmly believe in the doctrine of *organic evolution*. According to this doctrine all the organisms (plants and animals) which are existing today have descended form simpler pre-existing forms.

The term evolution means unrolling or unfolding of events and was used by *Herbert Spencer* for the first time. *Charles Darwin* desecirbed evolution as "descent with modification". In other words, evolution may be described as derivation of new species of plants and animals from those existed in the past. A hot of evidences that evolution has actually taken place have now been accumulated. A short account of these evidences is given below.

EVIDENCE FROM TAXONOMY

The second major category of evidence for evolution is taxonomy, the science of the classification of organisms. Classification would be absolutely necessary because of the seer numbers of species, even if no other purpose beyond that of facilitating study were to be served. There are something on the order of 1,000,000 species of animals and 250,000 species of plants described in biological literature. Large numbers of these may live even in very restricted localities. Thus, in Lake Maxinkukee, Indiana, Jordan identified sixty-four species of fish, eighteen species of amphibians, and 130 species of molluscs. This does not infer that other groups may not be as liberally represented, but simply that they are not cited. It is evident that no extensive and orderly study of the living world would be possible unless it be divided into categories about which generalizations could be made.

LINNAEUS AND BIOLOGICAL NOMENCLATURE

Modern classification is based upon the work of Carolus Linnaeus, a Swedigh botanist (1707-1778) who undertook the classification of the entire living world. Previously, scientific names of organisms had been short (or not so short!) descriptions written in Latin. Thus, Mark Catesby in 1754 referred to the red-headed woodpecker as *Picus capite toto rubro* and to the red-winged blackbird as *Saturnus niger alis suprene rubentibus*.

Linnaeus introduced the practice of giving each animal a binomial name, the first member of which is the generic name, and is shared with other closely similar species; while the second member, the specific name, differentiates the species from other members of the same genus. This binomial system of nomenclature is now universally accepted. Thus, the red-headed woodpecker becomes *Melanerpes erythrocephalus* and the red-winged blackbird becomes simply *Agelaius phoeniceus*.

The Species Concept

Basic to this is the species concept, the idea that there are definite kinds of plants and animals, the individuals of any one kind differing from each other only in minor traits, except sex; sharply separated in some traits from all other species; and mutually fertile, but at least partially sterile when crossed to other species. In the viewpoint of Linnaeus, such species were absolute. "There are just so many species as in the beginning the Infinite Being created." But many present-day biologists feel that the species is to some extent artificial, that is, the boundaries between closely related species are arbitrary rather than natural.

Toxonomic Categories

Linnaeus recognized that several species may have so much in common that they must be grouped together as a genus, distinct from other species-clustes (genera). Thus, the current edition of the American Ornithologist' Union Checklist of North American Birds lists two species of *Melanerpes* and three of *Agelaius*. But genera can be much larger. Thus, the Checklist gives twelve species of the genus *Puffinus* (shearwaters). And there are a few very large genera, for example the weed *Crepis,* of which Babcock recognizes 196 species!

The genera, also, Linnaeus found to fall naturally into larger cluster, based on similarity in rather fundamental characters. These groups of like genera he called orders. The genus *Melanerpes* belongs to the order Piciformes, along with ten other genera of woodpeckers and allied birds. *Agelaius* belongs to the great order Passeriformes, including the great majority of song birds. Finally, these orders were grouped into classes, the diverse members of which share only very fundamental characteristics. Thus, all of the orders of birds comprise the single class Aves.

Linnaeus grouped the classes into the two kingdoms, Plantae and Animalia, but did not feel the need of any category intermediate between class and kingdom. Ernst Haeckel, in the immediate post-Darwinian era, introduced the term *phylum* to include related classes. Phylum means, etymologically, a line of descent, and the term was chosen especially for its appropriateness to the new study of evolution. For a similar reason, Haeckel introduced the *family* as a category intermediate between genus and order.

Thus the full hierarchy of essential taxonomic categories is species, genus, family, order, class, phylum, and kingdom. Every organisms which is described must be assigned to each of these categories, either overtly or implicity. Returning to the two birds discussed above, their complete classification is as follows:

Kingdom	Animalia	Animalia
Phylum	Chordata	Chordata
Class	Aves	Aves
Order	Piciformes	Passeriformes
Family	Picidae	Icteridae
Genus	Melanerpes	Agelaius
Species	erythrocephalus	phoeniceus

All of the taxonomic categories are commonly subdivided for purposes of detailed study. The prefixes sub- and super- may be added to any of the standard categories to indicate subdivisions or larger assemblages. Other intermediate categories are sometimes used, but do not have official status in zoology.

T. H. Huxley has said, "That it is possible to arrange all the varied forms of animals into groups, having this sort of singular subordination one to the other, is a most remarkable circumstance." It is indeed. Linnaeus accounted for this by the theory of archetypes, which assumes that the Creator worked from a series of plans, the archetypes, which limited in number. These archetypes, like the plans in an architect's folio, were not all equally distinct, but fell into definite, classifiable categories. Thus each class would correspond to a major archetype, the various orders within a class would be lesser archetypes, and so on down the hierarchy. Thus, Linnaeus attributed the similarity between species of a genus not to descent from a common ancestor, but to the supposed fact that each is a more or less imperfect copy of rather similar archetypes, plans of the Creator.

SIGNIFICANCE OF THE TAXONOMIC HIERARCHY

Darwin's explanation of this "remarkable circumstance" is quite different. The taxonomic categories simply represent degrees of blood relationship. Thus, all members of the phylum Chordata have common ancestors, but they are exceedingly remote, and hence only the most fundamental chordate characters are held in common by extreme members of the phylum. Within any class, however, the degree of relationship is much closer, and hence more numerous and less fundamental characters are held in common by diverse members of a class. All birds, for example, share many characters in common. As one goes down the taxonomic scale this trend becomes stronger until finally members of a single species differ only in minor characters, and this because of their common inheritance. It is difficult to study any group of organisms in detail without feeling that this argument is a cogent one.

The Tree of Life

Taxonomists have always tried to summarize their studies with diagrams. A *tree* is the most successful type of diagram, but this has not always been evident. Linnaeus experimented with map-like diagrams, but it was found that no arrangement was possible which would always place similar forms together and always separate dissimilar forms. Later, Lamarck and others tried to arrange living forms on a ladder-like diagram, on the principle that the adaptation of living forms is progressive so that any particular animal should be proceeded by one somewhat lower in the scale of life and followed by another one somewhat higher. In a general ways, this can be done. It may be easily conceded that amphibians are more advanced than fishes, and that reptiles are more advanced than amphibians. But one cannot continue this series—birds, then mammals. For, while one mammals has advanced so far beyond all other animals that he alone studies the world in which he lives, nonetheless the majority of birds are in every respect quite as "high" as are the majority of mammals. And so it appears that two rungs are required at the same level on the ladder. This type of dilemma is not only repeated frequently among the major groups of organisms, but it becomes very much more frequent when the study is extended to the lower levels of classification. As soon as the need for parallel rungs at many levels became apparent, with parallel series above them, it immediately was evident that a tree would make a much better diagram, and this was generally accepted long before Darwin.

Now everyone understands that the parts of a real tree are related to one another in a branching fashion because the whole organisms is the product of growth from a single seed, growth accompanied by branching and differentiation. Independent creation and secondary union of the many parts would be unthinkable. The taxonomic tree is not strictly comparable to an actual tree, for here the processes of branching and differentiation are not generally amenable to direct observation. But the analogy cannot be avoided: the fact that no other type of diagram can symbolize the data of taxonomy so readily as a tree strongly suggests that, like a real tree, the tree of life owes its branching character to organic growth and differentiation—in other words, to evolution. Pre-Darwinian biologists were unable to understand why it was that classification seemed to fall into a tree-like pattern, yet they agreed that it was so. Long awareness of this fact undoubtedly helped to prepare the way for the ultimate acceptance of Darwinism.

A final a characteristic of the tree of life deserves special consideration. In any group, there are likely to be some members which are simpler and less specialized than the others. These may be known only as fossils, but many such archiac or primitive species are still living. In this case, the archaic species generally has its closest affinities with fossil rather than with living members of its group. Now it is a general rule that such primitive species resemble members of *other* groups more than do the more specialized members of the same group. To paraphrase this, species which are placed near the points of branching in the tree of life show especial resemblances to other species in both branches. If each species were created independently of all others, this fact would be inexplicable. Common characters should then be distributed among various groups without regard to their level of specialization. Yet, if the evolutionary theory be correct, then the most primitive members of related groups are those which have diverged the least from their common ancestry, and so primitive species should illuminate the relationships between groups. This is exactly what is found in nature.

Another suggestive aspect of taxonomy is the difficulty of distinguishing closely related species. In not a few cases, the ranges of variation of related species overlap. For example, in the genus *Dendroica,* the wood warblers, it is possible to set up a series ranging from the yellow warbler through magnolia, myrtle, and palm warblers to the chestnut-sided warbler in which the color pattern gradually shifts from predominantly yellow to predominantly black. The gradations are so fine that expert knowledge is needed to separate the species. Again, the fruti flies *Drosophila pseudoobscura* and *D. persimilis* are so similar that statistical analysis of populations is needed to distinguish them. Such situations strongly suggest descent from rather recent common ancestors.

Following examples illustrate homology:

A. Homology and Analogy

When we make a general survey of the animals or plants for structural evidence of evolution, we come across two types of characters, firstly those which are of common origin are known as *homology* and secondly, the adaptive features which are of unlike origin but are of similar functions, such are called *analogy*.

Those organs which are fundamentally unlike *i.e.* they have different origins but have the same function are called *analogous organs*. But the organs which are fundamentally the same in structure and have the same origin, irrespective of how greatly thy may be modified to perform widely different functions, are called *homologous organs*.

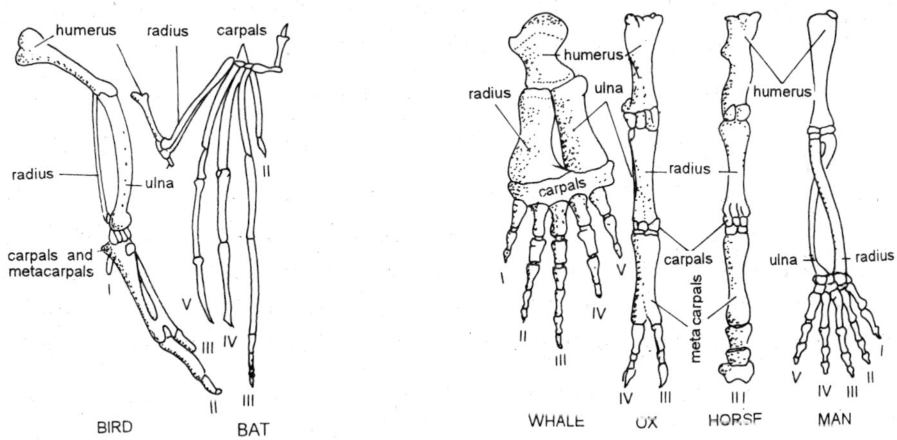

Fig. 2.1: Homology of Vertebrate fore-limbs.

EVIDENCE FROM COMPARATIVE ANATOMY

1. The Vertebral Column

The vertebral columns of all vertebrates originate from similar embryonic rudiments, four pairs of condensed mesenchymal masses in each somite, the arcualia, with associated looser mesenchyme. From these simple beginnings, the typical vertebra with its centrum, neural arch, spinous and transverse processes, and its various articular processes, is formed. Among the Cyclostomata, the most primitive of living vertgebrates, only the two dorsal pairs of arcualia formed, and these do not advance much beyond the embryonic condition, for they merely form two pairs of cartilage spikes flanking the spinal cord.

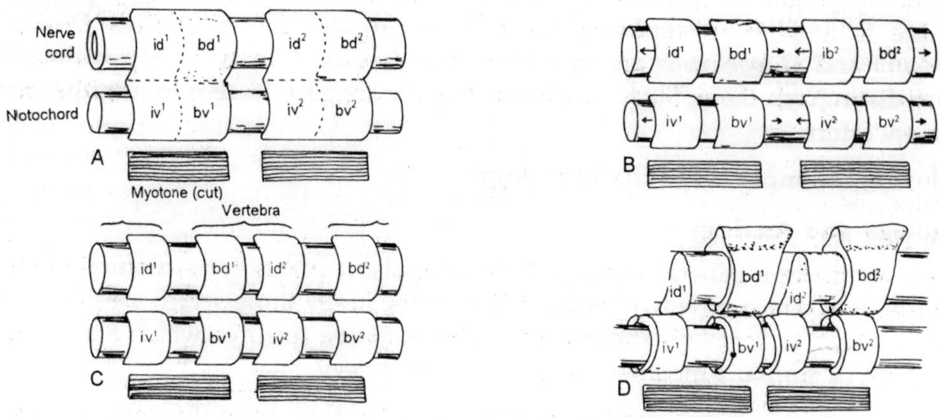

Fig. 2.2: The Arcualia which form the Vertebrae. A shows two body segments, each with four arcualia (actually four pairs, but only the left side is shown). *B* and *Ç* illustrate the separation of the anterior from the posterior arcualia in each segment. *D* shows how a vertebra comes to span two muscle segments because it is made up of arcualia derived from two adjacent body segments. The arcualia are named interdorsal, id; interventral, iv; basidorsal, bd; and basiventral, bv.

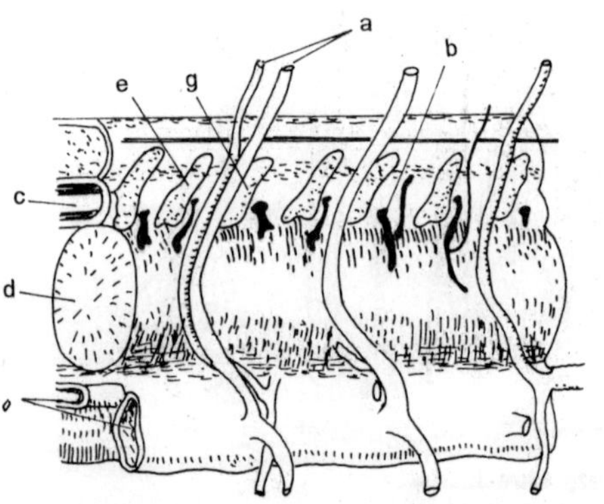

Fig. 2.3: The "Vertebrae" of the lamprey, consisting only of the two pairs of dorsal arcualia in each segment. a, shows intersegmental blood vessels; b, segmental nerve; c, neural tube; d, notochord; e and g, dorsal arcualia; and o, longitudinal blood vessels.

In the Chondrichthyes (sharks and their allies), a much more extensive development occurs. The two dorsal pairs of acrualia (= arch-formers) from neural arches and so-called intercalary over the spinal cord. The two ventral pairs fail to form arches in the trunk region, but in the caudal region the anterior ventral pair of arcualia forms a haemal arch in each segment, enclosing the caudal artery and vein. A centrum is formed around the notochord partly by the notochordal sheath, and partly by the bases of the arcualia and the surrounding mesenchyme. The articulation is strengthened by the notochord, which is still continuous in adult sharks. Ribs are formed in conjunction with these vertebrae. The entire skeleton of the adult is cartilaginous, a fact which is probably properly regarded as the retention of a basically embryonic condition.

The vertebrae of Osteichthyes (bony fishes) are not substantially different from those of the Chondrichthyes, except that they are ossified in whole or in part. In many fishes, *two* centra are formed for every body segment, an anterior *hypocentrum* based upon the anterior ventral pair of arcualia, an a posterior *pleuurocentrum* based upon the posterior dorsal pair of acrulia. This is exemplified by the Rhipidistia, an extinct group of lung-breathing fishes which is believed to have given rise to the Amphibia.

The most primitive fossil amphibians have vertebrae much like those of the Rhipidistia, but they change in two different lines of descent. In one, leading toward the modern Amphibia, the hypocentrum gradually replaces the pleurocentrum. In the other, leading to the reptiles, the pleurocentrum gradually replaces the hypocentrum. This replacement is not yet complete in the Rhynocephalia, the most primitive order of living reptiles.

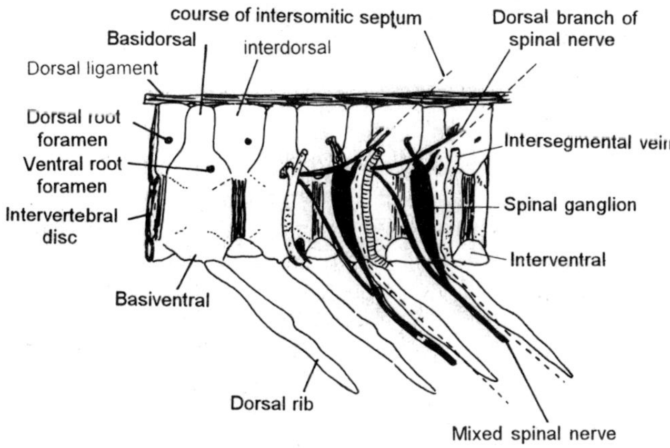

Fig. 2.4: Shark Vertebrae (*Squalus acanthias*).

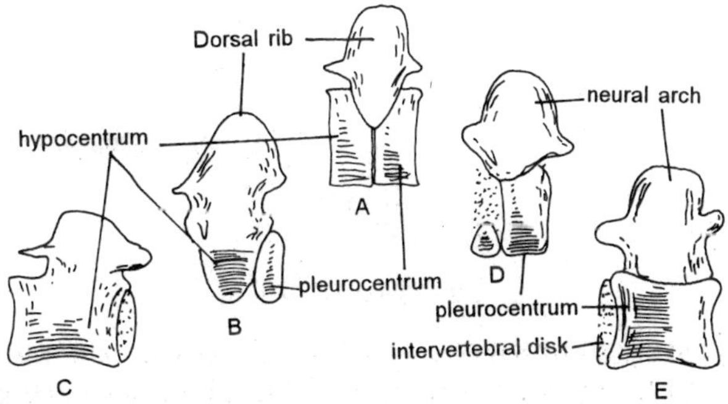

Fig. 2.5: Derivation of Amphibian and Amniote Centra. *A* is the primitive amphibian type. *A*, *B*, and *C* illustrate the transition to the type of vertebrae found in the modern amphibia, while *A*, *D*, and *E* illustrate the transition to the type found in amniotes.

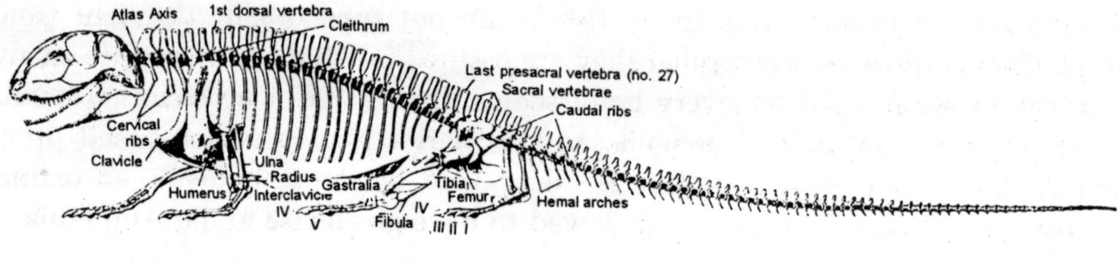

Fig. 2.6: Skeleton of a Primitive Reptile, Haptodus.

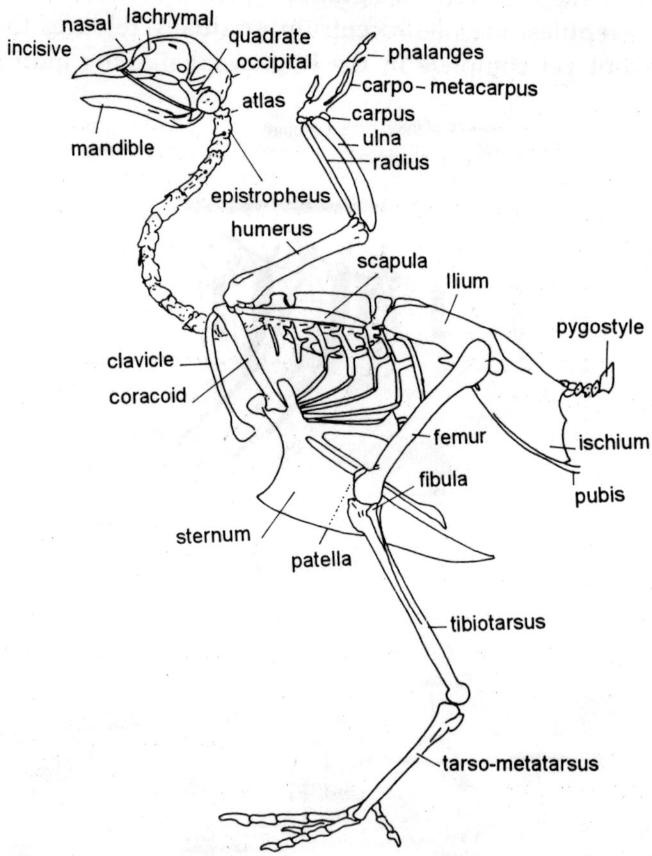

Fig. 2.7: Skeleton of a Bird.

The pre- and post-zygopophyses, articular processes on the neural arches, first appear among the Amphibia. They strengthen the joints between successive vertebrae. Another innovation is the addition of two more differentiated types of vertebrae in addition to the trunk and caudal vertebrae. One or two *cervical* vetebrae serve to make a freely movable joint between the skull and the vertebral column, a character of adaptive value to land-dwellers, but not to the strictly aquatic fishes. Between the trunk and caudal regions is a single sacral vertebra specialized by fusion with its ribs to attach the pelvic girdle to the axial skeleton.

The vertebrae of the Reptilia differ from those of the Amphibia mainly in that the centrum is a pleurocentrum rather than a hypocentrum (in extant forms). Also, the number of cervical and sacral vertebrae is increased. In the cervical region, the ribs are now fused to the vertebrae. The vertebral column of the birds differs from that of the reptiles chiefly in that a highly flexible articulation has been developed between the cervical vertebrae, and those of the body are largely fused. The caudal vertebrae are much reduced. The vertebral column of the mammals differs from that of the reptiles mainly by the development of curvatures associated with the mechanics of locomotion.

Thus, a survey of the vertebral columns of the extant classes of vertebrates leads back to the statement with which this discussion was opened, that, within a single phylum, any particular system seems to be based upon a prototype which is simply varied from class to class.

2. Serial Homology in the Crustacea

Structures, which are similar in different species because of common inheritance, irrespective of the diverse uses to which they may be put, are called *homologous* structures. There is no group of living things which does not present homologies in all of its structural system, but perhaps no set of homologous structures is more striking than the appendages of the Arthropoda, especially those of the Crustacea. In the typical crustacean, there is one pair of appendages borne by each segment of the body. These are all referable to a single structural plan, and so they are considered to be serially homologous. Serial homology is typical of all animals whose body structure is based upon a series of essentially similar segments. The major examples are in the phyla Annelida, Arthropoda, and Chordata. In the crustacean appendage, there is typically a basal *protopodite* of two segments. Distal to this are two parallel structures, a medial *endopodite* and a lateral expodite, each consisting of a series of segments. The appendages of various body segments have been modified in the most diverse ways, adaptive to widely different uses. Their condition in the crayfish is a familiar example. The first two pairs of appendages have been greatly modified to form antennae. Whether the eyestalk also represents a greatly reduced appendage is still subject to debate among specialists. The heavy, biting mandibles are formed by the shortened protopodite and endopodite of the third segment, while the expodite has been eliminated altogether. The fourth pair of appendages forms the first maxilla, in which the protopodite and the endopodite form a flattened plate which is used for handling food. The fifth pair are the second maxillae, but the exopodite is present here, and together with a dorsal out growth called the epipodite, it forms a vane to pass a current of water over the gills. This is the last of the appendages of the head region. The first three pairs of thoracic appendages are maxillipeds, the protopodites of which are flattened and serve for the handling of food, just as do the maxillae. The endopodite and exopodite are present on these appendages, but they are not very large, and they may be sensory in function. The first maxiliped bears an epipodite, while the second and third bear gills which extend dorsally under the carapace. The remaining five pairs of thoracic appendages are specialized as walking legs, with the exopodite absent and the endopodite terminating in a pincer. The first pair of walking legs forms the major pincers (or chelae) of the organism, serving as defensive and food-procuring organs. All but the last pair of walking legs bear gills. There are six pairs of abdominal appendages. The first is often much reduced or even absent in females, while in males it is modified together with the second abdominal appendage to form a copulatory organ. In females, the second appendage forms a swimmeret or pleopod in which there is a short protopodite followed by a more or less equally developed, filamentous exopodite and endopodite. In both sexes, the third, fourth, and fifth abdominal appendages follow this pattern, which appears to be the primitive appendage pattern. The final appendage consists of a short protopodite and a very broad, flat endopodite and exopodite, forming the telson, or terminal fan of the crayfish.

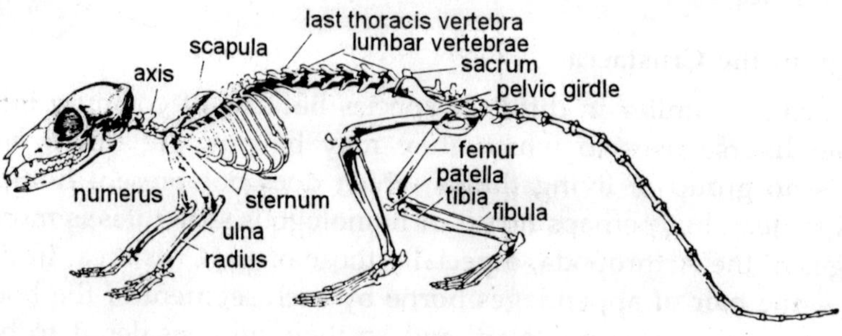

Fig. 2.8: The skeleton of a primitive mammal, the tree shrew, *Tupaia*.

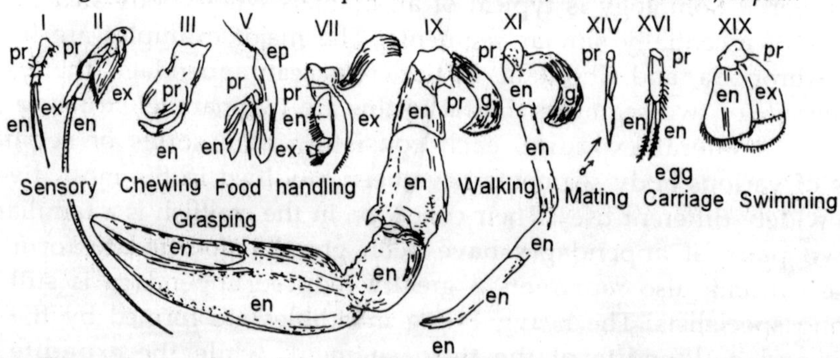

Fig. 2.9: Types of appendage of the Crayfish. Roman numerals indicate the body segment from which each was taken. *pr*, protododite; *ex*, exopodite; *en*, endopodite; *ep*, epipodite; *g*, gill.

Thus, within a single organism, the basic crustacean appendage is modified to serve no less than six to ten different functions (depending on how one may wish to classify them). If each of these appendages had been originally created for the function which it now serves, then it would be a strange circumstance that all of them are built upon the same pattern as are the legs. For most of these function are subserved in other groups by organs having nothing to do with appendages. Antennae of molluses, for example, are constructed on a completely different plan, yet there is no reason to suppose that they function less efficiently than those of the Crustacea. Nor is there any intrinsic reason why mouth parts, copulatory organs, and gills should be based upon a leg-like structure. Yet it is so in the Crustacea. These facts, so puzzling on any other theory, are easily understandable on the basis of the evolutionary concept. Given a primitive crustacean in which all of the appendages are in a simple condition, somewhat like the swimmerets of the crayfish, natural selection should favor diversification of function. There is a general tendency among the higher phyla for centralization of sensory functions in the head region; hence the sensory functions of the anterior-most appenages become intensified at the expense of other functions, and they become antennae and (possibly) eyes. Those appendages nearest the mouth naturally are used for feeding, and so they become specialized for chewing, biting, and food handling. Those appendages nearest the reproductive organs are, in the male, modified for the transfer of sperm to the female. Other appendages continue to serve the original locomotor function in various ways.

If the Crustacea as a whole be considered, then the range of adaptations of these appendages is still greater. In some of the Crustacea, mouth parts are more numerous than in the crayfish described above. In these, there are fewer legs. Why such a relationship should hold is utterly inexplicable except upon the theory that both the

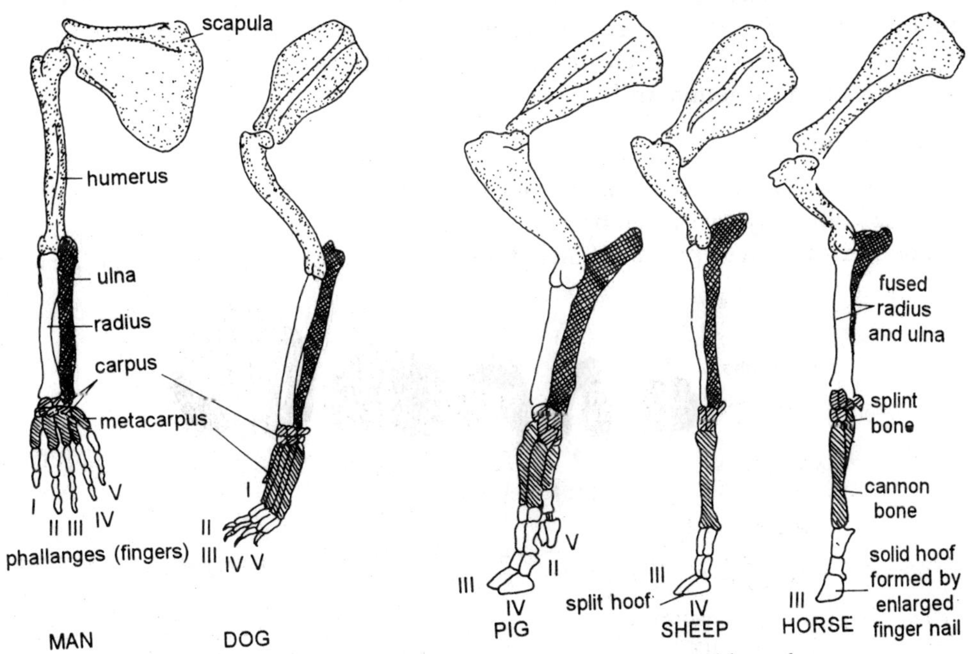

Fig. 2.10: Fore-limbs of few mammals showing serial homology.

mouth parts and the legs have been derived from primitive appendages by adaptive modification. In the barnacles, most of the appendages have been suppressed, ad the thoracic appendages have been specialized as plume-like cirri which sweep a current of food-bearing water toward the mouth. In the lobster, the abdominal appendages are flattened out to form broad, oar-like plates, effective as swimming organs. In the crabs, which normally hold the abdomen recurved against the ventral part of the thorax, the abdominal appendages are much reduced or missing entirely, excepting the first two which serve as copulatory organs. But through all of this great range of variation, a single pattern is discernible, a fact which bespeaks an organic relationship of all of the Crustacea.

3. Limb Structure of Vertebrates

The forelimbs of a mole, wing of bat, forelimb of man and terrestrial horse which look so much different and perform different functions are homologous structures because developmentally all of them possess a pentadactyle plan. This structural divergence or modifiction from a fundamental plan of structure can be explained only on the basis of adaptation to different environmental conditions. Thus biologists are convinced that animals have inherited homologous structures from an ancestor which they shared incommon in the remore past.

4. Eye Structure of Vertebrates

The eyes of different vertebrates, though may differ in their muscles and lenses and required by different degrees of accommodatrion, but they are all homologous. The eyes of vertebrate from cyclostomes to mammels, are constructed of similar parts like sclerotic, choroid, retina and lens.

5. Brain Structure

The brain of vertebrates raning from fishes to mammels consists o similar series of parts, viz., olfactory lobes, cerebral hemishperes, optic lobes, cerebellum and medulla oblongata and other less prominat structures. As we progress through the series from

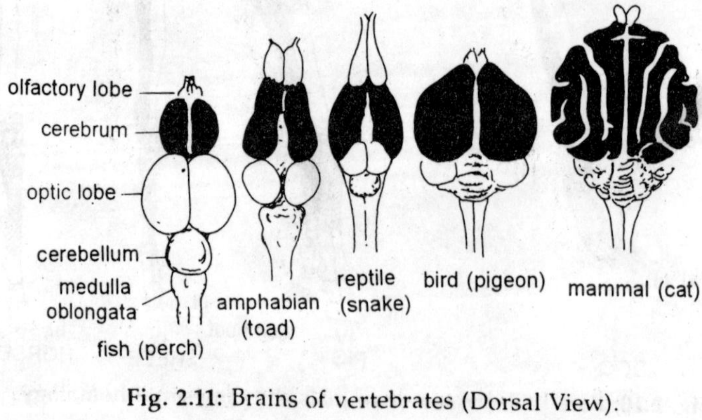

Fig. 2.11: Brains of vertebrates (Dorsal View).

fishes to mammals some lobes become more prominant thant other like cerebral hemispheres which are smaller in fishes but highly developed in mammals. It is due to adaptations and evolutioanry changes.

Catagories of homology: These are three categories of homology:

(i) *Phylogenetic homology*-which is existed between different species (e.g. pentadectyle limbs of terrestrial vertebrates).

(ii) *Sexual homology*-existed between the sexes of same species.

(iii) *Serial homology* is existed between the organs of same individual occupying different levels of the body (e.g. annelids, arthropods and chordates). In the crustacean all segments of body carry paired jointed appendages, each consists of a basal *protopodite* and the distal *endopodite* and *exopodite.* The appendages of the various parts of the body are variously modified depending upon the use for which one is meant. This can be explained using the example of prawn.

Following examples illustrate analogy:

The analogous organs have almost similar appearance and perform the same function but thier origin is different.

1. The most suitable example is that of 'wings'. Wings are found in bat (mammal) and birds and they are also found in the cae of insects. The purpsoe of these wings (in bat, birds and insects) is flight. The wings in case of bat and bird are similar in origin, structure and function and are thus considere as homolgous organs. But an insect's iwng is not similar in origin and structure through it performs the same function. Therefore, it is considered as analogous organ.

2. The other example is furnished by the fins of the fishes and flappers of aquatic mammals like whale and seals. These structure have similar appearahce and function but their structure found distinctly different and therefore, are analogous. The fins have very siple skeleton while the flappers resembles a pentadactyle limb.

3. The stings of honeybeen and scorpion perfrom similar function and similar in thier structure. But their orgin is different, in honey bee it is a modification of ovipositor while in scorption it is the modification of last abdominal segment. Hence are analogous.

Vestigial Organs

Another Very persuasive aspect of comparative anatomy concerns *vestigial* or *rudimentary* organs. These are dwarfed and generally useless organs which are found in many plants and animals, relatives of which may have the same organ in a fully developed and functional condition. Perhaps the most widely known example is the vermiform appendix of man. This small structure, without any known function in man, is notorious as a seat of disease. In other Primates, however, this organ is considerably larger. And in mammals which eat a coarse diet, involving considerable amounts of cellulose, the appendix and cecum form a large sac in which mixtures of food and enzymes can react for long periods of time. The appendix of man is easily understandable as a degenerating legacy from ancestors with a much coarser diet; but it is inexplicable why a useless and disease-ridden structure should have been created especially to plague him.

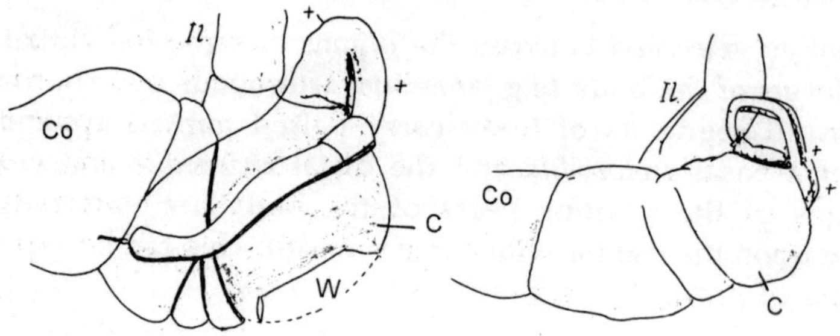

Fig. 2.12: The Appendix in the Orang-Utan and in Man. *il,* ileum; *co,* colon; *C,* caecum; *W,* window cut in caecum;+,appendix.

Weidersheim has listed nearly 100 such vestigial characters in man, and a few of these may be mentioned. In the inner corner of the eye of all vertebrates there is a transparent membranous fold, the nictitating membrane. In most vertebrates, this "third eyelid" can be swept clear across the eyeball to cleanse the latter, much like the blinking of a mammal. In birds, it is particularly well developed. It s use can be easily observed if a captive owl is watched by daylight. But in mammals, it forms a mere crescentic fold at the inner corner of each eye, with no known—or probable-function. It is understandable only as a degenerating inheritance from an ancestor to which the nictitating membrane was actually useful, as it now is to the majority of vertebrates. The ear muscles of man present a similar situation. Many mammals move the external ear freely in order to detect sounds efficiently. The complete muscular apparatus for these movements is present in man, but quite vestigial. While school boys sometimes pride themselves on their ability to wiggle their ears, the ability has no real usefulness, and even this limited ability is not shared by everyone. The presence of these muscles, then, suggests descent from an ancestor to which they were really useful.

Similarly, most mammals have a well-developed tail, but this is lacking in all of the higher Primates. It is represented in them by vestigial caudal vertebrae, usually three to five in man. Rarely, a fleshy tail extends a few inches beyond the caudal vertebrae. Whether or not an external tail is present, the muscles which move the tails of other mammals are also present in the Primates.

A final example from man concerns the wisdom teeth. The wisdom teeth, or third molars, are the posterior-most teeth, as well as the last to erupt. In other Primates, these teeth are as sound and as fully developed as the rest of the dentition. But in man, they are far more variable than are the other teeth with respect to size and time of eruption. Frequently, they fail to erupt altogether. And when they are present, they are far more subject to all types of dental defect than are the other teeth. Thus it is probable that these teeth should be regarded as vestigial, and, in view of the frequency with which they fail to erpt, that they will in time be completely lost to man.

Many examples of vestigial characters may be found among lower animals. The external ears of whales are completely of the type found in terrestrial mammals, but they are much reduced in size, and it seems unlikely that they are efficient auditory organs. Also among whales, the hind limbs are completely missing, yet in some species rudiments of the pelvic girdle still remain, but have lost their connection to the vertebral column. In ungulates (horse, deer, and other hoofed animals), the smaller bone of the lower rear leg, the fibula, has been reduced to a mere splint on the larger bone, the tibia. A similar reduction of the fibula has occurred in the birds. Perhaps no feature of the anatomy of snakes is so generally known as their leglessness. No snake shows any vestige of the forelimbs, but some (pythons and boas) have small, ineffective rudiments of the hindlimbs. These are capped by claws which show externally, but they are so reduced that they appear at a glance to be scarcely more than raised scales.

Many animals, both vertebrates and invertebrates, have become adapted to life in deep caves, where the light of the sun never reaches. Living thus in perpetual darkness, there is no adverse selection against degenerative changes of the eyes, and in fact blindness is a general characteristic of such cave dwellers. Their eyes show all degrees of degeneration from just short of the typical functional condition to complete absence of the eyes. Examples include the blind, cave-dwelling salamander of central Europe, *Proteus anguineus;* the many species of cave-dwelling fishes of the United States as well as other parts of the world; and the blind cray-fishes. The latter have eyestalks which do not, however, bears eyes. While such degenerated eyes are easily understandable on the basis of descent from ancestors with functional eyes, their presence is inexplicable, indeed it is contradictory, on any other theory.

Typical beetles, as every school boy knows, are strong fliers. The island of Madeira, a wind-swept island about 600 miles off of the coast of Portugal and 400 miles to the west of North Africa, has a rich beetle fauna. But a large proportion of these Madeiran species are either wingless or they have much reduced wings which can be of no use for flight. If one makes the assumption that the reduction or loss of these wings has been a result of natural selection, then the selective force may be easily surmised. For beetles in flight would be quite likely to be blown out to sea and lost. Thus a strong selective force should favor any variations toward reduction of wings. Those species which have been on the island and subject to this selection longest should be most numerous in the list of flightless species. That this is probably true is indicated by the facts that almost all of the endemic species are flightless, while most of the flying species are also represented in Europe or Africa, or have close relatives on these contiments. Only the evolutionary explanation for these facts can be rational.

When structures undergo a reduction in size together with a loss of their typical function, that is, when they become vestigial, they are quite commonly considered to be degenerate and functionless. But Simpson has recently pointed out that this need not be true at all: the los of the original function may be accompanied by specialization for a new function. Thus the wing of penguins has become reduced to a point that will not permit flight, but at the same time it has become a highly efficient paddle for swimming. The wings of rheas, ostriches, and other running birds are also much reduced, and have been described as "at the most still used for display of the decorative wing feathers!:" But Simpson has observed that the rheas, when running, spread the wings and use them as balancers, especially when turning rapidly. It seems quite probable that this is true of the running birds generally.

A comparable case is afforded by the pineal gland. This small structure grows dorsally from the forebrain, and its histological structure shows a mixture of glandular and nervous characteristics. In the lampreys and in many reptiles, including the very primitive, lizard-like *Sphenodon* of New Zealand, this instructure bears a third eye, located on the dorsal surface of the skull. Fossil evidence indicates that this is also true of the Crossoptery-gii, the group of fishes from which land vertebrates appear to have been derived. Thus the possibility is open that the possession of a pineal eye may have been a primitive character of the vertebrates, and that the pineal gland of most extant vertebrates is in fact a vestigial eyestalk. But if so, it may well have an important new function, for it is commonly regarded as an endocrine gland, although the evidence is incomplete.

Connecting Links

There are certain animals living or extinct which possess characters intermediate between two major groups of animals. Thus, lung fishes (living) connect fish with amphibians, *Seymouria* (extinct) connects amphibians with reptiles. *Archaeopteryx* (extinct) connects reptiles with birds and Prototheria (egg laying mammals—living) connect reptiles with mamals. *Neoplina* frms a connecting link between annelids and molluscs. *Peripatus* connects annelids and arthropods. These connecting links may not be direct ancetors of the next higher group but they do show that evolution has taken place. The lower animals during the course of time have changed into higher animals through the bridge of animals which possessed intermediate charactes.

Atavism or Reversion

It has been observed that some individuals are born with some such abnormal features which are supposed to have been present in their ancestors, but have lost during the course of evolution. The sudden reappearance of characters are called atavism. Examples are additional nipples in thoracic region, presence of short tail in newly born, presence of thick body hairs etc.

Thus the study of comparative anatomy reveals that all the animals in fact are related either closely or distantly to each other. They have diverged from their ancestral groups during the geological years in much the same way as the larger branches and twings of a tree diverge from its main trunk.

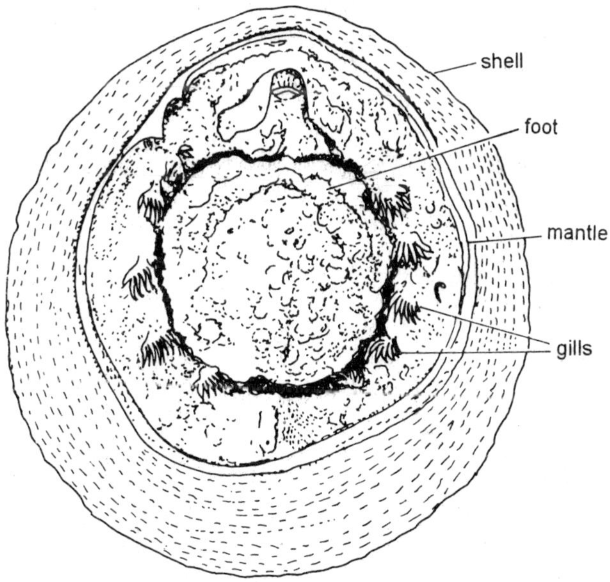

Fig. 2.13: Neoplina.

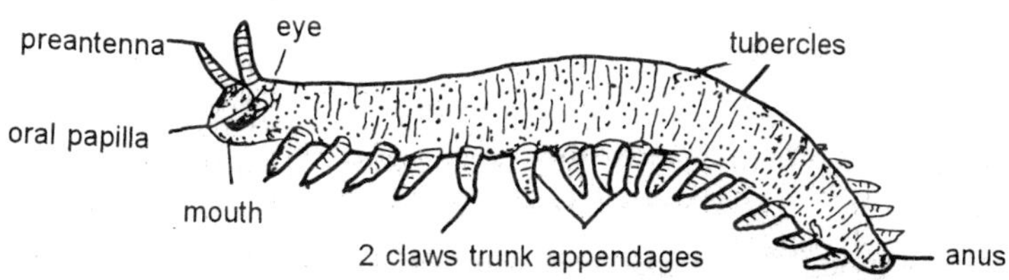

Fig. 2.14: Peripatus.

EVIDENCE FROM EMBRYOLOGY

Comparative embryology is a specialized branch of anatomy which furnishes evidence for evolution regarded by Darwin as "second to none in importance." Ernst Haeckel brought this field into prominence in the immediate post-Darwinian period with his Biogenetic Law, which states that "Ontogeny recapitulates phylogeny." He believed that embryonic stages corresponded to ancestral adults, and hence provided direct evidence of lines of descent. Recapitulations do occur, but not as Haeckel thought, for resemblances are chiefly between embryos, not embryos and adults, and embryos, too, have adaptive problems.

A striking example occurs among crustaceans. A series of six larval types strongly resemble adults in a sequence from primitive to advanced. The larvae pass through these stages at successive molts, with larvae of primitive crustaceans stopping early in the series and advanced crustaceans going through most of the stages, as indicated in the chart. Failure of the Cypris larva to appear in the development of higher crustaceans simply means that *Cypris* and barnacles represent a side branch of crustacean evolution.

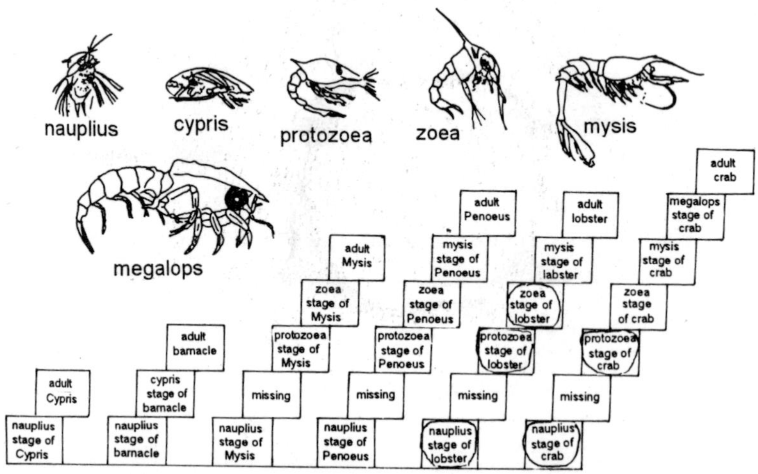

Fig. 2.15: Developmental stages of crustaceans. At the top are a series of larval types, several of which resemble adults of corresponding groups of crustaceans. The sequences of squares at the bottom indicate the developmental histories of the major crustacean types. Stages in ovals are completed in the egg, but others are free-swimming larvae.

Examples Among Vertebrates

The ontogeny of man, considered biogenetically, indicates a long, complicated history. The fertilized egg corresponds to a protozoan ancestor, but it soon becomes multicellular, indicating a primitive metazoan grade. Gastrulation forms a coelenterate-like embryo, but this soon becomes triploblastic, like a flat worm. Fundamental chordate characters (dorsal nerve tube, notochord, and pharynx specialized for respiration) are then developed. Fish-like characters, such as gill slits and aortic arches, appear, followed by tetrapod characters, such a the pentadactyl limb and metanephric kidney. Finally mammalian, then primate, and at last specifically human characters appear.

The details of the development of specific systems are sometimes very impressive. The kidneys of vertebrates are all developed from the *nephrotome,* a segmented mass of mesoderm lying on either side of the somites, from which so many of the serially homologous structures of the body are developed. Yet there are three distinct types of kidneys among the vertebrates. All vertebrate embryos first develop a pronephric type of kidney, utilizing only the anterior-most part of the nephrotome. Only in the hag-fishes and a few of the bony fishes does this remain as the functional kidney of the adult. In all other vertebrates (including all of the bony fishes), a mesonephric type of

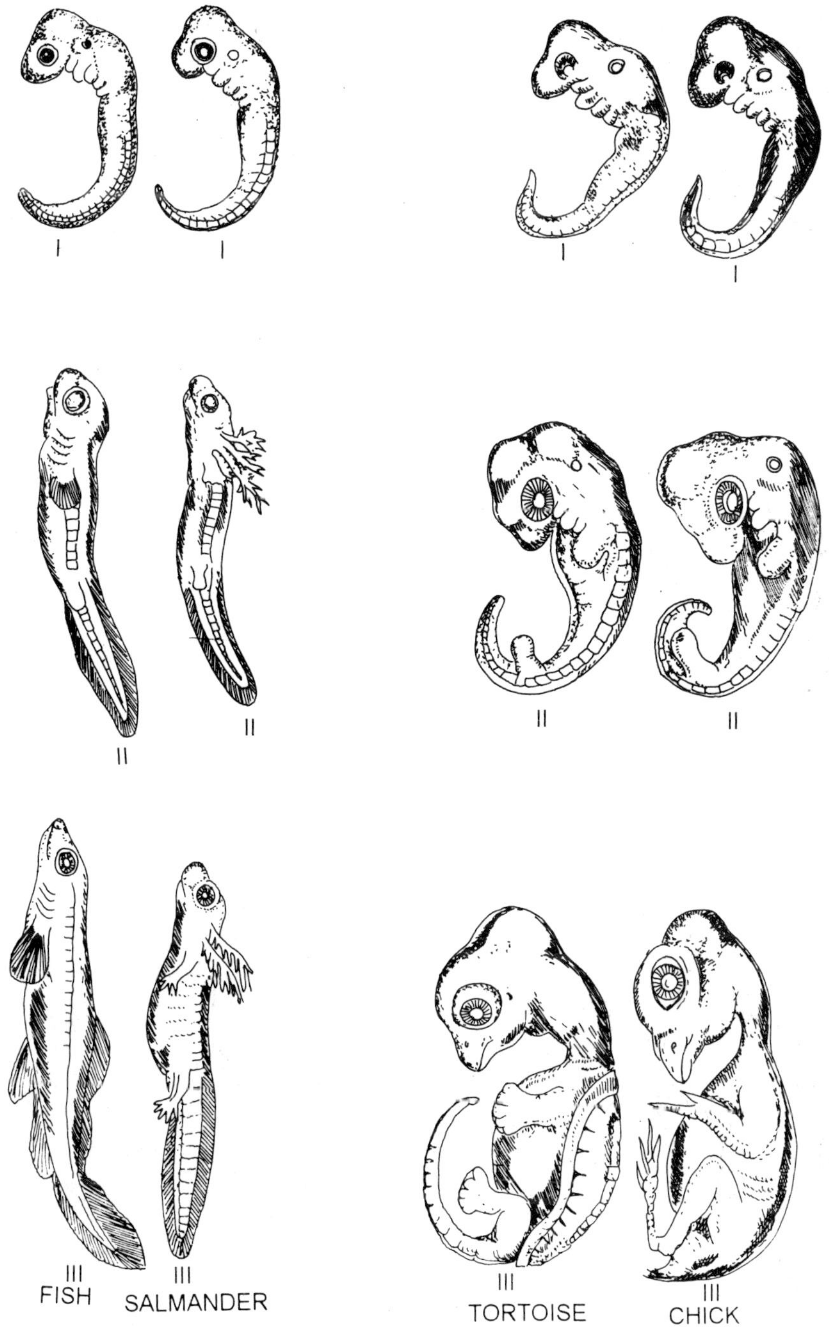

Fig. 2.16: (Continued on opposite page)

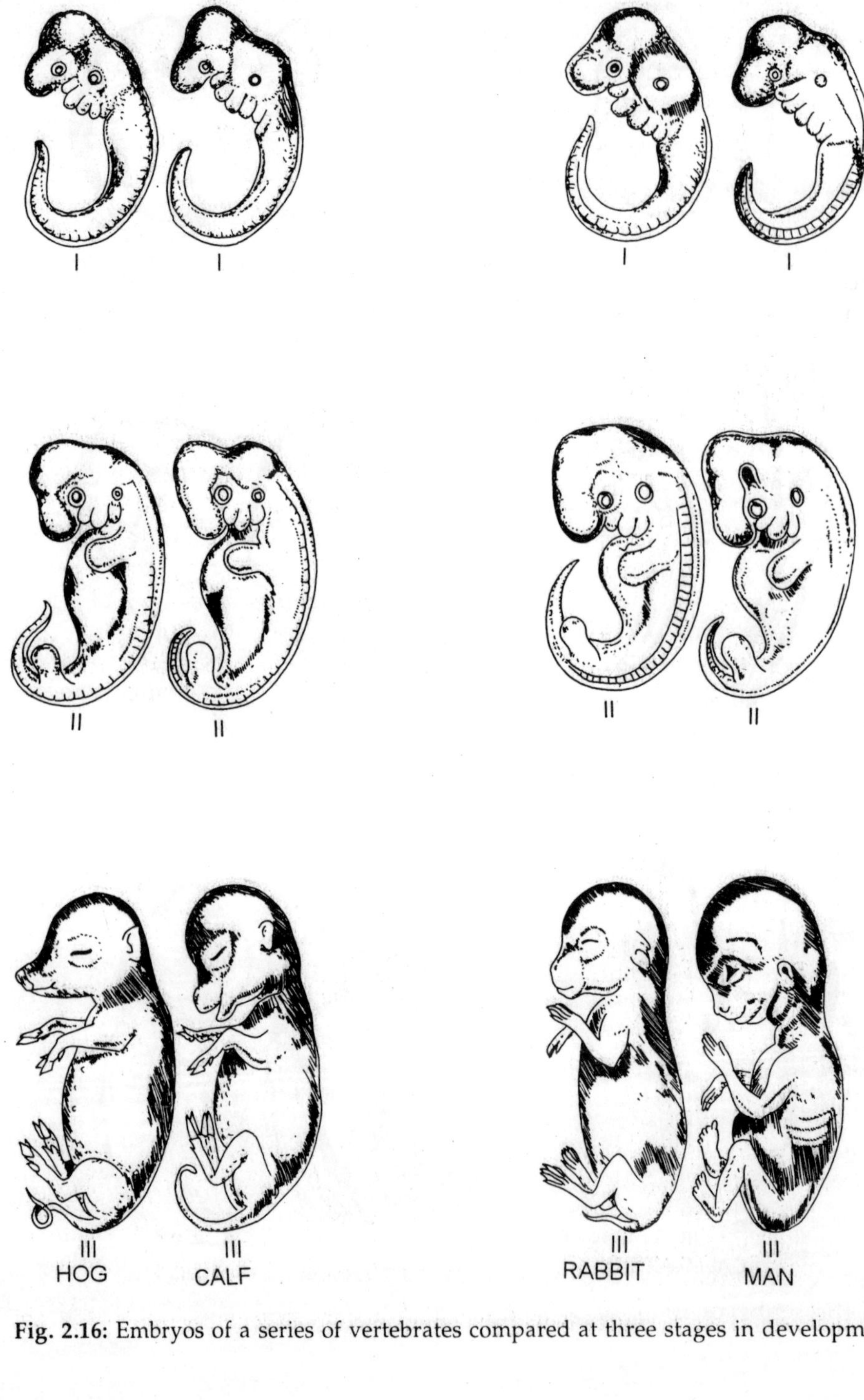

Fig. 2.16: Embryos of a series of vertebrates compared at three stages in development.

kidney is developed posterior to the pronephros, and the pronephros either degenerates or is partly incorporated into the mesonephros. This mesonehros is the functional kidney of adult fishes and amphibians, and of the embryos of reptiles, birds, and mammals. But in all of the last mentioned classes, a third type of kidney, the metanephros, is formed posterior to the mesonephros, and serves as the functional kidney of the adult organism.

Similarly, all vertebrate embryos develop a series (most commonly six) of aortic arches, each of which runs unbroken from the ventral aorta to the dorsal aorta, much as in adult Amphioxus. In the fishes, these arches are modified in several ways, all of which involved the separation of each aortic arch into a ventral afferent branchial artery and a dorsal efferent branchial artery, the two being connected by a capillary network in the gill filaments. In the Choanichthyes, the group of fishes most closely related to the Amphibia, the first arch drops out, and is largely missing in the adult, but its ventral and dorsal roots, together with new growths from them, form the major arteries of the head (the external and internal carotids). The sixth arch has given rise to a *pulmonary* branch which supplies the lungs. This tendency for parts to drop out after having been formed in the embryo, and for the remaining parts to be diverted to completely different functions from the original purely respiratory function, is the principal factor in the embryology of this part of the circulatory system of all tetrapods. Among the urodeles, the main portions of the first and second arches drop out, so that now the carotids arise from the third arch. The third arch is broken by a capillary network early in development, but it soon becomes continuous again. The dorsal connection between arches three and four disappears, with the result that the ventral connections now appear as common carotid arteries on either side, while the two fourth arches now supply the major circulation to the body. The fifth arch becomes reduced in size and may be lost altogether, while the sixth arch again gives rise to a pulmonary branch. In the Anura and in the reptiles, this process goes a little farther, with the fifth arch being lost completely and with the dorsal part of the sixth arch being lost, so that all of the blood entering the sixth arch must go into the pulmonary artery. The birds have essentially the reptilian system, with, however, the left fourth aortic arch degenerating, thus leaving the right fourth arch to carry the entire systemic circulation. In the mammals, it is the right fourth arch which degenerating, and the left one which persists. Thus, of the six original pairs of aortic arches, only three persist in the highest classes of vertebrates. Arch three serves the head region, arch four (but only one of the pair) serves the systemic circulation, and arch six serves the lungs. Yet all six pairs are developed in the embryos of birds and mammals.

A similar story could be told with respect to almost any organ system in any major group. The details differ, but the general facts are the same. When, in the course of embryonic development, a new organ system is formed, its structure is closely similar even in the most widely dissimilar species of the same class, or even phylum in many instances. As differentiation proceeds, detectable differences first become apparent between the embryos of those species of which the adults are most widely different.

Thus the very early embryos of fishes and of mammals are quite similar, but they soon become recognizably differentiated. As development proceeds, the similarities of the embryos become progressively restricted to smaller and smaller taxonomic groups until finally the characters which distinguish the adults of closely related species are formed. In many instances, this process is not completed until after birth (or hatching). Thus juvenal robins have the speckled breast which is typical of the adults of most species of the thrush family, to which the robin belongs.

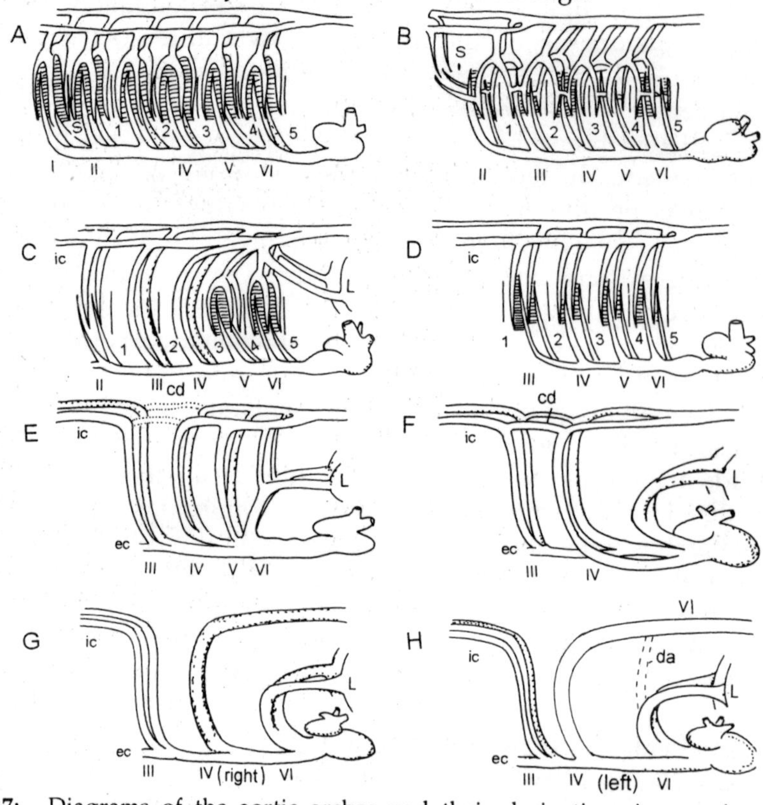

Fig. 2.17: Diagrams of the aortic arches and their derivatives in a series of vertebrates. *A*, hypothetical ancestor; *B*, shark; *C*, the lungfish *Protopterus*; *D*, teleost; *E*, salamander; *F*, lizard; *G*, bird; *H*, mammal. Roman numerals indicate aortic arches; *s*, spiracle; Arabic numerals, gill slits; *cd*, carotid duct; *da*, ductus arterisous; *ec*, external carotid artery; *ic*, internal carotid artery; *L*, lung.

Many examples indicate relationship within as well a between classes. Thus the whalebone whales, which feed by straining minute organisms out of theses water, have no teeth in the adult stage, yet their embryos have a set of tooth buds which are resorbed without ever erupting. Whale embryos also have a coat of hair which is entirely lost to the adults. The absence of birds' teeth is proverbial, yet their embryos also have ephemeral tooth buds. Such facts are readily understandable on the principle that the basic factors of embryology are determined by heredity, and so are common to related species and groups. Both whales and birds are descended from ancestors which had teeth. The hereditary factors which initiate tooth development are still present and active. But an additional hereditary change (mutation) which acts later in development has been independently acquired in each group since its origin. And this change causes the tooth buds to abort.

Much more radical changes in developmental pattern may also occur, so that only the early embryonic stages can give a clew to the true relationships of the organisms involved. Some extreme examples occur among the cirripedes (barnacles). Typical cirripedes show many typical crustacean characters, so that it is not difficult to understand that they belong in this subphylum. Nonetheless, they are sufficiently aberrant as to have deceived so competent a zoologist as Cuvier, who treated them as a class of the Mollusca. Yet the larvae of all cirripedes are unmistakably crustacean larvae, and it was this fact which finally established the correct taxonomic position of the branacles. A much more extreme example is presented by *Sacculina,* a very aberrant barnacle which parasitizes crabs. This organisms goes through all of the typical developmental stages of a barnacle until it settles down upon the abdomen of its host. At this point, other barnacles would develop the usual adult structures of the group, but *Sacculina* undergoes a degenerative development. Appendages are lost, and the parasite becomes sac-like, with branches invading the tissues of the host like a tumor and absorbing nutriment. The only organ system which is well developed is the reproductive system. This parasite resembles no other organism, and so it defied classification until the larvae were studied. The larvae showed clearly that the parasite could be nothing else but a much degenerated barnacale. The case of the enteroxenid snails which parasitize echinoderms is quite similar. They are worm-like externally. Internally, they have only a much reduced gut and an hermaphroditic reproductive system. Yet they produce typical snail larvae.

Another interesting case which was originally studied by Darwin is that of the loggerhead duck, or steamer duck, *Tachyeres cinereus.* This bird is widely distributed in southern South America, the Straits of Magellan, and the Falkland Islands. The young ducks are fairly good fliers, but the growth of the wings does not keep pace with the growth of the body, and the adults are completely flightless.

EVIDENCE FROM COMPARATIVE PHYSIOLOGY AND BIOCHEMISTRY

Many modern biologists are of the opinion that evolution is basically a biological phenomenon. Now it has been fully realised that any morphological characteristic is always an outcome of some developmental processes which are all essentially physiological and chemical. *Haldane* (1937) stated, "our final theory of evolution will see it largely as a biochemical process". For this purpose, physiological processes and chemical composition of various cells and tissues of different animals have been compared. Physiology and biochemistry have given some of the following most important and dependable evidence to support the idea of evolution.

1. Protoplasm Chemistry

The protoplasm is considered as "the physical basis of life". A biochemical study of the living matter in the protoplasm suggests that protoplasm from variety of sources has the same biochemical constitution. It mainly consists of substances like proteins, carbohydrates, fats, water etc. This would suggest that during evolution the most fundamental property of living things has remained intact, while variations in essential respects produced the variability according to the needs of varied forms.

2. Chromosome Chemistry.

Like protoplasm, another remarkable similarly at the biochemical level is found in the chemistry of chromosomes. The chromosomes of all living organism basically consist of nucleic acids (DNA, and RNA) and proteins (histones and protamines). The molecules of these chemical substances remain arranged in all chromosomes in, a almost identical fashion. Such a uniformity in the composition of chromosomes again suggests a common origin of most living beings.

3. Enzymes

A large number of enzymes (bio-catalysts) are same in all the animals. For example *Trypsin,* a protein splitting enzyme is present from protozoans to mammals and hence called an *ancient enzyme.* Similarly *amylase* is found sponges to mammals.

4. Hormones

There are a number of hormones which throw light on the organic evolution. For example *thyroid hormone* is common in all vertebrates and this hormone from one class of organisms can be substituted for that in another class of organisms. For the treatment of human thyroid deficiency, beef thyroid has been successfully used. Also in case thyroid gland of frog is surgically removed at a young stage, frog will not metamorphose, but if it is again fed with thyroid tissue from mammals, the metamorphosis can be easily brought about.

Another examples is that of M.S.H. or *melanophore stimulatin, hormone* of amphibians. This hormone is found in mammals also, but has no known effect there. If the hormone, extracted from mammalian pituitary glands, is supplied to amphibians, it is as effective as the amphibian hormone itself. It suggests that this enzyme in mammals does not have any known function and is only found as a vestigial of the amphibian ancestry.

5. Excretion

In most of the aquatic animals, the nitrogenous waste is ammonia, in terrestrial forms it is urea, but in insects, reptiles and birds, it is the semisolid uric acid. This is due to economy in the supply and intake of water. Ammonia being toxic is eliminated soon in aquatic forms which have ample supply of water while in other forms it is converted into harmelss urea or uric acid. An interesting example is that of embryo of birds. In birds, in the embryo for the first five days the nitrogenous wastes are in the form of ammonia, for the next few days (till 9th days) urea, and finally uric acid.

6. Haemoglobin

Haemoglobin, a respiratory pigment, when treated with glacial acetic acid precipitates in crystals called *haemin* crystals. The shape, size and structure of these crystals is similar in all the member of a species. They are different from those of other species. The degree of differences agrees with the classification of animals.

7. Blood Plasma

The similarity in the composition of blood plasma of different animals is very characteristic. Plasma contains proteins in the solute form. Besides, it also contains salts of sodium, potassium, calcium and magnesium.

8. Comparative Serology

If small amount of blood serum from an animal is injected into the blood of guinea pig or any other animal, *antibodies* against the *antigens* of the injected blood will be produced. These antigens will be destroyed now by antibodies thus produced in the blood stream as well as in a test tube. The strength of precipitate depends upon the concentration of antigen. The precipitate is the *precipitint* and the test is *precipitation test*. *Dr. George H.F. Nuttal* employed the precipitation method for finding out the interrelationship among different animals. Some blood of man is taken and allowed to clot. The clot is removed and the remaining yellowish fluid is left called *serum*. It is now injected in a small dose into the rabbit. The proteins of this serum act as *antigens* in the body of rabbit. Due to it *antibodies* are formed in the blood of rabbit. These antibodies clump and destroy the human blood. After the last injection of heavy dose, the rabbit is allowed to live for some days and then its blood is drained away and allowed to clot, its clots is removed and the left serum is collected called *anti-human serum*. If the serum of man is mixed with this anti-serum, a white precipitate is formed. When it is mixed with the blood of dogs, cats, primates etc. With the blood of dogs and cats precipitation does not take place but precipitation occurs very frequently with blood of primates (monkeys, apes, gorilla etc.). This test shows the relationship between man and the other primates and not with the dogs and cats.

Different types of serum can be prepared in this manner and relationship can be tested *Nuttal* and *Graham Smith* concluded the following observation after conducting thousands of such tests.

(i) *Anticarnivora serum* given maximum reactions among the bloods of carnivora than others.

(ii) *Antipig sera*. It gives maximum reactions only with blood of species belonging to pig family, moderate reactions with ruminants and camel. *Anti-Llama serum* gives moderate reaction with the blood of camel.

(iii) *Anti-whate serum*. It gives maximum reactions only with bloods of other whales and slight reactions with that of pigs and ruminants.

(iv) *Anti-turtle serum*. It gives maximum reactions with the bloods of turtles and crocodiles and negative with lizards and snakes.

(v) *Anti-lizard serum.* It reacts with the bloods of lizards and snakes.

(vi) *Anti-mammalian serum.* It gives positive test of interrelationship with all the mammals, but degree and time taken varies in every case.

Thus the serological tests prove the relationship. In general the species of a single genus exhibits very close serological similarity, genera of the same family show moderate reaction and families of the same order show slight similarity.

9. Phosphagens

The phosphagens play a key role in muscle contraction and are the sources of energy for the resynthesis of ATP, once they are brokedown. In the muscles of most vertebrates, phosphagen is always a specific compound called *creatine phosphate,* while in most invertebrates it is *arginine phosphate.* Hemichordates, the most primitive

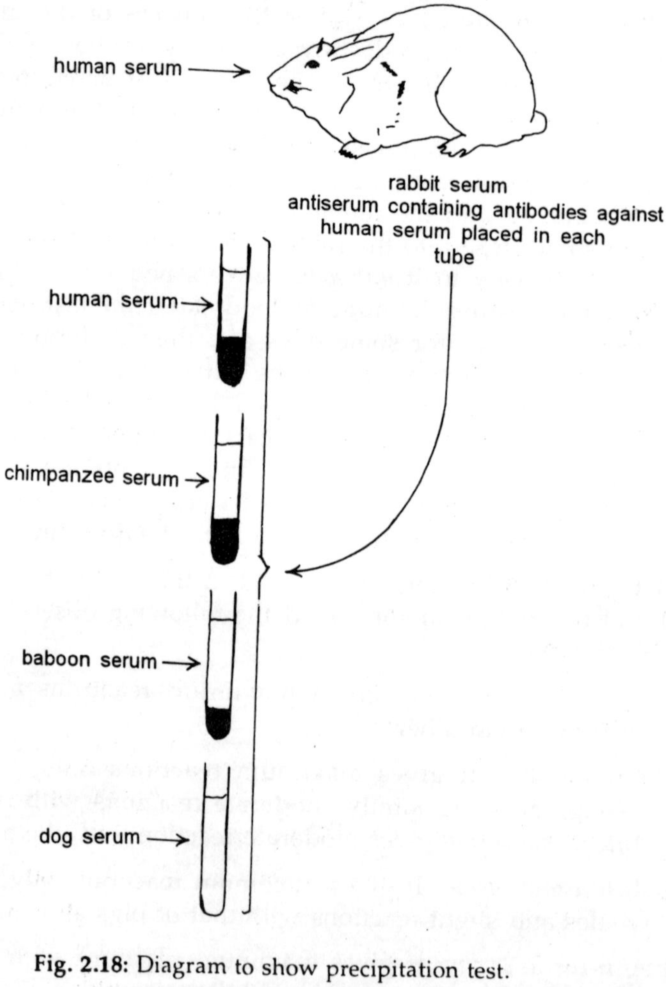

Fig. 2.18: Diagram to show precipitation test.

chordates, have both the phosphagens, the creatine phosphate as well as arginine phosphate, such a situation is also found in echinoderms and on morphological grounds echinoderms have been considered close to the ancestor of chordates.

10. Cytochrome 'C'

It is a respiratory enzyme present in the mithochodria. It is made up of 104-112 amino acid monomers. The sequence of amino acid in its molecules has been determined by *Dickerson* (1972) and other workers in diverse forms such as man, rabbit, fishes, snakes etc. Variations found in the sequences are in agreement with accepted phylogenetic relationship.

EVIDENCE FROM PALEONTOLOGY

The various lines of evidence discussed thus far all indicate that the species now populating the earth must have been produced by evolution, but generally the only possible direct evidence for a specific line of descent is a series of fossils leading stepwise from an ancestral to a descended species. Hence the science of paleontolog, which deals with fossil remains, has unique importance for evolution.

THE GEOLOGICAL TIME SCALE

Before beginning the discussion of paleontological evidence for evolution, however, it is necessary to introduce the problem of geological time. Fortunately, there are methods of determining the ages of rock deposits in the earth's crust. The oldest method is applicable only to sedimentary rocks—those successive layers or strata of rocks which are formed by slow settling out of sediments from the oceans or other large bodies of water. The use of the method for dating purposes is based upon the assumption that those geologial processes which are observable in action now are the same ones which have determined the past history of this earth, and that they have in the past acted at rates comparable to those now observable. As applied to the problem at hand, this simply means that the sedimentary rocks of the past were deposited at rates comparable to those which are now being deposited. In a purely comparative way, dating by this means is fairly satisfactory. The deepest strata are the oldest, while the most superficial strata are quite recent. Thick strata represent long-continued deposition, while thin strata represent short periods of deposition. Thus some relative concepts of geological time are readily gained from an inspection of the sedimentary rock deposits. But more exact ideas are difficult to obtain because it is clear that sedimentation is now progressing at very different rates in different parts of the world, and there is nor reason to doubt that the variation in the rate of sedimentation has been as great over much of the world's history. It may well have been greater at times. Hence calculations of age based upon the thicknesses of stratified rocks lead to such statements as that the Mesozoic Era began somewhere between 1900,000,000 and 240,000,000 years ago. But the difference between these figures is more than 25 per cent of the smaller one. In addition to this difficulty, the strata have been changed by such geological processes as folding and erosion, so that often the record is fragmentary and confused. There

Table 1. Geologic Timetable

Era	*Period*	*Epoch*	*Duration in millions of years*	*Time from Beginning of period to present (millions of years)*	*Geologic conditions*	*Plant life*	*Animal life*
Cenozoic	Quaternary	Recent	0.025	0.025	End of last ice age;	Decline of woody plants;	Age of man
(Age of Mammals)		Pleistocene	1	1	climate warmer Repeated glaciation; 4 ice ages	rise of herbaceous ones Great extinction of species	Extinction of great mammals; first human social life
	Tertiary	Pliocene	11	12	Continued rise of mountains of western North America; volcanic acitivity.	Decline of forests; spread of grasslands; flowering plants, monocotyledons developed	Man evolving; elephants, horses, camels almost like modern species
		Miocene	16	28	Sierra and Cascade mountains formed; volcanic activity in northwest U.S.; climate cooler		Mammals at height of evolution; first manlike apes
		Oligocene	11	39	Lands lower; climate warmer	Maximum spread of forests; rise of monocotyledons, flowering plants	Archaic mammals extinct; rise of anthropoids; forerunners of most living genera of mammals
		Eocene	19	58	Mountains eroded; no continental seas; climate warmer		Placental mammals diversified and specialized; hoofed mammals and carnivores establishe Spread of archaic
		Paleocene	17	75			
Rocky Mountain Revolution (Little Destruction of Fossils)							
Mesozoic (Age of Reptiles)	Createaceous		60	135	Andes, Alps, Himalayas, Rockies formed late; earlier, inland seas and swamps; chalk, shale deposited	First monocotyledons; first oak and maple forests; gymnosperms declined	Dinosaurs reached peak, became extinct; toothed birds became first modern birds, archaic mammals common
	Jurassic		30	165	Continents fairly high; shallow seas over some of Europe and western U.S.	Increase of dicotyledons; cycads and conifers common	First toothed birds; dinosaurs larger and specialized; insectivorous marsupials

	Triassic		40	205	Continents exposed; widespread desert conditions; many land deposits	Gymnosperms dominant, declining toward end; extinction of seed ferns	First dinosaurs, pterosaurs and egg-laying mammals; extinction of primitive amphians
	Appalachian Revolution (Some Loss of Fossils)						
Paleozoic (Age of Ancient Life)	Permian		25	280	Continents rose; Appalachians formed; increasing glaciation and aridity	Decline of lycopods and horsetails	Many ancient animal died out; mammal-like reptiles, modern insects arose
	Pennsylvanian (Carboniferous)		25	255	Lands at first low; great coal swamps	Great forests of seed ferns and gymnosperms	First reptiles; insects common; spread of ancient amphibians
	Mississippian		25	280	Climate warm and hu-	Lycopods and horsetails	Sea lilies at height;
	(Carboniferous)				mid at first, cooler later as land rose	dominant; gymnosperms increasingly widespread	spread of ancient sharks
	Devonian		45	325	Smaller inland seas; land higher; more arid; glaciation	First forests; land plants well established; first gymnosperms	First amphibians; lung-fishes, sharks abundant
	Silurian		85	360	Extensive continental seas; lowlands increasingly arid as land rose	First definite evidence of land plants; algae dominant	Marine arachnids dominant; first (wingless) insects; rise of fishes
	Ordovician		65	425	Great submergence of land; warm climates even in Arctic	Land plants probably first appeared; marine algae abundant	First fishes, probably freshwater; corals, trilobites abundant; diversified molluscs
	Cambrian		80	505	Lands low, climates earliest rocks with abundant fossils	Marine algae	Trilobites, brachipods dominant; most modern phyla established
	Second Great Revolution (Considerable Loss of Fossils)						
Proterozoic			1500	2000	Great sedimentation; volcanic activity later; extensive erosion, repeated glaciations	Primitive aquatic plants —algae, fungi	Various marine protozoa; towards end, molluscs, worms, other marine in vertebrates
	First Great Revolution (Considerable Loss of Fossils)						
Archeozoic			???	???	Great volcanic activity; some sedimentary deposition; extensive erosion.	No recognizable; indirect evidence of living things from deposits of organic material in rock	

are a few instances in which stratified rocks appear to have been laid down in definite annual layers, the varves, more or less comparable to the annual growth rings of trees. In such cases, the extent of the periods of sedimentation can be determined with great exactness, if the varves actually do represent annual layers. But this is by no means certain. And even if it were certain, the number of known examples is so small that it could have little importance for the general problem of dating geological history.

Although the study of sedimentary rocks has not led to a satisfactory dating system, it has been possible to determine the sequence in which the various strata of the earth's surface have been laid down, from very ancient rocks right up to those of very recent origin. Particular strata are identifiable not only by their position and their physical characteristics, but also by the fossils which they contain. Thus it has been possible to divide geological time into a series of eras, the sequence of which is undoubted. The first two eras, the Archeozoic and the Proterozoic, are not of great interest for the present discussion, because the rocks deposited in these eras contain very few fossils, and those are generally of doubtful character. During the Paleozoic Era, fossils were deposited in great abundance, but only archaic types were present. At first, only invertebrates were represented, but fishes, amphibians, and finally reptiles made their appearance during the Paleozoic Era. The next great era was the Mesozoic, or Age of Reptiles, during which birds and small mammals also arose. Finally, the Cenozoic Era, which is still in progress, has been marked by the rise to dominance of the mammals and man.

The eras are, however, immensely long expanses of time, characterized by progressive differences of flora and fauna, and not infrequently of climate and other physical characteristics. The eras are therefore divided into *periods* of shorter, but still very long, duration. Thus the Paleozoic Era lasted for about 300,000,000 years, but it is subdivided into seven periods, the durations of which vary from 25,000,000 years to 80,000,000 years. Finally, it is sometimes desirable to break up the periods into still smaller divisions, the *epochs*. The Tertiary Period of the Cenozoic Era, for example, lasted for about 74,000,000 years. This is subdivided into five epochs the durations of which varied from 11,000,000 years to 19,000,000 years.

While this system of time measurement is not quantitatively accurate, it is, on the whole, quite workable. For, if a fossil is found in strata from the Cretaceous Period, it is possible to state with complete assurance that it followed Jurassic forebearers and preceded Paleocene descendants, if any descendants were left. The geological time scale, together with some of the characteristics of life at each level, is summarized in Table 1.

The Lead Method

In 1907, Boltwood introduced a method for dating geological strata based upon radioactive elements. The conclusions to which the new method led indicated that the earth was vastly older than had been generally believed, and the method was received with skepticism. But it has since become the standard by which the accuracy of other methods of dating is judged. The method is based upon the fact that uranium 238 will slowly disintegrate to produce lead with an atomic weight of 206 and helium. The rate at which this occurs is calculable. With any definite amount of uranium, one half of the molecules will break down, forming lead and helium, in the course of 4,510,000,000 years. As this figure is independent of the actual quantity of uranium originally present, it is called the "half-life" of the element. Now, if a uranium-bearing rock is found, the ratio of uranium to lead 206 can be determined, and from this, utilizing the half-life, the interval since the formation of the rock can be calculated.

Recently, additional geochemical methods have been developed. Potassium 40 yields calcium 40 and argon 40; rubidium 87 yields strontium 87; thorium 232 yields lead 208; and uranium 235 yields lead 207. Each of these parent elements has its characteristic half-life, ranging from 126,000,000 years to as much as 60,000,000,000 years.

While the "lead" method is now universally accepted, it has serious limitations. Uranium is not a common element, and it is often found in geological formations which are not readily fitted into the geological time scale. The introduction of additional geochemical methods helps to overcome these limitations, and so there are good grounds for hope that an accurately dated scale of geological time may be available in the near future. For the present, it may be said that the oldest dated rocks are more than 3,000,000,000 years old, and the oldest which might have borne life are about 2,000,000,000 years old. Around three quarters of that 2,000000,000 years during which life might have existed passed before the beginning of the Cambrian Period, with which the useful fossil record begins, for a lead measurement of the age of a late Cambrian deposit gave a figure of 440,000,000 years. The next exact determination is in the early Permian Period, at an age of 230,000,000 years. Thus the entire Paleozoic Era probably lasted about 300,000,000 years; the Mesozoic about 130,000,000 years; and the Cenozoic about 75,000,000 years up to the present. There is an accurately dated deposit from the beginning of the Eocene Epoch which places this at 58,000,000 years ago. The dates are few, but fortunately they are widely scattered in geologic time. But as yet, accurate determinations are not available for the extent of any of the periods or epochs.

The Radio-Carbon Method

Another radioactive method has been developed which shows promise for shorter range determinations of age, up to perhaps 40,000 years. It has been found that living organisms utilize a small, but constant proportion of their organic carbon in the radioactive form. The half-life of radioactive carbon is 5568 ± 30 years. Hence, remnants of bone, wood, or other carbon-containing remains of dead organisms can be assayed for their radio-carbon content. The difference between the average amount in fresh tissue and in the fossil may then be treated as being due to radioactive disintegration, and the age of the fossil calculated from the half-life. This method has proved very useful in the study of late Pleistocene and Recen remains. It has been checked against historical objects of known age, such as wood from Egyptian pyramids, and has been found reliable.

FOSSILIZATION

Any remnant of living forms from the remote past may be regarded as a fossil. While there are many ways in which fossils may be formed, most of them involve the burial of the dead organisms. As more and more sediment is laid down above, the depth of the fossil increases, and hence, in a general way, the oldest fossils may be expected to be found in the deepest layers of the earth's surface, while the more superficially located fossils may be regarde as recent in origin. Thus, ideally, one ought to be able to read the story of life in the correct historical sequence by examining the fossil record in sequence from the deepest to the most superficial strata.

But the vast majority of plants and animals are not fossilized after death. Decay and destruction ordinarily await the dying organisms. Predators and scavengers may not only eat away the soft parts of the body, but they may also break up skeletal structures beyond all hope of preservation or recognition. It is only the unusual instance in which the organism is rapidly buried or in some other way protected from scavengers and from oxidation that a fossil may be formed. Thus the fossil record, even if completely known, would have to be very fragmentary, because the majority of organisms never take the first step toward fossil formation.

Much the most common method of fossilization is burial in the sediments which are continually deposited on the floor of the oceans and of other large bodies of water. When aquatic organisms die, they may fall into deep sedimentary deposits in which the bodies are protected from scavengers and from oxidation. The soft parts of the body gradually decay and are carried away by the seepage of water. Bones and other hard parts may remain as such, or they may be replaced particle by particle by minerals in the water. As this process continues, the layer of sediment which is being deposited grows ever thicker, and its lower portions gradually harden into rock, the sedimentary, stratified rock which is characteristic of the beds of marine or aquatic deposits everywhere, and which is the hallmark of ancient seas in areas which are now dry land. However, this is not the only means by which organisms may be buried. Dust storms can have the same effect, and will be effective in causing fossilization of terrestrial organisms. Again, volcanic ash may also rapidly bury organisms and thus preserve them as fossils. Pompeii, which was buried by volcanic ash from Mount Vesuius in 79 A.D., has been extensively investigated in modern times. Whole families, together with their domestic animals, have been preserved as cavities in the ash, from which casts can be made. Desert forms may be dried out by the hot, dry, desert winds, then buried under the shifting sands.

Some special methods of burial are also occasionally effective. If a petroleum spring should occur, evaporation of the more volatile oils will produce first a pool of sticky tar and then one of viscous asphalt. This has happened during the Pleistocene epoch at Rancho La Brea in southern California. Many Pleistocene and Recent mammals and birds have been trapped in this asphalt, and they are among the best preserved of fossils. It appears to have worked in the following ways. Small mammals, herbivores, and birds try to reach the rain pools which occur on the surface of the asphalt. In so doing, they become stuck in the soft asphalt, and predators are then ensnared while attempting to catch the former. Water birds may alight on the water pools and then become entrapped in the asphalt around the edges. Thus Rancho La Brea is one of the richest sources known for well preserved fossils of recent mammals and birds. Because the city of Los Angeles has grown up around it, it no longer entraps the wild fauna of the region, but the Los Angeles fire department is still occasionally called out to rescue a child who has gotten his feet stuck in the asphalt.

Another unusual method of burial is the entrapment of insects in amber (Fig. 2.19). Such fossils are sometimes preserved almost perfectly, so that even histological details are comparable to those of freshly fixed specimens.

Fig. 2.19: Termites in Mid-Cenozoic Amber.

Lastly, organisms may be petrified, that is, their actual tissues may be replaced, particle by particle, by minerals in solution in the waters of the locality. The principal minerals utilized in this type of fossilization are iron pyrites, silica, calcium carbonate, and other carbonates. The most widely known example of petrifaction is afforded by the petrified forests of southwestern United States; however, animal remains may also be petrified. Generally, this preserves only the hard parts of the body, but occasionally soft parts are so well preserved that even fine details of cells can be made out in thin sections. Most of the fossils from sedimentary rocks are of this type: the original material has been replaced by minerals from the surrounding medium.

(i) **Petrified fossils.** In this type of fossils, as a result of petrification the organic matter a dead animal or plant is replaced particle by mineral matter. In petrification, quartz, limestone or iron pyrites are used.

(ii) **Dead and preserved bodies.** The actual dead and preserved bodies, or parts of bodies, of animals and plant, with the original tissues intact; and enclosed either in ice or in amber or else mummified in various ways, form one class of fossils. From the arctic ice, almost fresh bodies of mammoths, dead thousands of years ago, have been recovered. Several specie of ants and insects have been preserved in amber especially found on the shallow coats of the Baltic sea. Several mollusc-shells, teeth of sharks, bones of animals have been found in well preserved condition.

(iii) **Casts and impressions.** Some times an animal or plant that has lain in mud or clay long enough to have left its impression; the mud hardens about the body and forms a mold; the organic matter disintegrates and the mold is filled with hard mineral matter; and the matrix may then be removed so as to leave the perfect cast.

(iii) **Trails and footprints.** Fossils are also found in the form of trails and footprints. Animals when they move from one place to another leave impressions, which are preserved when sand and mud become stone. Such fossils have been found of organisms such as worms, molluscs, arthropods and vertebrates.

(v) **Cropolites.** These are the structures, which represent fossil casting or excreta.

(vi) **Gastroliths.** These are present in the body cavity of certain reptiles. These structures are believed to have been of some use in grinding the stomach contents of the extinct reptiles.

Fig. 2.20: Process of Fossil Formation.

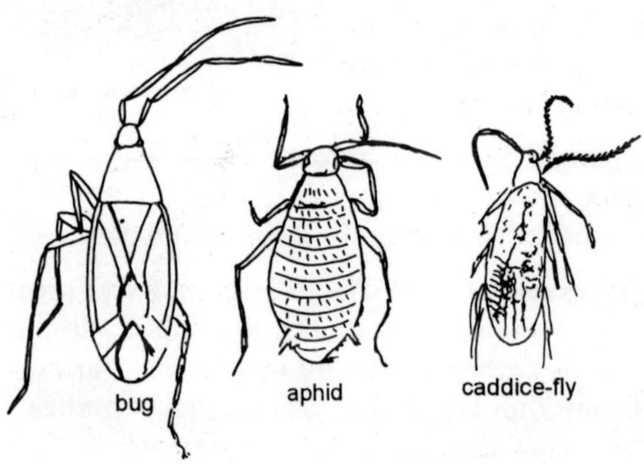

Fig. 2.21: Insects preserved in Amber.

(vii) **Pseudofossils.** Many objects of inorganic origin closely resembles the forms of organic origin, and are found in the sedimentary rocks. These are called pseudofossils.

Conditions for fossilization

There are two important conditions for fossilization (i) hard parts and (ii) immediate burial.

The Geological Record

On the basis of their studies palaentologistis have divided the geological history of earth into five major *eras;* these are in turn divided into *periods* and periods into *epoches*.

Importance of Fossil Record

1. Most primitive forms of life are present in the oldest rocks.
2. From older to recent strata or layers, there is a succession of higher and more complex forms.
3. In many cases, a group arose in one era or period and remained scare. However, in the next period or era it became dominant due to adaptive radiation.
4. Large groups became extinct but after establishment of all the major phyla, some species of each phylum remained upto the present.
5. The past forms of life are not exactly similar to any of the living forms.
6. The main trend in evolution has been a gradual progression from simpler and more generalised types towards more complex and specialized forms, as one proceeds from the lower towards the upper strata.
7. It is seen that a new group arises near the close of a geological period during which great climatic changes were taking place. Such a incipient group become dominant and characteristic group of next period.
8. Many complete fossil pedegrees have been worked out which establish the most convincing evidence at our disposal that evolution has taken place.

Transitional or Intermediate Fossil Forms

Transitional fossil forms are known between different existing groups from fishes to amphibia, from amphibia to reptile, from reptiles to birds and from reptiles to mammals. For example.

Stegocaphalus is a form linking fishes and amphibians. *Archaeopteryx* and *Archaeornis* are the toothed birds connecting reptiles and birds. *Inostracevia* was a intermediate form between reptiles and mammals. *Manchurosaurus* was also a link between reptiles and mammals.

Incompleteness of the Geological Record

Unfortunately the geological record does not give a complete picture of all the plants and animals which existed in the past. If it had been complete no further proof of organic evolution, would have been required.

The incompleteness of the geological record is due to various reasons, some of which are listed below.

1. In the earlier time plants and animals were soft and usually not fossilized. Most of them, after the death, were destroyed by micro-organisms, predators, etc. In some cases some part fossilised and some not, as in the case of plants where wood is preserved and flowers, leaves etc. are destroyed. In this way we find that most plants and animals are poorly fossilised.
2. Several examples of the cracking of rocks have been found. The cracking may be caused by tension. Many such rocks containing fossils have been destroyed in the world's history.
3. Fossil formation generally takes place at the bottom of sea. Many land animals and plants could find no chance to reach the sea.
4. Sometimes the superficial rocks are flown away with the moving glaciers due to which the deeper strata are exposed and these exposed strata are eroded by wind and water destroying the fossils preserved there. This incompleteness of fossils record does not give a complete picture of progressive evolution.
5. Fossils lying at great depths have been destroyed by pressure of the overlying rock layers and heat.

Evolution of Horse

The history of horse dates back to about 60 million yeas in the eocene period, involving about 20 genera. The horse phylogeny starts with *Hyracotherium* (including *Eohippus*) a member of Equidae. This animal was a little less than a foot high, the teeth were 44 and the check teeth were only moderately specialised for grinding. The front foot had four toes and a splint and hind foot with three toes and two splints. The line of descent perhaps passed from *Hyracotherium* through *Mesohippus, Miohippus, Parahippus, Merychippus* and *Pliohippus* to *Equus,* which represents the modern horse and its allies.

In the above transition, changes were brought about from the size of fox to that of horse. The teeth became specialised for grinding coarse siliceous grasses. These changes were associated with elongation of jaws. The limbs were elongated and the toes were lost. The metacarpels and matatarsals were fused. The neck became much longer and mobile.

EVIDENCES FROM GEOGRAPHICAL DISTRIBUTION

The geographical distribution or *biography* of animals and plants on the earth give a clear understanding of the organic evolution theory (*Wallace*). Geographical

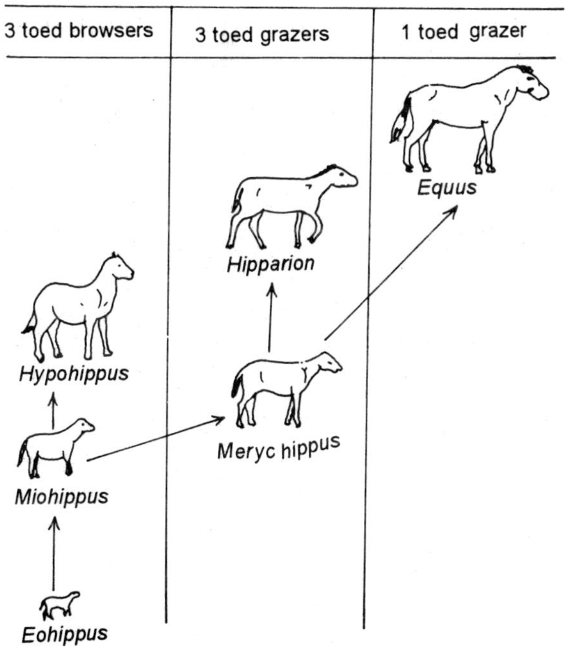

Fig. 2.22: Evolution of horse.

distribution may also be called a study of the *horizontal distribution* of organisms known as *spatical distributions* of the earth's surface at a given time. The distribution of animals is governed by many factors such as environment, temperature, availability of required type and amount of food, salinity, enemies, topography and barriers.

On the basis of similarity and differences of fauna and flora in different regions, the world has been divided into following six biogeographical regions.

(a) Nearctic—North America down to the Maxican Plateau.

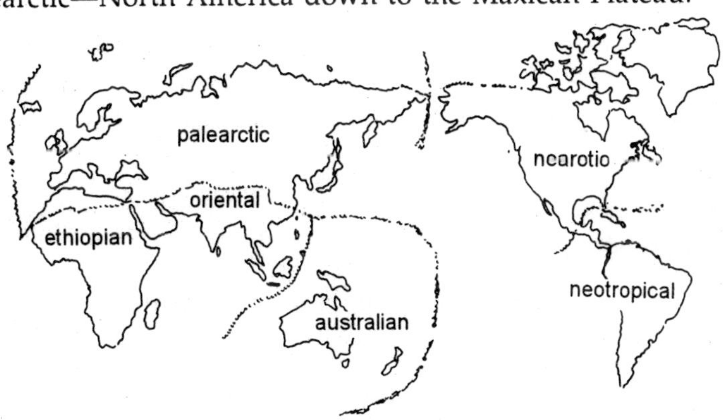

Fig. 2.23: Geographical Regions

(b) Palearctic—Asia North of the Himalayas, Europe and Africa, North of Saharta Desert.

(c) Neotropical—Central South of America.

(d) Oriental—Asia, South of Himalayas.

(e) Ethiopian—Africa, South of Himalayas.

(f) Australian—Australia and associated islands.

Most of these biogeographical regions are separated from one another through barriers of sea, deserts, mountains, etc. and these barriers are quite ancient in time. The physical conditions of one region may be similar to that of the other, and one would expect that the plants and animals of one region should have no difficulty in inhabiting the other region. Yet these different regions have little in common so far as their fauna and flora are concerned. Thus, elephants are found in tropics, but in India and Africa alone, and not in other parts of the earth between the same latitude. Conditions at the two poles are almost identical but their fauna are strikingly different: penguins in South Pole and polar bears in the North Pole. Climatic conditions of British Isles is very much like that of New Zealand, but the fauna of two are very different.

Marsupials are restricted to Australia and South America (with the exception of North American *Opossums*). Anthropoid apes occur in isolated areas in West Africa and South East Asia. On the other hand, dogs, mice among mammals, thrushes, swallows and plovers among birds and many families of insects are examples of almost universal distribution. The fauna of West of Africa (elephant, antelopes and gorilla) is entirely different from that of Brazil (tapirs, sloths and monkeys with prehensile tails). Distance also do not determine distribution. Great Britain and Japan separated by long distances posses similar fauna while the islands of Bali and Lombock separated by only a sea of 15 miles distance contain entirely different faunas.

The above facts cannot be explained b the theory of special creation; but the explanation lies in the doctrine of organic evolution. The dissimilarities in the fauna and flora of regions having identical environmental conditions, can be interpreted as under:

1. Each species has originated only once. The originate in all habitats at all times.

2. Each species tends to expand from its centre of origin and this migration is blocked by physical or climatic barriers.

3. New species can arise only insulation and hence related species tend to fond in adjacent areas.

4. Due to geological changes or action of enemies in the course of time, extension took place in large areas. This resulted in *Discontinuous distribution*.

EVIDENCES FROM CYTOLOGY

Animals are made up of the aggregrates of cells. The fundamental structure of a cell is similar throughout the animals kingdom. It consists of nucleus, cytoplasm and the surrounding cell membrane. Process of cell division (mitosis and meiosis) are also the same throughout the animal kingdom. The only explanation is that all the animals had a common origin.

EVIDENCES FROM GENETICS

The final line of evidence for evolution is drawn from Genetics, the science of heredity and variations. Hereditary variations on which the whole idea of organic evolution is based will result only through the application of certain basic principles of genetics. Thee are mainly two sources of hereditary variations namely, recombination and mutation. While recombinations after hybridization will yield new combinations, mutations will create new genetic material which never existed earlier.

Examples of hybridization and mutations are available which show that evolution has taken place. Man has tried to cross different species of animals to give rise hybrid. Besides artificial hybrids, there are some naturally occurring hybrids in plants.

A mule is the offspring of a jackass (*Equus asinus*) and a mase (*Equus clabus*) is sterile hybrid. Natural hybrids in animals are among butterflies, moths, crickets, fresh

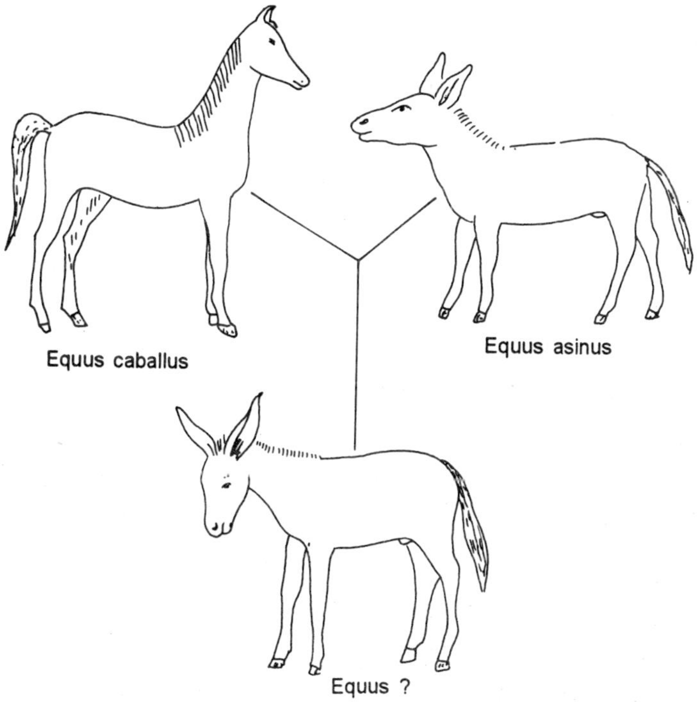

Fig. 2.24: Mule, an offspring of mare and Jackass

water fishes, toads and ducks. Hybridization is uncommon in reptiles and birds. Changes in the basic number of chromosome can occur due to breakage and reunion of chromosomes.

Different kinds of mutations-namely gross mutation due to variation in chromosome number (polyploidy and aneuploidy), and point mutations introduce different kinds of variations in plant and animals are consequently result in speciation. Plant breeders have employed induced polyploidy methods in producing numerous economically important varieties of plant. Geneticists have produced many new strains of micro-organisms, plants and animals (*Drosophila*) by artificially inducing point mutations in them.

Mutations form the raw materials for evolution. Experimental data with induced mutations proves beyond any doubt "descent with modification". The cross breeding in animals and plants has frequently resulted in the production of sudden and unpredictable new types from older parent types.

Chapter—3

Lamarckism & Ned-Lamarckism

Lamarck's Laws.

Buffon, Erasmus Darwin, and Lamarck, as we saw in the first chapter, all developed evolutionary theories which had one thing in common, the inheritance by offspring of the modifications impressed upon the parent during its lifetime. Buffon believed that these modifications were the result of the direct action of environmental influences; Erasmus Darwin that they sprang from within through the reaction of the organism to outside conditions, applying his theory to both animals and plants; while Lamarck's belief was a compromise, setting forth views comparable to those of Erasmus Darwin as applied to animals and of Buffon with reference to plants.

Lamarckism

The first extensive and coherent theory of evolution was put forward by *Jean Baptiste Lamarck* (French) in his *Philosophic Zoologique* in 1816.

Lamarck was very much impressed by the effect of environment on plants and animals. His conclusions are summarized as follows:

(1) The internal forces of life tend to increase the size of an organism, i.e. the whole body and also the different parts upto a limit which it brings about itself. (There is gradual increase in size).

(2) The production of new organ in an animal body results from the arisal and continuance of a new need, and from the new movement which, this need brings into being and sustains. (New need produces new organ).

(3) The degree of development and their forces of action are directly proportional to their use, continuous use strengthening the organ while disuse have reverse effect.

(4) All that has been acquired, imprinted or changed in the organization of the individual during the course of its life is preserved by generation and transmitted to the new individuals that descend from the individual so modified. (Acquired characters are inherited).

The fourth principle, that is, acquired characters are inherited is the most attractive and controversial law, now known as *'Lamarckism'* or *'Lamarckian doctrine'*.

Analysis of Lamarck's Theory

(i) The first of his law is the statement of a supported fact, which is regarded to be universally far from truth. While the evolutionary trend in certain groups may be associated with increase of size, there are a number of cases, where evolution proceeded not only without any increase in size but rather through a reduction in size. In several groups of plants such a reduction in size can be seen.

(ii) According to 2nd law of *Lamarck*, the development and efficiency of an organ is in proportion to the use of that organ. If an organ is used constantly, it will tend to become highly developed and if it is not used by the organism, then it would atrophy or may disappear altogether in the course of future generations.

(iii) The law of use and disuse of organ appears satisfactory and is supported by the following evidences:

1. Giraffe is found in forests of Africa. In an attempt to catch the leaves of the branches, high above the ground, its neck went on elongating and attained the present day form. In this effort, his front legs were also used along with the neck, as a result of which the legs were lengthened with the stretching of neck. This change was inherited to the offsprings gradually and caused the evolution of present giraffe.

2. The snakes live in bushy places and therefore they have got a long body. *Lamarck* explained that the ancestors of the snakes were 4 limbed and tried to conceal themselves in the bushes. This required creeping habit and elongated body. These efforts resulted into aborted limbs and elongated body of snake. The locomotion was carried by the abdomen.

3. The flat fishes (flounder) have their both eyes on one side of the body. It is assumed that the ancestors of these fishes used to turn over to one side of the body in order to swim over in shallow water. They, therefore, made efforts to bring both the eyes on the upper side. This adaptation was inherited and all the flat fishes now have their eyes on one side.

4. Ducks and allied birds, when did not get enough food on land, they were forced to search it in water. Here they spread their toes and stretched their legs for wading purpose, resulting into webbed feet and long legs.

5. An athlete or a wrestler has at the peak of his training more powerless leg muscles due to his protected illness. A black-smith develops a large and strong *biceps* muscle to his constantly pounding on his anvil.

6. The fore limbs in birds have developed into wings with which they fly, but then continuous disuse in Ostrich and Kiwi has resulted in their reduction and weakness.

The third law of *use* and *disuse* of organs is true but it does not form a principle of evolution except in conjuction with the fourth law of *inheritance of acquired characters.*

(iv) The inheritance of the characters acquired during the life time of an individual is the fourth postulate and is an essential requirement, if the modifications produced due to changes in environment, have to have an evolutionary significance.

Lamarck recognized that such structural changes, irrespective of whether they are progressive or retrogressive, acquired by the individual during its life span are inherited or transmitted to the offspring.

Fig. 3.1: A=Original short necked ancestor of Giraffe., B=Keep stretching the neck., C=and stretching, D=Until neck becomes progressively longer.

Origin of New Species. Lamarck considered that existing individual is the sum total of characters acquired by its ancestors in the preceding several generations, so that the present form is so markedly different from is ancestors, that it has become a new species.

The evolutionary theory of *Lamarck* was strongly supported by *E. Geoffroy* (1772-1844)

ACQUIRED MODIFICATIONS

Acquired modifications are of such interest that it is well to enumerate the various sorts in some detail.

Restriction of Size

Food supply profoundly affects the ultimate growth of an animal. A number of house-flies of the same species will be seen to differ materially in size around a certain average. This does not mean that the smaller ones are younger than the larger, for an adult insect when once it has attained the power of flight has ceased to grow. What it does mean is that the small fly, unless it belong to a different variety, was unable to secure sufficient food during its period of larval life, resulting in a permanent dwarfing of the individual. Nothing is recorded, however, of any permanent effect of such dwarfing upon the race, while silk-worms, if meagerly fed for one generation, fail to attain the full optimum of size as adults for three generations, even though the larvae are amply fed. With the succeeding generations, however, the moths become larger and soon attain their normal size. These temporary dwarfings are called inductions.

Permanent dwarfing is known to occur, as in the case of the dogs owned by the Indians around the Hudson Bay trading posts, which have been in their possession about forty years and are now much smaller than better fed dogs belonging to the same original stock, but owned by the white men of the posts. Other instances of permanent dwarfing are seen in the Shetland ponies, of which the average height is about 10 hands, though many do not exceed 9 hands. Whence the ancestral stock of these ponies reached the Shetlands is unknown, some writers suggesting a Scandinavian, others a Scottish origin, but some of them are cart-horse-like in build, others more slender and Arab-like. Whatever their origin, they are unquestionably derived from a much larger stock, and the diminution of size is apparently a direct response to circumscribed surroundings together with hard conditions and meager fare.

In the islands of Malta and Cyprus have been found dwarf races of elephants, the adult individuals ranging in height from three to seven feet, relics of the old armies of migration when these Mediterranean islands were part of a broad highway of communication between Africa and Europe. And here again the same causes which have dwarfed the Shetland ponies seem to have had their influence, for the small size and innumerable variations of these elephants are ascribed to the struggle for existence that such a reduced and unfavorable feeding ground would entail. These dwarf elephants are now entirely extinct, but seem to have been of the same stock as the African elephants of to-day, which in the fullness of their growth possess a stature second to no living terrestrial form.

The effect of these restrictions of food is in the nature of transmission of the maternal condition rather than true heredity, and doubtless for many generations the dwarfing was simply the result of ontogenetic repetition and meant nothing more; ultimately, however, the repeated starving seems to have made itself felt in heredity, with the result of a permanently dwarfed race.

Mutilations have been one of the principal lines of research of those who would prove the inheritance or non-inheritance of acquired characters. Even when practiced for many generations or even for thousands of years (circumcision in man) these apparently fail to influence the unmutilated progeny in the slightest degree. Weismann, for instance, experimented on white-mice, producing no fewer than 901 young from five successive generations of artificially mutilated parents, and yet there was not a single example of a vestigial tail or any other abnormally in this organ.

Hence we may safely say that variations due to mutilation and to disease are not inherited, otherwise in all probability none of us would exist without some trace of hereditary crippling. Of the effect of climate, on the other hand, we are not so sure, for many of the observed changes which an animal undergoes as a result of the influence of cold or heat or humidity or dryness may be of an ontogenetic nature.

Effects of Cave-Dwelling

The strange modifications of cave-dwelling animals, are the direct result, first, of lack of light in the loss of pigment and of organs of vision, and, second, of scarcity of food which gives rise to depauperated bodies and attenuated limbs. Certain organs, such as tactile or gustatory structures, have hyper-trophied as those of sight have diminished, and while natural selection may be invoked to account for these well-developed structures, it cannot account for the atrophy of the others nor for the depauperation. Panmixia might *permit* these changes, but is of questionable value as a *cause*. Both of the principal American students of cave faunas, Packard and Eigenmann, feel the necessity of some other factor that the Darwinian, and Packard has expressed himself as follows with regard to the causes of production of cave faunæ:

1. "Change in environment from light, even partial, to twilight or total darkness, and involving diminution of food, and compensation for the loss of certain organs by the hypertrophy of others."
2. "Disuse of certain organs."
3. "Adaptation, enabling the more plastic forms to survive and perpetuate their stock."
4. "Isolation, preventing intercrossing with out-of-door forms, thus insuring the permanency of the new varieties, species, or genera."
5. "Heredity, operation to secure for the future the permanence of the newly originated forms as long as the physical conditions remain the same."

"Natural selection, perhaps, expresses the total result of the working of these five factors, rather than being an efficient cause in itself; or at least constitutes the last term in a series of causes. Hence Lamarckianism in a modern form, or, as we have termed it, Neo-Lamarckianism, seems to us to be nearer the truth than Darwinism proper or natural selection."

Inheritance of Instinct

Instinct has been defined as "the natural unreasoning impulse by which an animal is guided to the performance of any action, without thought of improvement in the method". It usually implies an action based upon inherited knowledge, but, judging from experimentation upon living animals, that knowledge is the result of trial and error upon the part of individuals which ultimately becomes part of the heritage of the race.

Birds are particularly interesting in this regard, for they possess so many instinctive traits, not only of food-getting and defense, but nest-building, migration, learning to fly in the mode characteristic of the species, and song. Many of these things have been attributed to actual instruction on the part of the parents or, in the case of the nest-building, to a recollection of the natal structure, but William Beebe tried the interesting experiment of raising a number of wild birds from incubated eggs. These, while never having had parental care or example, nevertheless manifested in each instance all of the instinctive actions of their race, although, as Beebe says, they were in some cases slower than they should have been about learning to fly, the maternal insistence being an aid in overcoming a very natural timidity on the part of the fledgelings.

So many of the curious manners and customs of domestic dogs, such as turning around three times to stamp out a lair, are doubtless relies of formerly valuable instinct, of little present worth to the house-dwelling associate of mankind, but perpetuated by heredity.

Ontogenetic Variations

It is often a matter of great difficulty to determine whether characteristics shown by an animal are those of the race (phylogenetic) or those of the individual (ontogenetic), even though observed in a number of successive generations where the life conditions remain the same. An interesting example is the brine-shrimp (*Artemia salina*) which has a wide-spread distribution from Great Salt Lake, Utah, to Central Asia. This creature is remarkable in several ways, as it can live in a 27 per cent brine solution on the one hand or in fresh water on the other. In some places the colonies seem to be altogether female, and parthenogenesis is the rule, the eggs developing without being fertilized. In other localities males are common and reproduction is biparental. Sometimes the brine-shrimp is viviparous, the eggs hatching within the brood-sac of the mother, and again it is variable in its form, especially as regards the end-lobes of the tail and the bristles they bear: all of which features seem to be correlated with the chemical diversity of the various habitats within which it is found. The eggs can survive being dried and may be blown about by the wind or carried by the feet of birds from one salt pond to another. They have also been bred from an English commercial product known as "Tidman's Sea Salt," of which a solution was allowed to stand for several days. A fresh-water ally, Branchipus, is said to be merely another phase of *Artemia,* the distinctions, as in most of the peculiarities of the latter genus which have been described, being largely due to the degree of salinity of the medium. If this be true, most of these variations are ontogenetic, as the transference of eggs to other conditions than those under which the parents lived would at once give rise to variation from the parental type.

Experiments by Loeb upon the unfertilized eggs of sea-urchins have shown that artificial parthenogenesis can be caused to occur by chemical stimulus, either by adding or suppressing certain salts in normal sea-water, which is suggestive of the parthenogenetic conditions of certain colonies of *Artemai* mentioned above. Owing to the brine-shrimps' method of dispersal and the fact that salt ponds are often isolated and of various degrees of salinity, the change of habitat is apt to be an abrupt one which renders individual adaptatbility a highly necessary asset.

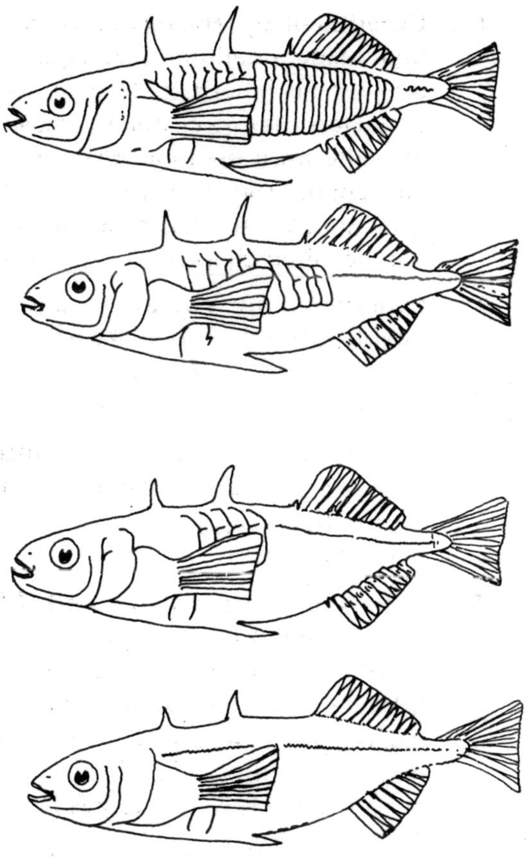

Fig. 3.2: Stickleback fishes, *Gasterosteus cataphractus*, showing ontogenetic variation of plates and spines as a response to varying salt content of the water. Uppermost figure collected in salt water, the next in brackish water, the third at a river mouth, and the fourth in fresh water.

The little fishes known as sticklebacks (*Gasterosteus catapharactus*), which have such curious nest-building habits, also show an ontogenetic variation dependent upon the chemical content of the water. Those living in salt water have from twenty to thirty bony plates along the back, in brackish water these are reduced to from fifteen to three, while in fresh water there are none at all.

As we have seen, it is often difficult to say whether the adaptive characters which are so often taken as criteria of species are racial or individual; if the latter, even though we may not know it, the form is an ontogenetic and not a genuine species, and it is highly probable that many of the observed instances of what was taken to be inheritance of acquired characters are not such in the true sense of the word but are simply due to the repetition of cause and effect in each individual generation. There is apparently an heredity control, however, for the adaptation to a change of habitat is not at random but always according to a definite plan, and gives rise to a predictable result.

CRITICISM OF LAMARCKISM

All experiments performed to test this postulate gave negative results with the possible exception of McDougall's experiments on rats. Following are some objections to Lamarckism.

1. **Experiment by *Weismann.*** *August Weismann* was the first who for the first time made a definite distinction between heritable changes and those which can not be inherited. In 1890, he performed some experiments to test if characters disappear due to disuse. This he did by cutting the tails (*mutilation*) of white mice for more than 20 generations to see if this has any effect on tail length. The measuring of tail length of the offsprings of 20 successive generations, revealed that on average, the tail were not shorter. It means that acquired character (cut tail) was not inherited.
2. **Experiment by *Pavlov.*** *Povlov,* a Russian scientist, wanted to show the inheritance of learning. He trained the mice to come for food on hearing the bell. But in later experiments it was found that training was necessary for each and every generation.
3. **Experiment by *Loeb.*** *Loeb* produced artificial parthenogenesis in the eggs of sea urchin by certain chemical stimuli. A generation was obtained from these unfertilized eggs, the members of this generation possessed no parthenogenesis.
4. **Experiment by *Castle* and *Phillips.*** *Castle* and *Phillips* performed transplantation experiments to shoe that environment has no effect on heredity. In one of the experiments they transplanted the ovary of a black female guinea pig into the body of white female guinea pig and the recipient female was mated with a white male guinea pig. They found that all the individuals from this pair were black. This shows that the environment does not effect the heredity as has been suggested by *Lamarck.*
5. **Indian Tradition.** Boring of ears and nostrils in Indian women has been continued from centuries but their female offsprings do not show any trace of holes in ears and nostrils.
6. Chinese women use iron shoes to keep their feet short but young ones at birth have normal size.
7. Muslim and Jewish boys have been circumcised for so many years, yet this has not resulted in a tendency towards the reduction of the produce in them.

Besides the above there are evidences which support that acquired characters are inherited from generation to generation. These are as follow:

1. ***Guyer* and *Smith's* experiment.** *Guyer* and *Smith* crushed the crystalline lens of the eye of rabbit and inoculated it into fowl which developed anti bodies. The antiserum of fowl (the fowl serum containing antibodies) was injected into pregnant rabbits. Some of the offsprings of rabbits were born with defective eyes and this character was inherited in successive generations.
2. **Experiment by *Mc Dougall.*** *Mc Dougall* trained the rats to escape the water tank only through the dark route as against the lighted route. This training

was continued through several generations in the progeny which make fewer and fewer mistakes in succeeding generations.

Agar et at (1954) also repeated the experiments of *Mc Dougall* for 50 generations, covering a period of 20 years. They got clear evidence against the hypothesis that effects of training are inherited.

3. **Experiment by *Brown Sequard.*** Various diseases like *Exopthalmia, Haematoma* and *Dry gangrene* are inherited from generation to generation by injuries in the restiform body of the brain shown by *Brown sequard.*

4. **Melanism in moth**. *Haslop Harrison* claims to have induced the formation of melanic mutants by feeding larvae of moth (*Selenia bilunaria*) on hawthorn leaves which had absorbed small quantities of salts of magnesium and lead. This has reference to the spread of melanic forms in the industrial areas of England.

5. **Experiment by *Lindsey.*** *Lindsey* subjected certain cold-blooded animals, warm-blooded animals and plants to unusual environmental conditions and found that environment affect certain changes in these organisms and these acquired modifications were inherited upto some extent in their offsprings.

6. **Experiment by *Summer.*** *F.B. Summer* reared the white mice at 20°C to 30°C. He found that they developed longer bodies, tail and feet. When these mice were bred at lower temperatures their offsprings showed their abnormal proportions of size from generation to generation.

7. ***Kammerer's* experiments on Salamander.** The skin of *Salamandra maculosa* is black with yellow sports. He kept some of them for four years in yellow box and found that yellow sports become much larger and united together to form a yellow band on the back. These salamanders were then mated with each other and the resulting offsprings were found to be more of yellow colour.

In another experiment the salamanders were kept in black box due to which yellow sports reduced and their offsprings were black.

8. **Inheritance of food habit.** *Sladden* (1934-38) produced the inheritance of the acquired ability of stick insects (*Dixippus morouns*) to subsists upon ivy (plant) instead of their normal diet privet (plant) after repeated forced feeding trials through six generations.

9. ***Tower's* experiment of potato beetles.** *Tower* found that exposing the young stages of potato beetles to extreme conditions of temperature and moisture, produce marked changes in offsprings and these modifications were inherited from generations to generations.

10. **Experiments *by Griffith and Detlefson.*** *Griffith* and *Detlefson* reared the rats in cages placed on rotating tables for several months. The rats were adapted

to rotating conditions to such an extend that when the rotation was stopped, they showed signs of dizziness (*nystagmus*). This condition was inherited for several generations.

NEO LAMARCKISM

In the present century, the scientists have again started thinking on Lamarckism and supporting in on the basis of many evidences. They are trying to fit in modern science. These scientists are called *Neo Lamarckians* and the modernised theory of *Lamarck* is known as *Neo Lamarckism*. Among the neo-Lamarckians notable supporters are *Herbert Spencer, Eimer, Cope, Hyatt, Naegeli, Dall, Haeckel* and *Gadow*.

At present the concept of Neo-Lamarckism is:

(i) Developed characters (whether acquired or not) are never transmitted by heredity and the hereditary constitution of the germ is not changed by changes in such character.

(ii) Possibly the environmental stimuli act upon germ cells at an early stage in their development. This may rarely cause changes in hereditary constitution; but changes, produced in somatic cells, by environment, do not cause corresponding changes in the hereditary constitution of germ cells.

(iii) Germ cells may undergo modifications which are not hereditary. Such influence may be carried over to next generation without becoming hereditary. All such cases are called *induction*. And many instance of supposed inheritance of acquired characters come under this category.

(iv) Environment may profoundly modify individual development but it does not generally modify heredity.

Thus neo-Lamarckism is neither a complete denial nor the complete acceptance of the old Lamarckism. It is merely, a slight modification of inheritance of acquired characters and it explains the appearance of acquired characters in a few succeeding generations by after affect or induction.

Chapter—4

Darwin's Natural Selection

Natural selection still seems to be the most important factor in evolution and has been defined as "The survival of the most fit wit the inheritance of those species-forming adaptations wherein fitness lies" (Jordan and Kellogg). Crampton says of it:

"Natural selection proves to be a continuous process of trial and error on a gigantic scale, for all of living matter is involved. Its elements are clear and real; indeed, they are so obvious when our attention is called to them that we wonder why their effects were not understood ages ago. These elements are (1) the universal occurrence of variation, (2) an excessive natural rate of multiplication, (3) the struggle for existence entailed by the foregoing, (4) the consequent elimination of the unfit and the survival of only those that are satisfactorily adapted, and (5) the inheritance of the mutations or recombinations that make for success in the struggle for existence. It is true that these elements are by no means the ultimate causes of evolution, but their complexity does not lessen their validity and efficiency as the immediate factors of the process."

Really great ideas in science are sometimes remarkably simple, so much so that they occur to more than one person independently. Natural selection is a case in point. The idea is basically that healthy and vigorous individuals have better chances of surviving and leaving progeny than ailing and frail ones. Empedocles (5th century B.C.) speculated that living things first arose as disjointed body parts—as heads, trunks, and limbs. These parts then combined at random, with only viable combination surviving. Lucretius Carus (1st century B.C.) reiterated the myth of Empedocles, but added some less bizarre biological reflections. Between 1835 and 1837 Edward Blyth in England discussed some examples of what we would now call normalizing natural selection. Although Darwin's failure to acknowledge Blyth as a precursor of his theory has puzzled some modern writers, we think the explanation is simple. Normalizing natural selection is a conservative rather than a creative factor in evolution. Darwin was interested in what changes biological species, Blyth in what keeps them stable.

In 1831 Patrick Matthew published a work in which natural selection was described as a factor of evolution. However, the title of the work was *Naval Timber and Arboriculture*, and the discussion of natural selection was overlooked by most readers.

In 1844, Robert Chambers published *Vestiges of the Natural History of Creation* anonymously because he feared, justifiably, adverse reaction from the Victorian establishment. This was the most widely read pre-Darwinian statement of evolutionary views. However, Chambers followed Lamarck in ascribing a greater role as an evolutionary factor to the inheritance of acquired traits, rather than to natural selection.

Darwin and Wallace considered natural selection, i.e., the differential survival and reproduction of organisms, the chief directing agent of evolutionary change. So do most evolutionist today. However, biology has advanced during the century and more since the first announcement of the Darwin-Wallace theory, and it is important to understand how the theory of natural selection had developed since then.

Darwin started his first notebook by compiling evidence on the transmutation of species in 1837. At that time he had no satisfactory explanation for transmutation. An explanation was suggested to him in September 1838, when he read "for amusement" the work of the sociologist T. R. Malthus, `*An Essay on the Principle of Population*. In this essay, first published in 1798, Malthus argued that since human populations can increase exponentially, they must inevitably outgrow their means of subsistence. The increase is sooner or later checked by hunger, disease, or war. Darwin saw that the potentiality for exponential population growth is quite universal in the living world. Yet most of the time populations of most species remain approximately constant in numbers. It follows that only a part of the progeny survive, and the rest are eliminated by death. It is a matter of chance alone what survives and what dies? No. In every species some individuals are stronger or in some ways better adjusted to the environments in which they live than other individuals. The offspring of the stronger are more likely to be represented among the following generations than the offspring of weaker individuals. The result of this differential survival is, in Darwin's words, that "natural selection is daily and hourly scrutinizing, throughout the world, the slightest variations, rejecting those that are bad, preserving and adding up all that are good; silently and insensibly working, whenever and wherever opportunity offers, at the improvement of each organic being in relation to its organic and inorganic conditions of life."

The same insight was suggested to Wallace in 1858 by the same essay of Malthus, under unlikely circumstances—when he was suffering from a malarial fever on the island of Halmahera in Indonesia. Although Darwin's insight antedated that of Wallace by almost 20 years, their discoveries were announced together in 1858. Darwin's great work, *Origin of Species*, was published in 1859; the sixth edition revised by Darwin appeared in 1872.

Some of the concepts of the Darwin—Wallace theory require comment. Artificial selection, practiced by breeders of agricultural plants and domesticated animals, has commonly been used as a model of the action of natural selection. However, in Lerner's words, "Natural selection has no purpose. . . . For any given generation, natural selection is a consequence of the differences between individuals with respect to their capacity to produce progeny. . . . Artificial selection, in contrast, is a purposeful process. It has a goal that can be visualized" (1958). Natural selection can and does take place in domesticated and laboratory organisms, and in mankind, under all sorts of natural and artificial conditions. Artificial selection is man-made, however. Natural selection has no selector, it is a self-generated outcome of interactions between organisms and their environments.

Darwin described natural selection as a consequence of the "struggle for life" or "existence." Although he carefully pointed our that the word "struggle" was used metaphorically, Darwin's phrase easily lent itself to abuse. So-called social Darwinists claimed that Darwin's theory justified war, aggression, and hostility between races and classes, and unrestrained economic competition. "Nature red in tooth and claw" was an adage that appealed to many in the heyday of capitalism and imperialism. Yet in nature the struggle for life does not necessarily take the form of actual combat between individuals of the same species.

Among their animal combat within species is often ritualized. Victory or submission may be achieved without the opponents inflicting physical harm on each other. Plants "struggle" against aridity by developing devices to protect against excessive loss of water, not by sucking water from each other. It is no paradox to say that under many circumstances the most effective "struggle" for life is mutual help and cooperation. This idea was advanced by K. F. Kessler in 1880 and by Peter Kropotkin in 1902.

Herbert Spencer described natural selection as the "survival of the fittest." Not without hesitation, Darwin accepted this phrase in one of the later editions of *Origin of Species*. The use of the superlative is gratuitous, however. What survives is not the "fittest" individual, but simply tolerably fit individuals. Who are the fit, or fittest, in the process of natural selection? This question could not be answered scientifically until the middle of this century and the development of population genetics, which made possible a rigorous definition and measurement of Darwinian fitness.

Darwin freely admitted that some critically important parts of his theory could not, in his day, be fully clarified. He took it for granted that individuals composing a species vary in their chances to be winners in the struggle for life. What is the source of the variations on which natural selection operates?

Darwin did not always distinguish between what we now call hereditary or genotypic variation and environmental induced phenotypic variation. Although in his day it seemed possible that both the environment and selection may change the hereditary materials, it was not until 1909 that the Danish geneticis W. Johansen experimented with pure lines of beans, obtained by self-pollination from a single progenitor. Most individuals belonging to a pure line are genotypically identical, but show some phenotypic variation, e.g, in seed size and seed weight. Yet the averages of the progenies of large and small beans in the same pure line were identical. Thus the selection was without effect. By contrast, the progenies of large beans from a mixed population were on the average large, those of small beans were likewise small. The selection extracts from a mixture of genotypes the desired ones. The source of the genetic variability is the mutation process.

It was believed in Darwin's time that the heredities, or "bloods," of the parents mix and blend in the offspring. This "blood theory" of heredity created another difficulty, pointed out to Darwin in 1867 by Fleeming Jenkin, by training an engineer rather than a biologist. Suppose that some useful variation, for example a lighter coloration, appears in a species composed of dark individuals. The light individual will likely mate with a dark one (unless many light variants appear simultaneously). The progeny will be intermediate in coloration, and in turn will mate mostly with dark ones. In a few generations the light variant will disappear as completely as a drop of ink in a sea.

The "blood theory" of heredity makes sexual reproduction a destroyer of genetic variability. In fact the opposite is true, but this could not be understood until the "blood theory" was invalidated by Mendel's discoveries. The heredities of parents are not bloods that mix, but arrays of genes that segregae during the processes of sex-cell formation.

Pordigality of Production

Perhaps one of the most impressive things in nature is the teeming abundance of living creatures. The swarms of gnats dancing in the sunlight, the great number of birds on certain oceanic islands, the immense collection of individuals in a great school of fishes: all are example of what Thomson has called the "insurgence of life," and when one realizes that he sees but the smallest fraction of the total numbers which occur, he is the more impressed. Speaking of the splendor of oceanic luminescence which is often met with at sea, Thomson says:

"There is a cascade of sparks at the prow, a stream of sparks all along the water level, a welter of sparks in the wake, and even where the waves break there is fire. So it goes on for miles and hours—a luminescence due to the rapid vital combastion of pinhead-like creatures (*Noctiluca* and others), so numerous that a bucketful contains more of them than there are people in London...

"On the night before the new or full moon in the middle or latter half of December there occurs the remarkable swarming of the Japanese Palolo worm. It invariably takes place about midnight just after floodtide. At 1 A.M., Akira Izuka relates, the worms 'covered the whole water as with a sheet' and were thick down to a depth of a fathom. By 2.15 A.M., there was not a single worm to be seen; the reproductive orgasm was over. The phenomenon appears to us to be a dramatic instance of the abundance of life, of the crisis-nature of reproduction, and of the precise way in which internal rhythms may be related to external periodicities."

The productivity of all living organisms is far beyond the ultimate numbers which can possibly survive, and the reason is this; organisms at their least rate of increase reproduce in geometric ratio, whereas the space they may occupy and the available food supply remain constant. Hence without some very efficient check the slowest breeders would soon exhaust the possibilities of food supply and space. For example, the elephant is the slowest breeder among mammals, but Darwin calculated that a single pair beginning to breed at thirty year and continuing to do so until a century old would produce on the average six young would have in 750 years, barring accident, nineteen millions of descendants. A rabbit, on the other hand, may have six young in a litter and four litters in a year, and the young may begin to breed at six months, a vastly more rapid rate of increase than that of the elephant.

Among the lower vertebrates where no parental care is given to the young the potential productivity is necessarily enormous. In four herring the number of eggs varied from 20,000 to 47,000; in cod there were 6,000,000; a turbot, 9,000,000; and a ling, 28,000,000, and yet despite the enormous number of offspring which might *possibly* be produced from a *single pair in one generation,* the ultimate number of herring or cod or ling remains on the average about the same. The chance of survival, therefore, of a ling's egg is one in fourteen million.

The vertebrates, however, are relatively slow breeders, for there are as a rule but one or at most a half a dozen generations in a year. With invertebrates, on the other hand, the actual number of generations may greatly exceed this, and this is what Linnaeus meant when he said: "Tres muscae consumunt cadaver equi, aeque cito ac leo." Huxley estimated that the descendants of a single green fly, if all survived and multiplied, would at the end of one summer weight down the population of China. Common house-flies would in the same time—six generations of three weeks each—occupy a space of about a quarter of a million cubic feet, allowing 200,000 to a cubic foot. An oyster may have 60,000,000 eggs, and the average American yield is 16,000,000. If all the progeny of one oyster survived and multiplied and so on until there were great-great-grandchildren, these would number 66,000,000,000,000,000, 000,000,000,000,000,000, and the heap of shells would be eight times the size of the earth!

Professor Woodruff in his experimental study of *Paramecium* has maintained a pedigreed race since 1907, the descendants of one wild individual. In five years there were 3029 generations, the mean rate of reproduction being three divisions in 48 hours. They were as healthy at the end as at beginning of the culture and had given evidence of the potentiality of producing a volume of protoplasm approximately equal to ten thousand times the volume of the earth ! It has been estimated that at the end of the 9000th generation, now long since passed, the mass would exceed the confines of the known universe and the rate of growth would be extending its circumference into space with the velocity of light!

With such extraordinary productivity on the part of all living matter, the efficiency of the check upon every species of plant or animal is at once apparent. This check is that which Darwin and Wallace both recognized, and called the *struggle for existence.*

Struggle for Existence

This struggle for existence is the competition between all organisms and between each individual and the physical environment. The struggle is threefold, although in the long run it is all against what may be called the environmental complex, which includes all surrounding nature, whether due to physical conditions, to plant or to animal life.

The *intraspecific struggle* is the struggle against the organisms's own kind, the internecine strife. In some cases this is the most severe check of all, for each one's needs are precisely similar and the competition, instead of touching at one or two points, is absolute. In human warfare, the hatred is more bitter the nearer the contestants are related, as shown by Germany's "Gott strafe England!" so it is with the organic world.

Examples of this intraspecific struggle are the young trees in a forest. As seedlings they may spring up over a devastated area in great abundance. Some soon die from lack of sufficient soil or moisture or due to other causes, but they are still numerous until they become so tall that their branches begin to mingle and a leafy canopy is formed which shuts out light and air from the trees of less vigorous growth. Then the weeding out of the less fit begins and the number of trees in the area rapidly diminishes until ultimately the relatively few great trees in a mature forest are the result.

In artificial lobster culture, experiments have shown that it is better to turn the newly hatched young at once into the sea rather than retain them within the limits of the aquarium for any length of time, for they are their own worst enemies and the results of what may be called cannibalistic selection are more destructive to the race than competition with the natural environment.

Interspecific struggle is the familiar struggle between members of different species, often in the nature of competition, but perhaps more frequently because the one may afford food for the other. Mankind is just as much concerned in this interspecific struggle as any other form of life, but in general it is the lower organism, be it plant or animal, which is worsted in the struggle and must make good its losses or perish.

Instances could be multiplied *ad libitum,* as the struggle is universal and no creature is immune, although in more favored communities, with rare exceptions, only parasites and disease germs make their direct attack upon man, but *every animate form* depends upon some other organism, be it animal or plant, for its food, so none is exempt from direct or indirect participation in the universal strife.

Environmental struggle is that against the physical environment —against excess of moisture or of drought, against extreme heat or cold, against lightning and tempest, earthquake and volcanic eruption. The eruption of Mont Pelee in May, 1902, slew practically every inhabitant of the city of St. Pierre, there being but one lone survivor out of a population of perhaps 28,000 souls; while in August, 1883, the volcano of Krakatoa as the culmination of a series of increasingly violent explosions threw back the inpouring sea and drove a wave of water high upon the neighboring coasts of Java and Sumatra, engulfing more than 36,000 people with their villages and lands (Idings). The Sicilian earthquake of December 28, 1908, caused the death of at least 77,283 people in Messina and the near-by city of Reggio, while the official figures for the Japanese earthquake of 1923 are: killed 103,000, injured 125,000, missing 235,000, a total of 463,000 souls! Again, as a result of neither earthquake nor volcano, the great tropical storm of September, 1900, piled up the waters of the Gulf of Mexico, almost overwhelming the city of Galveston, with a loss of about 5,000 lives. The destruction wrought by our western cyclones and tornadoes is well known though rarely is there so appalling a loss of life as in the instances that have been given. Lightning is said to be the greatest single cause in the destruction of ranch cattle in Nebraska, due, however, in large part to the conductive power of the wire fences against which the animals drift before the storm. Thus the lightning bolt which would in a state of nature be local in its effect has its danger zone very largely increased through human interference. River floods such as those of China take their annual toll of thousands of human lives and doubtless of many animals as well. The Cenozoic sediments on our western plains, formerly supposed to be those of extensive lakes, are now interpreted very largely as river flood-plain deposits. The fossil animals which they contain, and which must number many thousands, are largely the accumulations of flood victims of ancient days.

Drought is another potent cause of destruction, not only of plants but of animals as well, for the drying up of a water hole is frequently the forerunner of tragedy. Excessive cold for which the animals are not prepared is another cause, as in the case of the numerous skeletons of guanacos, a wild species, allied to the domestic llama of South America, which have been observed in Patagonia. Of these Hatcher says: "During the winter storms these animals would be driven from the surrounding plains to seek shelter in the river valleys and there, beneath embankments or in clumps of bushes, would be found the remains of such as, through old age or disease, were unable to survive the rigors of the storm they had sought to escape."

All these aspects of the struggle are, strictly speaking, with the conditions of life, and as well shall see, the interrelations of various organisms with these conditions are generally very intricate, and over against the record of a vast host of all living beings nature writes a sentence comparable to the handwriting on the wall: "Thou art weighed in the balances, and art found wanting." But the destruction is not without distinction, for although the boldest and best of a race are sometimes the first to be destroyed, and often the slaughter is utterly indiscriminate like that in the trenches of Flanders, or when a great Greenland whale rushes through an immense school of the delicate sea-butterflies which form the bulk of its food, during which thousands are engulfed and swallowed, no perfection of detail nor harmonizing color nor activity nor any other usually advantageous variation availing one any more than the other; nevertheless, taking nature as a whole, the fittest do survive in the long run and the least fit are the first to perish.

Survial of the Fittest

The survival of the fitted is therefore the result of natural selection. While the same conditions persist, specific change is very gradual, but with changing conditions placing a premium new or different characters, species also undergo a change. Natural selection therefore enforces adaptation, and of two forms in competition, the adaptable will crowd out the in-adaptable.

This is forcibly illustrated by the Tasmanian wolf or thylacine (*Thylacynus cynocephalus*) which is confined to-day entirely to the island of Tasmania but whose remains are found in the superficial deposits of the Australian mainland, showing it to be but recently extinct. The thylacine is one of the native pouched animals or marsupials which form the indigenous mammalian population of Australasia and is characterized among other things by being dull of wit as compared with a true placental dog. Australia has such a dog, the dingo (*Canis dingo*), the origin of which is doubtful, but it is supposed to be of Asiatic extraction and introduced into Australia by human agency. We have here the adaptable placental placed in competition with the in adaptable marsupial, with the inevitable result—the marsupial holding its own when it had only the ancient enemies of its race to contend against, but powerless in competition with the representative of a more vigorous stock.

Nature's Balance

Competition is inseparable from life and is really the source of all progress, but this very competition leads to a nicety of adjustment between rival organisms and a linking together of other beings into a marvellous web of interrelationships. This nicety of adjustment is known as nature's balance, and, although often unobserved, man has inadvertently sometimes upset the adjustment, causing dire results which he knows to his cost.

Not all organisms are enemies, but in many instances, although generally among unrelated forms, there is a marked interdependence, all of which adds to the complexity of the web of life, and at first sight the extremes of series of interrelated organisms, when the links in the chain are unknown or disregarded, make them seem very far-fetched as illustrations of the case in point. Such, for example, is the linkage suggested by Darwin between domestic cats and red clover, but when all of the links are known, the connection is at once evident. Darwin found by experiments that humble-bees are almost indispensable to the fertilization of the heartsease (*Viola tricolor*), for other bees do not visit this flower. He also found that the visits of bees are necessary for the fertilization of some kinds of clover; "for instance, twenty heads of Dutch clover (*Trifolium repens*) yielded 2290 seeds, but twenty other heads protected from bees produced not one. . . . Humble-bees alone visit red clover, as other bees cannot reach the nectar. . . . Hence we may infer as highly probable that, if the whole genus of humble-bees became extinct or very rare in England, the heartsease and red clover would become very rare, or wholly disappear. The number of humble-bees in any district depends in a great measure on the number of field-mice, which destroy their combs and nests; and Col. Newman, who has long attended to the habits of humble-bees, believes that 'more than two-thirds of them are thus destroyed all over England.' Now the number of mice is largely dependent, as every one knows, on the number of cats; and Col. Newman says, 'Near villages and small towns I have found the nests of humble-bees more numerous than elsewhere, which I attribute to the number of cats that destroy the mice'. Hence it is quite credible that the presence of a feline animal in large numbers in a district might determine, through the intervention first of mice and then of bees, the frequency of flowers in that district!" Huxley had added a link to each end of this chain of relationship by the supposition that the cats are very largely harbored by the unmarried spinsters on the one hand, whereas the clover affords sustenance for the cattle which in turn produce the "roast beef of Old England" which nourishes her valiant sons which are the source of England's might, thus making the number of maiden ladies in a community productive in this very roundabout way of the prowess of Great Britain.

It will readily be inferred from the foregoing how wonderfully delicate this balance of nature is in its adjustment and how easily it may be deranged by human interference. A few notabel examples of such derangement of the balance may be given. The English sparrow was introduced into a city of Connecticut for the purpose of eradicating the measuring worms which were defoliating the elm trees at the time so characteristic of the town. It may be said to its credit that the sparrow did largely abate the nuisance for which it was imported, but owing to its fecundity, greediness, and quarrelsome disposition, it has become a widespread pest, driving out the native song and insectivorous birds and offering in exchange only its own unattractive personality.

Another instance is the Colorado potato beetle (*Doryphora decemlineata*), a native of the Central West, where it fed upon the nightshade and was kept in check by its natural enemies. The introduction of the potato plant, a close ally of the nightshade, offered the beetle a new and abundant food-supply, to which it took with great avidity, multiplying rapidly, for the lessening of the competition for food rendered the other natural checks of relatively little avail. The beetles then began an eastward migration in an ever widening pathway, going from one potato plantation to the next, until they finally reached and extended along the Atlantic seaboard and now extract an annual toll equivalent to thousands of dollars from the raisers of the esculent tuber. In thus case, it was not, as usual, the introduction of an animal into new environmental conditions that upset the balance, but the reduction of one of the most severe checks to survival, the struggle for food.

Another very remarkable instance the entire history of which is known is that of the gypsy moth (*Porthetria dispar*) which was accidentally liberated in Medford, Massachusetts, in 1869. Professor Leopold Trouvelot, a French naturalist, was experimenting with silk-spinning caterpillars, especially the American species, to see if any could be made of economic importance. He also imported specimens of the insect under discussion from Europe. Evidence seems to show that some of the egg clusters or young caterpillars escaped from his place, and as he was aware of the dangerous nature of the insect in its native home, he destroyed all the caterpillars he could find, but soon seeing that he could not fight them single-handed, he reported the matter to the authorities. Little notice was taken of the insects, however, although after a few years they did become exceedingly troublesome to the people of the neighbourhood by defoliating their shade and fruit trees. During the summer of 1889, however, their depredations and numbers increased to such an extent as to become a public menace, depreciating the value of property and causing an exodus from the infested districts, as they swarmed everywhere and many trees and orchards died as a result of the repeated defoliation. The legislature of the commonwealth then took up the problem of extermination and appropriated the sum of $25,000 for their elimination. Additional appropriations were requested and granted until 1900, when the work of the state commission ceased. In the meantime another similar pest, the brown-tail moth (*Euproctis chrysorrhoea*) had also been introduced from Europe and, together with the gypsy moth, which spread alarmingly with the cessation of the work of extermination by the state, has become a problem of national importance. Hence the United States Government in cooperation with that of Massachusetts has taken up the fight. Actual extermination seems now out of the question, as every available method, mechanical, by poison, or by fire, has been tried, and what the authorities are striving to do is to restore nature's balance by introducing the parasites and other natural enemies of the two insects from their native home, with the hope that thereby they may be controlled and not become too great a burden, for their total destruction seems impossible.

Yet another noted instance is the introduction of the carnivorous mammal, the mongoose (*Herpestes griseus*), into Jamaica. "Rats brought by ships became plague in Jamaica. To cope with them the mongoose was imported, and it made short work both of the Old World rats and the Jamaican cane-rats. But when these were gone, the appetite of the mongoose remained, and the poultry and various ground birds began to suffer. Useful insect-cating lizards were also eaten, and another cloud rose on the sky—there was a multiplication of injurious insects and ticks, so that plants and animals began to be affected through an ever-widening circle" (Thomson).

One instance where balance has been partially restored after being upset by human interference is in the case of a scale insect accidentally introduced into California from Australia on some young lemon-trees. This multiplied until it became a most pernicous pest which various mechanical remedies failed to control. Search was made in Australia, and a natural enemy, a lady-bug, was brought over to California, with the result that not only was the scale reduced but almost completely eliminated. It was then found that the lady-bug depended upon the scale for food to such an extent that it died in turn, and now protected colonies of scale and lady-bug are kept in readiness to control future outbreaks of the pest!

Survival of the Existing

We have spoken of the survival of the fittest which controls the existence not only of a race, or the individuals which make up the race, but also of the characteristics which compose the individual. There are, however, certain characteristics which seem to be non-important and, so far as we can judge, have nothing to do with an organism's change for survival. These non-essential character are, as we say, not of selection value, and persist through heredity. They are minor traits such as peculiar color patterns, relative proportions of parts, or vestigial structures, like the red color seen in the fins of certain deep-sea fishes. These forms live in a habitat where, as no light exists, the color can not be seen and hence is of no possible utility to the animal. The spots on the coat of a lion-cub are also instances in point. Usually these indifferent traits may be interpreted in the light of historic vestiges and point to former conditions in racial history where they were of importance in determining the chances for survival. To this category belong the great number of vestigial organs of which Weidersheim has enumerated 180 in the human body alone!

OBJECTIONS TO SELECTION

1. Darwin in a letter to his friend Hooker (January 11, 1844) expresses his contempt of Lamarck's ideas in the following words: "Heaven defend me from Lamarck's nonsense of a 'tendency to progression,' 'adaptations from the slow willing of animals,' etc... Lamarck's work appeared to meto be extremely poor; I got not a fact or idea from it."

In spite of these views Darwin's *Origin of Species* is interlarded with Lamarckian explanations. Whenever the author feels the short-comings of the selection factor he lapses into an explanation involving the idea that the effects of use and disuse of organs are inherited. Followers of Darwin, especially Weismann, felt this to be the chief defect in the fabric of Darwinism and bent their efforts chiefly toward purging Darwinism of all taint of Lamarckism.

2. Darwin insisted upon the idea that minute fluctuating variations, which we now know are to a large extent non-heritable, were the principal, if not the sole, materials for natural selection to work upon. He knew of a considerable number of "sports" or "saltatory variations" (now called mutations), but considered these too infrequent to furnish the necessary basis for selection. We now know that mutations may be as small as fluctuating variations or as large as "sports" and that they are of much more frequent occurrence than Darwin supposed.

3. Darwin considered all variations as heritable. He did not distinguish between *somatic* variations and *germinal* variations. In fact, as we learn from a study of his pangenesis theory, he considered all variations as in the first instance somatic, and subsequently transferred by means of gemmules to the germ cells. Every somatic variation, whether induced by use, disuse, in response to environmental stimulus, or through mere spontaneous variability, was supposed to be able to give off gemmules into the blood stream that would carry to the germ cells the physical basis of the varying character. The pangenesis mechanism is now known to have no basis in fact.

4. The natural-selection theory is based upon a mistaken conception of the methods of articial selection. Darwin believed, without having any proof for this belief, that the way in which domestic varieties had been so profoundly modified at the hands of man was by the conscious or unconscious selection of slight fluctuating variations in favorable or desired directions, and that this resulted in the cummulative improvement or enhancement of the desired characters over a long series of generations. Darwin supposes that the radically changed conditions of domestication hasten and stimulate variability, thus offering a better opportunity for selection. Transferring this idea to nature, he thinks that changed natural conditions stimulate variability, just as does domestication, and that this is seized upon by natural selection to make for adaptation to the new environment and the resultant origin of new species.

Our modern experimental studies have shown that somatic modifications due to environmental changes are not hereditary, and that all of the recent domestic varieties whose origin has been observed have been the result of suddenly appearing germinal variations or mutations, that arrive fully formed and cannot be improved by selection, except that they usually need to be selected out or isolated in order to prevent swamping out through intercrossing with the parent-type.

5. Objection has frequently been made to Darwin's idea of the purely fortuitous or chance character of variations. According to this view variations occur in all structures and in all directions at haphazard, so that there would be the widest possible opportunity for a given adaptive variation to occur just when the circumstances would demand. It now appears that variations do not occur in all directions in random fashion, but that they tend to follow certain definite paths of change; in other words, variations are, to a considerable extent at least, orthogenetic. If variations really tend to follow certain definite lines, owing to purely internal causes, natural selection would be unnecessary, at least until orthogenesis went too far for the good of the species, or far enough to be a real importance in the struggle for existence.

6. The difficulty of explaining how natural selection could make use of the initial stages of adaptive structures is obvious. It is inconceivable that the first, almost imperceptible variation in a favorable direction could be of selective value, so as to effect the survival of the individual or the relative number of its offspring. What would be the advantage of the first few hairs of a mammal or the first steps toward feathers in a bird when these creatures were beginning to diverge from their reptilian ancestors? This objection is, of course, based on the fluctuating-variation idea. If the mutation idea were substituted, the difficulty would, to a great extent, clear up; for a mutation might be of sufficient importance in one generation to have selective value from the very first.

7. Natural selection is said to be incapable of explaining the origin of coadaptive and highly complex adaptations whose effectiveness depends upon the perfection of their adjustments to one another. For example, we may refer to some of the perfected adaptations. In the case of the electric organs of certain fish, the Darwinian assumption would be that the first step in the direction of an electric organ would be a very small one, and that it was built up little by little by means of natural selection. But, say the critics, the electric organ would be of no value until it became powerful enough to impart an effective shock to the intruder, and this would not be possible if the character began in a small way. The whole phenomenon of protective resemblance is open to the same type of criticism. As a specific example of this we may cite the case of the dried-leaf butterfly, *Kallima,* previously described. In its present condition this animal has a strikingly detailed resemblance to a dried leaf, which is therefore doubtless of some value. But of what value would be the first tiny change in the direction of resemblance? Until its resemblance became close enough actually to deceive the enemies of butterflies, the critics claim, there would be no chance for selection to act.

8. It is frequently objected that a vast number of characters of organs are useless or non-adaptive and, as such, could not have arisen through the instrumentality of natural selection. If these useless characters, which are sometimes quite large and prominent, are independent of natural selection, why do we need natural selection to explain adaptive characters? It is also claimed that a vast number of specific peculiarities are useless and therefore could not have helped in the differentiation of species. It should be said in defense that Darwin realized this difficulty quite as clearly as do his critics and was greatly puzzled by it. His idea of correlated variability, however, helps to answer it, for it may well be that many of these apparently useless characters are correlated, or linked in inheritance, with characters of supreme selective value such as general hardines or great fecundity. Darwin also points out that we are not in a position at present to pronounce judgment on the value of many structures or functions that have been adjudged non-adaptive.

9. Certain characters in organisms, past and present, have been interpreted as over specializations, organs that have evolved beyond the range of usefulness or that are more elaborate than is demanded for survival under the conditions of life. The case of the extinct Irish elk is often cited as an example of over specialization. This group of animals went to extremes in the development of size and elaboration of horns far beyond the range of usefulness, so that it is said to have brought about the extinction of the race. Natural selection, which is supposed to have brought the horns up to the point of adaptive perfection, should have kept them within the bounds of usefulness.

Again, the enormously overgrown and over specialized dinosaurs of long ago are thought of as having followed their lines of evolution far beyond the point of greatest effectiveness and adaptability.

10. The rudimentation of structures, which is such a common phenomenon in nature, is said to meet with no adequate explanation on a selection basis. The case of the whale's vetigial hind limbs is a case in point. Darwin's explanation would be that under aquatic conditions the first whale ancestors would be handicapped by hind legs and that any decrease in their size, which would be enhanced by disuse, would be of advantage. This might seem reasonable during the main period of limb reduction, but, after the limb is reduced to a subcutaneous rudiment, there could be little advantage in carrying the rudimentation still farther. Some whales have the hind limbs much more profoundly reduced than others, although they are all thoroughly out of the way and involve no hindrance in swimming. Any number of similar cases of the same kind might be cited. Darwin had no explanation to offer except a resort to Lamarcksm; but Weismann, the ablest neo-Darwinian, offered the theory of panmixia to cover this objection.

11. It is objected that, unless favorable variations occur in a large number of individuals at the same time, the character would be swamped out by inter crossing with individuals not possessing the favorable variation. The probability that such a swamping-out would occur was shown mathematically by various critics. By way of answer to this objection there arose a number of "isolation theories," according to which favorably varying individuals would be protected from back-crossing with the non-varying individuals. We might also point out that the Mendelian laws of dominance and segregation would serve to prevent loss of any new favorable character.

12. It is objected that natural selection might explain the "survival, but not the arrival, of the fittest." But Darwin met this perfectly when he said: "Some have even imagined that natural selection induces variability, whereas it implies only the preservation of such variations as arise and are beneficial to the being under its conditions of life."

13. Criticism has been directed against natural selection because of the fact that some of the supporters of Darwinism, notably Weismann, have made the claim that natural selection is the sole cause of evolution. This idea of the *Allmacht* or all-sufficiency of natural selection was not Darwin's, as is clear from the following statement; "I am convinced that natural selection has been the most important, but not the exclusive means of modification."

14. It is objected that many, if not most, of the fluctuating variations with which Darwinisms deals are purely quantitative or plus-and-minus variations; whereas the difference between species are qualitative. This is a serious objection and difficult to meet, yet a fair defense has been formulated by leading neo-Darwinians.

15. There is a growing skepticism on the part of biologists as to the extreme fierceness of the struggle for existence and of the consequent rigor of

selection. It may be answered that no very obvious fierceness is implied in the theory. So long as overproduction and a shortage of space and food exists the struggle for existence is inevitable.

16. Special objections are offered to the subsidiary theory of sexual selection. It is said that the type of sexual selection involving active rivalry and battling for mates needs no special theory, in as much as this is a mere phase of the struggle for the maintenance of the full life, including the chance to leave offspring. It is against the other side of sexual selection, which involves passivity on the part of the male and active choice on the part of the female of the more beautiful or otherwise attractive male, that objection is raised. It is claimed that such choice implies too high aesthetic powers in animals of relatively poor vision and mentality. Experiments have been performed with moths, in which the male and female coloration is strikingly different, in order to determine whether females actually do exercise any choice of mates that is based on considerations of appearance. The result proved conclusively that color patterns have no value in mating, but that the female is passive and mates with the first male to present himself, while the male finds the female through his exquisitely effective sense of smell.

We know now, however, that secondary sexual characters are intimately bound up in a physiological way with the functioning of the sex glands and are therefore doubtless to be interpreted as mere non-adaptive correlative variations or else as examples of obliterative coloration.

DEFENSE OF SELECTION

Not all of the sixteen objections are considered serious at the present time, for some are based upon lack of a full knowledge of what Darwin actually wrote; others are largely academic in character and fail to stand up under actual test; still others have been more or less adequately met by subsidiary or supporting theories which have been advanced by various neo-Darwinians.

Most of the special objections raised in this chapter have received the attention of various able Darwinians, and the student of evolution would doubtless be interested in the expert and fair-minded defense of Darwinism at the hands of Professor V. L. Kellogg as it appears in his book *Darwinism To-Day.*

A much briefer and considerably more general defense is that of J. L. Tayler, which is as follows.

GENERAL DEFENSE OF SELECTION

To realise how far the theory of selection is capable of explaining the facts of organic evolution, it is necessary to bear in mind the postulates in which the theory is founded.

1. It is obvious that natural selection can only act by preserving or eliminating the complete organism. Selection must therefore be organismal. This Darwin and other selectionists have clearly recognised.
2. As the whole organism must survive, if the favourable variation or variations are to be preserved, it follows that certain minor unfavourable variations may also be preserved if they happen to exist an individual which survives on account of its major favourable variations. And since no individual is completely adapted to its environment, it follows that there must be always a variable amount of residual unfavourable variability in every organism.
3. This residual unfavourable variability may be of considerable utility under changed conditions.
4. Complementary specialisation of parts, as Spencer has shown, is favourable to successful competition, and as it is the whole organism that is selected or eliminated, it follows that any weakness of one specialised part, since it would disturb the balance of all, would be detrimental. The more complex the organism, the more specialised the structures, the more dependent one part will be on the others for its existence, hence a complementary specialising tendency will be favoured by selection, and therefore all struggles of one part of an organism with another will be reduced to a minimum.

It is clear that there must be some underlying criterion which determines whether any given organism shall be selected or not, and that criterion must be the net result of its adaptability to its environment. One organism may conceivably survive, by its possession of a large number of small favourable variations, while another may survive in virtue of a single valuable one, but in each case it would be the whole value of that organism which determined its survival. This fact is continually disregarded by opponents of the neo-Darwinian position, yet this selection of the organism as a whole is the fundamental postulate from which the theory of selection starts. Thus it is not uncommon to read criticism bearing on the early development of some organ, in which the inadequacy of selection is supposed to be proved by the write demonstrating, or believing he has demonstrated, the fact that the particular variation in question must have been too small to be by itself of selection value. In many cases the particular variation would, no doubt, if taken alone be, as the objector asserts, too unimportant to be selected, but as it is the whole organism that is selected, it is not logical to make an artificial separation and study the development of one organ or structure irrespective of the other organs with which it is in nature associated. *Every organ in its evolution must be considered in relation to the whole of the particular organism in which that particular stage of development of that organ is found.*

Starting, therefore, with this fact that the net value of adaptability of the whole organism to its environment must be the basis which determines selection or elimination, it will follow that certain lines of development will result from the application of this criteron. In a series of organisms placed under new conditions, elimination will proceed along lines essential to bring about a proper adjustment to the new conditions. It the offspring of these adjusted organisms merely repeated in their generation the characters of the exterminated as well as of the surviving organisms, that temporary adjustment would be permanent as long as the conditions were unchanged. But since the offspring are produced only by the surviving organisms, selection is continually raised to higher and higher planes of adaptation, and, therefore, as long as conditions remain constant, the tendency of selection must be, as Darwin clearly saw, cumulative. He did not, however, apparently see that from this cumulative tendency definite variability must arise out of indefinite.

Selection in direct relation to climatic conditions is, therefore, of very minor importance, while selection among the members of a species and all forms of inter-organismal selection is of infinitely more importance, since it is this interaction, produced by the offspring in different degrees inheriting the advantages of both parents (both of whom have survived on account of certain advantages), that leads to the cumulative development and never-ending struggle for survival. Darwin came very near to this conception of definite variability when he pointed out that "if a country were changing, the altered conditions would tend not cause variation, not but what I believe most beings vary at all times enough for selection to act on." Extermination would expose the remainder to the "mutual action of a different set of inhabitants, which I believe to be more important to the life of each being than mere climate," and as "the same spot will support more life if occupied by very diverse forms," it is evident that selection will favour very great diversity of structure.

Bearing in mind this cumulative action of selection it will follow that under constant or relatively constant conditions the struggle for successful living will become more and more selective in character, even if the actual number of inhabitants remain more or less the same as when the struggle first commenced. The selection of variations will thus tend to pass through certain more or less ill-defined, but nevertheless, real states. In proportions the struggle becomes intense, either from the number or from there increasing adaptability of the organisms, or both, certain major essential adaptations, which were necessary for the climatic and other more or less comparatively simple condition, will be supplemented by minor auxiliary variations which in the earlier stages would not have appeared. And still later, is more and more rigorous conditions of life were imposed, the advantage would tend to rest with those organisms which possessed highly co-ordinated adaptations, since this would entail more rapid responsiveness to environment.

As evolution advances from the unspecialised to the specialised, and higher and higher forms of life come into being, with increasing complexity and specialisation of parts entailing and increasingly delicate adjustment of those parts to each other's needs, the relation of each part to the whole organisms becomes of more and more importance, and it follows that selection must become more and more generalised in its action. No single variation could be service to any of the higher forms of life unless it was in more or less complete harmony with the whole tendency of the individual. The adjustment of parts and their mutual interdependence make it essential for adaptation that the relation of parts be preserved; consequently, correlated minute favourable variations will tend to be more and more selected as evolution passes from the unspecialised to the specialised forms of life. This response of the whole organism should be still more delicate in those forms of life that are continually subjecting themselves to changed conditions; hence this delicacy of adjustment is far more necessary in the higher forms of animal life than in more stationary plant organisms, and in the developing nervous systems of animals we have just the central adjusting system that is required for these conditions. *With evolution of type there will thus be an increasingly definite tendency given to organic, especially the animal, forms of life, if the acting principle of evolution has been selectional. Selection is, therefore able to account for the steadily progressive tendency of life as a whole without calling to its aid any unknown and doubtful perfecting principle.*

To summarise: Natural selection, acting on the whole organism, tends to produce more and more definite tendencies in all surviving forms of life, which tendencies are progressive and continuous in character. Variable conditions, by partially altering the line of selection, induce a temporary indefiniteness. And lastly, the process of selection being itself able to be the indirect, though not the direct, cause of those favourable variations, which it subsequently selects from, is able to dispense with any subsidiary factors, provided it has a certain number of elementary properties of life which afford sufficient material to work with.

EXPERIMENTAL SUPPORT OF THE EFFECTIVENESS OF NATURAL SELECTION

Weldon's experiments with the shore-crabs of Plymouth Sound.—These experiments seem to show that under changed environmental conditions natural selection acts upon minute fluctuating variations of linear or quantitative type so as to reduce an alteration in the species; exactly as Darwinism would hold. A large breakwater was so placed near the mouth of Plymouth sound that the rate of flow of the river water was greatly slowed down in certain regions. This allowed an increased settling of the fine china-clay sediment that is carried by the river, and the changed condition caused the death of numerous crabs of the species *Carcinus maenas*. The question arose as to whether the survivors and those that had pershed showed any consistent difference son the basis of which selection could be operating. Careful measurements of hundreds of individuals showed that the mean breadth of frontum is slightly less in the survivors than in the perished. Measurements were repeated in two subsequent years and it was found that there was a progressive narrowing of the frontum. As an experimental check upon these conclusions Weldon placed a number of crabs in a large aquarium, in which china-clay was kept partly in suspension, and fond that about half of them died. Again the survivors were compared statistically with the perished and the same relation was found to hold: that the survivors had a mean frontal breadth distinctly narrower than that of the perished. Weldon conclude that his experiments "have demonstrated two facts about these crabs; the first that their mean frontal breadth is diminishing year by year at a measurable rate, which is more rapid in males than in females; the second is that this diminution in frontal breadth occurs in the presence of a material, namely, fine mud, which is increasing in amount, and which can be shown experimentally to destroy broad-fronted crabs at a greater rate than crabs with narrower frontal margins. . . . and I see no escape from the conclusion that we have here a case of Natural Selection acting with great rapidity, because of the rapidity with which the conditions of life are changing."

Cesnola's experiments with Mantis.—To test the selective value of color makings Cesnola fixed specimens of the brown and green *Mantis religiosa* on plants, some of which were against harmonious, others against disharmonious backgrounds. The result was that most of those which were inconspicuous because of a harmonious background escaped, while most of the others were eaten up by birds.

Poulton's and Sanders' experiments with butterfly pupae.-Numerous pupae of various colors were palced under conditions favoring protective coloration and others under opposite conditions. The conclusion was that protective coloration is a real survival factor, and one that operates so as to give the protectively colored individual a decided advantage in the struggle for existence.

Davenport's experiments with chickens.—A number of chickens, some black, some white, and some barred or checkered in color, were allowed to wander free in the fields. Hawks killed most of the whites and may of the blacks, but spared, to a large extent, the less conspicuous checkered and barred types which are harder to detect against a mixed background.

All of these experiments merely tend to show that discriminate survival actually occurs, but only the experiment of Weldon has a bearing on the possibility that mere quantitative changes of small dimensions might under certain conditions be of selective value. We badly need more experimental evidence of this sort and until this is forthcoming we shall have to admit that there is very little experimental evidence in favour of the type of natural selection that Darwin stood for.

NORMALIZING NATURAL SELECTION

Of considerable genetic and evolutionary significance is the interaction of mutaton and selection. A mutation is basically an accidental change in some component of the genetic system, comparable to a misprint or error in copying some word or paragraph. It is not surprising that mutations are rarely useful to their carriers when the latter live in the "normal" or usual environments of their species. One may conjecture that most useful mutations have already occurred and have become incorporated in the species gene pool. Useful mutations are more likely to be found when a population is transferred to a novel environment. Some mutations are favorable in heterozygous but unfavorable in homozygous condition. A majority of mutations range from neutral to deleterious to lethal. The accumulation in the gene pool of mutants lowering the fitness is impeded by normalizing natural selection.

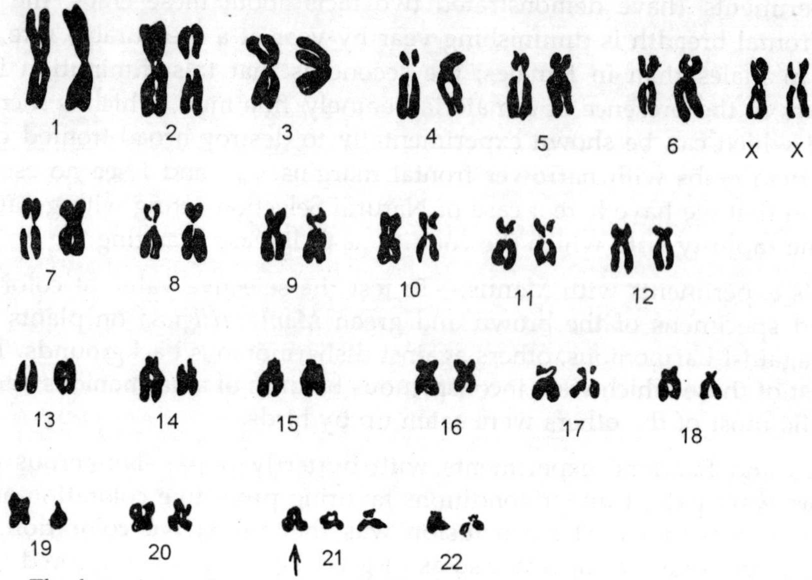

Fig. 4.1: The karyotype of a woman with Down's syndrome, in which three rather than the usual two chromosomes number 21 occur.

The situation is simplest for dominant mutations. A carrier of dominant mutant shows its effects in the phenotype, and is therefore exposed to.the action of natural selection. Let the mutation rate to a dominant condition be u per gamete per generation, and the Darwinian fitness of a carrier of the mutant gene $1 — s$. The equilibrium incidence of the dominant gene in the gene pool of the population will then be u/s. For a dominant lethal ($s = 1$) this will equal the mutation rate, $p = u$.

Mongoloidism, or Down's syndrome, in man is close to being a dominant lethal. Although most affected individuals survive, they seldom reproduce—their Darwinian fitness is very low. Mongoloidism is due to duplication of chromosome 21, so that most affected individuals have 7 chromosomes in their cells instead of the usual 46 (Fig. 4.1). The incidence of Down's syndrome at birth is about 15 per 10,000 (it is greate among children of older mothers). The mutation rate (u) is 7.5×10^{-4}, and is opposed by selection not far from $s = 1$. The frequency of Down's syndrome phenotypes of in a population is 2 U/s (since one mutant gamete is enough to produce an affected diploid individual).

Chondrodystrophic (achondroplastic) dwarfism in man is once of the many dominant mutants that appreciably reduce Darwinian fitness, but not to zero level. The dwarfs have heads and trunks of normal size, but relatively short legs and arm. A study made in Denmark has recorded the birth of eight achondroplastic infants among some 94,070 infants in families in which neither parent was an achondroplast. The mutation rate for achondroplasia is thus estimated approximately $8/(94{,}070 \times 2)$ or 4.2×10^{-5}. In families in which one parent is a dwarf, the number of children is no the average 0.25, while for normal families it is about 1.27. The Darwinian fitness of the achondroplasts is thus 0.25/1.27 = 0.2, and the selection coefficient discriminating against achondroplasia 1 - 0.2 = 0.8. The frequency of achondroplasts in the population should be $2u/s$ = (8.4 X 10^{-5})/0.8, or approximately 1 per 10,000, which is about what is observed.

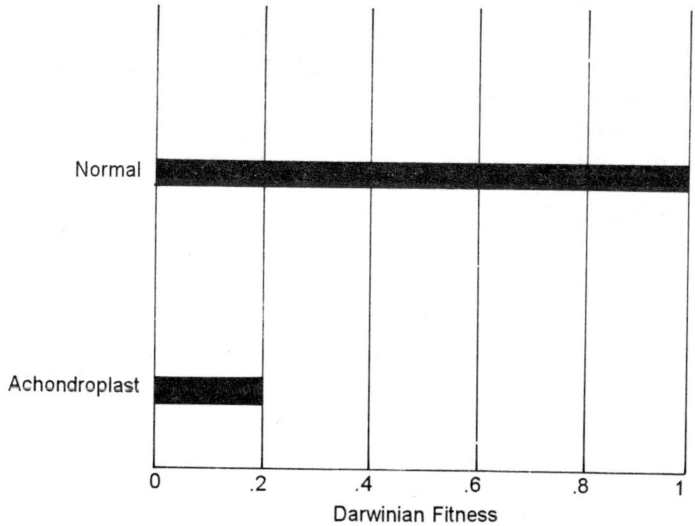

Fig. 4.2: Relative fitness of normal and achondroplastic humans. Achondroplastic dwarfs have heads and trunks of normal size, but relatively short legs and arms.

The situation is more complex with deleterious recessive mutations. A recessive mutant that is deleterious or even lethal when homozygous is sheltered from natural selection when it is carried in a heterozygote with a "normal" dominant allele. Recessive mutants will therefore accumulate in the population, until matings of heterozygous carriers are frequent enough to start yielding the ill-adapted homozygotes. An equilibrium between mutation and normalizing selection will then be reached when the number of the mutant alleles eliminated equals that of new mutants arising per generation. This will happen when $sq^2 = u$, or $q = \sqrt{u/s}$ (where s is the selection coefficient against the recessive homozygote, q the frequency of the deleterious recessive in the gene pool, and u the mutation rate).

Consider phenylketonuria and albinism, two examples among hundreds of more or less deleterious recessive mutants found in human populations. Both of them are rare—between 1:10,000 and 1:20,000—except that in some small and isolated populations albinism is more frequent. Phenylketonuria is a disorder of the metabolism of the amino acid phenylalanine. If untreated, it results in brain damage, gross mental retardation, and a Darwinian fitness of zero. If discovered early, however, and the affected infant is placed on a diet nearly free of phenylalanine, serious damage can be avoided, and the fitness increased to near normal. This is a fine example of dependence of Darwinian fitness on the environment. Albinism causes failure to synthesize pigment in the hair, skin, and iris of the eyes. Albinos are somewhat handicapped compared to normally pigmented persons, but most of them survive and reproduce.

Why do phenylketonuria and albinism persist in most human populations? The simplest hypothesis is that they arise repeatedly by mutation, but that normalizing selection prevents their undue accumulation. If the frequency of a recessive homozygote is $q^2 = 0.0001$ (on 10,000), the frequency of the recessive gene the gene pool of the population will be $q = 0.01$, and that of heterozygous carriers (according to Hardy-Weinberg formula) $2pq = 2 \times 0.99 \times 0.01 = 0.0198$. That is, almost two per cent of healthy persons in the population are carriers of the phenylketonuria gene, and two per cent carriers of the albinism gene.

The fitness of untreated phenylketonurics is zero, hence the selection against them is $s = 1$. Assuming that the fitness of heterozygous carriers is one, the same as for noncarriers, we can estimate the mutation rate from the normal to the phenylketonuric allele, using the equation $u = sq^2$. This is $1 \times 0.01^2 = 0.0001$, or 10^{-4}. The fitness of homozygous albinos is not known; assuming it is about 90 per cent normal ($s = 0.01$), we obtain an estimate of the mutation rate $u = 0.1 \times 0.01^2$, or 10^{-5}. Both estimates have a margin of uncertainty because we have assumed the fitness of the heterozygous carriers of phenylketonuria and albinism to be one. If it were lower, normalizing selection would be more important against the heterozygotes than against the homozygotes because the former are much more numerous than the latter. Our estimates of the mutation rates would then be too high. Similar problems arise also with respect to other deleterious recessive mutants, which are individually rare and yet are encountered in most human populations.

GENETIC BURDENS OR LOADS

Some examples of deleterious genetic traits in human populations have been used in this chapter to illustrate the action of normalizing natural selection. Indeed, genetic variability and normalizing selection are universal in the living world. Mutants arise in any species and many or most of them are deleterious. Some of these mutants may persist in the populations, i.e., one or several generations intervene between the origin and the elimination of a deleterious genetic variant. Hence, populations of any species, including the best adapted and most successful ones, carry burdens or loads of genetic defects. Unhappily, mankind is no exception.

One can distinguish between expressed (overt) and concealed genetic loads. The former comprise individuals with diseases, malformations, or constitutional weaknesses that are manifestly genetic. Concealed genetic burdens are due to recessive genes or gene complex carried in heterozygous condition in individuals who are themselves healthy or "normal." When homozygous, however, these genes or gene complexes are also debilitating or lethal. Concealed loads are uncovered when heterozygous and show the defect. Inbreeding increases the probability of uncovering concealed genetic malformations In human populations some rare recessive diseases are found mainly in progenies of parents who are relatives. It is an open question whether incest taboos were first instituted because of experience with the genetic risks. There is no doubt, however, that avoidance of incest is genetically a sound practice.

Inbreeding experiments have been made with many species of normally outbreeding animals and plants. Inbred lines of corn (*Zea mays*) have been studied most extensively. Almost without exception progenies obtained after several generations of self-pollination are strikingly less vigorous than outbred progenies. In genetically well-studied species of *Drosohila* techniques are available for uncovering and recording the frequencies of the constituents of concealed genetic loads. The essence of these techniques is a series of crosses designed to obtain individuals that carry in double dose (are homozygous for) certain chromosomes extracted from natural or experimental populations. The results obtained are qualitatively similar, although quantitatively variable, in all species. When chromosomes that are present in "normal," healthy flies are made homogygous, many of them become lethal (kill all the homozygotes), semilethal (kill more than 50 but less than 100 per cent of the homogygotes), subvital (kill fewer than 50 per cent), cause sterility of the females or males, or induce visible structural defects or abnormalities.

Very few, if any individuals in natural populations are free of chromosomes that produce more or less serious defects when indouble dose. Only a minority of the chromosomes give "normal" homzygotes, and a very small minority give supervital homozygotes. In point of fact, special experiments have shown that even these minorities are really handicapped. Normality (the adaptive norm) is defined as the average viability, fertility, development rate, etc., of heterozygous individuals. The heterogygotes conserve their normality when exosed to variations of temperature and other conditions. They possess developmental homeostasis, which makes them able to adjust to environmental changes. By contrast, the normal and supervital homozygotes deteriorate and become subvital, even semilethal, in changed environments. They are deficient in homeostatic ability to adjust to changed environments.

Components of concealed genetic loads become expressed not only as a result of inbreeding but also of accidental mating of not closely related individuals who happen to be heterogygous for the same recessive defect. The recessivity of genes that are lethal or grossly deleterious when homozygous may be incomplete; the heterozygous carriers may then be weakened compared to non-carriers. Extensive data on expressed genetic loads, though nor as precise as one would like them to be, are available for human populations. According to Stevenson (1961) and Trimble and Doughty (1974), in British populations some 12 to 15 per cent of pregnancies that continue more than four or five weeks end in spontaneous abortion, and two per cent in stillbirths. An appreciable part, although doubtless not all, of the abortions and stillbirths are genetically caused. Schull and Neel (1965) found 3.55 and 3.42 per cent childhood mortalities in Hiroshima and Nagasaki, respectively, in families in which the parents were not known to be related. What proportion of morbidity in human populations stems from genetic causes is even more difficult to determine than for mortality. Stevenson estimates that in the population of Northern Ireland some 7.9 per cent of all consulations with medical specialists and 6.4 per cent of those with general practitioners are sought by genetically handicapped persons, and 26.5 per cent of all hospital beds are occupied by such individuals.

What, if any, biological function do genetic burdens serve in living species? Genetic variability is a prerequisite for evolutionary change, and the mutation process is the ultimate source of this variability. You may deplore the fact that most mutatins are deleterious, instead of only useful ones being produced exactly where and when they are needed. One must realize, however, that to do so a living species would have to anticipate its future needs and be able to act purposefully to satisfy them. Only mankind can do that, and, lamentably, only to a very limited extent. Deleterious mutations are the price the species pays for having some useful ones.

Moreover, deleterious and useful mutants are not fixed categories. What is deleterious in one environment may be useful in another. Insecticide resistance is imperative for survival when an insect population is being treated regularly with an insecticide, but mildly deleterious in the absence of any treatment. A gene may be useful in combinations with some, neutral with others, and harmful with still other genes. Biologicla species usually differ in numerous genes. In the short run, the store of genetic variability carried in the gene pool of a population is a genetic burden. Yet in the long run, genetic variability is a treasure house enabling the population to adapt to changing environments.

POLYMORPHISM AND BALANCING NATURAL SELECTION

It has been seen than normalizing natural selection is pre-eminently a conservative force. It purges the gene pool of a population of deleterious genetic variants and thereby tends to keep the species constant. Its efficiency is far from perfect, however, since populations carry genetic defects, which lower the Darwinian fitness of homogygous individuals. Although these defects are of many kinds, most are rare. Among hundreds or even thousands of genetic diseases and malformatinos known in man, most have an incidence of 1:10,000 or less. Taken all together, these deficits nevertheless add up to a genetic load which is by no means negligible.

Sometimes two or more genetic variants occur in a population with such frequencies that they cannot be regarded as anomalies. Such populations and species are called *polymorphic*. Species in which at lest two phenotypically or genotypically discrete classes cannot be distinguished are *monomorphic*. Some polymorphsms are more apparent than others. For example, polymorhisms for eye and hair color in human populations are clearly visible, whereas blood group, enzyme, and taste-blindness polymorphisms require special tests to be revealed.

Polymorphism and continuous genetic variation are distinct in reality, although not in theory. Thus, variations in stature, weight, or head shape in human populations, though they doubtless have genetic foundations, are not usually described as polymorphisms. Tall and short, fat and slim, longheaded and roundheaded people do not form discrete classes. Not only do intermediates exist, but they are more frequent than the extremes. Nevertheless, if we could carry genetic analysis of these traits down to the gene level, we would find discrete variants in the gene pool underlying the continuous observed variation of the phenotypes. Eye color polymorphism, which at first sight seems discontinuous, is to a certain extent an instance of continuous variation, due to one gene with major and several with minor phenotypic effects. The more detailed the study of genetic variation in a species, the more likely it is to disclose polymorphisms.

What genetic mechanisms are responsible for the origin and maintenance of polymorphisms? This is a difficult and controversial problem in evolutionary genetics. Several mechanisms can be considered. A polymorphism can be balanced, so that two or more gene alleles or chromosomal variants are kept in a population at more or less stable equilibrium frequencies. Several forms of balancing natural selection can bring such a situation about. A transient polymorphism may be a stage about midway between the origin of a new useful variant by mutation and its eventual fixation in the population by directional natural selection. In addition, polymorphsms may have nothing to do with selection, i.e., the genetic variants concerned may be neutral, or equal in Darwinian fitness. They would then be due to mutation and random genetic drift. It is certain that all these mechanisms exist in reality; what is still controversial is their relative frequency in nature and their roles in the evolutionary development of the living world.

Consider the effects on Darwinian fitness of two gene alleles, A_1 and A_2. If the fitnesses of the homozygotes A_1A_1 and the heterozygotes A_1 A_1 are equal, but either lower or higher than that of the homozygotes A_2 A_2, the allele A_1 is said to be fully dominant and A_2 fully recessive. If the heterozygote A_1A_2 is intermediate in fitness between A_1A_1 and A_2 A_2, the dominance is incomplete or absent. The heterozygote A_1A_2 may also be superior, or inferior, in fitness to both homozygotes. If it is superior, we have a case of over dominance, or hybrid vigor, or *heterosis*. Inferior heterozygotes betoken some degree of hybrid inviability or hybrid sterility (the term "negative heterosis" has also been proposed, but it is rather misleading and should be avoided).

Let the superior fitness of a heterotic heterozygote be one, and the fitnesses of the inferior homozygotes 1 - s_1 and 1 - s_2 where s_1 and s_2 are selection coefficients. In a sexual and outbreeding population there will be:

	A_1A_1	A_1A_2	A_2A_2	*Total*
Fitness, *w*	$1 - s_1$	1	$1 - s_2$	W
Frequency before selection	p^2	$2pq$	q^2	1
Frequency after selection	$p^2 - s_1p^2$	$2pq$	$q^2 - s_2q^2$	$1 - s_1p^2 - s_2q^2$

Traditionally, since Darwin's day, one has usually assumed that natural selection eliminates what is less fit and puts in its place what is more useful. Overdominance means however that neither A_1 nor A_2 is best by itself, because the fittest heterozygote, A_1A_2, has them both. The outcome of selection will therefore be a balanced equilibrium, conserving the polymorphism in the population. The frequencies of A_1 and of A_2 at equilbrium will be:

$$p = s_2/(s_1 + s_2) \quad q = s_1/(s_1 + s_2)$$

If the advantage of the heterozygote is the same over both homozygotes ($s_1 = s_2$), the variants A_1 and A_2 will be equally frequent at equilibrium ($p = q = 0.5$). This works regardless of whether the advantage is small (for example $s_1 = s_2 = 0.01$) or great. If both homozygotes are lethal or sterile ($s_1 = s_2 = 1$) we have what is called a *balanced lethal system*. In many laboratory strains of *Drosophila* both chromosomes of a homologous pair have different mutant genes, which may or may not produce viable changes in the appearance of the flies when heterozygous but which are lethal in double dose. The mating A_1A_2♀ × A_1A_2♂ yields only A_1A_2 progeny; the strain is said to "breed true," but it can be shown that one half of the eggs deposited fail to develop to the adult stage. Similar in principle are some species of evening primroses (*Oenothera*) and other plants. They are "permanent heterozygotes" for one or several translocations between different chromosomes; all viable seeds are translocations heterozygotes like the plants that produce them.

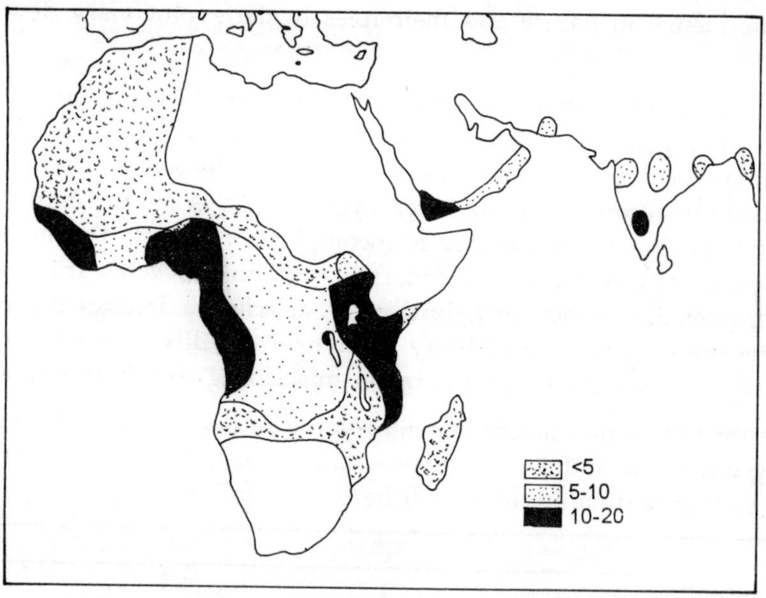

Fig. 4.3: Geographic distribution of the sickle-cell gene, *s*, which in homozygous condition is responsible for sickle-cell anemia. The frequency of *s* is high in certain parts of the world where falciparum malaria is endemic (see Fig. 4.4). This association may be due to the high resistance to malaria of individuals heterozygous for the *s* gene.

One of the homozygotes may have a fitness greater than the other, while both are less fit than the heterozygotes ($A_1A_2 > A_1A_1 > A_2A_2$). At equilibrium. A_1 will be more frequent than A_2 in the population. The classical (and some people think, the only well-established) case of balanced polymorphism in man is that of sickle-cell anemia, a disease that is fairly common in some populations in Africa and in populations of African and Asiatic origin. The anemia is due to homozygosis for a gene *s*, which produces an abnormal hemoglobin S_1 instead of the normal hemoglobin A produced by the "normal allele S. The two hemoglobins differ in a single amino acid, glutamic acid in hemoglobin A being replaced by valine in position 6 in the beta chain of hemoglobin S_1. The homozygotes *s/s* usually die before reaching sexual maturity; their Darwinian fitness is only slightly greater than zero. Let us take it to be w = 0.1, so that the selection coefficient against *s/s* homozygotes is 0.9. The heterozygotes *S/s* not only will survive, but, as first shown by Allison, will be relatively more resistant to, or less affected by, a form of malaria widespread in the tropics (*Plasmodium falciparum*) than are the "normal" homozygotes *S/S*. The heterozygotes will thus be more fit than the homozygotes *S/S* in countries where the falciparum malaria is widespread; where this malaria is absent there is no advantage in the heterozygotes.

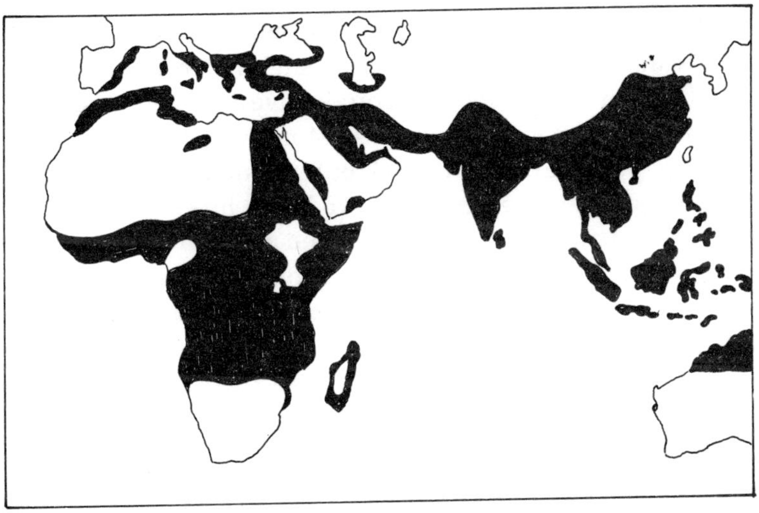

Fig. 4.4: Old World distribution of malignant malarias caused by the parasite *Plasmodium falciparum*.

Let us suppose that in some country where malaria is moderately common the fitnesses of the homo- and heterozygotes are $S/S = 0.8$, $S/s = 1$, $s/s = 0.1$. At equilibrium, the heterotic balancing selection will establish the frequencies of the alleles $S = p = 0.9/(0.9 + 0.2) = 0.82$, and $s = q = 0.2/(0.9 + 0.2) = 0.18$. In every generation $q^2 = 0.18^2$, that is, 3.24 per cent of the infants born will be the semilethal homozygotes *s/s*, most of whom will die before adulthood. However, in a malarial environment the average fitness of the population will be greater than that of a population consisting only of "normal" homozygotes, *S/S*. This is because $2pq = 2 \times 0.82 \times 0.18 = 0.295$; 29.5 per cent of the individuals born will be heterozygotes that are relatively resistant to malaria and have a fitness of one, whereas the fitness of "normal" homozygotes in malarial environment is only 0.8. If the population migrates to a malaria-free country, or if malaria is eradicated, a population having the *s* gene in its gene pool will be slightly less fit than one free of *s*. Normalizing natural selection will begin to act to reduce the frequency of *s*, and eventually eliminate it entirely. On the other hand, if malaria becomes more frequent, balancing natural selection will tend to increase the frequency of *s*.

Although it has not been proven, some hereditary diseases and defects may be maintained in human populations by balancing selection if the fitness of the heterozygous carriers is greater than that of the noncarriers. The same holds for some of the constituents of the genetic loads in *Drosphila* and other organisms. Reliable evidence bearing on this problem is difficult to obtain because small selective advantages or disadvantages require impractical amounts of experimental or observational data.

Suppose that the Darwinian fitnesses of the three genotypes A_1A_1:A_1A_2:A_2A_2 are 0.99:1:0, i.e., the heterozygous carriers of A_2 have a one per cent advantage over the noncarriers, while the homozygote A_2A_2 is lethal. How shall we go about discovering this one per cent advantage? It may be due to a slightly lower childhood mortality, greater fecundity, earlier marriage, longer reproductive life, better social adjustment, or a combination of these and other factors. Moreover, the advantages may be pronounced in some physical or cultural environments and absent in other environments. Finally, the heterozygotes may have had a fitness advantage in the past, in environments before the advent of civilization and modern technology, but may no longer be advantageous. For all these reasons, pinpointing small selective advantages and disadvantages in man is a formidable task. It is laborious even with laboratory animals and plants. The question of how many recessive genetic defects are maintained in human populations by balancing selection, rather than by mutation pressure, is wide open.

Heterotic balancing selection can be demonstrated and measured in situations where selective advantages and disadvantages are large. Populations of some animal and plant species are polymorphic for variants of the chromosome structure, inversions, and translocations. As a rule these chromosomal variants produce no visible effects on the external appearance of their carriers, but they can be detected by microscopic examination of the chromosomal complements. This is easiest in many representatives of Diptera, which have giant banded chromosomes in the cells of larval salivary glands. The inversion heterozygotes (heterokaryotypes) have two chromosomes of a homologus pair with gene arrangements that differ in one or more chromosome segments (block of genes) placed in inverted orders—ABCD and ACBD. The pairing of such chromosomes give rise to an easily visible loop.

Drosophila pseudoobscura is one of many species the natural population of which are polymorphic for chromosomal inversions. For reasons that are still unknown, most of the inversions occur in only one of the five chromosomes the species has—the third chromosome. Around 1940 the surprising observation was made that the relative frequencies of the inversions in some populations change from month to month. These changes are not only quite significant statistically but are cyclic with the seasons of the year, i.e. are repeated in successive years (Fig. 4.5). The hypothesis was framed that the changes are due to the action of natural selection. To test the hypothesis, experimental populations are set up in the laboratory in so-called population cages. A population is started with known frequencies of two or more gene arrangements differing in inversions. The population is left to breed freely for as many generations as needed. At desired intervals, samples of larvae are taken, and the chromosomes in the salivary glands examined.

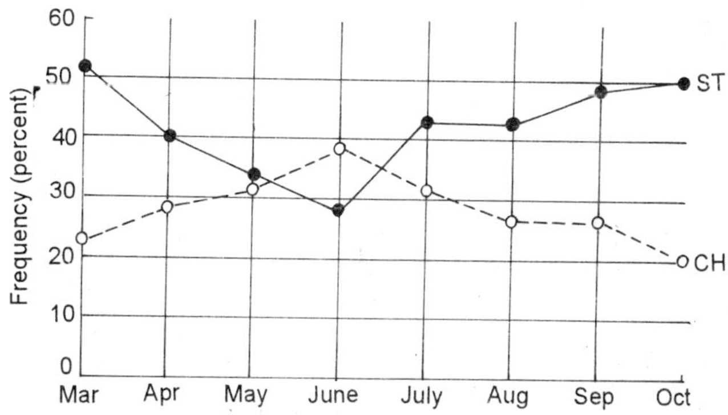

Fig. 4.5: Frequencies of two gene arrangements in a natural population of *Drosophila pseudoobscura*. The frequencies of the two chromosomal arrangements, CH and ST, change throughout the year. CH has its highest frequency at the beginning of summer, when the frequency of ST is lowest.

Typical results are shown in Fig. 4.6. In the early generations (a generation takes about a month under the conditions of this experiment) the frequencies of the inversions change rapidly. Later the changes slow down, and eventually an equilibrium is reached when the frequency of the inversions remains constant. None of the competing inversions are eliminated and none reach 100 per cent frequencies. This is prima facie evidence that a balancing natural selection is acting. The most probable explanation is that the fitness of the heterokaryotypes is higher than that of the homokaryotypes. Taking into consideration the slope of the selection curve and the equilibrium values, one can compute the adaptive values of the karyotypes. These turn out to be one for the heterokaryotypes and 0.89 and 0.41 for the homokaryotypes.

This is very strong selection, considering that the selection is not for some kind of genetic disease but for "normal" constituents of natural populations. One of the two homokaryotypes has a fitness less than one half that of the other! To be sure, the fitness is remarkably sensitive to environment (temperature, nutrition, etc.). The results in Figure 4.7 are from an experiment done at 25° C. Lowering the temperature by only 10°C renders the adaptive values of the karyotypes so nearly equal that no significant changes in the chromosome frequencies are observed for about 10 generations.

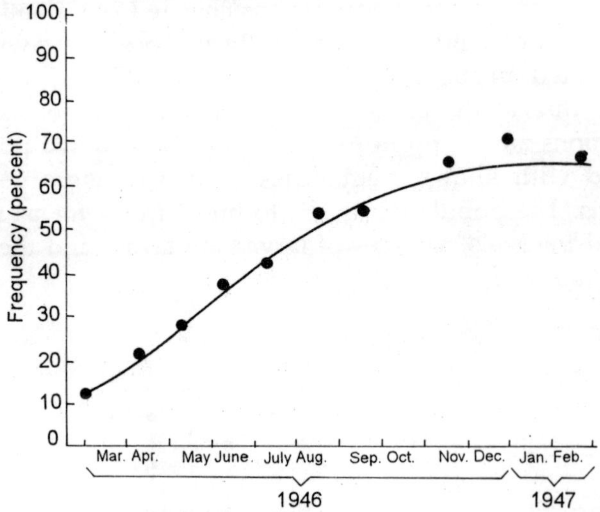

Fig. 4.6: Changes in chromosomal frequencies in an experimental population of *Drosophila pseudoobscura*. Two chromosomal arrangements, CH and ST, compete in the cage. The frequency of ST gradually increases from its initial frequency of 12 per cent, to reach an equilibrium frequency around 70 per cent. Correspondingly, CH decreases from about 88 per cent to an equilibrium frequency around 30 per cent.

These experiments throw some light on cyclic seasonal changes in the frequencies of different chromosome sin nature. Such changes are expected if one of the homokaryotypes (ST/ST) is superior to the other (CH/CH) in summer (June-September), but inferior in spring (March-May). Just what physiological and ecological properties of the karyotypes are responsible for the changes is not well known. Some experiments suggest that ST is superior under conditions of crowding, and CH superior in uncrowded populations. Other characters, such as viability and fecundity, are probably also involved. Nevertheless, the evidence shows unambiguously that *Drosophila,* an organism that produces several generations per year in at least some of its natural habitats, undergoes genetic evolutionary alterations under the influence of seasonal climatic or biotic changes in its environments. There is also evidence that natural selection induces genetic alterations in different years, presumably because of the year-to-year environmental fluctuations.

Another question is why an inversion of a block of genes should change the Darwinian fitness, either in homo-or heterozygous condition. After all, a chromosome with an inversion carries the same gene loci, though arranged in a different sequence, that it had before the inversion. The answer is suggested by the fact that gene recombination is reduced or suppressed in chromosomes in which the two members of a pair differ by an inversion. Suppose that a pair of chromosomes carry gene alleles $A_1B_1C_1D_1$ and $A_2C_2B_2D_2$, respectively. The two gene combinations interact favorably, giving heterosis in the heterokaryotype $A_1B_1C_1D_1/A_2C_2B_2D_2$. Other combinations of the same alleles, for example $A_1B_2C_1D_2$, have adaptively less favorable effects. An inversion that suppresses recombination guards against disintegration of a favorable gene pattern, and is therefore favored by the balancing natural selection.

Inversion polymorphism is really an especially clear case of a much more general phenomenon of linkage disequilibrium. Genes do not act in development independently of other genes. They interact in concert, and natural selection favors placing together in the same chromosome an assemblage of gene alleles that confers a high fitness on the possessors. Such assemblages of genes are frequently referred to as supergenes. The role of chromosomal inversions in evolution may then be conservation of adaptively valuable supergenes.

How widespread inversion polymorphism are in organisms other than *Drosophila* and its relatives is uncertain. In the absence of giant chromosomes the detection of inversion is difficult—only those inversions that change the position of the centromere or those that produce certain unusual configurations at meiotic divisions are detected. Nevertheless, inversion heterozygotes are known in many animals, e.g., species of grasshoppers; in plants they are found chiefly in interspecific hybrid. In human populations inversions seem to be rare, but this may only mean that their disvoery is uncommonly difficult. By contrast, polymorphism for translocations is found in populations of many plant species and relatively few animals. The biological function of translocations and inversion polymorphisms is probably similar. Establishment of adaptively valuable linkage disequilibria and supergenes.

The adaptive superiority of hetercozygotes, hererosis, is the most frequently discussed, but not necessarily the most widespread, genetic mechanism underlying polymorphisms in natural populations. Frequency-dependent selection may be at least equally or more important. In the foregoing discussion we tacitly assumed that the adaptive values and selection coefficients are constant for any given genotpye, no matter what its frequency in the population relative to other genotypes. This simplifies the mathematical treatment of selection processes but it is often unrealistic. After all, the adaptive value, Darwinian fitness, is a function of the environment in which an organisms lives, and other organisms of the same and different species living in the same neighbourhood are important constituents of the environment.

Suppose that the fitnesses of two competing genotypes, say A_1A_1 and A_2A_2, are high when a given genotype is rare in a population, but diminish when that genotype becomes frequent. Natural selection will enhance the frequency of the initially rare genotype. As its frequency increases, its fitness dwindles, while the fitness of the alternative genotype grows as it becomes less frequent. It may be that at certain frequencies the carriers of A_1A_1 and A_2A_2 are equally fit. If so, a stable equilibrium will be established at these frequencies, and the population will become stably polymorphic. Very extensive experimental evidence on frequency-dependent selection in some species of *Drosophila* has been obtained by Kojima and his colleagues. The evidence is even more abundant in cultivated plants. The yield of a planting of a proper mixture of genotypes may exceed that of equal areas planted to each of these genotypes singly.

There are at least two mechanisms that bring about frequency dependence. One is diversity of environments. Some environments may be more suitable, or preferred, by carriers of one genotype and others by competing genotypes. This may lead to diversifying natural selection. Another mechanism is interaction of the carriers of different genotypes. The interaction may be mutually beneficial (facilitation, synergism), or may result in one or both genotypes being harmed (disoperation).

Dawood and Strickberger (1969), as well as Anxolabehere and Periquet (1972), have adduced evidence of such interactions between larvae of different mutants and species of *Drosophila*. It should be noted that heterotic and frequency-dependent balancing selections are not mutually exclusive—they can operate simultaneously. In fact, one of the first instances of frequency dependence was found in experimental population cages of *Drosophila pseudoobscura* polymorphic for chromosomal inversions. We have seen above that polymorphisms for inversions are consequences of superior fitnesses of heterokaryotypes; nevertheless, analysis of the selection process in experimental populations has shown that the relative fitnesses of the karyotypes is dependent on the kinds and frequencies of the chromosomes present.

DIVERSIFYING NATURAL SELECTION

Biologists often speak about the environment in which a species or a population lives. Yet even carefully controlled laboratory environments are hardly ever completely uniform. Natural "environments" are mosaics of more or less similar or disparate habitats or sub environments. Some of these habitats may be more suitable for some genotypes, and others for competing genotypes. How is a species to cope with the variety of habitats that it meets both in space and time? One stragegy is natural selection favoring a genotype in which the carriers possess a well-developed homeostasis, such that they have well-adapted phenotypes in all, or at least in most, subenvironments. In a way, this strategy is universal. Not only every species but also every individual meets with a range of environments; if it were specialized to live in a single environment it would not live long. Another strategy is polymorphism. Natural selection diversifies the gene pool in such a way that two or several classes of genotypes having optimal adaptiveness in different subenvironments are produced. The two strategies are of course not mutually exclusive. Conditions under which each of them is expedient have been investigated mathematically by Levins (1968).

The converse of diversifying selection (also called "disruptive selection") is stabilizing selection. If the environment is rather uniform in space and time, it may be advantageous for a population to limit the range of its variability, so that only, or mostly, phenotypes tested by natural selection are produced and deviatins eliminated. In a sense, stabilizing selection is related to normalizing selection, and so understood it is really universal. Every species has no the average a certain body size, weight, and proportions. It is doubtful that giants above seven feet tall or dwarfs below five feet would fare well in human populations. In fact, neonatal deaths are more frequent among abnormally heavy and abnormally light infants that among infants or average weight.

Diversifying selection may be closely related to frequency-dependent selection. Suppose that a population inhabiting a certain territory has two or several groups of genotypes and meets two or several subenvironments. A genotype that is rare but is well adapted to a certain habitat will be promoted and increased in frequency by selection when its habitat is not fully occupied. Conversely, it will no longer be promoted when its habitat is saturated. A frequent genotype may saturate its optimal environment, and its excess population may spill over into subenvironments that are stressful for them. Diversifying selection will then establish a balanced polymorphism. Frequencies of the different genotypes will be reached such that the exploitation of the available mosaics of subenvironments will be as complete and efficient as possible. Levene (1953), Maynard Smith (1970), and others have explored the conditions under which the polymorphism will be stable. Owell (1971) and McDonald and Ayala (1974) have shown experimentally that laboratory populations or *Drosophila* placed in environments deliberately made heterogeneous support more genetic polymorphisms than do populations in more homogeneous environments.

Diversifying selection is a process that enhances the adaptedness of populations that live in heterogeneous environments. By and large, genetic uniformity is a drawback, except in ideally constant environments, and diversity is an advantage. Does diversifying selection play any role in mankind? Although upbringing and training are the chief factors that qualify persons for different occupations and social roles, individual genetic predispositions do facilitate training for one or another vocation. Insofar as this is so, diversifying selection renders important services in the human species.

SEXUAL SELECTION

In many species of animals females and males are strikingly dimorphic. Males are often (though not always) large, stronger, more brightly colored, and possessing various "adnorments," some of which may seem to be handicaps rather than advantages to their possessors. Females and males always differ in behaviour to some extent, and sometimes strikingly so. Could sexual dimorphisms originate in evolution by natural selection? Darwin found this improbable as a general explanation, and proposed, in 1871, sexual selection as a more plausible solution. In his words, "sexual selection depends not on a struggle for existence in relation to other organic beings or to external conditions, but on a struggle between the individuals of one sex, generally the males, for the possession of the other sex."

During the century since it was proposed by Darwin, the theory of sexual selection has met with even more opposition than the theory of natural selection. At present, sexual selection is usually regarded as a special case of natural selection. After, all, Darwinian fitness is reproductive fitness. The carriers of some genotypes transmit more of their genes to the following generations than do other genotypes. Higher fitness may be achieved because of some advantage in the "struggle for existence in relation to other organic beings or to external conditions"; it may also be achieved because of greater proficiency in securing mates. Components of fitness may or may not be correlated. Greater sexual prowess may gc together with lower survival rate, or vice versa. Darwinian fitness is a resultant summation of advantages and disadvantages. Too conspicuous a coloration, overgrown antlers, or horny outgrowths, present in some recent and fossil species of mammals, may have been nuisances to their possessors, but if they enabled them to sire more offspring they were promoted by selection.

Sexual selection has been experimentally demonstrated. For example, yellow body color, one of the classic mutants in *Drosophila,* is selected against in experimental populations containing mixtures of yellow and normally pigmented flies. The lower fitness of yellow is due largely to the fact that yellow males are seldom accepted by normal females, while normal males are accepted about equally by both yellow and normal females. Seual selection in nature is a complex issue. Do females really discriminate in favor of more "handsome" males? In some species, among both vertebrates and invertebrates, males engage in agonistic displays or actual fights, and the winner gets the female as the "prize". Here the females play passive roles. Persistence in courting females varies, and the more preservering males are accepted on the average more often than more sluggish ones.

Some modern authors have broadened the concept of sexual selection to mean any departure from random mating among inhabitants of the same neighbourhood. So defined, sexual selection is certainly widespread. In the first place, different species, even closely related ones, usually exhibit ethologicla isolation, i.e., a preference for mating with conspecific individuals. To exercise such a preference, both females and males must be able to recognize the species and sex of individuals whom they meet. The recognition marks may be visual, auditory, olfactory, or kinesthetic. Selection is certainly involved in development of the recognition marks. The bright colorations and othe adornments of males may then be understood either as appealing to "aesthetic perceptions" in the females or as species recognition marks. The fact that such adornments are found more often in males than in females in consistent with more active sexual role observed of males in many species.

Moreover, deviations from randomness of mating are by no means uncommon among conspecific individuals in the same neighbourhood. A most interesting situation, combining sexual and frequency-dependent selections, was discovered independently by Petit and by Ehrman in several species of *Drosophila*. In Ehrman's experiments equal numbers of females and males of two kinds of *Drosophila pseudoobscura* were introduced into an observation chamber, but the proportion of flies of each kind varied from experiment to experiment. Whichever males were in a minority had a relative advantage in mating with females of either kind. In one experiment the two kinds of flies differed in geographic origin—the ancestors of one strain were collected in a locality of California, and those of the other in Texas. Table 4-1 shows the results.

Table—4.1

Observed and expected numbers of matings of two strains of *Drosophila pseudoobscura* in mating chambers with different proportions of the two kinds of flies.

	Males mated		*Females mated*		*Chi-square*
Proportions	*C*	*T*	*C*	*T*	*for males*
12C:12T					
Observed	55	49	50	54	
Expected	52.0	52.0	52.0	52.0	0.39
20C:5T					
Observed	70	39	84	25	
Expected	87.2	21.8	87.2	21.8	16.96
5C:20T					
Observed	39	65	30	74	
Expected	20.8	83.2	20.8	83.2	19.91
23C:2T					
Observed	77	24	93	8	
Expected	92.9	8.1	92.9	8.1	34.75
2C:23T					
Observed	30	70	12	88	
Expected	8.0	92.0	8.0	92.0	65.76

It can be seen in Table 4.1 that when the numbers of males of California and Texas origins in the mating chamber were equal, the matings occurred in random proportions. This was not the case when either California or Texas males were a minority. The observed number of matings per minority male turned out to be greater than expected on the assumption of randomness, while the majority males mates less frequently than expected. The differences between the observed and expected number of males mating was statistically quite significant—a chi-square greater than 6.63 is expected to occur by chance in fewer than one per 100 trials. By contrast, the numbers of majority and minority females mating were nearly in proportion to their abundance. Mating advantages of minority males were also observed by Ehrman when the two kinds of the flies in the chamber differed inchromosomal inversions, in mutant genes, or in having developed at different temperatures. Some observations indicate that mating advantages of rarity occur also in *Tribolium* beetles and in certain butterflies. Nevertheless, this phenomenon can hardly be universal. Grossly deleterious mutants, such as hereditary diseases or malformations, surely do not confer mating advantages.

Mating advantages of minority genotypes may lead to stably balanced polymorphisms. Of course, a mating advantage tends to increase Darwinian fitness.

However, the advantage decreases as the frequency of a genotype increases. At some frequencies the competing genotypes mate at random, and an equilibrium is reached. It is worthwhile to reiterate that different kinds of selection may operate simultaneously. It may be that the frequencies of chromosomal polymorphisms in natural populations of *Drosophila* are determined by interactions of heterotic and frequency-dependent balancing selections.

DIRECTIONAL SELECTION

Physical, e.g. climatic, and biotic environments of a species do not remain constant for long periods. Environmental changes do not necessarily induce genetic changes; the genetic variants composing a species may be sufficiently homeostatic so that adaptedness remains adequate in the altered environments. On the other hand, the Darwinian fitnesses of variant phenotypes and genotypes may be shifted, some in positive and others in negative directions. A *directional natural selection* then comes into operation. It transforms the gene pol of a species or population toward the highest level of adaptedness (given the genetic materials available) that can be reached in the new environment. Apart from environmental change, origin of new and favorable mutant genes or supergenes brings into action directional natural selection, even if the external environment continues without change.

Rates of changes in gene frequencies under directional selection depend of course on the relative Darwinian fitnesses of the favored genotypes. In principle, the rates are the same as under normalizing selection. The similarity of the selection rates under normalizing and directional selections is not surprising. If selection favors a certain allele of a gene, it necessarily discriminates against the other alleles of the same gene.

Experimental models of directional selection have been made. A known mixture of different genotypes is introduced in a population cage, or sown or planted on an experimental field, and left to breed freely. In every generation, or at other covenient intervals, samples of the population are taken, and the frequencies of some genetic variants determined. From the observed rates of changes in these frequencies, the adaptive values of the variants can be computed. It should be stressed that what is observed in such experiments is natural selection, no matter how "artificial" the experimental environments may be.

Artificial selection, practiced by breeders of domestic animals or plants, can be used as a model of directional natural selection. It was so used by Darwin in *Origin of Species*. Experiments on artificial selection have also been made by geneticists on a variety of characteristics of many species. Perhaps the most notable result of these experiments is that at least some change in the desired direction can be obtained with almost anything one selects. In Lewontin's words, "There appears to be no character—morphogenetic, behavioural, physiological, or cytological—that cannot be selected in *Drosophila*" (1974). The significance of this must be appreciated. Natural populations, at least of sexual and outbreeding species, contain stores of genetic variability much greater than classical geneticists, and even Darwin, dared to suppose. Furthermore, this variability is pervasive, in the sense that it affects every aspect of the organism's constitution. This disposes of the disingenuous objections against natural selection as a directing agency of evolution that alleges that to progress in a given direction selection has to wait for the occurrence of novel mutations. To be sure, to quicken the progress of artificial selection practical breeders sometimes take steps to augment the genetic variance of the materials with which they work. Yet striking, even spectacular improvements in the productivity of agricultural plants and animals have been achieved by artificial selection without conscious steps to enhance the genetic variances. Call to mind the far-reaching differences between domesticated species and their wild ancestors!

Examples of the action of directional selection in nature can be seen most easily in genetic changes that occur in nondomesticated species because of transformation of their environments by man. The evolution of resistance of insect species to pesticides has in recent decades been a spectacular demonstration of the ability of living species to undergo genetic changes in response to challenges of the environment. The story is always the same: when a new insecticide is introduced, a certain relatively small amount or concentration is sufficient to achieve a satisfactory control of the insect pest abundance; the necessary concentration gradually increases, until it becomes totally ineffective or economically impractical. In 1947 Sacca was the first to report that a population of the house fly, *Musca domestica,* had become resistant to DDT. Since then, resistance to one or more insecticides has been recorded in at least 225 species of insects and other arthropods. Usually the resistance is specific to one or a group of chemically related insecticides, but occasionally instances of multiple resistance are reported. An interesting parallel is found in a mammal species, the grey rat. A substance known under the trade name of Warfarin was destroying rats, chiefly through its action as a blood anticogagulant and interference with vitamin K. Warfarin-resistant rats appeared first in a locality in England, and have since been spreading and replacing the ordinary Warfarin-sensitive rats.

In a few species resistance to insecticides has been induced experimentally by selection in the laboratory. It has been ascertained that resistance can be due either to a single gene with a major effect or to a co-acting set of genes with individually minor effects. The ubiquity of genes that confer adaptednes to synthetic chemicals only recently invented by man may seem strange or even mysterious, however, the mystery can be dispelled by studies of the physiological bases of specific resistances. It is usually found that in the resistant strains the insecticide is detoxified by enzymes and metabolic processes that also exist in nonresistant strains, although with a lower level of activity. The emergence of pathogenic and other bacteria resistant to antibiotic and chemotherapeutic agents is due to directional selection quite analogous to that for insecticide resistance. Resistance in pathogenic bacteria has ushered in practical problems that are even more troublesome than those created by insect resistance to insecticides, in as much as in the former problems it is human health and even survival that are involved.

Directional natural selection has also been observed in the spread of so-called industrial melanisms in many species of moths, best studied in England. Since the middle of the last century the appearance and spread of darkly pigmented variants of moths has been recorded in industrial regions where the vegetation is blackened because of pollution from soot and other wastes. In some localities the dark varieties have almost completely replaced the formerly "normal" lightly pigmented forms, while in nopolluted areas the light ones still occur. Altogether about 10 species have shown industrial melanisms. In a majority of the species examined genetically, the difference between the dark and light variant is due to a single gene, the black usually ·being dominant. Interestingly enough, further evolution of the melanics has also selected a system of gene modifiers with relatively minor phenotype effects, which intensify the phenotypic effects of the main gene. Observations and experiments have shown clearly that predation of the moths by birds is the main selective factor promoting industrial melanisms. The melanic varieties are protectively colored when they rest on blackened vegetation; they are conspicuous on nonpolluted vegetation, wher elight varieties are protected.

Plants have also produced some clear-cut instances of directional selection. Bradshaw, Gregory, Antonovics, and their colleagues have studied varieties of several species of plants able to grow on soils contaminated with heavy metals (copper, zinc, lead). Such contaminated soils are often found near old mines where ores of these metals were exploited. The resistant varieties maintain themselves on soils with concentrations of the pollutants that are lethal or at least stressful for ordinary members of the same species. A conspicuous feature is that the resistant and nonresistant varieties may grow at distances only a few meters from each other, but rigorously confined to their respective contaminated and uncontaminated soils.

NATURAL SELECTION AND ADAPTEDNESS

Darwinian fitness is a function of reproductive efficiency. The carriers of some genotypes transmit their genes to the next generation more often than do carriers of other genotypes. By definition, natural selection promotes Darwinian fitness. However, is high Darwinian fitnesss always beneficial to an individual or population? Does it always go together with adaptedness, good health, improved survival ability, enhanced adaptability in varied environments? By and large it does; if it were otherwise, life on earth would have become extinct long ago. One must nevertheless be wary of projecting the vernacular meanings of "fitness" onto Darwinian fitness. Natural selection is not a benevolent genie, but a consequence of certain combinations of physical and biological variables. A hypothetical situation in which natural selection could have mischievous consequences is easily imaginable.

If intelligent people had fewer children then less intelligent ones, and provided that human intelligence is in part genetically conditioned, the average intelligence would exhibit a downward trend. In our system of values high intelligence is a positive quality, and low intelligence a negative quality. Moreover, deterioration of the intelligence may conceivably endanger the survial of the species. High Darwinian fitness may thus be a property of genotypes that are not only socially undesirable but injurious to the species as well.

Genetic research has uncovered several instances of hereditary variants promoted by natural selection despite their being, at least in the long run, harmful to the population. Dunn and his collaborators found most unusual mutants at the *t*-locus in the house mouse. A male mouse heterozygous for a *t*-mutant and for its "normal" allele generates more, sometimes may more, functioning spermatozoa with the *t*- mutant than with the normal allele. This deviation from the usual 1:1 segregation ratio confers upon the mutant alleles a selective advantage. If not counterbalanced by some disadvantage, the mutant *ts* will spread and crowd out the normal alleles. Actually, most mutant *ts* are lethal when homozygous. For this and perhaps other still undiscovered reason, *t* alleles do not outnumber normal alleles, although they are found in many mouse populations.

Another example of a genetic variant that has subverted the normal processes of meiosis and sex cell formation, and thus gained a selective advantage, is the so-called chromosomal "sex ratio" found in natural populations of several species of *Drosophila*. A male with a "sex-ratio" X chromosome produces functioning spermatozoa most or all of which carry X chromosomes, and few or none with Y chromosomes. Crossed to any female, such males give progenies consisting of daughters and few or no sons. Ordinary males pass their X chromosomes to only one half of their offspring, because the other half are males that inherit Y chromosomes from their farthers and X chromosomes from their mothers. Unopposed by some other agency, the "sex ratio" X chromosomes would take over the population; in a species incapable of parthenogenesis, a population consisting only of females would become extinct. In fact, however, this does not happen in nature in populations in which "sex ratio" chromosomes are found. The reason for this is still conjectural.

Species now living are descendants of only a minority of the species that existed in the geological past, and increasingly smaller minorities as we look back to more distant times. Indeed, the usual fate of an evolutionary line is extinction. Yet the evolution of extinct lines is also controlled by natural selection. The solution to this apparent paradox is rarely found in the appearance of "subversive" mutants, like those of the *t*-locus in the mouse or the "sex ratio" in *Drosophila*. Extinctions may often be due to the fact that natural selection promotes genetic variants with high Darwinian fitnesses, here and now, regardless of what their fitnesses may be in future environments. Artificial selection is goal-directed, but natural selection is not. Having no information about the future, natural selection is opportunisitic and blind. A genotype does not have high or low Darwinian fitness in the abstract; its fitness is a function of the environments in which it is placed or to which it can migrate.

A genotype only passably successful now may flourish in some future environments. Such a genotype is often said to be "preadapted" to what it will meet in the future. The concept of preadaptation is a fallacy if it is taken to mean that evolution somehow contrives adaptations for use in as yet nonexistent environments. Nevertheless, in building adaptedness for present environments, natural selection can only operate with what is inherited from the past. Bipedal gait was surely adaptive in mankind's remote ancestors, and it was selected for the benefit of those ancestors. Yet its upshot was the development of hands capable of manipulating tools, machines, and surgical instruments. This ability made possible the development of technology and culture, and is highly adaptive in human environments created much later by culture.

GROUP AND KIN SELECTION

In a population under selection the frequencies of some genes increase and those of others decrease as generation follows generation. The reproductive efficiency of individuals who carry the "successful" genes is higher than that of the carriers of "unsuccessful" ones. What survives or dies—remains sterile or engenders offspring—are however not genes themselves but living individuals, who are their carriers. The question can nevertheless beasked whether, in addition to natural selection on the individual level, there can be *group selection,* i.e. selection of groups of individuals, such as Mendelian populations. Related species compete for resources that both are in need of, and one species may outbreed and crowd out another. Thus, species introduced from foreign countries sometimes become pests or weeds, and eliminate or greatly reduce the abundance of native species. This is particularly striking on ocenic islands. Island endemic species often seem to have reduced competitive abilities, and they are given short shrift by introduced forms.

A more complex and interesting problem is whether group selection is also important intraspecifically. Wright, in his classic essay "Evolution in Mendelian Populations" (1931), devised a model of a species subdivided into semi-isolated colonies. The gene exchange between the colonies is presumed to be rare enough that the colonies may diverge genetically, owing to differential selection and random genetic drift. He compared the colonies to scouts or trial parties "exploring" the field of gene combinations. Evolutionary changes in some colonies may result in superior adaptedness, and in other colonies in inferior adaptedness and even extinction. The colonies that flounder and fail may then be replaced by migrants from the successful ones. The whole species will then rise to a higher level of adaptedness.

Variants of this model were later utilized to explain some evolutionary phenomena that are difficult to account for by the more usual selection of individual contributions to the gene pool of following generations. Consider, for example, altruistic behaviour. For the purposes of the present discussion altruism can be defined as a behaviour that benefits other individuals at the expense of the altruistic. If altruistic behaviour is genetically conditioned, the Darwinian fitness of the altruists will be diminished; that of individuals who refuse to act altruistically will, on the contrary, be augmented. It would seem that natural selection would discriminate against altruism and favor self-preservation and egotism. Yet a population, colony, or tribe that includes many altruists may prosper more than a population burdened with egotists. Altruism and related behaviours may be promoted by group selection. But they may also be promoted by natural selection when the altruistic behaviour of an individual benefits its close relatives. This form of natural selection is known as *kin selection.*

It is arguable whether it makes sense to speak of altruism and selfishness except in species able to choose freely between different possible courses of action. The only such species is man. Many kinds of behaviour are equivalent to altruism, whereas they are really automatic and stereotyped. The behaviour of honeybees, ants, termites, and other social insects has excited the curiosity and admiration of many observers for a long time. Individual "workers" toil selflessly and are ready to sacrifice themselves for the benefit of the colony to which they belong. The evolutionary development of this ostensible altruism is easily understood. The "workers" and "soldiers" do not reproduce. They are sterile females, or males with underdeveloped sex organs. The prepetuation of an insect society is the task of a sexual caste, the so-called queens and fertile males, which are a minority, sometimes single individuals, in the colony. Sexual individuals do little or no work after the colony is founded, and evince no particular readiness for self-sacrifice. The workers share genes with the sexual members produced in the colony. The result of this arrangement is that natural selection operates not in individuals but in colonies consisting of reproductive and sterile castes as units. The ostensible altruism of the workers enhances the probability of transmission to the next generation of the genes these workers share with the sexual caste of their colony.

Behaviours analogous to human altruism are by no means confined to social insects. Production of progeny exacts, as a minimum, an "investment" of nutrient materials on the part of the parents to produce. Sex cells. Where parental care is developed the progeny is, for a shorter or longer time, fed by the parents. More than that, parents defend their progeny from aggressors and predators, and in so doing expose themselves to dangers of injury or death. Among herd-living animals the large and powerful males may protect other members of the troop, particularly females with young from predators, and do so at a risk to themselves. One should not expect natural selection to promote selfish behaviour in every individual under all conditions. Hamilton (1964), Trivers (1972), and others advanced a theory of kin selection, which provides a simple and reasonable explanation of ostensibly altruistic behaviour in animals. Altruism may promote the "inclusive Darwinian fitness" of the altruist, instead of his individual well-being. Indeed, each child shares 50 per cent of his genes with siblings and with each parent. Therefore, a parent who is disabled or killed defending his progeny may "save" more of his genes than if he remained alive while his progeny perished. This is especially true of older parents, whose reproductive life is finished or nearly so. The closer the genetic relationship between the altruist and the beneficiaries of his conduct, the greater the biological, selectional warrant for altruistic behaviour. Members of a herd, e.g., baboons or wild horses, are in general more or less close relatives. Altruistic and cooperative behaviours are, as a rule, more developed the closer the genetic relationships.

Group selection has also been invoked to explain some forms of behaviour that do not fit the kin-selection model. In some species, physiological and behavioural phenomena result in decreasing reproductive rates of members of populations that have grown beyond optimal number. Wynne-Edwards (1962) sees in this the result of group selection that benefits populations rather than individuals. Indeed, excessive population growth may destroy the food supply and cause extinction of the population. Some species have evolved social hierarchies of dominant and subordinate individuals within populations. One alleged consequence of this is greater efficiency in the utilization of food. Another is that some weaker or "surplus" individuals are excluded from the population, and effectively condemned to die from starvation or predation. Group selection supposedly makes the surplus individuals stoically accept their fate because it benefits the species, even though at their expense. Most evolutionists find such explanations questionable. There is clearly much to be learned about the roles of different kinds of group and kin selection.

THE PRESENT STATUS OF NATURAL SELECTION

It has come to be rather generally believed that the natural selection that Darwin himself believed in stands almost unscathed as one very important causal factor. In fact it is the only explanation ever offered for adaptation that even approaches adequacy. As an explanation of the origin of new types or new species it falls far short of adequacy, and I think Darwin evidently realized this, although he was unfortunate enough to entitle his book *Origin of Species*. As an explanation of the origin and perfection of adaptation natural selection has only one rival, the far less satisfactory Lamarckian theory of the inheritance of acquired characters. There is a strong tendency among geneticists to conclude that the modern germ-plasm hypothesis, with the aid of mutations and the mechanism of Mendelian inheritance, furnishes all the necessary explanation of the causes of evolution. There is, however, marked dissent to this extreme position. In his critique of De Vries's rather extreme position that the mutation theory needs no aid from natural selection, Weismann shows in most able fashion the inadequacy of mutations to account for adaptation, and, in contrast, how well natural selection accounts for them.

In a very recent paper Professor C. C. Nutting attempts to show that natural selection is still an important factor in evolution and quite in harmony with both the mutation theory and Mendelism. We perhaps can close the present chapter no more fittingly than by quoting Professor Nutting's paper.

THE RELATION OF MENDELISM AND THE MUTATION THEORY TO NATURAL SELECTION

Two marked tendencies are evident in the history of any important theory after its publication.

First

The followers of the discoverer carry the theory too far and attempt too universal and application. This is manifestly true of Wallace and Weismann who out-Darwined Darwin in their claims for natural selection; of the followers of Mendel, such as Morgan and Pearl; and of many mutationists who make much greater claims for that theory than does De Vries himself.

Second

Each generation of biologists is so occupied with its own work and contemporary theories that it makes no real effort to understand preceding theories.

This second tendency seems to me most marked in the attitude of present workers along genetic lines towards natural selection. They reveal an apparent lack of understanding of what Darwin really meant and of what he claimed; and when criticising that theory they are often engaged in the classic, but unprofitable, exercise of "fighting windmills."

Darwin presented the following three factors in natural selection.

First

Heredity, by which the progeny tend to resemble their parents more than they do other individuals of the same species.

Second

Individual variation, by which the progeny tend to depart from the parental type and sometimes from the specific type.

Third

Geometrical ratio of increase, by which each species tends to produce more individuals than can survive.

Each of these factors is practically axiomatic, so little is it open to argument.

No one doubts the fact of heredity, whether pangenesis, Weismannism or Mendelism be the correct expression of the mechanism involved. These do not affect the fact of heredity nor invalidate it as a factor in natural selection.

No on doubts the fact of variation; whether it is the "individual variation" of Darwin, the "fluctuating variety" or the "mutation" of De Vries. All that is necessary for Darwin's purpose is that there be heritable variations. That there are such things all parties agree and it matters little what you call them. They are adequate to act as a factor in the Darwinian scheme.

No one doubts the fact geometrical ratio of increase. It is a proposition easily capable of mathematical demonstration, and that is sufficient for Darwin's purpose.

These three factors, then, are not debatable as facts, whatever their mechanism or causes.

A moment's reflection will show that geometrical ratio of increase is a quantitative factor, giving an abundance of individuals from which to select; that individual variation is a qualitative factor, giving the differences which make a selection possible; and that heredity is a conservative factor, holding fat those characters which better fit the organism to its environment.

Now it seems to me that there is no possible outcome of the necessary action and interaction of these three factors that would not be a selection of some sort. Darwin thought it comparable in a large way to the selection by which the stock-breeder improves his herd, and therefore called it "natural selection," carefully guarding the phrase from misinterpretation from the teleological angle as well as from a too close parallelism between artificial and natural selection. And I believe no one has suggested a more acceptable term for the process of selection resulting from the interpaly of natural laws.

Three outstanding theories have been advanced since the publication of the *Origin,* each involving an advance in our knowledge of the mechanism of heredity on the one hand and the origin of variations on the other.

Weismann's theory of the continuity and stability of the germ plasm was of immense importance in its discussion of the mechanism of heredity, and his amphimixis gave a plausible explanation of the origin of variations. His results were almost universally regarded as confirming and greatly extending the scope of natural selection.

Mendel's theory regarding the purity of the gametes, their segregation in the sexcells, and the whole complex Mendelian mechanism so admirably described by Morgan; all of these, fascinating and important as they are, deal with the mechanism rather than the fact of heredity. In my opinion their acceptance or rejection does not affect the status of natural selection as a theory of organic evolution.

But it is the theory of mutation that has furnished most of the ammunition for the opponents of natural selection; and this in spite of the fact that De Vries, the originator of the mutation theory, expresses himself with great clarity as follows:

> "My work claims to be in full accord with the principles laid down by Darwin and to give a thorough and sharp analysis to some of the ideas of variability, inheritance, selection, and mutation which were necessarily vague in his time."

In 1904, when these words were published, there did seem to be a sharp distinction between the ideas of Darwin and those of De Vries. The former believed that natural selection acted upon many small variations and accumulated them until the differences were sufficient to constitute new species; while De Vries claimed that new species were formed by the sudden appearance by mutations of forms specifically distinct from the parents. That mutants were new species!

It seems evident that Darwin did not regard "saltatory evolution" as the common method, while De Vries did.

Darwin believed that individual, usually small, variations furnished the material on which selection acts; while De Vries thought that mutants, usually large variations,

furnished the material. Both, however, believed thoroughly that natural selection was *vera causa* of evolution.

But things have changed greatly since 1904. The work of Morgan, Castle, Jennings and a host of others has shown that many mutations are so small, from a phenotypic standpoint, that they are quantitatively no greater than the individual variations of Darwin; and that they are heritable in the Mendelian way.

Castle produced a perfectly graded series of hooded rats which exhibits almost ideally the steps by which a new form might be produced by natural selection. He says:

> "If artificial selection can, in the brief span of a man's lifetime, mould a character steadily in a particular direction, why may not natural selection in unlimited time also cause progressive evolution in directions useful to the organism?"

Jennings says:

> "Sufficiently thorough study shows that minute heritable variations—so minute as to represent practically continuous gradations—occur in many organism: some reproducing from a single parent, others by biparental reproduction. . . . It is not established that heritable changes must be sudden large steps; while these may occur, minute heritable changes are more frequent. Evolution according to the typical Darwinian scheme, through the occurrence of many small variations and their guidance by natural selection, is perfectly consistent with what experimental and paleontological studies show us; to me it appears more consistent with the data than does any other theory."

Many believers in mutation have been needlessly befuddled by the diverse meanings of "varaitions" as used by Darwin and De Vries. Darwin included in his "individual variations" both the "fluctuating varieties" and the "mutations" of De Vries. Phenotypically they cannot even now be distinguished. De Vries himself candidly admits that this was Darwin's attitude, thus providing himself more clear-sighted than many of his followers. All that Darwin needed for his purpose was proof of variations that are heritable, and these are found in mutations, be they large or small.

Just as Mendelism has to do with the mechanism and not the fact of heredity, so the mutation theory deals with the nature and not the fact of variations. Neither, in my opinion, has any implication that is antagonistic to the theory of natural selection.

The statement has been made that natural selection "originates nothing" because it does not explain the origin of variations. I must confess scant patience with this point of view. As well say that the sculptor does not make the statue because he does not manufacture the marble or his chisel; or that the worker in mosaic originates nothing because he does not make the bits of stone which he assembles in his design!

The material corresponding to the bits of stone in the mosaic is furnished by heredity and variation, and its quantity by geometrical ratio of increase. Natural selection acts in selecting and putting together this material in the formation of new species. Thus, in a true sense, it seems evident that something new has appeared—something that is, but was not.

Another favourite figure, is "Natural selection acts only as a sieve" determining which forms shall be retained and which shall be discarded. This also seems to me to fall short of a complete statement of the truth. If the material subjected to the sifting process be regarded as changing with each generation by the addition of variations, or mutations if you prefer, some of which are favourable to a nicer adjustment of the species to its environment, the figure would be more nearly correct. To make it complete, however, the mesh of the sieve must change from generation to generation so that a quantitative variation which would be preserved in one generation would be discarded in a later one. But in this case natural selection would do more than a sieve could do. It would combine a number of favourable variations in the production of something new, a new species!

In conclusion it seems to me that we are justified in maintaining that Mendelism an the mutation theory, while forming the basis of the most brilliant and important advances in biological knowledge of the last half-century, have neither weakened nor supplanted the Darwinian conception of the "Origin of species by means of Natural Selection."

WEISMANN'S GERMPLASM THEORY

August Weismann (1834-1914) was, a well-known neo-Darwiman, German biologist. His germ-plasm theory was published in 1892 in a volume *Das Kemplasma.* His views were just opposite to those of *Lamarck* and in a sense to those of *Charls Darwin.* According to him all the organisms are made up of two types of substances '*germplasm*" and '*somatoplasm*'. The germplasm is the protoplasm of the germ cell and is responsible for inheritance of the characters of the organism. The somatoplasm constitutes the body of organism and is not concerned with the inheritance. In reproduction, a portion of germplasm is derived from each parent i.e. paternal sperm and maternal ovum, each of which has an equivalent share in the inheritance. These combine to form zygote. When the zygote divides, in the first or subsequent cleavages each daughter cell receives an equal share of germplasm and this holds true for all cells which go to form the adult body. None of the somatic cells, however, has any share in subsequent generations from the original zygote. Thus, through germ cells a continuing of germplasm is maintained from generation to generation. The somatoplasm is not inherited and dies with the organism. During the division, the zygote, which is purely germplasm, produces somatoplasm and which makes the body of organism while the germplasm the gonad, as usual. According to this theory, only those variations which appear or which are introduced in the germplasm can be inherited and not those which appear in the somatoplasm.

According to *Weismann* the germplasm is composed of '*ids*' i.e. equivalent portions of germplasm contain all kinds of determinants present in the organisms, which are responsible for the development of characteristics in the off-springs.

Weismann's Theory was widely accepted by biologists. There are several evidences which supported his theory. An albino female guinea pig was selected whose ovaries, before attaining maturity, were replaced by the ovaries of a young black female guinea pig. It was mated with a white male and it was found that all the offsprings produced were black. The white young ones were not produced as it should be in the normal albino female. This shows that the heredity is determined by the germ cells and not by the body cells.

Bornier made some experiments in support to Weismann's theory. He transferred some plants from the plains to alpine region. Due to environmental change a number of new characters developed. After 18 years plants were transferred back to the plain where they again restored the original characters and lost the alpine characters. This shows that environment does not affect the genotype of an individual and the environmental variation is not inheritable.

This theory is not very correct and does not hold true with all organisms. It may be difficult to explain certain cases of asexual reproduction or such abnormal reproduction as regeneration.

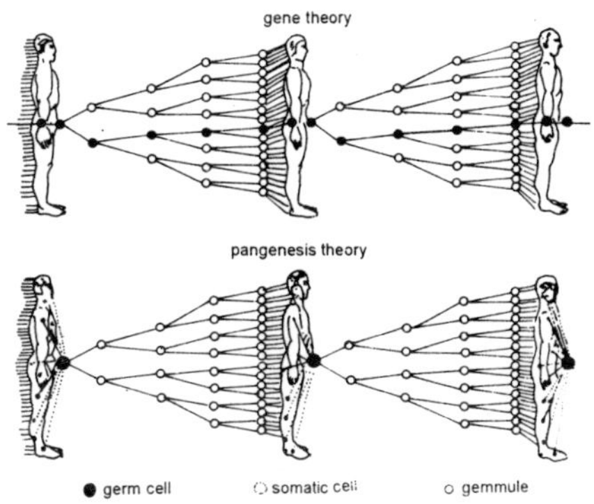

Fig. 4.9: Diagram to illustrate the continuity of germ-plasm.

In certain protozoans like *Amoeba* not only the germplasm is handed down, but the entire body of the organism is divided between the offsprings. Yet other protozoan (*Ephelota*), which reproduce by budding, give up only the sometoplasm and a portion of nucleus. In *Bignonia*, the vegetative reproduction is accompanied by the division of sometoplasm. In the earthworm, the reproductive organs reside in the anterior segments and yet, no matter where the worm is cut, if enough of tail remains, it will reproduce the entire organism, including the reproductive bodies with all the hereditary details.

Heredity deals with one more problem. The latent (recessive) and potential (dominant) characters are present almost in all plants and animals. When a tall plant is crossed with a dwarf one, all the plants produced are tall. Here tall is the dominant character. In these plants the dwarf character is present in the recessive form which only gets a chance to appear in the next generation. One more example may be taken from bees. The known drone bees are produced *parthenogenetically* from the unfertilized egg and hence have no father. But they have all the male like characters e.g., its size, reproductive organs etc., which distinguish them form females. This is proved on the basis that the male characters, which the queen had received from her father were present in recessive form and reappeared in drones in the next generation. This phenomenon of reappearance of reversion of ancestral characters is termed 'atavism'.

Thus the theory of *Weismann* is purely speculative. It has no practical support and can not be proved by experiments.

Chapter—5

Sexual Selection

Darwin in 1871 published his work entitled, *"The Descent of Man"* in which the proposed this theory. It is based upon the competition or struggle for mating. This theory is to supplement his theory of *Natural Selection*. *Darwin* proposed this theory to account for the secondary sexual characters found in animals which could not be explained by the theory of Natural Selection. Before discussing the theory of Sexual Selection, it is necessary to know the sexual characters. These are two types of sexual characters: Primary sexual characters and Secondary sexual characters.

PRIMARY SEX DISTINCTIONS

The primary sex distinctions are the functional ones which naturally differ in the two sexes. They are the reproductive organs, ovaries and testes, with their essential glands and ducts, and the orans in the female whereby the young are nourished and developed before and after birth, the placenta and mammary glands.

SECONDARY SEX CHARACTERS

The secondary sex characters, on the other hand, are such as are often not directly concerned with procreation but may nevertheless be of considerable importance to the organism. They often enable us readily to distinguish the sexes, as in the case of the pea fowl (*Povo cristatus*), whereas in other instances sexes may be only recognizable by a microscopic examination of the reproductive glands, as in the American oyster (*Ostrea virginica*).

Special Organs for Mating

The secondary sex characters may be grouped under several headings for the purpose of discussion. First among them are special organs for mating, structures sometimes found in the male in addition to the normal devices for coition, which are primary sex organs. Such, for example, are the holding organs in the male frogs, curious pad-like structures on the thumb of the hand, by which the female is firmly clasped on either side of the hip girdle. Its purpose is to keep the two sexes together, as there is generally no internal impregnation but the eggs are fertilized in the water, which must be done immediately after extrusion before the gelatinous envelope has time to swell, since after that impregnation is impossible. The shark-like fishes, with one living exception, have a portion of each pelvic fin in the male modified into what is known as the clasper for use as an intro-mittent organ in mating, while in the related chimaeroids or silver sharks there is in addition to the pelvic claspers a curious device on the top of the head, which looks not unlike a door-knocker and is provided with hooks derived from the skin. The actual use of this knocker-like organ has never been observed, but it is also supposed to be a clasping mechanism.

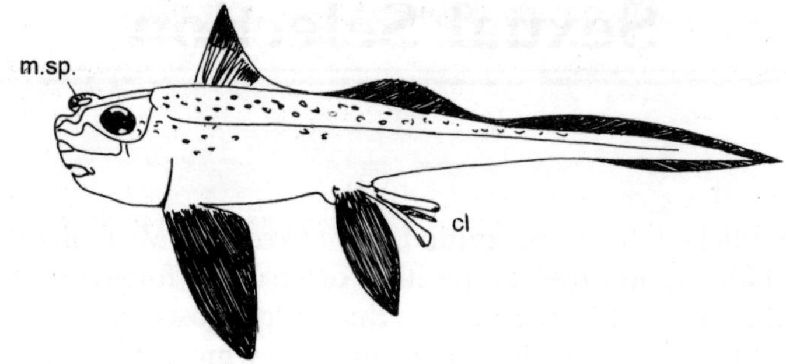

Fig. 5.1: Chimaeroid fish, *Chimaera colliei*, male, showing secondary sex characters. *cl*, claspers; *m. sp*, frontal spine (clasper).

The cephalopods show a remarkable seasonal alteration of one of the tentacle-like arms in the male which is used in mating, a modification known as hectocotylization (Gr. *εkarov*, hunded, and *korvλn*, hollow vessel). The arm is filled with the spermatophores prior to coition and is detached from the male during the latter process, a new one being developed to take its place.

Fig. 5.2: *Argonauta argo*, mate, *hc*, hectocotylized arm.

Brooding Organs

Another kind of secondary sex characters are the brooding organs found especially in the female but sometimes in the male. Under this head comes the pouch of the female marsupial mammal, either one transverse or two longitudinal folds of skin which cover the teats and serve to protect the very immature and helpless young until they are able to fend for themselves. A comparable device is sen in the sea-horse (*Hippocampus*,), a small but highly specialized fish in which the ventral fins have been modified into a brood pouch to carry the eggs until time for them to hatch. In this case, curiously enough, it is the male which, as Thomson says, "carries the eggs about in his breast-pocket." The same is true of some related pipefishes (*Syngathus*).

The frogs and toads also show marvellous brooding devices, especially the famous Surinam toad (*Pipa americiana*) in which the skin of the back of the female forms growths for the reception of the eggs and in these the young undergo their whole metamorphosis. Each receptacle consists of a cavity with a lid; the origin of the latter is not understood but it is probably produced by the remnant of the egg-shell itself, which, after the larva is hatched, remains on the top of the cup. Gadow says of this group: "The greatest charm of the Anura (frogs and toads) lies in their marvellous adaptation to prevailing circumstances; and the nursing habits of some kinds read almost like fairy-tales."

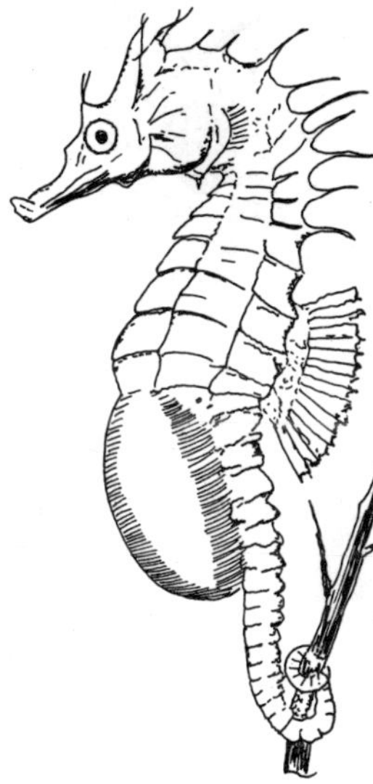

Fig. 5.3: Sea-horse, *Hippocampus antiquorum*, male, showing brood-pouch formed from combined pelvic fins.

Special Sense and Sound-producing Organs

A third instance of secondary sex characters is represented by special sense and sound-producing organs. Here it is the insects which from the best means of illustration, for being as a rule active locomotor types, their special senses are proportionately developed. Male moths, especially the so-called giant silk-worm moths, the cecropia, polyphemus, luna, and others, have the feathery antennae enormously developed compared with those of the female. They are the seat of a remarkable sense which, for want of a better under standing has been compared with the human sense of smell. It is used by the male in searching out the female for the purpose of mating, and is wonderfully effective, the male moths assembling around a newly emerged female from a radius of several miles. Among the scale insects (coccids, Fig. 5.4) both sexes are degenerate sedentary type during their young condition and often obscurely colored so as to be discerned with difficulty. The female remains scale-like throughout life but the male upon metamorphosis changes into a delicate two-winged fly in sharp contrast to his lowly mate. Abortion of the mouth parts in adult insects is not infrequent, as in the Chinese silk-worm (*Bombyx mori*), may-flies, etc., but in the male coccid they are replaced by an extra pair of eyes on the under side of the head as a necessary aid in finding the obscure female!

Vocal organs are almost exclusively confined to the vertebrates, with the exception of the insects, but in the latter the males only are provided with sound-producing organs. With the crickets and katydids they are a modification of the fore wings, which have sonorous structures that are thrown into vibration by being rubbed past each other, the sound being amplified by tightly stretched membranes analogous to a banjo head, or by resonant chambers formed by the wings, and comparable to the body of a violin.

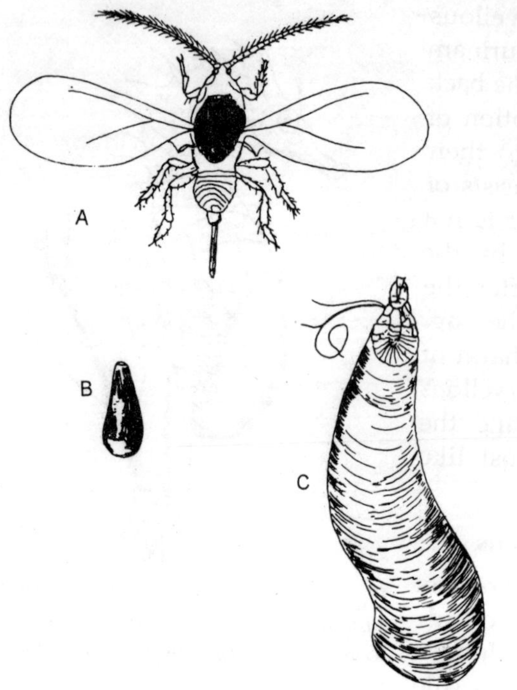

Fig. 5.4: Scale insects, *Mytilaspis pomorum*. A, adult male; B, young male scale; C, adult female scale and egg capsule. Greatly enlarged.

The grasshoppers rub the hind leg across the edge of the wing, and the cicadas have remarkable pair of kettle-drums at the base of the abdomen which are made to resound by the alternate stretching and relaxing of the membrane which covers them. They lie beneath a pair of backwardly projecting flaps comparable to the skirts of an eighteenth-century waistcoat.

Special locomotor organs, such as occur in the male scale insects and certain moths, are perhaps not so much a development in the male as degeneracy in the female, for absence of locomotive powers on her part renders increased activity upon that of the male imperative for the sake of future generations. In the famous gypsy moth, the female, although still possessing wings, has lost the power of flight because of the great wight of her body, while in the canker-worm moths ithe female's wings have utterly disappeared.

Organs of special use usually take the form of weapons such as the antlers of the male deer, the tusks of the boar and of the male musk deer the tusks of the boar and of the male musk deer and chevrotains, and among insects the huge mandibles of the stag-beetles and the spines on the head and thorax of the rhinoceros-beetles. Among bees and wasps, on the other hand, the sting is confined to the females or to the so-called neuters or workers which are in reality undeveloped females. The sting is a modification of the ovipositor or egg-laying organ characteristic, for instance, of the parasitic Hymenoptera which are near relatives of their stinging allies.

Special characters for exciting or attracting the opposite sex are such as appeal to the senses of sight, hearing, and smelling, taking the nature of cries and antics, of color and plumage, and of special odors. They may be either seasonal or persistent and are in many ways the most inexplicable of all the secondary characters of sex. They are of course confined almost entirely to the male. Instances may be multiplied indefinitely, as they have been observed by all—the gorgeous tail of the peacock, the pompous dignity of his carriage and even his strident voice are all supposed to make an irresitible appeal to the female. The strutting of the common turkey and especially the drumming of the ruffed grouse are further instances. The beard of man and the mane of the lion may also be included under this head.

Reciprocal organs are found in a areduced state in the female, as horns in the goats, antelope, caribou, and reindeer. Stridulating organs, though undeveloped, are present in female crickets and kaatydids, while in male mammals the mammae are present in a rudimentary state. In the male, the mammary apparatus normally becomes absorbed, though occasionally at birth and at puberty milk is produced in the human subject. Male goats nad castrated sheep have also been known to give emilk (Wiedersheim). A functionless rudimentary brood pouch is present in the male Tasmanian wolf (*Thylacynus*).

Indifferent Characters

Under the head of indifferent characters are included such as are of no vital importance to their owner and yet are sex distinctive. They are mainly the negative reciprocals of certain structures already mentioned as characteristic of the opposite sex, as of instance, the reduced wings of certain female insects, the vestigial gut in the male rotifers or wheel animalcules, the slight differences in size or color in many birds and insects, or the differences in the proportions or the pattern and arrangement of the veins of insect wings and in the number of tarsal or antennal segments.

Atavistic characters are those distinctive of ancestral conditions, like the hairy chest and greater bodily strength which often occur in men but rarely in women.

These are the secondary sex characters to explain which the theory of sexual selection was proposed, but it will be seen that with the exception of those under the caption of indifferent characters, and perhaps the last three, they *may* all be regarded as the result of natural selection, as they are useful in the struggle for existence. The characteristics included in the special characters forexcitation, etc., on the other hand, are such as make their owner more conspicuous and doubtless expose him to dangers from which a more obscure animal would be immune. Hence their development is opposed to the principle of natural selection, as the results are a handicap and not an aid in the struggle for existence.

THEORY OF SEXUAL SELECTION

The theory of sexual selection is based upon the competition or struggle for mating, and while natural selection presupposes a passive female, sexual selection implies presumptive choice on her part of the most desirable male to be the father of her progeny, with the consequent inheritance by the offspring of his estimable characteristics. Darwin also included in his theory the competition or struggle for mating where rival males fight for the possession of a female as in the case of the deer or the sea-lion, the female being in each instance the passive reward of the victor. But this fighting on the part of the males is merely a phase of the intra-specific struggle, although it is not a fight for food and space but for a chance to mate. This chance, however, often depends upon a life and death struggle and no theory of *sexual* selection is necessary to account for the development of the weapons or greater prowess which enables one animal to become the victor over another.

Postulated Bases of the Theory

As Darwin originally proposed it, the theory was based upon the following assumptions, the truth of which was apparently attested by a varying number of facts:

"First, many secondary sexual characters are not explicable by natural selection; they are not useful in the struggle for life. Second, the males seek the females for the sake of pairing. Third, the males are more eabundant than the females. Fourth, in many cases there is a struggle among the males for the possession of the females. Fifth, in many other cases the females choose, in general, those males specially distinguished by more brilliant colours, more conspicuous ornaments, or other attractive characters. Sixth, many males sing, or dance, or otherwise draw to themselves the attention of the females. Seventh, the secondary sexual characteristics are especially variable.

"Darwin believed that he bad observed certain other conditions to exist which helped make the sexual selection theory probable, but the conditions noted are sufficient if they truly exist" (Kellogg).

Difficulties

Some of the difficulties which stand in the way of our acceptance of Darwin's tenets are thus enumerated by Kellogg: The theory can apply only where the males are more numerous or polygamous, otherwise even if rejected by the first female each male, however undesirable, would sooner or later find his mate and thus the unornamented males would leave as many progeny as the ornamented, which would prevent any cumulation of ornamental variations by selection. Among the higher vertebrates, where a great number of ornamented males occur, the proportion of the sexes is about equal, and where polygamy exists there is always a struggle for master, not an active female choice.

In most species the mating female is observed to be wholly passive and propinquity is seen to be the greatest factor indetermining which two shall pair.

Ornamental colors are as often characteristic of the males of species in which there is no real pairing as among those which pair. Sexual choice certainly cannot account for the remarkable ornamental colors of the males of many fishes in the breeding season, for the female may not even see the male which fertilizes her already laid eggs. Eigenmann, on the other hand, notes the utter absence of such ornamentation among cave fishes which live in the dark, and argues that where they do appear, therefore, they must be due to visual selection. This may, however, be explained in a different way, as we shall see.

A high degree of the aesthetic sense on the part of the females of animals would be necessary for choice on a basis of ornament and attractiveness. For this we have no other proof, and we can not imagine its existence on the part of invertebrate animals in which ornament is often so highly conscpicuous. Even among the higher forms wherein the aesthetic sense may be developed to a certain extent, we have no assurance of any one standard of beauty for individual taste varies greatly with men and probably also with animals. Even if we may attribute a certain aesthetic sense to mammals and birds, the question naturally arises whether it can be so keenly developed as to lead the female to make a choice among slight differences of color or song. Yet this assumption is necessary if the theory of sexual selection be accepted. Would not the evolution of this aesthetic sense upon the part of the female be just as difficult to account for? In all probability he rreactin is purely emotional rather than discriminative.

Very few cases of actual choice by the female have ever been observed, for even so earnest an observer as Darwin himself, who had every reason to record the greatest possible number, failed to note more than eight cases among birds, and but half a dozen more, all doubtful, are mentioned in the literature from his day until 1907.

Many of the so-called attractive characters of males have been found to be during life of such a nature that the females could not note them; as, for instance, the brilliant colors and curious horns of the male dung-beetle which are so obscured by filth that they cannot be used for display, while in melodious or luminous insects like the crickets or fireflies, nearness will make a relatively feeble song or light seem much finer than that of a distant but much more musical or brilliant insect.

That dances and serenades do occur during the breeding season is well known, but they may be for the purpose of excitation of the usually passive female and may not imly the necessity of any discriminating choice upon her part among a number of males. That rivalry would be developed among the dancers or singers is only natural and may or may not lead to further perfection of the art.

Experimental Evidence

There is as yet comparatively little evidence based upon actual experiment, but such as we have is strongly opposed to the sexual selection theory. In this connection the experiments of Mayer are of remarkable interest. He worked with the giant silk-worm moth *Callosamia promethea,* in which there is decided sexual distinction in color and pattern. The females have a reddish brown ground color, while the males are blackish, and the ornamentation in the two sexes is distinctly different. If there is any moth species in which the colors and general pattern of the male ought to be readily obvious to the female, and in which sexual selection might be presumed to have been the influence in producing a pronounced male type of preferred pattern, it is this species. The experiments were varied, but is was found that sight had nothing to do either with the finding of the female or choice of mate, for reversal of wings—male of female or female on male— or any other disguising of the sexes made no difference whatever in the mating. The assembling power, as we have already noted, was resident in the antennae of the males.

From the experiments it is concluded that the mating instinct on the part of these two insects is a phenomenon of chemotaxis (Lat. *chemicus,* chemic, and Gr. *τάξις,* arrangement), as sexual selection on the ground of color alone does not affect it, and there is no associative memory connected with it.

Alternative Explanations

It will be seen from the foregoing that the Darwinian factor of sexual selectionrests upon very meager evidence, the great majority of secondary sex characters being explicable by natural selection, and on the other hand even where natural selection cannot be invoked, sexual selection does not stand the test of experiment. Some alternative explanation becoming necessary, the following have been offered:

"First, the secondary sexual characters are produced as the result of the immediate stimulus (naturally different) of the sexually differing primary reproductive organs, this stimulus being usually considered to result from an internal secretion of the genital organs acting on certain tissues of the organism; and second, that the males in most species possess an excess of energy which manifests itself in extra-growths, extra-development of pigment, plumage, etc., and that displays by the males of special movements, sound-makings, etc., are direct effects or manifestations of sexual excitation" (Kellogg).

In explanation of the first of these alternative theories it should be remembered that there are within the body numerous glands, some of which, like the salivary glands, liver or pancres, have ducts connected with them, and these ducts bear away a definite secretion which, like the saliva, pancreatic juice, and bile, has a particular and well known function. In addition there is the group of so-called ductless or endocrine glands, such as the spleen, the thymus and thyroid bodies, and the suprarenal capsules. These are sometimes of considerable size and very constant in a large number of vertebrated animals; hence their importance is manifest, but is not in every instance clearly understood. They seem, however, to have a very vital reglating function of one sort or another and their serious derangement is always followed by or concomitant with disease. The reproductive glands, especially those of the male, are often large, seemingly unnecessarily so if procreation is their only function. Ductless glands known as interstitial have been demonstrated in connection with the testes or ovaries as the case maybe, so that, in addition to the "normal secretion" of sperms and ova, there is poured into the blood stream an "internal secretion" consisting of the chemical messengers or hormones, which are known to control or stimulate the development of certain secondary sex characters, such as antlers, mane, or beard. Castration which removes the sex gland prevents the appearance of these characters; on the other hand, the injection of male hormones into a castrated animal restores its secondary sex deficiencies.

Steinach's experiments in which he reversed the reproductive organs in the two sexes of guinea pigs gave rise to amazing results, as the male then developed female traits, both physical and physiological, and vice versa, although they were unable to produce young. (Parker).

The excess of vigor in the male is due to the vastly greater task which the female has of providing nourishment for the offspring, either in the form of yolk in bird or reptile, or of interuterine nourishment and milk in the mammal. With approximately equal vitality to begin with, the drain upon the female would lead to relative excess of energy on the part of the male which could manifest itself in the development of the male characteristics. As we shall see in a later chapter cave animals are characterized by depauperation, lessening of stature due to starvation. Excess of energy is inconceivable on the part of a cave salamander or fish and therefore could not give rise to exuberant growth. This may answer Eigenmann's argument, at least in part. In order that the duty of procreation may not be shirked, it is necessary for the male to have increased activity and sexual instinct in proportion as the female is passive, which would account for the greater sexual excitation on his part.

A final explanation for secondary sex distinction is that of Emery, who believes that "many cases of secondary sexual differences are explained by the sudden appearance (mutation) of another form of male or female, the persistence for a while of the two forms side by side, as now exists in numerous dimorphic species (especially among insects), and then the gradual dying out (killing out by natural selection) of one of the two old original forms (the one like the other sex), thus leaving the other, or aberrant form" (Kellogg). This explanation is also offered to account for the development of mimicry among butterfiles and will be further discussed under that head. Kellogg seems to think that as an explanation for the development of sex characteristics it is rather far-fetehed, but in the promethea moth which we have discussed the larger and more brilliantly colored individual is the female, which makes it difficult to apply the excess of male vigor theory, to say the least.

Criticism of Sexual Selection Theory:

Although *Darwin* gave several evidences in the favour of his theory *Kellogg* pointed out the following difficulties in accepting this theory:

(1) Sexual selection can apply only where males are more numerous than the females or they are polygamous. In the polygamous forms, there is always struggle for mastery and not an active female choice.

(2) Mating female, in most species, is completely passive.

(3) Ornamentation is completely absent in cavefishes, living in dark, therefore, it is regarded that the ornamentation is due to visual selection.

(4) There is no proof for the aesthetic sense on the part of the females for choice on the basis of ornamentation and attractiveness.

(5) Only a few cases of actual choice by female have been observed.

(6) Many attractive characters of males are not noted by the females, as for instance, the brilliant colours and curious horns of the male dung-beetles.

(7) The dances by males for the excitation of females may be only natural during the breeding season.

There is as yet comparatively little evidence based upon actual experiments, but such as we have is strongly opposed to the sexual selection theory.

Alternative Explanation

It is thus clear that Darwinian factor of sexual selection has very meagre evidence. The great majority of secondary sexual characters can be explained by natural selection. Even where natural selection cannot be invoked, sexual selection does not stand the test of experiment.

Therefore, alternative explanations can be suggested as:

1. Secondary sexual characters are formed as a result of the immediate stimulus of the sexually differing primary reproductive organs. This stimulus is due to internal secretion of genital organs acting on some tissues of the organism.
2. Males in the majority of species possess an excess of energy, which results in extra-growths and extra-development of pigment, plumage etc.

There is then very little evidence which prove the theory of sexual selection. To some extant it has been seen that even the choice of male, on the part of female, may be significant because a male with more effective patterns of courtship will have greater success in gaining the favour of female. Now it is realised that sexual selection is simply one aspect of natural selection. Now-a-days natural selection is described in the terms of reproductive fitness. Genes which are responsible for fitness either by contributing or to mating success, tend to increase their frequency in subsequent generations. Thus sexual selection can be compared in its effects to differential viability, longevity or fecundity and can be grouped with them as one of the elements in natural selection.

Chapter—6

Recapitulation Theory and Extinction

RECAPITULATION

The celebrated German savant, Ernst Haeckel, conceived and set forth in the form of a law the wonderful similarity which exists between the life history of any organism and the evolutionary history of the race of beings to which it belongs. This principle, usually known as Haeckel's biogenetic law (morphogenesis of Hyatt), may be stated as follows: The life history of the individual (*ontogeny*) gives a brief resume of the evolutionary history of the race (*phylogeny*). Or, more briefly, ontogeny repeats phylogeny.

While this is in the main true, the phylogenetic record may be falsified in ontogeny in several ways, just as any historical document may lack certain important portions through the accidental or international multilation of the volume, or may have spurious chapters added thereunto by a later hand. In some instances the results remind one of the palimpsests—ancient parchments from which the work of an older scribe has been erased and over the almost indecipherable traces of the ancient writing a new penman has engrossed an inscription of later date. Where the record is essentially correct in recounting in their orderly sequence the historical events, we may compare it to *palingenesis* (Gr. πάλιν, again, and γενεσις, production), in which truly ancestral characters conserved by heredity are reproduced in development. The introduction of spurious matter may be likened to caenogenesis (Gr. καινός, recent), in which non-primitive characters make their appearance in consequence of secondary adaptation of the young to the peculiar conditions of its environment. The elimination of certain chapters finds its parallel in tachygenesis (Gr. ταΧύs, swift), the acceleration of characters which are crowded back further and further into the embryonic life or out of the life cycle entirely.

Palingenesis

When one consider the millions of years of evolutionary life which have fallen to the lot of all organisms, if, as we believe, they have all evolved out of one primal creation of life, and compares those untold ages with the brief span of individual existence, he will see that the record is necessarily so greatly abridged as to make a perfect recapitulation practically impossible. Nevertheless the metamorphosis undergone by the common frog may be taken as a fairly typical instance of palingenesis, as there is little evidence in its life history, except in the development of certain larval organs, such as the adhesive suckers, of any material falsification of the record.

The same may be also true of the ontogeny of the lower insects, but as soon as forms like the moths are considered, especially such as have highly adapted larvae, as in certain of the mimicking forms, it is at once evident that caenogenesis has been at work.

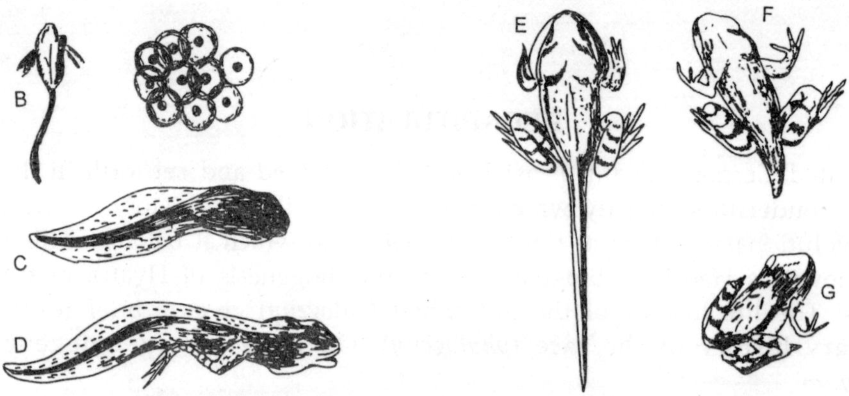

Fig. 6.1: Metamorphosis of the frog, *Rana temporaria*. A, eggs, greatly enlarged; B, tadpole with external gills; C, hind legs appearing; D, hind legs well developed; E, tadpole with all limbs free; F, G, stages in which the tail is resorbed, F being a miniature adult.

Cænogenesis

Among the inch-worms, or geometrid larvae, protective mimiery is common, the creature being elongated and twig-like in form, in color, and in its attitude when disturbed, for it throws itself out rigidly at an angle with the supporting branch to which it is attached by the hook-bearing proplegs at the hinder end of its body. This is clearly an adaptation to meet the vicissitudes of larval life, and the whole insect is very different from the form of a primitive species, such, for instance, as the fish-moth, *Lepisma*, or *Compodea*, "the simplest living insect." Then, too, the pupa state of the geometrid, in which the insect passes into a condition of quiescence and the marvellous transformation into the adult takes place, can have no precise parallel in the history of the race, for it is inconceivable that a long period of racial dormancy, during which the evolution of wings was accomplished, could ever have occurred.

Tachygenesis is the rule in ontogeny, and numerous instances might be cited, but certain of the frogs and toads may again be taken as typical instances, for while in most of them the eggs hatch out in water and the young undergo the typical palingenetic development mentioned above, others show all degrees of the suppression of larval stages until metamorphosis is practically eliminated and development from egg to frog is direct.

"In some the eggs hatch on land, having been laid in holes, on grass or leaves, and when the tadpoles are hatched, they wriggle into water or are washed into pools by the rain. In others, again, the eggs are laid on land, and the tadpoles have lost their gills before they are hatched, but the metamorphosis is completed later on. In a few the complete change occurs inside the egg, and when hatching takes place, little frogs appear, sometimes, however, with a stump of the tail still left. In others the eggs are carried by the parent, and here, too, they may be hatched as tadpoles or as perfect frogs" (Mitchell).

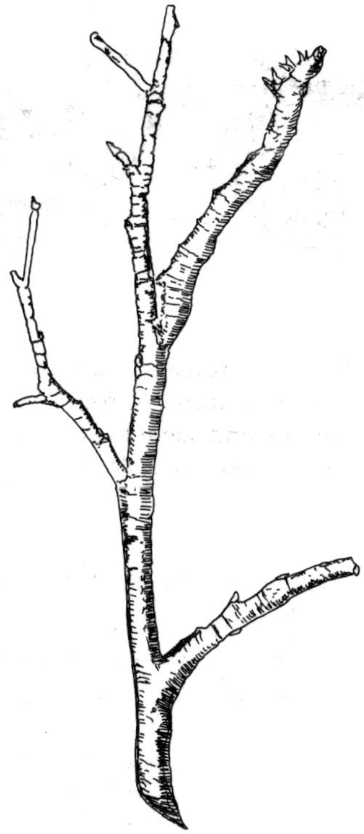

Fig. 6.2: Geometrid larva on a branch. The twig-mimicking caterpillar is the upper right-hand projection from the stem.

Yet another remarkable life history wherein the strict phylogeny is departed from is that of *Ichthyophys* one of the Gymnophiona of the class Amphibia. These are curious, burrowing, snake-like forms with neither tail nor limbs, but which, despite their degenerate specialization, seem to have inherited more of the characteristics of their ancient stegocephalian forebears than have any other living amphibians. The female of a species which lives in Ceylon and breeds just after the spring monsoon, digs a hole close to the surface near running water where the ground is moist. Here she lays about two dozen eggs, around which she coils herself, probably to protect them against other burrowing lizards and snakes, which are very numerous. During the period of incubation, if such it may be called, the eggs swell to twice their former size, and the mature embryo weighs four times as much as the newly laid egg. The embryo has external gills, which, however, have lost their primal function of respiration for that of nutrition, for they move up and down in the fluid of the egg. The lateral line sense-organs develop, organs of prime importance to all aquatic vertebrates but here functionless while within the egg, which may also be said of the fin developed by the short tail. When the embryo has reached a length of about seven cm., the gills begin to shrink, and at the same time one pair of gill-clefts appears at the base of the third external gill. When the larvae are hatched, the gills are lost and the young animal takes to the water in a gilless state, although at the bottom of the aperture on either side two gill arches may be seen, and the larva frequently comes to the surface to breathe. The lateral sense-organs and tail fin now fulfill an important function. The creatures seem to live a long time in the larval stage, but at last the gill-

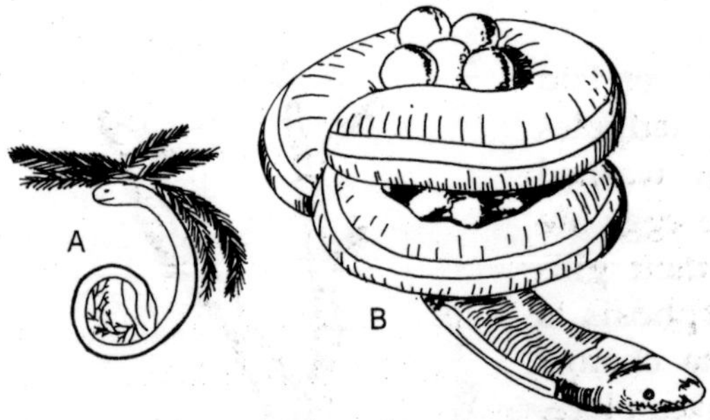

Fig. 6.3: Limbless amphibian, caecilian, *Ichthyophys glutinosa.* A, nearly mature embryo, with gills, tail-fin, and some food yolk; B, female guarding her eggs, coiled in a hole underground.

clefts close, the tail fin is lost, the skin assumes a totally new structure, and the fish-like larva turns into a burrowing creature which readily drowns when forced to remain in the water. In this instance the most striking tachygenetic features are the crowding of the gills, which might well be of material service to the aquatic larva, back into the embryonic stage, and their assumption of a different role, and the total elimination of any trace of legs from the life history, although the Gymnophiona must have been descended from ancestors possessing limbs.

THE RACIAL CYCLE

Students of fossil forms, especially of the molluscs, have come to recognize a series of definite stages in their phylogenetic. These are as follows:

	Ep-acme		*Acme*	*Par-acme*
Ontogeny	{embryonic nepionic	adolescent neanic	adult ephebic	senescent gerontic
Phylogeny	{phylo- nepionic	phylo- neanic	phyloephebic	phylogerontic

If the life cycle be represented by a curve, the ascending limb will include the nepionic and neanic or what may be called the epacme. During the acme the summit of the curve is reached and the organism is in the full fruition of its powers. The par-acme is the period of decadence, when the organism's failing strength is represented by the descending curve terminating in death.

Shells of molluscs, especially of gastropods and cephalopods, preserve the ontogenetic characters, often in a single specimen, better perhaps than the remains of any other group, and have as a consequence lent themselves especially to students of this interesting problem, notably the American paleontologists Hyatt, Beecher, Grabau, Jackson, and J. P. Smith. Not only is this true of molluses, but in a more or less complete degree of brachiopods, echinoderms, and corals.

In many of the gastropods and cephalopods especially, a single shell of an old individual may have preserved a record of all of the changes it has undergone during the animal's lifetime. Thus we can find at the apex of the coil the tiny embryonic shell or protoconch, then the neanic portion, formed during the animal's youth, simpler perhaps in ornamentation than the adult or ephebic section when the height of development of all of the features, ridges, bosses, spines or complexity of suture characteristic of the species is attained. Then the gerontic or senescent portion is seen, recognizable by an increasing simplicity, comparable to that of the neanic shell but retrogressive rather than progressive in the assumption of characters. Furthermore, it has been proved by paleontologists that the several stages shown in the development of a given shell reflect the adult condition of more ancient forms, presumably ancestral. Thus, the living *Nautilus,* the sole survivor of a formerly abundant group of cephalopods, has a closely coiled shell, but the earlier cephalopods had not, and coiling was gradually assumed and sometimes secondarily lost in phylogerontic types. In the course of its development, however, *Nautilus* passes through arcuate (L. *arcuatus,* shaped like a bow), loose-coiled, then close-coiled stages directly comparable to the adults of the Paleozoic genera *Cyrtoceras* and *Gyroceras,* and the later *Nautilus* representatives of its own group.

Another highly evolved Cretaceous cephalopod, the ammonid *Placenticeras pacificum,* is characterized in the adult by complex sutures, the lines of juncture of the transverse partitions or septa which separate the chambers of the shell and its outer wall. In the development of its sutures the individual shell passes through simpler stages which are comparable to the adult structures of nautilid and goniatitic forms, followed by stages in which the septa are comparable to those of early *Ammonites* before it assumes its adult generic features.

In the lamp-shells or brachiopods, the spirally wound buccal (mouth) arms which serve for food-getting usually have an internal limy support which is an outgrowth of the inner wall of the shell. Beecher and others have shown that the stages in development of the exterior and interior of the shell and the brachial (arm) supports can be closely correlated with adult characters of more primitive representatives in the group. As has been said, while stages in development from the young to the adult are all progressive, in senescence the stages that appear are in the main retrogressive; Nautilids and ammonids, which are characterized by close-coiled shells, build loose-coiled or even uncoiled additions; specialized *Ammonites,* with complex septa, in senescence build simple septa, thus assuming simpler characters comparable to those seen in their own youth, and also comparable to the characters of adults in regressive (degenerating) series in their own groups.

While shell-bearing invertebrates thus lend themselves admirably to the study of recapitulation, vertebrates as a rule do not, for the skeleton, changing as it does with age, gives only the characteristics of the owner at the moment of death, and one can rarely learn much of the ontogeny from a single individual or the portion of an individual such as is usually available for study. Nevertheless where several individuals of a specie are known, or the annually shed antlers of a captive deer, enough may be learned to show that the law of recaptitulation holds with them as with the invertebrates.

Persistently Primitive Types

Both plants and animals in nature, as under domestication, show a remarkable variation in plasticity, so that while in the great majority time has wrought wonderful evolutionary changes, with a few it is as though the world stood still and the ceaseless centuries passed over them without effect. In some cases isolation in a remote place, where inter-specific competition has largely diminished, is doubtless one causes of their changeless survival; others are in the thick of the strife but seem to be immune to the influence of changing conditions.

Such, for example, are *Amaba* and the simpler Protozoa and unicellular plants, doubtless relics from the remote Proterozoic age, unless, as seem hardly probable, life evolved from the lifeless more than once and these are the primitive stages of a later creation. *Orbulina* and *Globigerina,* two foraminiferal Protozoa, are known from the Ordovician and doubtless existed long before, while among the Brachiopoda one relic type is *Lingula* and another *Crania,* both dating also from the Ordovician, and persisting practically without change until the present. *Nautilus,* mentioned above, has persisted from the Tertiary, and the family to which it belongs from the Jurassie. Of the vertebrates, an extremely old type is the lung-fish *Ceratodus,* of which a modern derived genus, *Neoceratodus,* is now found isolated in certain Australian rivers, and which dates from the Triassic. The ancient Port Jackson shark, *Cestracion,* has persisted since the Jurassic and members of its family are found in Lower Carboniferous rocks. In many respects the most interesting relic of all is the tuatera (*Sphenodon*) resident on certain small islands bordering the mainland of New Zealand—a Permian type although somewhat modified from its Paleozoic ancestors. *Sphenodon* is the sole survivor of an important order of reptiles and is of great value to those who would revivify the creatures of other days because of the flood of light which its structure on their probable anatomy.

These ancient forms are what are called generalized or primitive, as opposed to specialized types, for they alone can tarry in their evolution while time rings in its changes. High specialization, on the other hand, means a relatively brief career.

PHYLOGERONTIC CHARACTERS

Just as senility may readily be recognized in the individual by certain characteristics such as the graying of the hair, loss of teeth, of upright carriage, of vigor and elasticity of step, so to the trained eye characteristics are discernible which may point to racial senility. These have been recorded by the English paleontologist, Sir Arthur Smith Woodward, who arranged them under several heads, as follows:

Relative Increase of Size

One of the characteristics recognizable as belonging to racial senescence, although having no parallel in that of the individual, is relative increase of size far beyond that which is usual in the group to which the animal belongs. In certain recorded instances among the prehistoric animals, such an increase was followed by extinction, and in several living examples racial death is certainly threatened if not a very real probability, for, as we shall see, great increase of size is accompanied by slow maturity and consequent lessening of the rate of increase, which severely handicaps the species in the struggle for existence. An example of phylogerontic immensity is the genus *Productus*, of which *P. giganteus* is the largest brachiopod known, sometimes attaining a width of nearly one foot, while most species hardly exceed one or two inches. *Productus* was very abundant in the Carboniferous and Permian and then was blotted out. The cephalopod *Ammonites* was another relatively huge form, while the living giant squid, *Architeuthis*, is by far the largest known invertebrate, reaching a recorded length of 50 or more feet. Certain of the smaller cephalopods, on the other hand, measure but an inch or two, while between one and two feet would be a fair average for squid.

Among vertebrates the Mesozoic dinosaurs were truly gigantic, *Brontosaurus* with 67 feet of length and 30 or more tons of weight, but especially *Gigantosaurus*, 80 or more feet long and proportionately heavier, occurring just before the extinction of their suborder; while of carnivorous dinosaurs *Tyrannosaurus*, the culminating form of its race, was the largest and most terrible flesh-eating terrestrial animal whose existence is thus far revealed. Of living types, we have the giant whales, the sperm, Greenland, and sulphur-bottom, all gigantic compared with the more conservative dolphins and porpoises, and rapidly nearing their extinction. The hippopotamus, exceeding by far all other swine-like animals, is also restricted in numbers and habitat compared with former times, while the elephants are very few to-day, though formerly world-wide in distribution and extraordinarily abundant. Of the old African elephants which reach a stature of 11 of 12 feet, very few are now alive, and the most majestic species of all, *Elephas meridionalis* of Europe and *E. imperator* of America, are extinct. Of primates, the gorilla exceeds all others in size, with the exception of occasional men, and he is now making his last stand against fate in the dismal fastness of the dark continent.

Spinescence

Another gerontic character found occasionally in all skeleton-bearing animals is spinescence, that is, the tendency on the part of the shells of molluscs and of brachiopods, the external mail of Crustacea, and even the internal skeletons of vertebrates to produce a superfluity of dead matter. In some instances spines and horns are undoubtedly of genuine protective value, they may be ornamental structures, although the extent to which purely ornamental characters without other practical value to the organism develop in nature is open to question. In general, such excrescences seem like growth-force run riot, as though with the lessening vitality incident to racial old age, it is no longer adequately controlled.

Examples are, among brachiopods, *Spirifer*, in some species of which the hinge-line becomes elongated into spine-like processes, and *Productus horridus*, whose valves are liberally bedecked with spines. Among molluscs there are a number of highly spinescent types, such as the gastropod *Murex* and the bivalve *Spondylus*. Among vertebraes there is a group of lizard-like reptiles, Car-boniferous to Permian in distribution, members of the order Pelycosauria, related to the living relic *Sphenodon,* mentioned above. These creatures toward the close of their career were characterized by the enormous elongation of the vertebral spines, which in extreme cases also bore transverse processes like the yard-arms on the masts of a square-rigged ship. *Edaphosaurus* was perhaps the most remarkable vertebrate from this point of view. There seems to be evidence that these lengthening spines were independently acquired in more than one line of descent within this group, but in all cases the appearance of spinescence heralds the extinction of the line.

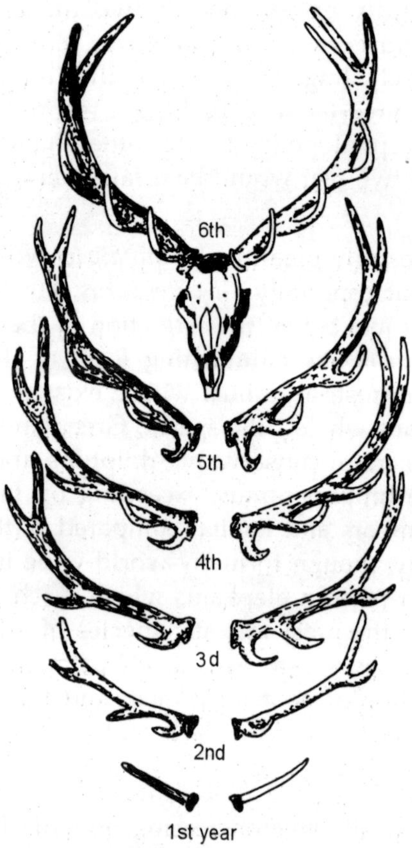

Fig. 6.4: Antlers of stag, showing ontogenetic increase in complexity. Series in the British Museum (Natural History).

The dinosaur *Stegosaurus* shows phylogerontic characters in at least two ways—rapid increase of size over that of allied forms, and a marvellous overgrowth of armor plates and tail spines which heightens the bizarrerie of this most grotesque of beasts. *Stegosaurus* again is the last of its phylum, for no trace of the genus has ever been found in rocks of later date than the Morrison formation which produced it.

The deer are the most familiar recent instance of spinescence in the bony growths known as antlers which surmount the brow. These antlers, which are periodically shed, increase annually in weight, complexity, and number of points until the deer is old, when they begin to simplify again. An interesting recapitulation is shown, for the antlers of a young animal are comparable to those of fully adult Miocene deer, those of somewhat later age to those of the deer of the Pliocene, while a "stag royal" in his prime bears antlers comparable to shoe of the Pleistocene and Recent deer. Moreover, as one would expect, the deer which attained the maximum development of horns, the great Irish stag or "elk", is now extinct, and we know of no other reason for their extinction than the racial senescence which the antlers imply. The not infrequent interlocking of the horns of two fighting bucks, resulting in their destruction by starvation or the attacks of other animals, shown that the point of greatest utility as weapons of defense and offense has apparently been exceeded, so that the structures are an actual menace to their possessor.

Degeneracy

Physical degeneracy is another phylogerontic trait, this time paralleling certain senile characteristics of the individual. Among these is *loss of teeth,* which is recorded several times: first, among the fishes such as the sturgeon and certain deep-sea forms like the "gulper" eel, *Macropharynx,* which evidently feeds upon the bottom oozes. The trutles, which are among the oldest of living reptiles, had lost their teeth by Triasic time, when they first appear in the rocks; the birds, which were toothed during the Age of Reptiles, have also been toothless since its close. In each of these instances the jaws are sheathed with a horny beak variously modified, so that while the turtles are doubtless fewer than of old, the birds, except for man's interference, can hardly have begun to wane as a whole, although many races are extinct. Among dinosaurs, three phyla were, at the time of their extinction, rapidly losing their dental armament: the sauropod *Diplodocus,* 85 feet in length, whose teeth are slender structures no larger than lead pencils: *Genyodectes,* a carnivorous dinosaur of Patagonia: and *Struthiomimus,* one of the same group from the late Cretaceous of Alberta and Montana. Mammals occasionally show loss of dentition, notably the very ancient, egg-laying monotremes of Australia and Tasmania and the anteaters among edentates.

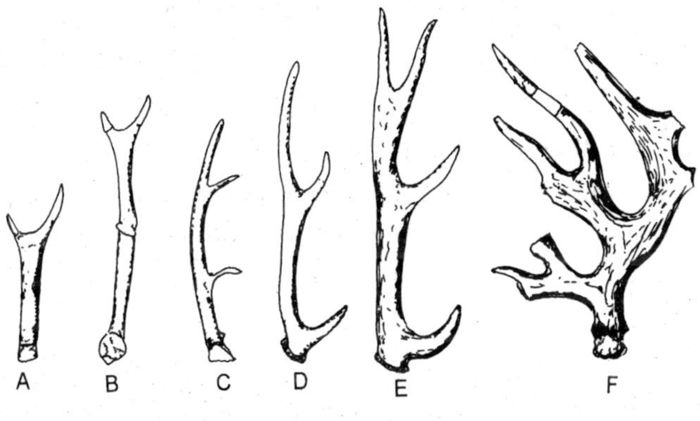

Fig. 6.5: Antlers of deer, phylogenetic series. A, B, *Cervus dicrocerus,* Mid-Miocene; C, C. *matheronis,* Upper Miocene: D, C. *pardinensis,* Upper Miocene and Pliocene; E, C. *issiodorensis,* Pliocene; F, C. *sedgwickii,* Pleistocene.

Another degeneracy is the assumption of an *ee-like, elongated form* which W. K. Gregory tells us has been acquired no fewer than forty-four distinct times among vertebrates alone. These have been the result of independent or homoplastic evolution, and the criteria which in almost every instance distinguish them are: anguilliform (eel-like) body with multiplication of vertebrae, gephyrocercal tail (tapering to a point), reduced pelvic limbs, and usually predatory habits. Gregory lists three groups of cyclostome fishes, one of sharks, a lung-fish, thirty evolutions among the Teleostomi or bony fishes, three among Amphibia, five among Reptilia, and one group of mammals, a remarkable record.

Modern views on Recapitulation Hypothesis

Dispite of above criticisms, the theory of recapitulation has continued to find favour with some embryologists and with most palaentologists. The most clearcut analysis of the relation between phylogeny and ontogeny comes from a Russian school headed by *A.N. Sewertzoff*. This he has described as *phylembryogenesis*. The studies have shown that the relation between phylogeny and ontogeny does not follow any one rule but is of several different sorts such as:

1. *Addition, prolongation or anaboly*. In this type, the development follows the ancestral pattern upto the last ancestral ontogenic stage to which further new stages are added. It may be represented by the following scheme in which small letters stand for embryonic stages, capitals for the adult stages.

a b c d D -Ancestral ontogeny

a b c d e E -Ontogeny of the individual.

In the above expression descendent follows the ancestral ontogeny through stage d and later adds on a stage e, which leads directly to the definet adult stage E. It implies that strictly speaking, the adult ancestor is not repeated in the ontogeney of the descendent.

2. *Deviation*. In this type the ontogeny of the descendent follows the ancestral pattern to a certain stage and then diverge along a new line thus.

a b c d e f F -Ancestral ontogeny

a b c d_1 e_1 f_1 F_1 -ontogeny of the individual

Deviation confirms *Van Baer's* third rule.

3. *Archallaxis*. The embroys of the ancestral and descendent species deviate from the early stages of embryogenesis. Therefore, no recapitulation is exhibited.

a b c d e f F (Adult)

a b_1 c_1 d_1 e_1 f_1 F_1 (Adult)

4. *Abbrreviations*. The end stages of the ancestral ontogeny are ommitted so that the definitive adult of the descendent resembles the adult of a remote rather a recent ancestor.

a b c d e f F

a b c d e E_1

Cases of neoteny full under this category. But such omission of the end stages of development, structures may also remain in a rudimentary or simple state or disappear altogether.

5. *Acceleration*. Here some of the middle stages of ancestral ontogeney are dropped out so that the later structures appear in the ontogeny of the descendent at an early stage that they did in the immediate ancestors.

a b c d e f F

a b e_1 f_1 F_1

6. *Omission*. Some of the early embryonic stages of ancestral development are omitted and deviation of the later stages.

a b c d e f F

c_1 d_1 e_1 f_1 F_1

Such omission of early stages may be followed by progressive evolution in the structure concerned or may result in rudimentations of an organ.

7. *Coenogenesis*. The embryonic stages of the descendent acquire certain special adaptive features, which are new to them. These have no ancestral significance and disappear during larval or embryonic life.

8. *Heterochronism*. It includes a variety of time displacements in development in which organs appear earlier or later in ancestral antogeny.

From the forgoing discussion it is evident that the developmental stages of an animal may resemble adult ancestors or the embryonic stages of an ancestor or may deviate somewhat from either of these or may bear no resemblace at all to its ancestor. Resemblence to ancestral forms, embryonic or adult is a common and widespread phenomenon and consequently the practice of drawing phylogenetic conclusions from the study of development is to a large extent justifiable (*Hyman*). Recapitulation in its narrow Haeckelian sense, as repitation of adult ancestor is not generally applicable, but ancestral resemblance during ontogeny is a general biological principle. It is advisible that the usage of word recapitulation should be altered.

EXTINCTION

Extinction in phylogeny has two aspects, each of which has its equivalent in ontogeny. Of these the one which first comes to mind is racial death—the actual cessation of continuity with the blotting out of the line. The other is the transmuting of a given species into one of higher or more specialized type. The first is comparable to childless death, the second to the passing on of life to offspring. In each instance, except among the potentially immortal Protozoa, the individuals die but the line continues to exist.

Illustrations of the two methods of extinction may be found among the Miocene three-toed horses. Some, like the browsing "forest" horse, *Hypohippus* died without issue, due to increasing aridity of climate and the consequent shrinkage of their natural environment and food supply, together with the specialization of their teeth for succulent vegetation only. Another horse, contemporary with *Hypohippus*, was *Merychippus*, the blood of which, for aught we know to the contrary, flows in the veins of all existing horses, although *Merychippus as such* has cease to be and may be reckoned as an extinct animal just as truly as *Hypohippus*..

In several instances among prehistoric forms extinctions have come with apparent suddenness so far as our records show, and in each instance without known competitors which could possibly have worsted the race in an interracial strife. As the representatives of these forms often showed in one way or another certain of the phylogerontic characteristics which have been mentioned, the inference is that they died a natural death. Instances are the dinosaurs, of which the first to be rendered extinct were the gigantic Sauropoda, *Brontosaurus* and its kind, most of which died out long before the final extinction of the dinosaurs as a whole; and when the race had well-nigh run its course, we see some forms gigantic, others spinescent, others toothless, and the marvel is, not that they died, but that they survived so long, for the years of dinosaurian dominance exceeded one hundred and forty million!

Another group which seems to have died a natural death were the so-called "fish-lizards" or ichthyosaurs which swarmed in the high seas of the Mesozoic almost from its beginning, but died out early in Upper Cretaceous time. Possible competitors were the mosasaurs or sea-lizards, whose known remains are confined to the Upper Cretaceous, but the ichtyosaurs were whale-like, not only in appearance but also in inferred habits, and were more numerous and widespread in their prime than were the mosasaurs. They certainly could not have been in competition with the whales, for a long interval separates the respective times of their existence during which, unless the mosasaurs filled the role, the seas were bereft of whale-like forms. The mosasaurs and pleisosaurs died abruptly without apparent cause at the close of the Mesozoic, as did the dinosaurs, and, like the latter, showed gigantic forms toward the close of their career.

CAUSES OF EXTINCTION

The causes of extinction (racial death) as applied to mammals have been admirably worked out by Professor Osborn (1906). As these laws probably covered nearly all extinction-causes in forms other than the mammals as well, they may be briefly summarized.

Changes in Physical Environment

These are the changes wrought by the elevation or subsidence of land masses, with the resultant formation or severance of land-bridges, either facilitating or preventing migrations, or, by permitting incursions of hostile animals, producing competition which cannot be withstood. Restriction of habitat, especially insular habitat, also increases competition at the cost of the weaker forms. On the other hand, insular animals which do survive are isolated from further competition and may live long after the general extinction of their kind elsewhere. Witness the monotremes of Australia and *Sphenodon* of New Zealand.

Changes in Climate

Increasing cold is apparently a very potent cause for extinction, for while certain forms like the musk oxen and the Siberian mammoth may become adapted to rigorous climate, many more will fail. The last glacial period was a time of wholesale extinction, either as a direct or indirect result of the devastating cold.

Increase of moisture diminishes the supply of harsher grasses which afford nourishment to the great host of grazing mammals, and by far the great majority of hoofed animals extant are of this nature. Moisture also fosters the growth of poisonous plants and increases the numbers of insect pests and the consequent misery which they cause. The latter are often the carriers of disease such as the surra sickness and the sleeping sickness, which are fatal to domestic horses and, in the latter case to men. Increase of moisture also gives rise to forest tracts which may form effective barriers to the migration of certain animals, and on the other hand afford migratory tracts to the semi-arboreal carnivores like the jaguar.

Decrease of moisture changes the character of the food supply, increases the length and severity of the dry seasons, removes forest barriers, thus increasing competition through the permitted migration of invading hordes, and actually causes the direct extinction of such forms as the primates, which depend upon forests for their livelihood. The extinction of primates in North America, with, and possibly as a result of, the increasing dryness toward the close of the Eocene is an historic fact. Diminution of the succulent herbage and increase of grasses, which are direct results of moisture diminution, favor the grazing but eliminate the browsing forms. An instance would be the extinction of the browsing titanotheres and forest horses an the great spread of grazing horses, camels, and other grass-feeding types in the Oligocene and Miocene epochs.

Changes in Living (Biotic) Environment

Competition is inseparable from life, and while the adaptable forms are impelled thereby to rapid evolution, the in adaptable often find the competition too severe to be met and consequently perish. This competition may be brought about by one of two ways: by rapid multiplication of certain local or native animals, or by the immigration of new animals, giving rise to the slow or rapid extinction of certain local forms. This may be due to direct competition, as when a mother marsupial is attacked by another animal; her young which she is forced to carry handicap her greatly in comparison with a placental mother, and her destruction means also that of her young. The food supply of larger forms may be destroyed by smaller grazing mammals. St. Helena, once heavily wooded, has been rendered a barren rocky island through the introduction of goats by the Portuguese in 1513, as they destroy the seedling trees and hence when the old ones died there were none to take their place. This must have had a profound effect upon the native forest-dwelling fauna.

Restrictions of *island life* result in the dwarfing of native forms through hard conditions and competition, as for example in the Shetland ponies, which are dwarfed far below the standard of their species. Certain islands in the Mediterranean, Cyprus, Malta, Sicily, and sardinia, have yielded the remains of Pleistocene dwarf elephants, all of which are now extinct. These islands are relics of old migratory routes.

The *introduction of higher carnivorous mammals* often brings extinction to lower forms of a less well-equipped type. The entrance of the dingo, a placental dog presumably of Asiatic stock, may well have been the cause of the extinction of the native marsupial "wolf," *Thylacynus*, in Australia. *Thylacynus* is now confined to Tasmania whither the dingo has not yet penetrated, but its remains are found in superficial deposits in

Australia, showing it to be only recently extinct. The migrations of the saber-tooth cats into South America during the Pliocene probably were a supplementary cause of extinction of the giant sloths, *Mylodon* and *Megatherium*.

Internal Causes

Inadaptive structures such as the *highly specialized teeth* of browsing quadrupeds like the titanotheres, conformable only to the needs of a browsing form and incapable of supplying subsistence when the forests shrank and the grasses became dominant, may well have been an important factor in their sudden extinction. *Unprogressive feet*, such as those of the Amblypoda, archaic ungulates, could not bear their owners to the evolutionary goals when the competitors were the swift-footed forebears of modern hoofed animals. *Large size*, aside from its indication of racial senescence, is in itself a menace, as it requires more food for its sustenance and there go with it slow breeding and a long period of adolescence which multiply the creature's chances for destruction before it can procreate its kind.

Extreme specialization always greatly increases the creature's risk, for the unspecialized frequently survive where the specialized, whose dominant organs may tend to over-development, perish.

Osborn gives four measures of *mental capacity* in extinct types: (1) absolute size and weight of brain, (2) convolutions, (3) proportionate size of frontal lobes or cerebrum, the seat of intellect, (4) ratio of brain to body weight. Among mammals especially a premium has been placed upon mentality ever since their initial evolution, and the shrewder of two competing groups generally wins. In the competition of dingo and thylacine mentioned above, relative mentality, which is notably deficient among marsupials, may well have been the deciding factor. In the old-time competition of archaic and modernized mammals the former were handicapped by inadaptive feet, teeth, and brain, and the last count especially was the one upon which they stood condemned.

Osborn thus concludes: "Following the diminution in number which may arise from a chief or original cause, various other causes conspire or are cumulative in effect. From weakening its hold upon life at one point an animal is endangered at many other points."

Chapter—7

Artificial Selection

Under the general heading of selection are three aspects—Artificial, Natural, and Sexual; the first being, as the name implies, a man-made process which is interesting as indirect evidence of evolution, for the plasticity of animals and plants in the hands of the breeder shows that remarkable changes in the character of organisms can occur, and if produced by Man, why not by Nature which is vastly more influential and has time without stint at her disposal?

The processes of artificial selection are undoubtedly more or less analagous to those of natural selection and aid us, as they aided Darwin, to understand more clearly the methods whereby Nature works to accomplish her results. It seems wise, therefore, to discuss this aspect of selection first as an introduction to the general theme.

Artificial selection is the process whereby the various breeds or races of domestic animals and plants have been formed by man. It is of the character of experimental biology upon which modern investigators rely so largely for their facts; but it has been found that laboratory experiments sometimes succeed where Nature fails, so that it is quite possible that the theory of natural selection, for instance, rests too largely upon an assumed analogy with the artificial.

Artificial selection is, in its simplest aspect, an extremely old process, judged by human standards, extending as it does over at least seven thousand years, and some of the changes wrought among animals and plants are truly marvellous. Certain types, on the other hand, like the peacock and guinea hen, lack plasticity and all the centuries of their domestication have failed to produce any very marked departure from the original stock. Such are static in contrast to the more plastic forms.

Nature of Product

The forms produced by artificial selection are often very far from a natural ideal and are not such as would be produced or even survive in the wild state. In nature, characteristics which make for success in the struggle for existence are the product of evolution; in domestic races the qualities or traits which are useful to man or conform with his fancy or with style are selected, many of which would prove an insuperable handicap in open competition. Another characteristic of artificial breeds is their instability, their proneness to revert to the original type in a few generations if the vigilance of the breeder in prohibiting promiscuous mating between different varieties is relaxed.

EXAMPLES OF ARTIFICIAL SELECTION

Pigeons

Among the most plastic of all domesticated animals are the pigeons, of which considerably more than 150 named varieties, all descended from a single known source and breeding, were known at the time of Darwin (1868); moreover, the range of variation is so great that some domestic races of the rock-pigeon differ fully as much from each other in external characters as do the most distinct natural genera (Darwin).

The rock-pigeon, *Columba livia,* which includes two or three closely allied subspecies or geographical races, is to be regarded as the common parent form. From this conservative bird there have been derived some remarkable types, such, for instance, as the pouter, in which the body and legs are elongate and the feet are fully feathered; but the most peculiar feature, which gives the name to the variety, is the enormous size of the cesophagus, which is, often inflated, giving the bird a remarkably pompous air. A second variety consists of the carrier pigeons, with elongated beak, neck and body but with the eyes surrounded by much naked, generally wattled skin. These pigeons are capable of the most sustained flight and are used abundantly as homing birds. They gave remarkable service as messengers during the World Wars.

In the fantail variety, the tail is expanded, directed upward and formed of many feathers, with no oil gland, and there is a rather short body and beak. The carriage of this bird is remarkable, the head touches the tail feathers, and habitually trembles very much, and the neck has an extraordinary backward and forward movement. Good birds walk in a singular manner as if their feet were stiff, and owing to their large tails they fly badly on a windy day.

The turbits have the feathers reversed along the front of the neck and breast while in the Jacobins they are reversed even more along the back of the neck, forming hood which, meeting in front of the neck, almost enclose the head. Tumblers have a curious habit of tumbling backward during flight, some, particularly the Dutch Rollers, tumbling to an extraordinary degree "Every few seconds over they go; one, two, or three summersaults at a time. Here and there a bird gives a very quick and rapid spin, revolving like a wheel, though they sometimes lose their balance, and make a rather ungraceful fall, in which they occasionally hurt themselves by striking some object" (Brent in Darwin).

In the Indian frill-back, as in the Jacobin, the feathers are reversed, but in the frill-back they curl backward over the whole body. The trumpeters have a tuft of feathers at the base of the beak curling forward, and a very peculiar voice.

These instances, the descriptions of which have been taken from Darwin, will suffice to show not only the extremes of variation but the impossibility of natural survival on the part of some of them, notably the tumblers and fantails, with their poor locomotive powers, and the frill-backs, whose feathers, instead of being an admirable protection against inclement weather, actually invite the destruction of the bird. The young of all of the highly improved fancy birds are extremely liable to disease and death owing to the detrimental effect produced on the constitution of an animal by too close inbreeding. Some of these breeds are very old, pouters, Jacobins and tumblers having existed prior to 1600. Nevertheless they are not yet stable and it is said that only two generations of promiscuous breeding will suffice to produce a mongrel in many ways an apparent reversion to the ancestral rock-dove type.

FACTORS OF ARTIFICIAL SELECTION

The means employed by the stock breeder in the production of new varieties have been enumerated by Jordan and Kellogg as follows:

First, unconscious selection with more or less complete isolation. What this really means is a weeding out of the unfit, a sort of lethal selection of the less desirable animals from every point of view. These are eliminated and the remaining animals permitted to breed. This improves the stock without giving the race distinctive qualities. Nevertheless where isolation is effective and environmental conditions differ, it leads in time to marked racial divergence. Among the sheep of England several markedly distinct breeds have been thus produced, as follows:

Hornless Varieties:

Southdown sheep of Sussex, tawny face and legs.

Hampshire sheep, black face, ears, and legs, black spot under tail.

Devonshire sheep, similar to preceding but without black spot under tail.

Cheviot sheep, face and ears white, head free from wool, ears erect.

Shropshire sheep, dun face, face more or less covered by wool.

Horned Varieties:

Dorset sheep, white face and ears, small white curved horns. Irish sheep, black horns.

Second, conscious selection of the more desirable individuals, (methodical selection) emphasizing such points as larger size, plumpness, earlier maturity, greater docility, and fertility. While unconscious selection tends to produce racial divergence, here the reverse is true, for these good qualities appearing in all individuals tend to obscure natural or racial traits and make all sheep looks like.

Third, conscious selection directed toward definite or special ends, that is, emphasizing certain individual characteristics rather than working toward a good, all-round animal. Variation is continually bringing new points to the breeder's notice; those which strike his fancy or increase the vale of the animal will be cumulatively emphasized in successive generations by breeding only from these individuals which show the characteristic best. In this way increased milk-flow or yield of meat or wool, or the disappearance of horns is obtained. Often the character which strikes the breeder's fancy is non-beneficial and hence opposed to a result which would be produced by natural selection.

Fourth, crossing or hybridizing. The crossing of two individuals is often of great benefit because it (1) increases the range of variation, (2) adds or combines certain desirable characters, (3) eliminates the undesirable. The offspring of such crossing may vary greatly, especially after the first generation, and the qualities they show may be either good or poor. They will diverge widely from either parent or show a combination of the characteristics of both. Sometimes traits appear that are distantly ancestral and again those that are entirely new. When this cross-breeding is accompanied by careful selection directed toward definite or special ends, remarkable results are accomplished, and it is mainly upon this series of processes that breeding as a fine art depends. In this way Luther Burbank has produced a marvellous assemblage of plant varieties, flowers, and fruits, attaining by his patience and skill results which are almost like those of wizardry.

LIMIT OF ARTIFICIAL SELECTION

These are not met with as soon among plants as among animals, and the first impediment is lack of fertility in making wide crosses. Among animals, crosses of varieties only are practicable and as a rule related species such as the horse (*Equus caballus*) and the ass (*Equus asinus*) can be bred only for one generation, as the offspring is almost invariably sterile. Among plants, on the other hand, not only are wider crosses practicable but many plants can be propagated from cuttings, while higher animals, including all domestic animals, cannot, and the only way of getting another individual such as a mule is toe make the original cross again. Mules can be improved therefore only by improving the parents or by crossing different breeds of the horse and ass respectively.

Cross-breeding Contrasts

Contrasting results are sometimes obtained in cross-breeding by reversing the sexes of the two varieties mated. Thus the common mule is the offspring of a male ass and female horse; from the father he inherits certain superficial traits such as the small hoofs, somewhat scantier mane and tail, voice, and some peculiar mental characteristics; while from the mother come the deeper-lying size and strength and symmetry. Some authorities claim that his intelligence is greater than that of either parent, although he lacks the dignity and poise of a good horse. Reverse the cross and the result is a hinny, more horselike in contour and appearance, but with the smaller stature of the ass. The voice is that of the horse, and the animal is more apt to be fertile than the mule. The superficial characteristics and voice are in each instance those of the father, the stature, strength and symmetry those of the mother.

Among plants the reverse hybridizing of the walnut gives another instance of contrasts. The California walnut (*Juglans californica*) male and the black walnut (*J. nigra*) female when propagated give trees which increase in size twice as fast as the combined growth of both parents, with clean-cut, glossy, bright green leaves from two to three feet long, an odor like apples, but no nuts. The hybrid is therefore sterile. From the male black walnut and the female California tree, however, there is produced a tree with larger nuts of a quality superior to that of either parent.

Mechanical Limits

The second limit to artificial selection is a mechanical one, for just as architecture and especially bridge construction have their limitations fixed by the strength of the materials at the disposal of the builder, so a limit may be reached in nature. Speed among animals requires among other characteristics long and slender lower segments to the limbs, which reach their highest perfection in animals of the size of the average African antelope. With a larger creature, such as a horse, the impacts and strains to which the limbs are subjected are increased, not only because of the greater force required to move a larger animal, but very rapidly with each added increment of speed. The modern race horse with a speed of over thirty miles an hour has just about reached the limit of strength on the part of bone and muscle, and a marked increase on the part of future animals is not to be looked for without proportionate risk of frequent injure to the horse. There are known cases in which a positive limit has been reached in attempting to modify organs through selection alone.

Variations are nature's contribution to artificial selection and are the basis of all man's work, for he can create nothing, but the variations which true provides, and there are definite limits to these variations.

Chapter—8

Orthogenesis and Kinetogenesis

Of the several hypotheses to account for the origin of species which are opposed or supplemental to the Darwinian factor of natural selection, two stand our sharply as of great importance, if they are finally found to be true. This proof can only be established by impartially conducted observations and experiments extending through years to come. At present the mass of evidence which can be offered in favour of each is to some extent nullified by other well-attested facts, and so the matter stands *sub judice*. Of these hypotheses that of orthogenesis stands first, although it is not an explanation of evolution but a statement of possible fact.

Orthogenesis

Orthogenesis (Gr. 'ορθος, straight, and γενσις, production) is the assumption that variations and hence evolutionary change occur along certain definite lines impelled by laws of which we know not the cause. Thomson says of it: "We must be on our guard, however, against the possible fallacy of concluding from the apparent orthogenesis in fossilized and surviving stages that there was no zigzagness and pruning in the process. Types may have had their waywardness gradually sifted out of them."

Orthogenesis has been proposed in explanation of several phenomena for which natural selection seemingly cannot account, such as:

1. The beginnings of advantageous modifications by the selection of individual variations occurring in every direction of change, for it is evident that unless a character appeared at once in a degree of development to give it "selection value" it would be difficult to account for its beginning. But the teaching of the Neo-Darwinians postulates the selection of minute random variations as well as such salutations as the idea just stated would imply.
2. Disadvantageous structures or degrees of development, such as structures which have apparently evolved along fixed and seemingly not advantageous lines, and the development of parts beyond the point of greatest usefulness,

even so far as to lead to individual death or racial extinction. Paleontology reveals to us repeated instances of the evolution of characters which lead onto extinction, such as the huge, unwieldy size of the sauropod dinosaurs; the growth of excrescences, observed in the armored and horned dinosaurs, and, later in time, in the Irish deer (*Cervus megaceros*); the uncoiling of the shell-bearing cephalopods (ammonites and nautilids): all of which are examples of development along disadvantageous lines or to disadvantageous degrees. When these are observed in a race a wholesale thinning of the ranks, if not final extinction, may be confidently predicted. Natural selection implies racial improvement, and the development of advantageous, not deterimental, structures; hence the theory of orthogenesis has been proposed to account for the latter as well as the former.

3. Degeneracy and the loss of structures. As we have seen, the reversal of natural selection would account for the elimination of a character formerly useful but which, in altered circumstances, has become a positive menace to survival. On the other hand, if the structure becomes merely valueless, cessation of selection would account for its diminution but not for its utter elimination. Orthogenesis has been offered as a supplementary explanation here.

EXPLANATIONS OF ORTHOGENESIS

The two principal explanations of orthogenesis, those of Naegeli and Eimer, have been stated by Newman:

> "Carl von Naegeli's ideas . . . involve a belief in a sort of mystical principle of progressive development, a something, quite intangible, that exists in organic nature, which causes each organism to strive for or at least make for specialization or perfect adaptation. . . . Naegeli believed that animals and plants would develop essentially as they have without any struggle for existence or natural selection. Hence this form of the orthogenesis theory is . . . alternative to natural selection."

Theodore Eimer's explanation is more scientific and less mystical than Naegeli's. "He believed that lines of evolution were not miscellaneous and haphazard but were confined to a few definite directions, determined at their initial stages not by natural selection but by the laws of organic growth, aided by the inheritance of acquired characters. A new character makes its beginning as would the first step in a slow chemical change, or series of such changes, and it must go through to a fixed end, under given conditions, just as surely as does the chemical process. Only when a given character or line of evolution results in the production of a very positive advantage or disadvantage to the species does natural selection step in to interfere with orthogeneisis. The causes of orthogenesis are said 'to be in the effects of external influences, climate, nutrition or the given constitution of an organism.' Natural selection is therefore subordinate but not dismissed."

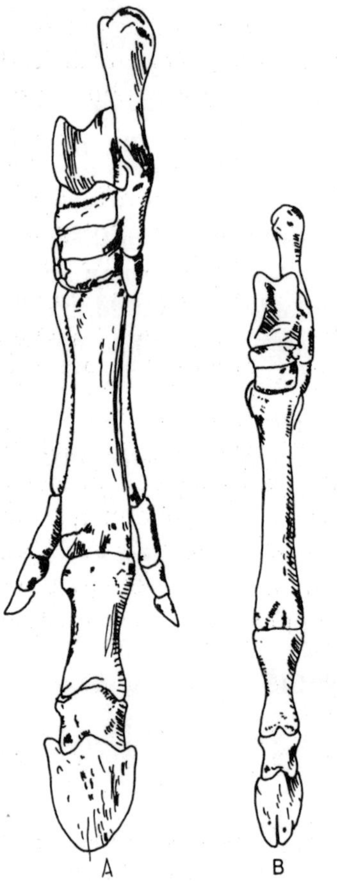

Fig. 8.1: Hind feet of pseudohorses. A, three-toed *Diadiaphorus,* Miocene of South America; B, one-toed *Thoatherium,* Miocene of South America. One-half natural size.

Orthogenesis Contrasted with Ortho-Selection

At first sight there seem to be no real distinction between orthogenesis and descent controlled by selection, as the latter must also lead to evolution along certain definite lines; in fact, it can produce no other kind of evolution. Where modification is the result of the elimination, through the selective struggle, of all other lines of variation, it may be called ortho-selection,. In orthogenesis, however, the lines of variation are presumed to be predetermined; hence the resultant lines of evolution are also.

POSITIVE EVIDENCES FOR ORTHOGENESIS

1. Parallelisms

Analogous or parallel variations are of frequent observation. They are the modifications of similar character which appear indifferent branches of the same large group or in unreleased groups, although in the latter case they are often called convergences. Comparative anatomy has revealed many examples to show that modification in a definite or determinate direction may be seen in all the sub-groups of a large family, although appearing in varying degree in different species. Examples are the total reduction of the side toes among the artiodactyls which has occurred in several unrelated genera, such as the giraffe, camel, and prong-buck (*Antilocapra*). A still more remarkable instance would be the parallelism which existed between the pseudo-horses of South America and the true horses of the northern hemisphere, both lines showing a three-and finally a one-toed condition owing to the progressive shortening of the lateral digits and proportional strengthening of the middle one. In this instance the relationship is quite remote, although both races belong to the ungulate group.

2. Over-specializations

Over-specializations are the numerous "excessive structures" which are developed far beyond the limits of usefulness. Such, for example, are the tusks of the wild hog (*Babirussa alfurus*), the huge horns of the "big horn" sheep, or the enormously elongated and slender neck of several weevil- or snout-beetles like *Apoderus tenuissimus*. One of the dolphins, *Mesoplodon*, has a mouth which can be opened but a little way when the animal is full grown, because a lower tooth grows around the upper jaw on either side. Such excessive structures, which have probably contributed to the extinction of many former species, include among others the tusks of the Jefferson mammoth and the antlers of the Irish deer.

3. Constitutional Limitations on Variation

The constitution, or sactual chemical composition of the body, permits, as a rule, changes only in a few directions. The breeder of animals or plants may not always produce any desired form or color. "No one has yet succeeded in producing a blue *Maiblume,* a grass with divided leaves, a hen with a parrot's beak." And we can be confident that a notochord can never appear in a beetle. The fact that an animal belongs to a given group renders the possibilities of variation distinctly, sometimes very narrowly, limited. "In this connection it should be mentioned that some biologists have seen in this restriction of the range of variation which inveitably accompanies specialization an important factor in the determination of lines of descent.

4. The correlations which bind each organ to others also restrict the range of variation.

5. Facts from Paleontology Supporting Orthogenesis

Many facts of paleontology seem to prove the existence of orthogenetic evolution. Wherever sufficient material permits the working out of a phyletic series, we always see a comparatively limited number of lines of development, which, except for occasional lateral branches, run essentially in straight lines, in gradational steps. Such for example are seen in the fossil horses.

The cusps of teeth in the mammals afford interesting instances of orthogenesis, as they emerge complete and do not depend upon use for their production; in fact, use destroys rather than perfects them, yet they appear independently but always in an adaptive position in many orders of mammals. From observing the initial trend the ultimate tooth pattern may be often accurately predicted.

Osborn's exhaustive study of the titanotheres, a group of huge Tertiary perissodactyls, has led to many important generalizations. He says:

> "Every known step of this transformation [of the titanotheres] is determinate and definite, every additional character which has been observed arises according to a fixed law and not according to any principle of chance.
>
> "In *eleven* principal branches which radiate from the earliest known forms (*Eotitanops gregoryi*) of this family, exactly similar new characters arise quite independently at different periods of geologic time which are separated by lapses of tens of thousands of years.
>
> "In certain branches the horns appear thousands of years later than in others and after their appearance may exhibit a singular inertia, or lack of momentum, over a long period of time. Every character has its own rate of velocity both in individual development and in racial development [evolution]."

"The phyletic series . . . of recent species also show, where we are able to trace them, distinctive single lines of development" (Kellogg).

KINETOGENESIS

Cope's theory of kinetogenesis or "mechanical genesis" has been accepted so widely, especially by paleontologists and pathologists, that it should be studied in some detail, with the vital objections which have been raised against it.

The word *kinetogenesis* (Gr. *κινητος* movable, and *γενεσις* production) is thus defined: "The doctrine or hypothesis that animal structures have been produced directly or indirectly, by animal movements." Cope's arguments, especially as applied to the vertebrates, are based upon the following: The vertebrates present two distinct advantages. First, we have a more complete paleontologic series. Second, we have the best opportunity for observation and experiment on their growth processes, since we ourselves and our companions of the domesticated animals belong to this branch of the animal kingdom.

MECHANICS OF THE VERTEBRATE SKELETON

Abnormal Articulations

Many of the data for experimental work are found in the records of surgery, and from this source inferences have been drawn as to the construction of normal articulations. Experiments have shown that in the case of unreduced dislocations or permanently bent joints certain predictable changes will always occur. That portion of the articular cartilage which fails to meet its fellow either throughout because of a limitation of its movement, or through dislocation so that it only comes in contact with the soft parts, is permanently destroyed and, under certain conditions, a new articulation may be formed where a permanent displacement of the joint has occurred.

The great plasticity of bone was the basis for the wonderful bloodless operation for the reduction of congenital hip dislocation invented by the famous Viennese surgeon, Doctor Lorenz, and demonstrated in America some years ago. In this operation, the dislocated thigh-bone itself was made the instrument, pulled from the abnormal socket which had been formed, and the head, being placed in the partially occluded normal socket, was forcibly ground into it. This always results in the destruction of bony tissue in the motionless bone. The femur was then bound firmly in place until, through the resorption of unnecessary material and the rebuilding of bone where it was needed, the socket was restored to its original form, and then induced movement and massaging completed the cure. The operation was successful in a high percentage of cases where performed upon children of tender years.

Even the pressure of soft parts can cause the absorption of bone, as shown by the interior cavity of the skull, which in many instances retain a faithful impression, not only of the general proportions of the brain, but even of its convolutions and of the blood-vessels in its membranous coverings.

Normal Articulations

Cope uses for illustration particularly the strong complex ankle-joint of the ruminating animals (ox, deer, camel, etc.) and the horse, in which the motion, while wide in its force-and-aft range, is restricted to one plane of space, resulting in a treble tongue-and-groove joint which resists dislocation although it may be broken by force. The articulation lies between the upper bone of the ankle, the astragalus, and the shin-bone or tibia, the former, which is convex antero-posteriorly, having two keels which fit into corresponding grooves in the concave end of the latter. The lower end of the tibia possesses a single keel which corresponds to a groove between the two astragalar keels. The formation of this joint has been thus explained:

> "In all bones the external walls are composed for dense material, while the centers are spongy and comparatively soft. The first bone of the foot (astragalus) is narrower, from side to side, than the tibia which rests upon it. Hence the edges of the dense side-walls of the astragalus fall within the edges of the dense side-walls of the tibia, and they have pressed into the more yielding material that forms the end of the bone, and causing bone absorption, pushed it upward, thus allowing the side-walls of the tibia to embrace the side-walls of the astragalus...

"The same active cause that produced the two grooves of the lower end of the leg produced the groove of the middle of the upper end of the astragalus. Here we have the yielding lower end of the tibia resting on the equally spongy material of the middle of the astragalus. There is here no question of the hard material cutting into soft, but simply the result of continuous concussion. The consequence of conclusion would be to cause the yielding faces of the bones to bend downward in the direction of gravity, or to remain in their primitive position while the edges of the astragalus were pushed into the tibia. If they were flat at first they would begin to hollow downward, and a tongue above and a groove below would be the result. . . . This inclusion of th astragalus in the tibia does not occur in the reptiles, but appears first in the Mammalia, which descended from them. . . . Every line of Mammalia commenced with types with an astragalus which is flat in the transverse direction, or without median groove. From early Tertiary times to the present day, we can trace the gradual development of this groove in all the lines which have acquired it. The upper surface became at first a little concave; the concavity gradually became deeper, and finally formed a well marked groove" (Cope).

The history of the wrist joint is similar, and in many instances the articulation between the bones of the palm (metacarpals) and sole (metatarsals) with the respective digits as well. This is especially true of forms which, like the speedier ungulates, have elevated the heel and wrist off the ground and walk upon the modified claws, the hoofs. Where the foot is supported by a yielding pad, as in the camels, the keels at the ends of the metapodials are incompletely developed, as their purpose, where present, is to keep the toes from spreading, and a yielding foot is a necessary part of the desert adaptation which the camels so admirably illustrate.

Vertebral Column

The vertebral column among vertebrates, especially among terrestrial types, is a marvel of mechanical designs. Aquatic adaptation relieves the backbone of the creature's weight, and hence the vertebrae tend to retain or re-acquire primitive simplicity of structure. The articulations of the vertebrae one with another are effected by the faces of the adjacent centra or by the additional articular facets, borne on the neural arch, known as zygapophyses. The reptiles exhibit the greatest variety of articulations, except those of the zygapophyses which are fairly uniform, while in the mammals the modification of each group of articulations is equally striking.

The forms which the articulations of the centra assume are four in number: first, the amphicoelous (Gr. *ἀμφι* at both ends, and *κοῖλος*, hollow, said of vertebrae in which both ends of the centrum are concave); the ball-and-socket, which may be procoelous (Gr. *πρό*, before, concave in front) or opisthocoelous (Gr. *ὄπισθεν*, behind, concave behind); the plane or amphiplatyan; and the saddle-shaped, in which the same face of a given vertebra is at once concave in one dimension and convex in the other. Of these various sorts the first is chiefly distinctive of fishes and certain Reptilia (ichthyosaurs), while among mammals it is imperfectly developed, being only a modification of the plane surface and usually occurring in comparatively few vertebrae.

The ball-and-socket type is mainly found in the neck of long-necked reptiles (dinosaurs), in the crocodiles, and in the mammals, where it permits the maximum degree of flexibility. Those mammals (perissodactyls and artiodactyls) in which the ball-and-socket articulation is found in the neck also show it, although in reduced degree, in the vertebrae of the loin, while the thoracic vertebrae exhibit a tendency in the same direction. The saddle-shaped articulation, while characteristic of the neck vertebrae of birds, is present in the mammals only in certain genera of monkeys. The majority of Mammalia have plane articulations on all of the vertebrae. In those forms in which movement of the vertebrae upon one another has become impossible, the centra coossify or fuse together, as in the region known as the sacrum, where a variable number of vertebrae, depending upon the length of the ilia or hip-bones, unite to form a firm structure which is solidly articulated with the pelvic girdle. Among birds this coossification is apt to extend still further, in some instances including practically the entire trunk, while in the whales, the form of whose body renders any independent movement of the head impossible, the neck is much shortened and the vertebrae, especially in the whale-bone whales, unite into a solid mass of bone.

There is, in all of this, abundant evidence of the effects of use and disuse, the ball-and-socket joints being developed where the greatest all-round flexibility is characteristic. This is therefore the prevailing type of articulation among Reptilia, the degree of its development being in direct proportion to the weakness of the limbs, for in the large and long-limbed terrestrial dinosaurs the articulations of the trunk and to a less extent of the tail vertebrae tend to become plane. In the mammals it is best developed in the most flexible regions, the neck and loin. The saddle-shaped articulation also permits considerable flexibility, but mainly in the vertical and horizontal planes. The fact is that in the ancestral whales the neck was considerably longer than in their modern descendants and had its centra distinct, the shortening and final coalescence of the center arose with the gradually increasing powers of locomotion through the water, which would enable the creature to overtake and capture its prey without the necessity of using a long, darting neck to seize it in the pursuit. The contrast is strikingly illustrated by the whale-like ichthyosaurs among reptiles on the one hand and the plesosaurs on the other. In the former as in the whales the tail became the principal organ of locomotion with, in all probability, a corresponding increase of speed, whereas in the pleisosaurs the more laborious method of propulsion by the paddle-like limbs made the speed of the creature as a whole considerably less and necessitated a proportionately longer and more flexible neck analogous to that of the fish-eating alligator snapping turtle (*Macroclemmys temmincki*). Of this vicious beast Agassiz says: "It does not withdraw its head and limbs on the approach of danger, but resorts to more active defence. It raises itself upon the legs and tail, highest behind, open the mouth widely, and throwing out the head quickly as far as the long neck will allow, snaps the jaws forcibly upon the assailant, at the same time throwing the body forward so powerfully as often to come down to the ground when it has missed its object."

The wonderful complication of the axial mechanism is shown in its highest perfection in certain of the dinosaurs, for never before nor since has nature produced such mighty animals unsupported by an external sustaining medium. In one of the most remarkable of these forms, *Diplodocus*, we have an animal of relatively short body borne on massive column-like limbs and with an extremely long neck and tail, the former of which was evidently a very mobile and self-sustaining organ, while the latter, though capable of considerable movement and self-support, probably was either largely water-borne—for the creature was at least semi-aquatic in its habits—or may have dragged on the ground when the animal came ashore. The entire fabric of the vertebral column is a marvel of lightness and ingenuity of design. The great mobility of the neck is indicated by the highly developed (opisthocoelous) ball-and-socket central articulations but especially by the extreme lightness of the centra themselves, which are pierced by deep lateral cavities leaving a median dividing wall so thin as to be readily broken through. This mobility is also shown by the complexity of the indicated musculature, for the points of muscle attachment are well developed and numerous keels and buttresses running obliquely in both directions across the centra and neural arches show the lines of stress not of few massive muscles but of numerous smaller muscles and tendons. Moreover, the neural spines, which are usually single, are here deeply cleft from the third cervical back to the sixth dorsal, indicating the pairing of the great muscles which run along the mid-dorsal line of the neck and back, and the independent action of the two members of the pair. This of course is indicative of a wide lateral sweeping of the neck and head.

In the dorsal region the faces of the centra flatten, indicating little flexibility unless the centra were separated by thick compressible pads of cartilage, a supposition which the articulation of the zygapophyses does not bear out. Here, except for the development of deep lateral cavities (pleurocoeles) in the centra, the vertebrae are relatively simple, indicating a similar simplicity of the musculature. The sacrum is a massive structure consisting of three closely coalesced vertebrae united not only by their centra but by the neural arches and even the dorsal spines, and a fourth vertebra the spine of which is free, although the centrum is well coossified with the others. This last vertebrae is interpreted as a caudal which age and lack of mobility have caused to unite with the vertebrae in front. The sacrum is the fulcrum of the whole wonderful lever, and the coalesced spines afford a firm anchorage for the long muscles and tendson which run forward toward the neck and backward to the tail—the tension members of the fabric. The sacrum is very solidly fastened to the hip bones or ilia by massive processes which extend outward and backward and unite distally into a heavy roughened plate forming the abutment against which the ilium bears. The hip socket is large and thoroughly braced by this bony plate and the bone (peduncle) extends downward in front of the socket in such a way as to meet the thrust of the thigh in ordinary standing posture, in walking when the hind limb is used to urge the animal forward, and also if the creature reared on its hind legs, evidently a very feasible act when the body was partially water-borne. The obliquity of the transverse processes which has already been described is such as to meet this unusual strain.

The tail lacked the great flexibility of the neck, but must have been capable of some laterla movement as well a as certain amount of elevation. The centra are again lightened but are by no means as complicated as are those of the neck. The transverse processes, especially on the anterior vertebrae, are widely expanded plates of bone, indicating powerfully developed lateral muscles. The tail must have had three uses: for swimming, somewhat as a serpent swims, a movement which requires a certain flexibility but does not perhaps necessitate the extreme range of the food-getting head and neck; for defense, for the only visible weapon of which the creature stood possessed was some ten feet of slender whiplash-like terminus to its tail, comparable to that seen in many modern lizards and serving as a very efficient flagellant for the punishment of would-be offenders against the owner's person; and for temporary support when the animal reared the fore quarters aloft. That this last need may have occasionally arisen is still further attested by the occasional abnormal coalescence of two of the caudals as in the great specimen in the Carnegie Museum at Pittsburgh. These vertebrae, the estimated seventeenth and eighteenth, are at about the point where the tail would meet the ground in such a manoeuver. To gain a full appreciation of all this, it should be borne in mind that *Diplodocus* was one of the giants of geologic time, stretching its mighty length through a span of eighty feet or more, of which some sixty-five lay without the pillar-like supports, and therefore, except for the resting of the tail upon the ground or the support given by the water when submerged, the neck or tail must each have been capable of being sustained by a single end.

There is a wonderful freedom of design in the construction of the individual vertebrae comparable to that seen in Gothic architecture, for not only is each vertebra different from its fellows but even the two sides of the same bone may be unlike. The whole structure of the *Diplodocus* skeleton is so perfect a response to the multitudinous stresses to which its various elements have been subjected that to one who can appreciate the design of a bridge or building or any other sustentative fabric the conclusion that it is the result of mechanical genesis is almost irresistible.

Tendons which have the minute structure of bone are often preserved in dinosaurs, notably in the iguanodons of Europe and their American relatives, camptosaurs and hadrosaurs and here the lines of tension are beautifully indicated just as the keels and buttresses of the vertebrae give evidence of lines of compression.

Limb Proportions

Limb proportions also follow definite mechanical laws which at first seem curiously contradictory, for both impact (compression force) and strain (extension force), while opposite in their action, have the same effect upon bone, that of causing it to elongate in the line of the stress. Transverse stress, on the other hand, would cause growth at right angles with the length of the bone.

Speed adaptation generally results in the elongation of the two lower segments of fore and hind limb and the relative shortening of the upper, while relatively slow progression as in the elephant gives rise to a lengthened proximal segment, that nearest the body, the distal one remaining relatively short. With the apes, such as the gibbon, in which most of the progress from branch to branch and from tree to tree is sustained by the arms, the latter are enormously elongated as compared with the legs, which are of relatively less use and are proportionately very short, so that when the ape stands with the body practically erect the knuckles still touch the ground. A regular gradation has been shown by Huxley to exist between these extremes and that in terrestrial mankind, in which the legs are the longer, the orang, chimpanzee and gorilla representing the intermediate stages in the order named, the last standing nearest to man. If Man's legs are longer than his arms as a result of their greater use, their growth stimulus was *impact,* whereas with the gibbon's arms which are the longer the stimulus was a *strain*. The tree sloths also have greatly elongated proximal segments to the limbs, the hands and feet being reduced to curious structures terminating in immense hook-like claws. The huge pleistocene ground sloths, on the contrary, although retaining enough arboreal characteristics to point to a partial tree-dwelling ancestry, had comparatively short and ponderous limbs, but the relative lengths of limb-segments were retained.

Aarchelon, the giant turtle from the Cretaceous of South Dakota, mounted in the Yale Peabody Museum, had suffered mutilation during life, for the right hind for is missing and the condition of the lower leg bones, which lack their distal end, is pathologic. The most interesting feature, however, is the difference in size of the two thigh bones, that of the perfect left limb being materially larger than that of the crippled right, showing the result of the cessation of growth-stimulus with the loss of utility. But this is an ontogenetic instance and could not, so far as our knowledge goes, become evolutional; the other instances which are constant and predictable are the result of evolutionary processes, whatever they may be. Cope and others have taken many other instances, such as the modification of form into radial symmetry for sedentary types and bilateral for the locomotor, the mechanics of the teeth, of muscular development, of the shells of invertebrates; but enough have been given to show that numerous modifications of animals *conform* with mechanical laws whether mechanics is the prime mover in their production or not.

OBJECTIONS TO KINETOGENESIS

Several arguments have been offered by the opponents of this theory of which the most important are: First, the apparently illogical and self-contradictory assumption that stimuli of different kinds produce similar results, while stimuli of the same kind may produce different results. Experiment, however, has proved the truth of this apparently paradoxical statement, for the irritation of bone will produce either bone deposits or bone absorption according to the degree of irritation. Thus moderate stimulus, such as the pressure and stretching mentioned above, may stimulate growth. Continued heavy pressure, on the other hand, causes bone absorption at the point of contact.

A second objection which has been made to kinetogenesis is that if growth-stimulus exist, how can there be a limit to increase, so long as the stimulus of use prevails? This objection is met by the assumption that the stimulus is stress due to disharmony between an organism and its environment, and that kinetogenesis is the result of the effort on the part of the organism to overcome this lack of harmony. When the organism is sufficiently adjusted to meet the requirements of the environment, equibrium is attained, the stress is reduced to the point necessary for the maintenance of the mechanism in working condition, and further growth ceases. In easy circumstances, where little or no exertion is necessary, there is not even sufficient stimulus to raise the mechanism to a state of efficiency, and the degeneracy of disuse results.

Despite Cope's arguments, this is one of the objections to his theory which is most emphasized by its opponents. It will at once be seen that even if the objections above are met, a third will yet remain, that of the acceptance of the Lamarckian factor of the inheritance of acquired or ontogenetic modifications, upon which the whole doctrine of kinetogenesis seems to depend for its inclusion among the potent factors of evolution.

Osborn's Theory of Coincident Selection

This objection Osborn has striven to avoid by his idea of "coincident selection," which he states as follows: "Individual or acquired modifications in new circumstances are an important feature of the adult structure of every animal. Some congenital variations may coincide with such modifications, others may not. The gradual selection of those which coincide (coincident variations) may constitute an apparent inheritance of acquired modifications."

Although these may occur they would hardly seem sufficient to account for the host of mechanical adaptations which exist. Many of these may be ontogenetic, recurring in successive generations through the influence on each individual of similar conditions with resultant similar adaptations. An illustration would be the tanning of the skin upon exposure to the sun in an otherwise fair individual. This would be repeated in each successive generation as an ontogenetic modification. Ultimately a germinal variation in the direction of greater swarthiness might occur. As this is a distinctly advantageous variation in those exposed to the harmful rays of the sun, Natural Selection would at once intervene to perpetuate it. Thus the ontogenetic modification, while not in itself heritable, would serve as a safeguard until the germinal variation coinciding with it could be established.

Chapter—9

The Mutation Theory

It will be recalled that Darwin, although depending upon the ever-present fluctuating variations as the material for natural selection to work upon, recognized the occasional occurrence of "sports" or "saltatory variations." These, however, seemed to him to be so rare in nature as to offer no adequate basis for selection. During the latter part of the nineteenth century several investigators, feeling the inadequacy of fluctuating variations to produce qualitatively new characters, decided to make a more careful examination of animals and plants in nature in order to discover whether saltatory variations might not be of more frequent occurrence than Darwin had supposed.

In England William Bateson collected a large number of instances of a type of variation which he called *discontinuous* in contradistinction to the *continuous* type which we have been calling fluctuations. Such variations, instead of being in a closely graded series with the typical variations of a species, were frequently quite sharply different from the majority. Although no experiments were conducted in order to test the hereditability of these "discontinuous variations," it is probable that some of them were "mutations" in the sense of De Vries.

At about the same time Hugo De Vries in Holland, partially as the result of his rediscovery of Mendel's work and his confirmation of the latter's laws of heredity, became convinced that new species arise not by the accumulation, through natural selection, of minute fluctuating variations, but by the sudden appearance in one generation of fully formed new elementary species. He began a systematic research for species of plants in nature that were giving rise to new species. Many species were examined in their natural surroundings and were then brought into the experimental garden for more careful observation, but for a long time the search for a species throwing off new elementary species was unsuccessful. Finally, however, in a field near Hilversum, in the vicinity of Amsterdam, he found what seemed to him to be just the kind of plant he had been looking for in the evening primrose (*Oenothera lamarckiana*).

"Lamarck's evening-primrose" says De Vries, "is a stately plant, with a stout stem, attaining often a height of 1.6 meters and more. When not crowded the main stem is surrounded by a large circle of smaller branches, growing upwards from its base so as to form a dense brush. . . . The flowers are large and bright yellow attracting immediate attention even from a distance. They open toward evening, as the name indicates, and are pollinated by bumble-bees and moths."

On account of the classic character of De Vries's mutants of *Oenothera lamarckiana* we shall follow his own detailed description of the more significant of these.

Fig. 9.1: *Ocenothera lamarckiana;* the original type used by De Vries in his experiments. This is the stock from Hilversum, from which arose in successive generations a series of mutants.

NEW SPECIES (MUTANTS) OF *OENOTHERA*
HUGO DE VRIES

This striking species (*Oenothera lamarckiana*) was found in a locality near Hilversum, in the vicinity of Amsterdam, where it grew in some thousands of individuals. Ordinarily biennial, it produces rosetles in the first, and stems in the second year. Both the stems and the rosettes were at once seen to be highly variable, and soon distinct varieties could be distinguished among them.

The first discovery of this locality was made in 1886. Afterwards I visited it many times, often weekly or even daily during the first few years, and always at least once a year up to the present time. This stately plant showed the long-sought peculiarity of producing a number of new species every year. Some of them were observed directly on the field, either as stems or as rosettes. The latter could be transplanted into my garden for further observation, and the stems yielded seeds to be sown under like control. Others were too weak to live a sufficiently long time in the field. They were discovered by sowing seed from indifferent plants of the wild locality in the garden. A third and last method of getting still more new species from the original strain was the repetition of the sowing process, by saving and sowing the seed which ripened on the introduced plants. These various methods have led to the discovery of over a dozen new types never previously observed or described.

Leaving the physiological side of the relations of these new forms for the next lecture, it would be profitable to give a short description of the several novelties. To this end they may be combined under five different heads, according to their systematic value. The first head includes those which are evidently to be considered as varieties, in the narrower sense of the word, as previously given. The second and third heads indicate the real progressive elementary species, first those which are as strong as the parent-species, and secondly a group of weaker types, apparently not destined to be successful. Under the fourth head I shall include some inconstant forms, and under the last head those that are organically incomplete.

Of varieties with a negative attribute, or real retrograde varieties I have found three, all of them in a flowering condition in the field I have given them the names of *laevifolia, brevistylis* and *nannella.*

The *laevifolia,* or smooth-leaved variety, was one of the very first deviating types found in the original field. This was in the summer of 1887, seventeen years ago. It formed a little group of plants growing at some distance from the main body, in the same field. I found some rosettes and some flowering stems and sowed some seed in the fall. The variety has been quite constant in the field, neither increasing in number of individual plants nor changing its place, though now closely surrounded by other *lamarckians.* In my garden it has proved to be constant from sed, never reverting to the original *lamarckiana,* provided intercrossing was excluded.

It is chiefly distinguished from Lamarck's evening-primrose by its smooth leaves, as the name indicates. The leaves of the original from show numerous sinuosities in their blades, not at the edge, but anywhere between the veins. The blade shows numbers of convexities on either surface, the whole surface being undulated in this manner; it lacks also the brightness of the ordinary evening-primrose or *Oenothera biennis*.

These undulations are lacking or at least very rare on the leaves of the new *laevifolia*. Ordinarily they are wholly wanting, but at times single leaves with slight manifestations of this character may make their appearance. They warn us that the capacity for such sinuosities is not wholly lost, but only lies dormant in the new variety. It is reduced to a latent state, exactly as are the apparently lost characters of so many ordinary horticultural varieties.

Lacking the undulations, the *laevifolia*-leaves are smooth and bright. They are a little narrower and more slender than those of the *lamarckiana*. The convexities and concavities of leaves are a useful character in dry seasons, but during wet summers, such as those of the last few years, they must be considered as very harmful, as they retain some of the water which falls on the plants, prolonging the action of the water on the leaves. This is considered by some writers to be of some utility after slight showers, but was observed to be a source of weakness during wet weather in my garden, preventing the leaves from drying. Whether the *laevifolia* would do better under such circumstances, I have, however, omitted to test.

The flowers of the *laevifolia* are also in a slight degree different from those of *lamarckiana*. The yellow color is paler and the petals are smoother. Later, in the fall, on the weaker side branches these differences increase. The *laevifolia* petals become smaller and are devoid of the emeargination at the apex, becoming ovate instead of obcordate. This shape is often the most easily recognized and most striking mark of the variety. In respect to the reproductive organs, the fertility and abundance of good seed, the *laevifolia* is by no means inferior or superior to the original species.

O. brevistylis, or the short-styled evening-primrose, is the most curious of all my new forms. It has very short styles, which bring the stigmas only up to the throat of the calyx-tube, instead of upwards of the anthers. The stigmas themselves are of another shape, more flattened and not cylindrical. The pollen falls from the anthers abundantly on them, and germinates in the ordinary manner.

The ovary which in *lamarckiana* and in all other new forms is wholly underneath the calyx-tube, is here only partillay so. This tube is inserted at some distances under its summit. The insertion divides the ovary into two parts: an upper and a lower one. The upper part is much reduced in breadth and somewhat attenuated, simulating a prolongation of the base of the style. The lower part is also reduced, but in another manner. At the time of flowering it is like the ovary of *lamarckiana,* neither smaller nor larger. But it is only reached by very few pollen-tubes, and is therefore always very incompletely fertilized. It does not fall off after the fading away of the flower, as unfertilized ovaries usually do; neither does it grow out, nor assume the upright position of normal capsules. It is checked in its development, and at the time of ripening it is nearly of the same length as in the beginning. Many of them contain no good seeds at all; from others I have succeeded in saving only a hundred seeds from thousands of capsules.

These seeds, if purely pollinated, and with the exclusion of the visits of insects, reproduce the variety entirely and without any reversion to the *lamarckiana* type.

Correlated with the detailed structures is the form of the flower-buds. They lack the high stigma placed above the anthers, which in the *lamarckiana*, by the vigorous growth of the style, extends the calyx and renders the flower-bud thinner and more slender. Those of the *brevistylis* are therefore broader and more swollen. It is quite easy to distinguish the individuals by this striking character alone, although it differs from the parent in other particulars.

The leaves of the *O. brevistylis* are more rounded at the tip, but the difference is only pronounced at times, slightly in the adult rosettes, but more clearly on the growing summits of the stems and branches. By this character the plants may be discerned among the others some weeks before the flowers begin to show themselves.

But the character by which the plants may be most easily recognized from a distance in the field is the failure of the fruit. They were found nearly every year in varying, but always small numbers.

Leaving the short-styled primrose, we come now to the last of our group of retrograde varieties. This is the *O. nannella*, or the dwarf, and is a most attractive little plant. It is very short of stature, reaching often a height of only 20-30 cm., or less than one-fourth of that of the parent. It commences flowering at a height of 10-15 cm., while the parent-form often measures nearly a meter at this stage of its development. Being so very dwarfed the large flowers are all the more striking. They are hardly inferior to those of the *lamarckiana*, and agree with them in structure. When they fade away the spike is rapidly lengthened, and often becomes much longer than the lower or vegetative part of the stem.

The dwarfs are one of the most common mutations in my garden, and were observed in the native locality and also grown from seeds saved there. Once produced they are absolutely constant. I have tried many thousands of seeds from various dwarf mutants, and never observed any trace of reversion to the *lamarckiana* type. I have also cultivated them in successive generations with the same result. In a former lecture we have seen that contrary to the general run of horticultural belief, varieties are as constant as the best species, if kept free from hybrid admixtures. This is a general rule, and the exceptions, or cases of atavism, are extremely are. In this respect it is of great interest to observe that this constancy is not an acquired quality, but is to be considered as innate, because it is already fully developed at the very moment when the original mutation takes place.

From its first leaves to the rosette period, and through this to the lengthening of the stem, the dwarfs are easily distinguished from any other of their congeners. The most remarkable feature is the shape of the leaves. They are broader and shorter, and especially at the base are broadened in such a way as to become apparently sessile. The stalk is very brittle, and any rough treatment may cause the leaves to break off. The young seedlings are recognizable by the shape of the first two or three leaves, and when more of them are produced, the rosettes become dense, and strikingly different from others. Later leaves are more nearly like the parent-type, but the petioles remain short. The bases of the blades are frequently almost cordate, the laminae themselves varying from oblong-ovate to ovate in outline.

The stems are often quite unbranched, or branched only at the base of the spike. Strong secondary stems are a striking attribute of the *lamarckiana* parent, but they are lacking, or almost so in the dwarfs. The stem is straight and short, and this, combined with the large crown of bright flowers, makes the dwarfs eminently suitable for bed or border plants. Unfortunately they are very sensitive, especially to wet weather.

Oenothera gigas and *O. rubrinervis*, or the giant, and the red-veined evening primroses, are the names given to two robuts and stout species, which seem to be equal in vigor to the parent-plant, while diverging from it in striking characters. Both are true elementary species, differentiated from *lamarckaiana* in nearly all their organs and qualities, but not showing any preponderating character of a retrograde nature. Their differences may be compared with those of the elementary species of other genera, as for instance, of *Draba*, or of violets, as will be seen by their description.

The giant evening-primrose, though not taller in stature than *O. lamarckiana*, deserves its name because it is so much stouter in all respects. The stems are robuts, often with twice the diameter of *lamarckaina* throughout. The internodes are shorter, and the leaves more numerous, covering the stems with a denser foliage. This shortness of internodes extends itself to the spike, and for this reason the flowers and fruits grow closer together than on the parent-plant. Hence the crown of bright flowers, opening each evening, is more dense and more strikingly brilliant, so much the more so as the individual flowers are markedly larger than those of the parents. In connection with these characters, the flower-buds are seen to be much stouter than those of *lamarckiana*. The fruits attain only half the normal size, but are broader and contain fewer, but larger seeds.

The *rubrinervis* is in many respects a counterpart to the *gigas*, but its stature is more slender. The spikes and flowers are those of the *lamarckiana*, but the bracts are narrower. Red veins and red streaks on the fruits afford a striking differentiating mark, though they are not absolutely lacking in the parent-species. A red hue may be seen on the calyx, and even the yellow color of the petals is somewhat deepened in the same way. Young plants are often marked by the pale red tinge of the mid-veins, but in adult rosettes, or from lack of sunshine, this hue is often very faint.

The leaves are narrow, and a curious feature of this species is the great brittleness of the leaves and stems, especially on annual individuals, for example, on those that make their stem and flowers in the first year. High turgidity and weak development of the mechanical and supporting tissues are the anatomical cause of this deficiency, the bast-fibres showing thinner walls than those of the parent-type under the microscope. Young stems of *rubrinervis* may be broken off by a sharp stroke, and show a smooth rupture across all the tissues, while those of *lamarckiana* are very tough and strong.

Both the giant and the red-veined species are easily recognized in the rosette-stage. The very young seedlings of the latter are not clearly differentiated from the *lamarckaian*, and often a dozen leaves are required before the difference may be seen. Under ordinary circumstances the young plants must reach an age of about two months before it is possible to discern their characters, or at least before these characters have become reliable enough to enable us to judge of each individual without doubt. But the divergencies rapidly become greater. The leaves of *O. gigas* ar broader, of a deeper green, the blade more sharply set off against the stalk, the whole rosettes becoming stout and crowded with leaves. Those of *O. rubrinervis* on the contrary are thin, of a paler green and with a silvery white surface; the blades are elliptic, often being only 2 cm. or less in width. They are acute at the apex and gradually narrowed into the petiole.

It is quite evident that such pale narrow leaves must produce smaller quantities of organic food than the darker green and broad organs of the *gigas*. Perhaps this fact is accountable partly, at least, for the more robuts growth of the giant in the second year. Perhaps also some relation exists between this difference in chemical activity and the tendency to become annual or biennial. The *gigas*, as a rule, produces far more, and the *rubrinervis* far less biennial plants, than the *lamarckiana*. Annual culture for the one is as unreliable as biennial culture for the other. *Rubrinervis* may be annual in apparently all specimens, in sunny seasons, which would allow a large part of the *gigas* to remain in the state of rosettles during the entire first summer. It would be very interesting to obtain a fuller insight into the relation of the length of life to other qualities, but as yet the facts can only be detailed as they stand.

Both of these stout species have been found quite constant from the very first moment of their appearance. I have cultivated them from seed in large numbers, and they have never reverted to the *lamarckiana*. From this they have inherited the mutability or the capacity of producing in their turn new mutants. But they seem to have done so incompletely, changing in the direction of more absolute constancy. This was especially observed in the case of *rubrinervis*, which is not of such rare occurrence as *O. gigas*, and which it has been possible to study in large numbers of individuals. So for instance, "the red-veins" have never produced any dwarfs, notwithstanding they are produced very often by the parent-type. And in crossing experiments the red-veins gave proof of the absence of a mutative capacity for their production.

[Besides the mutants just described three occurred two weak forms that could survive only if reared under protection and would have failed to survive in nature. Here we have a place for the action of natural selection, but operating with mutations instead of with fluctuating variations. These two mutants are "the whitish and the oblong-leaved evening-primroses or the *Oenothera albida* and *oblonga*."

All of the mutants so far mentioned are constant forms that breed true to type. Certain other types were either incapable of being bred or else were decidedly inconstant. *Oenothera lata* had only pistillate flowers and therefore could not be fertilized by pollen of the same mutant. *Oenothera scintillans* and *O. elliptica* are fertile to their own pollen, but produce progeny only partly like the parent, the rest reverting to the original type, *Oenothera lamarckiana.]*

As De Vries has pointed out, each mutation may be different from the parent form in only a slight degree for each point, although all the points may be different. The most unique feature of these mutations is the constancy with which the new form is inherited. It is this fact, not previously fully appreciated, that De Vries's work has brought prominently into the foreground. There is another point of great interest in this connection. Many of the groups that Darwin recognized as varieties correspond to the elementary species of De Vries. These varieties, Darwin thought, are the first stages in the formations of species, and, in fact, cannot be separated from species in most cases. The main difference between the selection theory and the mutation theory is that the one supposes these varieties to arise through selection of individual variations, the other supposes that they have arisen spontaneously and at once from the original form. The development of these varieties into new species is again supposed, on the Darwinian theory, to be the result of further selection, on the mutation theory, the result of the appearance of new mutations.

In consequence of this difference in the two theories, it will not be difficult to show that the mutation theory escapes some of the gravest difficulties that the Darwinian theory has encountered. Some of the advantages of the mutation theory may be briefly mentioned here.

1. Since the mutations appear fully formed from the beginning, there is no difficulty in accounting for the incipient stages in the development of an organ, and since the organ may persist, even when it has no value to the race, it may become further developed by later mutations and may come to have finally an important relation to the life of the individual.
2. The new mutations may appear in large numbers, and of the different kinds those will persist that can get a foothold. On account of the large number of times that the same mutations appear, the danger of becoming swamped through crossing with the original form will be lessened in proportion to the number of new individuals that arise.
3. If the time of reaching maturity in the new form is different from that in the parent forms, then the new species will be kept from crossing with the parent form, and since this new character will be present from the beginning, the new form will have much better chances of surviving than if a difference in time of reaching maturity had to be gradually acquired.
4. The new species that appear may be in some cases already adapted to live in a different environment from that occupied by the parent form; and if so,

it will be isolated from the beginning, which will be an advantage in avoiding the bad effects of intercrossing.

5. It is well known that the differences between related species consist largely in differences of unimportant organs, and this is in harmony with the mutation theory, but one of the real difficulties of the selection theory.

6. Useless or even slightly injurious characters may appear as mutations, and if they do not seriously affect the perpetuation of the race, they may persi.

LATER INVESTIGATIONS OF MUTATIONS

Since the publication of De Vries's classic investigations a large amount of attention has been paid both by botanists and by zoologists to the subject of mutations. Some of the investigators, notably B. M. Davis, went far toward discrediting the whole of the exceptionally careful work of De Vries by claiming that *Oenother lamarckiana* is of hybrid origin. It was pointed out that the form worked with is a domestic type escaped from cultivation and that there is nowhere in the known world any wild species comparable with it. It is supposed to have been brought to Europe from America many years ago, but there is no such species in America today. Davis claims that he has succeeded in producing, by crossing two American wild species, a hybrid from distinctly resembling *Oenothera lamarckiana*, and that when inbred this hybrid produces offspring showing various combinations of the two parent-species that are not unlike some of the mutants observed by De Vries. Jeffreys has also pointed out that the pollen grains of *Oenothera lamarckiana* exhibit a high percentage of sterility, which he believes to be a stigma of hybridity. The general tenor of this type of destructive criticism is to invalidate the whole mutation theory as developed by De Vries and to reduce his mutants to the level of mere Mendelian recombinations of characters once introduced from two or more parental species.

The large amount of work on the cytology of *Oenothera* by Gates and others has, however, served to show that the mutants of De Vries are more than hybrid segregates. Moreover, the beautiful work of Blakeslee on the Jimson weed (*Datura*) and the work of many other botanists, whose findings are reported by Gates in a contribution quoted below, serve to indicate that the type of evolutionary behaviour first observed in *Oenothera* is by no means exceptional, but is probably a common thing at least among plants and may be commoner than we at present know in animals. It may be said by way of anticipation of Gates' detailed account that nearly all of the mutations observed in various species of plants may be definitely correlated with observable changes in the chromosomes of the germ cells, involving changes in number or changes in arrangement of these nuclear elements.

While the botanists busied themselves with their type of mutations, the zoologists, especially T. H. Morgan and his able collaborators, were making discoveries of equal moment in connection with their studies of the mechanism of Mendelian heredity in

Drosophila. As has already been shown in previous chapters, hundreds of new hereditary types arose, apparently spontaneously, in pure-pedigreed stock. Each new type is designated a mutant, and the cause of the changed hereditary condition is not a gross chromosomal change, but an invisible change at a definite point in a definite chromosome, whose cause is unknown but whose location can be exactly determined. Such mutations are known as gene mutations. Like the mutants of *Oenothera*, these *Drosophila* mutants do not differ from the parent species in just one or two characters, but in several or many characters. Usually some one or two characters in any given mutant are especially characteristic, and these serve to give a name to each mutant and make it easier to identify them. Both morphological and physiological characters are involved in these mutants, and every part of the body may be involved. Sometimes the change is so slight as to require an eye sensitized by much training to detect them. It may happen, for example, that two mutants of the eye are so much alike that the human eye is not sufficiently keen to tell them apart, but they may be distinguished by differences in their hereditary behaviour. A large percentage of the mutants discovered in *Drosophila* are "lethals," which means that the change is decidedly for the worse, under the prevailing conditions of life, and that they render the individual unfit to live. Possibly under decidedly different conditions some of these lethal mutants might be better adapted than the normal invididuals. A further discussion of the role of mutants in evolution will be given in a later connection.

The following two rather technical, but very interesting, discussions as to the nature, causes, and significance of mutations are from two men who are recognized as perhaps the leaders in the two branches of mutations study. R. R. Gates had done a large amount of important work especially upon the cytology of *Oenothera*, and H. J. Muller has done and is still doing much to enrich our understanding of the mutational phenomena exhibited by *Drosophila*. While these two workers do not agree as to the relative emphasis that should be placed upon the two types of mutation, it is obvious that both types are of very great importance in evolution.

THE NEO-MUTATIONIST POSITION

Since the original work of De Vries, the subject of mutation in Oenothera has advanced in many directions and the explanation of the phenomena has taken on various aspects. Mutation has become essentially a cell problem. Cytological investigation have shown that many of the mutations are concerned with new chromosome numbers, and the precise nature of the change which has led to the appearance of forms with a new number is known with more or less certainly in the various cases. The new number is in each case present throughout the plant and will be found whether the chromosomes be counted in growing petals, root tips or anthers. These discoveries led in 1915 to the conception that each mutation with its new external characters is the result of nuclear changes transmitted by mitosis to every cell of the plant during development. A fundamental advance in the analysis of mutations has thus been made.

Mutations can now be classified into two types. (1) Mutations which are inherited as Mendelian differences, and which may be looked upon provisionally as the result of a chemical change in one gene or locus of a chromosome. The great majority of the mutations in Drosophila are of this type. On the other hand, very few Mendelian mutations are known in Oenothera. The two best known are *rubricalyx,* which is dominant in crosses, and *brevistlis,* which is inherited as a recessive. (2) Mutations resulting from a visible nuclear change involving a new number of chromosomes. Probably the majority of Oenothera mutations belong here. Similar mutations have been discovered in Drosophila and also in Datura. Examination of the chromosome numbers in related species of wild and cultivated plants shows that such changes have been of relatively frequent occurrence in the evolution of many plant genra, but they appear to be less common in animals.

CHROMOSOME MUTATIONS

The mutations with new chromosome numbers, which are particularly characteristic of Oenothera, may be classified into three types: (1) *trisomic forms,* i.e., mutations with one or sometimes two extra chromosome (2n+1); (2) *triploid* forms (3n); and (3) *tetraploid* (4n) forms. Other forms with different numbers, as 20 or 30, may be regarded as derivatives from these types.

Trisomic forms are known to arise through non-disjucntion, or a reduction division in which both members of one pair of chromosomes enter the same daughter nucleus. The best known mutation of this kind is *Oenothera lata* with 15 chromosomes. (*O. lamarckiana* having 14). But a number of others are now known, including *scintillans, albida, oblonga, subovata,* and more recently *cana, pallesccens, lactuca,* and *liquida.* Several other derivatives of *Oenothera lamarckiana* are also from their behaviour almost certainly trisomic. When pollinated from *O. lamarckiana* they give the two parental types of offspring having presumably 14 and 15 chromosomes. Since there are only seven pairs of chromosomes these trisomic types can not all be accounted for by duplication of a different chromosome in each case.

The genetic relationship between these trisomic mutants are also interesting and peculiar. Thus *lata* can give rise to *scintillans* in its offspring, and *scintillans* can similarly produce *lata.* Also *lata* (pollinated from *lamarckiana*) can give rise to several other trisomic mutations. Many of these relationships can be explained by the assumption of double non-disjunction, i.e., both members of one chromosome pair entering one germ cell while both members of another pair enter the opposite cell. It is highly probable that such cases occur. Thus germ cells would be produced with the chromosome content AACDEFG and BBCDEFG. It can be shown that by such a process one trisomic mutation could give rise to another in its offspring. It is very probable the *lata* is a primary trisomic mutation, e.g. $\frac{\text{AABCDEFG}}{\text{ABCDEFG}}$, i.e. with three a chromosomes, while some other trisomic (2n+1) mutations are probably secondary, e.g., $\frac{\text{AABCDEFG}}{\text{ACCDEFG}}$, i.e., with for example three A and three C chromosomes but only one B chromosome. All such irregular chromosome distributions are probably enabled to occur through a weakness in the attraction which normally leads to close pairing of homologous chromosomes during synapsis or on the heterotypic spindle. There is much evidence of variation in the strength of this attraction in Oenothera.

Cleland has recently found that the chromosomes in various Oenothera species retain their end-to-end connections even on the heterotypic spindle. He has also shown that in several species the arrangement of the chromosomes is more or less constant and characteristic during the stages immediately preceding the reduction division. Thus in *O. franciscana* four of the chromosomes from a ring while the other ten are arranged in five ring pairs which are at first linked to the circle of four in a definite way. In a form called *O. franciscana sulphurea,* which is derived from *O. biennis* × *francicana,* 12 chromosomes end-to-end form a circle and the other two form a pair which is at first linked round the larger chain of 12, and this arrangement is said to be constant. Thus in the derived form a rearrangement of the chromosomes with relation to each other has taken place. Cleland finds similar fixed arrangements in other species. Thus in a race identified with *O. muricata* the 14 chromosomes all form a single circle; in *O. biennis* they are usually in two interlooking circles, one with 6 and the other with 8 chromosomes; while in the mutant *oblonga* with 15 chromosomes the arrangement is more variable but frequently shows a ring of 5 single chromosomes with 5 pair attached to it. In a trisomic mutant from *O. rubricalys* × *hewettii* a varying number of ring pairs was found by the writer. If the relative constancy of these arrangements is confirmed, it will show an essentially new type of integration in nuclear structure, that is, a fixed positonal arrangement of the chromosomes in the nucleus with relation to one another. This will also confirm the hypothesis of the writer years ago, that the homologous maternal and paternal chromosomes in Oenothera are usually arranged alternately on the spireme thread before the reduction division. The fact that also, as observed by the writer in 1908, pairs of chromosomes are frequently detached from the rest of the spireme in certain forms at an early stage of diakinesis, would make it easier for double non-disjunction to occur on the heterotypic spindle in these forms. The further study of the chromosome arrangements in hybrids and mutants will throw more light on these rearrangements, and perhaps also on the nature of the forces which bring them about. Such rearrangements without change of number, if they prove to be constant, are to be considered as mutations of another kind.

Whether non-disjunction has played a part in the appearance of species in nature with a different chromosome number is as yet uncertain. But there are certain genera, such as Carex, in which the haploid numbers usually increase by one from species to species. Heilborn suggests that this may have happened by non-disjunction, as in *Oenothera lata,* followed by the division of the extra chromosome to form a pair.

Polyploidy

An increase in the chromosome number by multiples of the haploid number (polyploidy) is a phenomenon of considerable phylogenetic significance in plants, although it appears to be relatively uncommon in animals. There has been a burst of new knowledge on this subject in the last few years. Since the original mutant *Oenothera gigas* has been shown to be tetraploid, and *semigigas* mutations triploid, we have an experimental basis for the interpretation of all such cases. It appears probable that the triploid condition in Oenothera arises through the union of two male nuclei with the egg, as this condition has been actually observed in *O. lamarckiana* by Ishikawa, and in other plants as well.

A difference of opinion has long existed as to whether *O. gigas* arose from the fusion of two diploid cells or from a suspended mitosis with division of the chromosomes in the fertilized egg. I am still inclined to adhere to the latter view as more probable, with perhaps a sudden lowering of temperature just after fertilization as the cause. Although it is a fact that diploid germ cells do sometimes occur, diploid pollen grains are not known to be viable in plants.

The frequency of polyploidy in flowering plants, and also in other groups of plants, is one of the surprises of recent years. Not only does the condition occur in such cultivated plants as pineapples, bananas, mulberries, wheat, oats, sugar-cane, dahlias, and tobacco, but also in such genera of wild lants as the roses, maples, chrysanthemums, Erigeron, Hieracium, Rumex, Rubus, Crataegus, Spiranathus, and a number of others. The multiplication of chromosome sets runs as high as $8n$ in Rosa and in Acer, and even as high as $10n$ in certain species of Rumex (80 chromosomes) and chrysanthemums (90 chromosomes). Such a widespread phenomenon must be of fundamental significance in the evolution of the genera in which it occurs. The high degrees of polyploidy are probably often connected with hybridization, but there the higher chromosome numbers are usually accompanied by apogamous (asexual) reproduction, which renders constant even form with an unbalanced chromosome number. This is true even of the triploid mulberries and Erigerons, etc. In every case of polyploidy the higher numbers have not arisen gradually by the addition of single chromosomes, but one or more complete sets have been added each time and the process is a mutation involving considerable discontinuity.

That still other kinds of chromosome change occur, is shown both from experimental work and by comparison of the chromosomes of related species. Thus transverse segmentation of all chromosomes has taken place in *Primula kewensis,* and end-to-end fusion of certain pairs has evidently occurred in some species of Drosophila. In the Japanese violets there is some indication that the small number of large chromosomes in certain species may have been derived by the fusion of smaller chromosomes found in other species. A process suggesting transverse fragmentation of certain pairs of long chromosomes appears to have occurred in various genera of Liliaceae. Further study will no doubt throw light on the nature of these processes. It appears already that the passage from one genus to another has not infrequently been marked by a visible change in the chromatin morphology of the nucleus. Such alterations in the conformation of the nuclear material are to be regarded as germinal changes, even though they are not accompanied by external changes in the organism.

EVOLUTION BY SHULL

Mutations of Limited Kinds

A first step in circumscribing the course which evolution may take is the strict limitation placed upon the nature of mutations. In so far as these gene alterations furnish the primary changes for evolution, no change of species can go in directions for which there are no appropriate mutations. Certain discussions of evolution have been based on the assumption that mutations of every conceivable kind not only may be but are produced. Were this true, evolution *could* go in every conceivable direction. It is not true, evolution is limited to those directions which mutations throws open.

There are a number of indications that the nature of mutations is limited. The structure of the gene, if it is a protein molecule, should permit modifications only in certain ways. These ways should be numerous, but not of infinite variety. The more or less stable units making up the molecule resist alteration, while the bonds that connect them at least permit subsitution. Randomness of mutation would thus not be expected.

Several phenomena indicate positively that the changes of genes are not random. One is that mutations, in situations where a scale of possibilities exists, do not actually realize all those possibilities. Eye color in Drosophila should, if every change which is conceivable is also possible, mutate to any color in the spectrum. Actually, aside from white, the colors are all in the red range, none green or blue. Some other insects have green eyes; hence the absence of green in Drosophila must mean that mutations which would result in that color are impossible.

Parallel Mutation and Variation

Parallel mutation in laboratory animals and parallel variation in nature must also mean a limitation on the kinds of mutations. Two species of flies in the laboratory have produced identical mutations quite independently of one another. Both *Drosophila melanogaster* and *D. simulans* have experienced mutations of eye or color to prune, ruby, and garnet; of body color to yellow; of bristle shapes to forked and bobbed; and of wings to crossveinless, vesiculated, and rudimentary. That these mutations actually are identical is proved by crossing corresponding mutants of the two species and getting the same mutant phenotype in the offspring; if the parental mutations were not homologous, the offspring would be of wild type. These mutations cannot arise in one species and be transferred to the other by crosses for, while the cross can be made, the hybrids are sterile. The identical mutations arose independently. So many similar changes could not have been produced at random; something must direct mutation in particular channels.

With this proved example of parallel mutation at hand, instances of parallel variation in species that cannot be crossed must presumably be explained the same way. Ruby eyes with dilute fur color has appeared in mice, rats, rabbits, and guinea pigs. A recessive yellow fur color has arisen in three of these rodents, pink eyes with dilute fur color in three of them, brown agouti in three, a dominant black in two, and white fur with colored eyes in two. If variation were random, such equivalence could hardly exist. While homology of the mutations cannot be proved without crossing, it presumably exists.

Recurrent and Reverse Mutation

Repetition of the same mutation, which has often happened in genetically well-known organisms, and reverse mutation back to wild type, also indicate that change is guided. In X-ray experiments with Drosophila, the same mutation turns up again and again—garnet eye in 1 chromosome in 4000, forked, ruby, yellow, and white 1 in 8000, cut wing, singed, vermilion, carmine, and prune 1 in 22,000, and others. At the white-eye locus, while white mutates to the other alleles with about equal frequency, those other alleles mutate to white about three times as often as to one another. Among induced mutations in bacteria in one group of experiments, 27 changes destroying the capacity to synthesize tryptophane were all probably the same mutation.

With all conceivable changes open to them, the chance that genes would do the same thing a number of times or repeatedly undo what they have done must be considered negligible. The alternative possibility, and the one that must be regarded as probable, is that mutation is directed. Recurrent mutation is considered again later, not merely as proof that evolution is limited to certain channels, but as a positive factor promoting evolution along a given course.

Gene Frequency

The further steps which lead to the change known as evolution involve increase or decrease in the frequency of alternative genes at the various chromosomal loci. If gene A has only one mutant from *a*, which has spread to 0 per cent of the chromosomes of the population that carry that locus, the frequency of gene *A* is 0.9 that of *a* is 0.1. If something hapens to increase *a* to 11 per cent, reducing *A* to 89 per cent, that change in the frequencies is evolution. Now, frequencies of genes are determined by the frequencies of the individuals containing them. A gene is lost when individuals containing it are lost; it increases in frequency when more individuals possess it. The individuals possessing the locus of *A* and *a* are of three kinds, *AA*, *Aa*, and *aa*. Whether these three kinds of individual as are all different in phenotype or otherwise detectable respects depends on dominance. If one of the alternative genes is dominant over the other, there are only two recognizable kinds of individuals; if *A* is dominant, *AA* and *Aa* are alike. If neither gene is dominant, *Aa* is distinguishable from the other two, and there are three recognizable kinds of individuals.

In accordance with what is called the Hardy-Weinberg principle, if the frequencies of genes *A* and *a* in a large population are 0.9 and 0.1, respectively, the population should consist of 81 per cent *AA* individuals, 18 per cent *Aa*, and 1 per cent *aa*. This conclusion is arrived at on the theory of chance. The germ cells should be of two kinds, *A* and *a*, in the same fractions as are the genes. If one figuratively reaches into the supply of eggs and draws one out at random, the chance that the one darwin is an *A* egg is 0.9. A sperm similarly drawn at random from the supply has also a 0.9 chance of being *A*. Thus the chance that a fertilized egg (hence an individual of the next generation) will be *AA* is 0.9×0.9, or 0.81. The expected frequencies of the three kinds of individuals may be computed in one operation by merely squaring the binomial 008 A + 0.1 a. The result is 0.81 *AA* + 0.18 *Aa* + 0.01 *aa*. This product may be generalized as $p^2AA + 2p(1-p)Aa + (1-p)^2aa$, in which p is the frequency of one of the genes (*A* in the example), 1-p the frequency of its allele. The validity of this calculation rests on the assumption that fertilizations are made strictly at random.

The steps above are outlined on the assumption that gene frequencies are known, but in practice that is not immediately true. One determines the frequencies of genes by counting individuals in the population and reversing the calculation. One must know how the character in question is inherited, whether there is dominance and which gene is dominant. Suppose that 16 per cent of a population shows the recessive character *a*, hence that they are of genotype *aa*. The frequency of gene *a* must therefore be $\sqrt{0.16}$ or 0.4. That would mean the frequency of *A* is 0.6. Then, by squaring the frequency binominal, the expected frequency of the several genotypes would be found to be 36 per cent *AA*, 48 per cent *Aa*, and 16 per cent *aa* (the last having been determined by the census).

Genetic Drift

If both alternative genes *A* and *a* were abundant and well distributed through the population, there would be an "expectation" that in the next generation they would have the same frequency. But "expectations" are averages, and no average of many possibilities is ever likely to be exactly realized. The next generation is almost certain to have more than the expected fraction of one of the alternative genes, less than expected of the other. Either one may have gained. The shift may come from several sources. Not all individuals are likely to reproduce, and those which do may be more largely of one genotype than is expected. Those which do reproduce may not leave families of equal size, and the larger families may come from parents having a greater frequency of one of the genes. When the eggs of heterozygotes mature, one gene at meiosis may go more than half the time to the polar body. In fertilizations, one kind of sperm may enter with more than proportionate frequency. Not all fertilized eggs develop, and the losses may hit the two alleles unequally. Other accidents may help to change the proportions of *A* and *a* among the survivors.

The direction of change in frequencies may well be reversed the next generation. There is likely to be fluctuation in both directions over a period of generations. If these fluctuations of opposite sign balanced one another, there would be not *net* change, though all the fluctuations would be regarded as evolution. It would be *possible*, however, for the shift to go farther in the same direction for a number of successive generations. In the untold generations of millions of species of organisms, it would be surprising if there were not sometimes a succession of changes of the same sign. Small evolutionary changes may thus be purely accidental.

Reference was made in the preceding chapter to the decline in frequency of genes that are already in a small minority, particularly in small breeding populations. This sort of decline was excluded from the above example by stipulating that the two alleles were abundant and were both well spread through the population, so there would hardly be any portion of the range in which one gene was lost in one generation. This downward course of minority genes is part of the same general change referred to in this section-random change of frequency. Some evolutionists have applied the name "genetic drift" merely to the random loss of variability through complete elimination of genes that are already infrequent. The name "drift," however, is more fittingly applied to any kind of random change of frequency.

Mutation Pressure

The chief obstacle to the loss of variability through drift in frequencies is the tendency of mutations to occur again and again, already referred to as proof that mutation is directed. So many genes in Drosophila have mutated a number of times in the same way that one may well ask whether all genes are not thus mutable. Such a suspicion would be supported by the structure of the gene, discussed earlier. Some Drosophila mutations have occurred so often that geneticists have lost track of the number of times. After many repetitions, the recurrences were no longer reported. Other organisms furnish examples. One locus in the chromosomes of a particular strain of corn mutated in about 1 germ cell in 500, usually in the same way. Rats have produced a new repetitions, and the induced biochemical mutations of the mold Neurospora have sometimes been identical with earlier inductions.

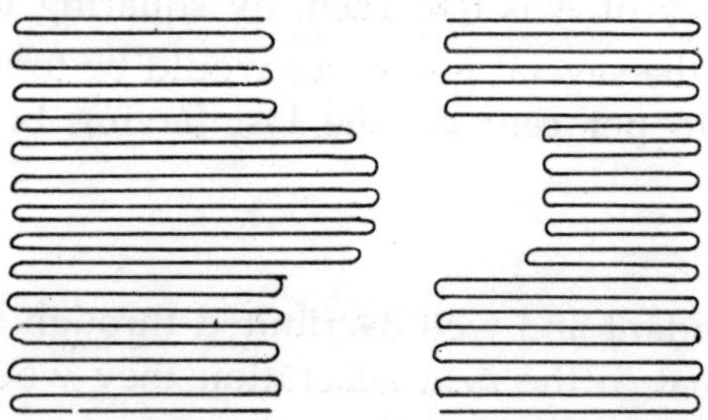

Fig. 9.2: Hypothetical folding of long protein molecule to form antibody having a configuration complementary to that of a gene.

Man experiences repeated mutations to hemophilia, chondrodystrophy, retinoblastoma, sickle-cell anemia, and other defects. Hemophilic men mostly die before maturity, so that a large fraction of the hemophilia genes are wiped out in each generation. Yet hemophilia seems to maintain its prevalence. To keep up its frequency, it would have to arise by mutation in 1 X-chromosome germ cell in about 35,000 in each generation. The rate for chondrodystrophy is computed at about 1 in 25,000, for amaurotic idiocy 1 in 25,000, for aniridia 1 in 140,000, for epiloia 1 in 165,000, for retinoblastoma 1 in 70,000, for thalassemia 1 in 2500, and for sickle-cell anemia 1 in 1000 in American Negroes. This last-named disease is due to one gene lacking dominance, in that the homozygote has sickle-cell anemia, a serious disease, while the heterozygote has no important pathological condition, though the red cells do sickle, less markedly.

The influence of repeated mutation of a given gene on evolution is called *mutation pressure*. If this pressure were unopposed by any other factor, the gene produced by the mutation would eventually replace the one from which it comes. Usually, however, such mutation is opposed by reverse mutation. If *B* mutates to *b* repeatedly, it is usually true that *b* sometimes mutates back to *B*. Under these circumstances *b* would not replace *B*. The two mutation rates are ordinarily not equal, the reverse mutations being the less frequent. If the mutation of *B* to *b* occurs at twice as great a rate as the mutation of *b* back to *B*, gene *b* would increase infrequency until it was in two-thirds of the pertinent chromosomes. Then the population would be in equilibrium, since the mutations in opposite directions would be equally numerous.

The net effect of recurrent mutation is usually to prevent permanent loss of variability even in the small populations where such loss is most likely to occur. Seldom is it powerful enough to effect any permanent change by itself. It is always opposed by something, and among the opposing factors may well be one which outweighs recurrent mutation many times over.

Other Mutational Guidance

Directing evolution by recurrent changes of a gene to some particular one of its mutant forms is limited to increasing the frequency of the mutant gene. There are suggestions that under some circumstances the effects of repeated mutation may be more varied. Genes that cause mutations were mentioned in the chapter on primary change. Sometimes these genes cause more mutations at certain loci than at others. A stock of Drosophila from Florida has a high mutation rate because of a certain gene in its second chromosome, but the effect of this gene is not spread evenly over all loci. Certain genes are "favored". The second-chromosome gene *could* exert quite as influence over long periods.

The effect of antibodies in causing mutations in Neurospora, mentioned in the account of artificial induction of mutations, suggests another internal control of evolution. The antibodies were produced in rabbit serum, against Neurospora extracts, and Neurospora was then exposed to these antibodies. Of the 25 modifications produced, and given the segregation tests to see whether they were mutations or some other sort of physiological change, 11 turned out to be mutations of the same gene. The explanation suggested is that antibodies were produced in relation to the configuration of this particular gene, and then because of their physical similarity the antibodies were able to modify the gene. The question arises whether, without antibodies artificially introduced, a gene might not occasionally create in the cell something nearly enough in its own image to react on the gene that produced it. The gene would in a way be guiding its own evolution.

Successive changes in the same direction have been reported for *Staphylococcus aaureus*, the bacterium of certain abscesses. Mutations providing resistance to sulionamide have occurred, and a succession of these mutations has furnished stepwise

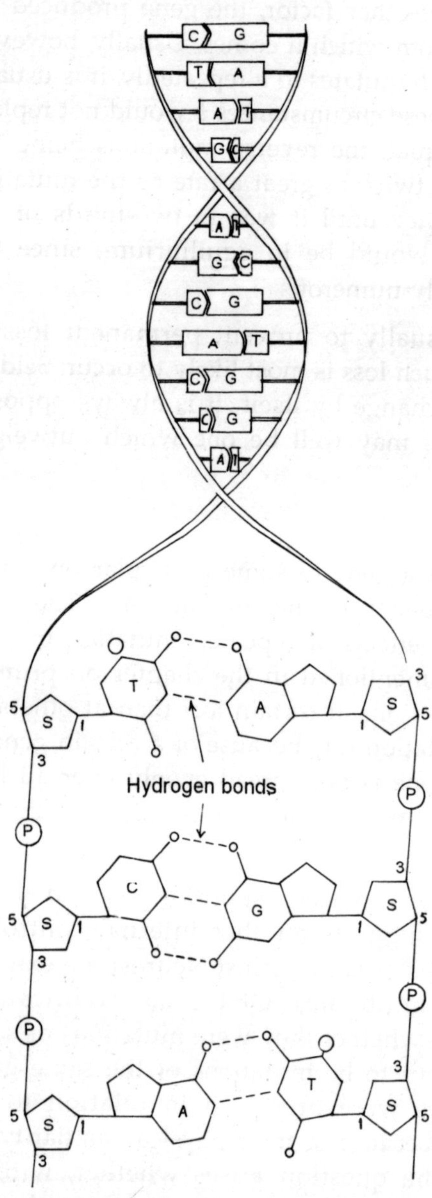

Fig. 9.3: Diagram showing the generalized structure of part of a molecule of DNA, the way in which the order of base pairs in the molecule is duplicated, and the way in which coding of messenger RNA determines the amino acid sequence of proteins. From Bonner and Mills, Heredity (Prentice-Hall, Inc.).

increases in the resistance. The rate of mutation was of the order of 1 per billion bacteria per generation. Some other modifications of bacteria could be interpreted as responses to the environment, but some of these might prove to be only selection of strains already in existence. Much interest attaches to the discoveries being made regarding the genetic mechanism of bacteria and the nature of their mutations. The interest centers in the question whether bacteria, unlike other organisms, experience environmentally produced modifications which fit them for the environment that produces the mutation—whether environment induces direct adaptation to itself. We return to the question of adaptiveness of mutation in bacteria in a later section.

The successive steps in one direction mentioned above might be regarded as a primary type of orthogenesis—evolution in a straight line, directed by an internal agency. One student of Drosophila thought he had discovered such mutational orthogenesis when, after heat-treatment, he obtained a succession of mutations causing darker and darker body color. Other investigators confirmed the production of mutations by heat, but did not observe successive changes in the same direction. It is not likely that any of the supposed instances of orthogenesis on a large scale, as conceived by some paleontologists—perhaps the evolution of the horse or of the dinosaurs—got their start from orthogenetic mutation.

STABBINS

The Nature of Mutation

Before discussing the role of mutation in evolution, we must understand fully the nature of the mutational process, both in respect to the type of change in the genetic material, and the effects of mutations on the phenotype. In its broadest sense, the term, MUTATION, has been used to cover all types of changes in the genes, chromosomes, or other hereditary particles which have a permanent effect on the genotype. Geneticists have, however, generally used the term only for changes at individual loci, which are called gene or point mutations. As our knowledge of gene mutations and chromosomal alterations has increased, different functions of the two types of changes have come to be recognized. Gene mutations have exclusively the effect of adding to the number of alleles available at a locus, and so increasing the gene pool. Chromosomal changes, on the other hand, though they sometimes produce additional genetic variation, have chiefly the effect of changing the linkage relationships between genes, causing certain pairs or larger groups of genes at different loci to be inherited partly or entirely as single units. Consequently, they will be discussed in the next chapter on the organization of genetic variability.

Point mutations can now be defined on the basis of the molecular structure of the hereditary material, or deoxyribonucleic acid (DNA). Since a detailed description of this basic molecular system of heredity is given in every modern textbook of biology or genetics, only those points most essential to our discussion will be brought up here. The molecule of DNA is a double ribbon containing thousands of units, or nucleotides, arranged in a linear order. These are of four different kinds, distinguished by the presence of one of four different bases; the two purines, adenine and guanine, and the two primidines, thymine and cytosine. When forming part of a DNA molecule these nucleotides are usually paired, since the molecular structure of adenine enables it to form a supra-molecular complex with thymine, while guanien can likewise unite with cytosine.

The specificity of a DNA molecule, or of any part of it, depends upon the order of these base pairs in the ribbon. Because this specificity is affected by the symmetry of the molecule, the pair adenine-thymine is significantly different from thymine-adenine. Consequently, the DNA molecule contains an informational "code," similar to the Morse code of telegraphy, in which the four nucleotide pairs containing the bases A-T, T-A, G-C, and C-G correspond to two kinds of "dots" and two kinds of "dashes." A "triplet" of three nucleotide pairs determines indirectly, through the medium of messenger ribonucleic acid (*m*-RNA), the position of a particular amino acid in a molecule of a protein.

Proteins have several functions. Most of them, the enzymes, catalyze the numerous chemical reactions which make the organism alive. Others, the structural proteins, determine the form and properties of various parts of living cells, particularly their membranes. Still other proteins, such as components of cilia and muscles, are responsible for animal movements. In many different ways, the regulation of bodily activities, including the action of genes in development, is carried out by various kinds of proteins. For all of these functions, the linear order of amino acids is of primary significance. The specific properties of a particular protein molecule depend upon the order of the twenty kinds of amino acids in the chain-like molecule of which it consists.

Once we understand these facts we can see that any change in the order of nucleotide pairs in the DNA molecule will bring about a corresponding change in the order of amino acids in the protein which it codes. It is likely to alter, to a greater or lesser degree, the properties and functions of the protein, and consequently of the cell of which it forms a part. A point mutation is such a change. Probably the most common type of point mutation is the substitution of one nucleotide pair for another, but deletions or various kinds of transpositions of one or a few nucleotide pairs would also represent point mutations. Most of these latter types of mutations would, however, be deleterious since they would be likely to interfere with some vital metabolic reaction.

Because the battery of enzyme systems which the organism produces during its life is responsible for all of its activities, we would expect to find among a large number of them, mutations which affect every character in a variety of ways. This is exactly what we find. In the fly, *Drosophila,* one of the best known organisms genetically, mutations are known which affect the color and shape of the eyes, the body color, the size of the fly, size and shape of wings, number and distribution of bristles on the body, number and structure of appendages, and many others. In various higher plants, mutations are known which affect the size, shape and lobing of the leaves, the structure of the inflorescence, the shape and color of the flowers, earliness of flowering, resistance to disease, branching pattern of the plant, and so on. In microorganisms, hundreds of mutations are known which alter or inhibit the synthesis of basic compounds such as amino acids, organic acids, components of the cell wall, and cellular organelles.

THE ADAPTIVENESS OF MUTATIONS

Perhaps the most important fact for students of evolution to realize about mutations is that all modern species and races of organism have existed as successful populations, well adjusted to their environment, for thousands or millions of generations. We would expect, therefore, that all of the potentially useful mutations would have occurred at least once during the evolutionary history of the species and have been incorporated by natural selection into the gene pool. Consequently, the theoretical expectation would be that all or nearly all of the mutations occurring in a successful population would lower its adaptation to its accustomed environment, and so would be rejected by natural selection unless the environment were changing relative to the needs of the organism. This is, in fact, what has been found in actual experiments.

In *Drosophila,* hundreds of mutations have been obtained by looking for those which occur spontaneously in laboratory cultures as well as by subjecting the flies to radiations and chemical substances which have specific effects on the DNA of the chromosomes. Scores of these mutation flies have been placed in competition with their wild-type alleles in laboratory bottles under standard conditions with nearly always the same result. After a greater or lesser number of generations the mutants are eliminated by the corresponding wild-type alleles. There are, however, a few experiments in which flies bearing mutant and wild-type alleles have been made to compete with each other under conditions different from those under which the fly is usually raised. Some of these have produced different results. For instance, a mutant known as *eversae* in *Drosophila funebris,* is less successful than the "wild-type" flies when raised at 16° or 29°C, but is better than wild-type when the flies are kept at 25°C.

Further evidence that some mutations may increase the adaptation of their bearers to a particular environment has been obtained from experiments with agricultural plants. One could reason logically that a cultivated field differs from natural habitats because it is man-made and supported by man, but that intrinsically, the differences between it and the surrounding wayside is no greater than that between two natural habitats. Consequently, if some mutations will make a plant more productive in the field, among the great variety of mutations which appear either spontaneously or following radiation or chemical treatment, there should be mutations which will make the plant more productive in a habitat not controlled by man.

For the past twenty years, hundreds of mutations have been produced and tested in cultivated barley. Most of them, as expected, are deleterious. Nevertheless, a small proportion, which has not been carefully estimated but is certainly less than one per cent, has increased productivity. Some mutations have given rise to other valuable characteristics of the barely plant, such as stiff stems which will support the weight of the ripening grain. These mutations, however, do not improve the plant under all conditions in which it might be grown but only under those in which they have been tested and produced. For instance, the best mutations produced by Swedigh geneticists are those characterized by early heading and stiff stems. In the moist climate of northern Europe stiff stems are particularly valuable since the moisture which accumulates on the plant during rainstorms tends to make the stems bend over and lie on the ground. In the dry climate of the western United States this happens less commonly, but the ripening grain is often exposed to hot, dry winds. If the stems are too stiff and brittle, these winds will break up the heads and cause the grain to be scattered over the field before it can be harvested. In the language of the agronomist the mutations which are valuable in Sweden because of resistance to lodging are disadvantageous in the United States because of susceptibility to shattering.

This example plus many others tell us that we cannot expect any single mutation to improve the overall fitness of an organisms, but only its adaptation to particular environments. Overall improvements in fitness do occasionally occur during the evolutionary history of a group. The conclusion reached there can be anticipated here. This is that such improved structures have evolved gradually, presumably through the accumulation of many mutations and gene combinations. Furthermore, early stages in the evolution of such structures are found in animals specialize for particular habitats and arise as specific adaptations. Their general value appears only later.

THE RELATIVE IMPORTANCE OF MUTATIONS WITH LARGE AND WITH SMALL EFFECTS

A question which geneticits often ask about the role of mutations in evolution is the following: Granted that some mutations have very great effects on the phenotype while the effects of other mutations are barely perceptible; which mutations have played the largest role in evolution, those with conspicuous effects or those with small effects? This question can be answered only by indirect means. A theoretical answer was given many years ago by the English biometrician, R. A. Fisher, in his well-known book. *The Genetic Theory of Natural Selection*. If an organism is well adjusted to its environment, slight changes in its genetic make up may adjust it better to modifications of that environment, but drastic alterations of one or a few characteristics are almost certain to make it function more poorly under any environment. The organism might be compared to a racing car carefully tuned up for making the most speed under a given set of conditions. Given a change in temperature, moisture, or condition of the track better performance might be assured by slight adjustments of the carburetor, the transmission, or the fuel, but any radical changes of these elements would certainly reduce the efficiency of the motor and might prevent it from running at all.

Indirect experimental evidence to show that most differences between natural populations have originated by the occurrence and establishment of many mutations, each one with small effects, can be obtained by making hybrids between differently adapted populations and studying their progeny. If segregation in the F_2 generation from such hybrids is according to simple Mendelian ratios, then we can assume that the adaptive characteristics by which they are differentiated are controlled by a small number of genes. Consequently they probably arose through the occurrence and establishment of a small number of mutations, each with a large effect on the phenotype. If, on the other hand, segregation for a particular character is not according to a well defined ratio but forms a blending pattern with most of the F_2 progeny falling at or near a mode intermediate between the parents, then we can infer that many genes contribute to the difference. It has, therefore, originated gradually through the accumulation of many mutations with each individual mutation having a small effect on the phenotype.

Many examples are now available of hybrids between different races in species of both animals and plants. One of the most extensive examples, which is quite characteristic of the usual situation, is that which the botanists of the Carnegie Laboratory at Stanford, California, Drs. Clausen and Hiesey, carried out on the Californian species related to the cinquefoils, *Potentilla glandulosa*. They studied two large F_2 progenies of hybrids between alpine and lowland races of the species which differ in respect to a large numbers of characters of both vegetative and reproductive organs. Most of the character differences between these races, such as winter dormancy, leaf size and shape, stem height, branching, hairiness, and seed weight (Fig. 9.4) are determined by so many gene pairs that their number could not be estimated. Others, such as the color and shape of the petals, are determined by two, three, or fourgene pairs. The only possible inference from these data and many others like them is that the races of *Potentilla* diverged gradually through the occurrence and accumulation of a large number of genetic differences. Evidence will be presented to indicate that the origin of distinct species has also been determined to a large extend by the accumulation of genetic differences with small effects.

There are, however, some conspicuous exceptions to this rule. Some of the best examples involve the origin of mimiery in animals and of different adaptations to pollination in plants. They indicate that although the establishment in populations of mutations with conspicuous effects is a very rare event compared to evolutionary change through the accumulation of small differences, nevertheless these rare events may be of great importance to evolution, far out of proportion to the rarity of their occurrence. They may occasionally orient the evolutionary line into an entirely new direction of adaptation.

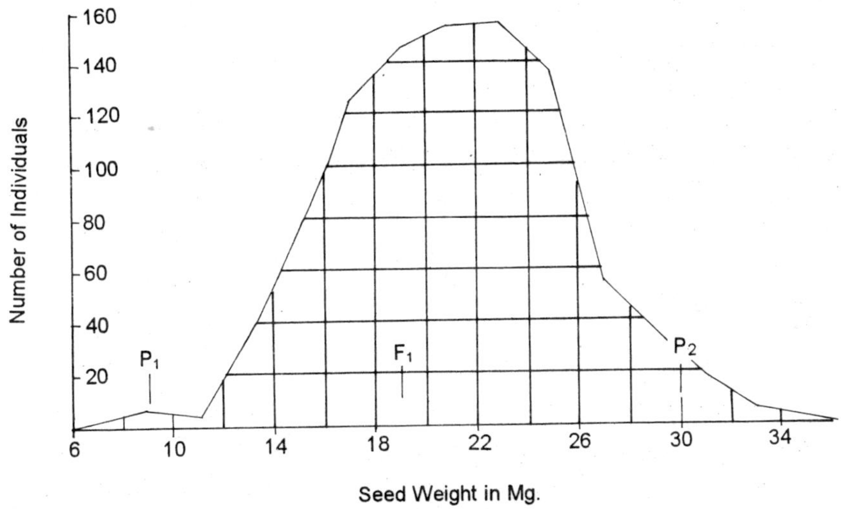

Fig. 9.4: Diagram showing the distribution of mean seed weights in two subspecies of the sticky cinquefoil, one from coastal Southern California and the other from timberline in the high Sierra Nevada, their, F_1 hybrid, and a progeny of 961 F_2 plants. Note that the value for the first generation hybrid lies about half way between that for the two parents and that the modal value for the F_2 progeny transgress those for both parents. From Clausen and Hiesey, Carnegie institute of Washington, Publication No. 615.

THE CAUSES OF MUTATIONS

A question often asked is, "What causes mutations?" In laboratory and garden populations kept under normal conditions for the organism in question they occur spontaneously with no apparent cause. Rates of mutation vary widely from one gene locus to another but are always low. Direct experiments to determine the cause of "spontaneous" mutations are almost impossible due to the low rate of their occurrence. Several kinds of radiations, such as x-rays, ultra violet rays, and neutrons have long been known to increase the frequency of mutations at rates proportional to their dosage. This has led some biologists to speculate that some of the "spontaneous" mutations

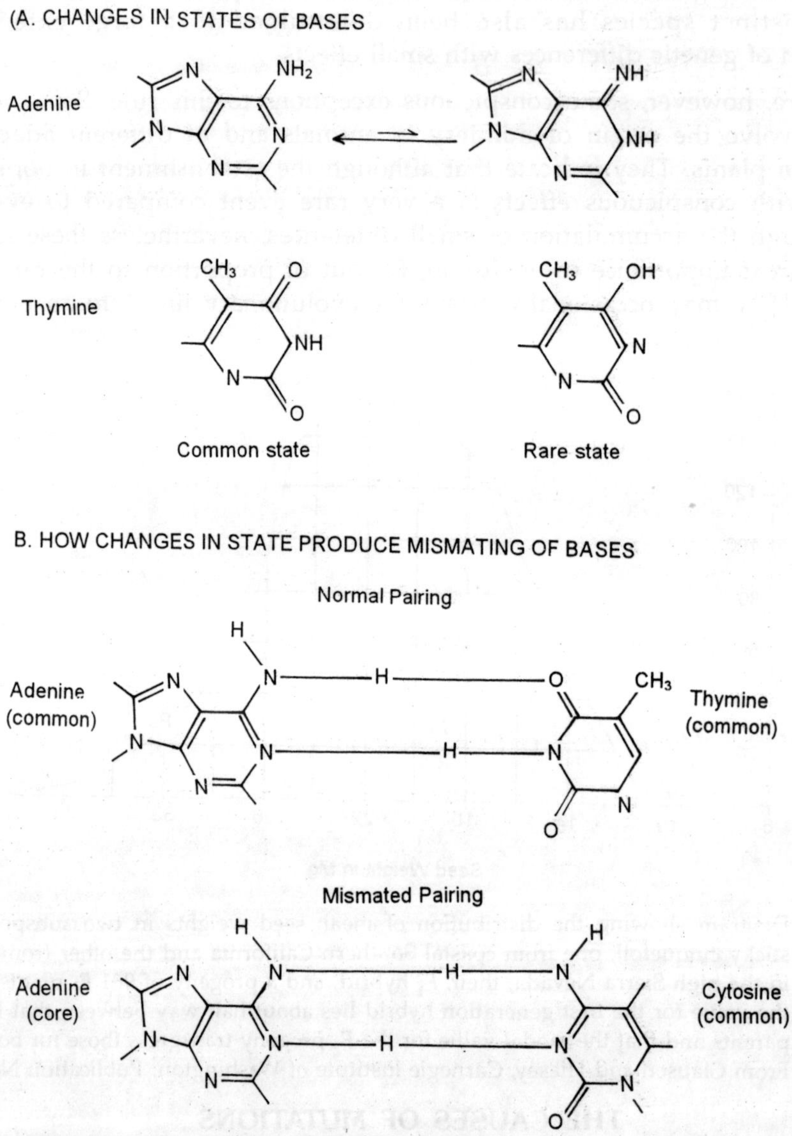

Fig. 9.5: Upper left hand right, normal and tautometic formulae for adenine; second row, the same for thyminel; third row, normal pairing of adenien and thymine in a typical DNA molecule; fourth row, abnormal pairing of tautameric state odenine with cytosine. Modified from Stahl, the Mechanics of inheritance (Prentice-Hall, Inc.)

are produced by natural radiations, such as cosmic rays. Yet when calculations are made, based upon known intensities of natural radiations compared to intensities needed to produce mutations, the results show that even where they are most intense, such as at high altitudes in the mountains, the intensity of natural radiations is so low that it could not account for more than a tiny fraction of the "spontanneous" mutations.

Recent studies of the chemical structure of DNA, the results of which have been outlined earlier in this chapter, have led to a new theory of the cause of mutations. This is well described by Dr. F. W. Stahl in his book, *The Mechanics of Inheritance.* The four bases (adenine, thymine, guanine, cytosine) which make up DNA can temporarily exist in rare states which have an affinity for the wrong base and form mismated base pairs. Thus, a rare state of adenine will mate with cytosine instead of thymine. When a DNA molecule containing such mismated paris occurs in a cell, it duplicates to give rise to one strand of normal DNA and one mutant strand, in which an adenine-thymine base pair has been replaced by one of guanine-cytosine. Experiments with both coli bacterium and bacteriophage have shown that many genetically identified mutants represent such substitutions of single base pairs.

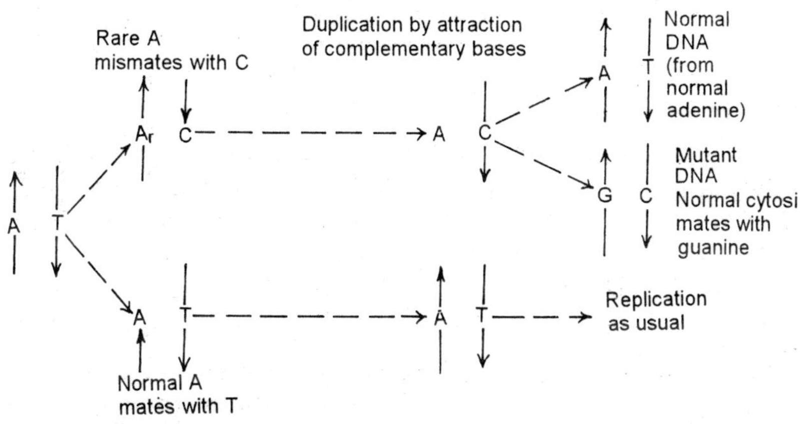

Fig. 9.6: Diagrammatic sequence showing how change of adenine from the normal to the tautameric state can produce subsitution of a G-C base pair from on A-T pair, this constituting a mutation. Modified from Stahl, *The Mechanics of Inheritance* (Prentice-Hall, Inc.).

Strong support for this theory is derived from the fact that a structural analogue of thymine: 5-bromouracil, induces large numbers of mutations in bacteriophage. This base molecule can become incorporated into DNA instead of thymine. Chemical evidence suggests that it changes more easily than thymine into a state which attracts guanine instead of adenine and so induces mutation by base pair subsitution. If 5-bromouracil induces mutations in this fashion, it should also speed up reverse mutation when cells containing mutations induced by it are again exposed to the chemical. This has proven to be the case.

Mutation by such rare states of base pairs may be only one of several ways in which slight chemical instabilities of cellular organization could produce alterations of molecular structure in chromosomal DNA and the compounds associated with it. Perhaps the safest generalization to make at present is to state that "spontaneous " mutations are by-products of accidents in nuclear or even cellular metabolism.

RATES OF MUTATION AND OF EVOLUTION

As was already mentioned in the first chapter, and will be emphasized in the next, mutations are rarely if ever the direct source of variation upon which evolutionary change is based. Instead, they replenish the supply of variability in the gene pool which is constantly being reduced by selective elimination of unfavorable variants. Because in any one generation the amount of variation contributed to a population by mutation is tiny compared to that brought about by recombination of pre-existing genetic differences, even a doubling or trebling of the mutation rate will have very little effect upon the amount of genetic variability available to the action of natural selection. Consequently, we should not expect to find any relationship between rate of mutation and rate of evolution. There is no evidence that such a relationship exists. Studies of mutation rates are of great importance in increasing our understanding of genetics, but they are of relatively little importance to the evolutionist. To understand the basis of differing evolutionary rates, we must learn as much as possible about the interrelationships between populations of organisms and their environment. This topic will be discussed in later chapters.

We must, however, ask the following question: has a scarcity of suitable mutations ever been a limiting factor in evolution? A few simple statistics will helps to answer this question. While rates of mutation vary enormously from one gene locus to another, and can also be greatly influenced by the environment, a rate of one mutation per gene locus in every 100,000 sex cells is a conservative estimate. Because all higher organisms contain at least 10,000 gene loci, and most of them contain many more, we can conservatively say that one individual out of ten carries a newly mutated gene at one of its loci. As already pointed out, the great majority of these mutations are deleterious, but a small proportion of them are beneficial. From various experimental studies we can arrive at a conservative estimate of the proportion of useful mutations as one in a thousand. On the basis of these estimates we can calculate that in any species about one in ten thousand individuals in each generation would carry a new mutation of potential value in evolution. Using conservative values of 100 million as the total number of individuals per generation and 50,000 as the number of generations in the evolutionary life of the species, we could expect that at least 500 million useful mutations would occur during this life span. We do not known how many new mutations are needed to transform one species into another, but five hundred is a reasonable estimate. On this basis, only one in a million of the useful mutations or one in a billion of all mutations which occur needs to be established in a species population in order to provide the genetic basis of observed rates of evolution.

This final calculation is based upon the assumption that any of the rare useful mutations which occurs is capable of being combined with other useful genes by crossing and segregation and spread through the species population by means of natural selection. Given the opportunity of free intercrossing, gene exchange, free recombination of genes among all of the individuals of a species, and the number of generations which occurs during the long evolutionary life span of a species, this is the ortically possible as the calculations of J.B. S. Haldane and Sewall Wright showed us many years ago. Actually, however, free gene exchange and recombination between all members of a species is never possible. In a widespread species the long distances which separate individuals indifferent parts of its range are themselves a partial barrier. Often instinctive preferences for mating between individuals which live near each other provide additional barriers to gene exchange.

These calculations lead to the conclusion that limitations on evolution are imposed not by the scarcity of available mutations, but by restrictions on the ability of the gene pool of a population to generate adaptive gene combinations that can meet the challenge of an altered environment.

Chapter—10

Colouration and Mimicry

COLOURATION

Everyone, whether a trained observer or not, has been struck with the wonderful range of colours borne by different members of the animal kingdom, and this is especially true in the tropics, where riotous colour is the rule. In New England, on the other hand, conspicuous colouring is relatively rare.

To the student of biology, colouration of animals is of striking interest, for much of it is intelligible as part of the great adaptation scheme of nature; but all colours are not adaptive and sometimes it is apparently impossible to account for the existence of certain hues from the standpoint of utility. Our first question therefore is as to the means whereby colour is produced, after which we may pass to a discussion of its significance.

COLOUR PRODUCTION

Colour in nature is the result of some sort of interference with the beams of white light, either through the absorption of certain of the component rays, allowing others to be reflected to the eye, or by some arrangement of surface sculpturing or prismatic glass which refracts a beam of light and breaks it up into its constituent rays. The first method is chemical, by means of the absorptive powers of a definite pigment, melanin, the oxidation of which produces the various colours, and the second is physical.

Chemical Colours

Pigment is found in nearly all portions of an animal's anatomy, not only on the surface but in the deeper-lying parts as well. In some instances the colour is merely due to the absorptive powers of the chemical substance of which the tissues are composed, as for instance, the haemoglobin, a compound of iron which gives the red colour to the blood of the verterates and certain worms, or haemocyanin, which gives the blue colour to the blood of the octopus. Both of these substances have a respiratory function, as they are the oxygen-carrying media of the blood, and their colour is determined just as that of any mineral or chemical substance is determined and has no other significance to the animal in which it is found.

Again, the pigment may be external and give a colour to the organism which may have a real value in the struggle for existence. Such pigment seems to be primarily for that purpose, and the cells which contain it are differentiated into plain pigment, which gives an unchanging hue to the animal, that is, one incapable of rapid temporary alteration, and chromatophores or changeable pigment spots, such as produce the flushes of colour which pass over the skin of a chamelon or of a squid. These are cells which have radiating fibers lying in a plane parallel to the surface of the skin. During a period of relaxation the mass of pigment lies deep and thus presents but a small visible area: upon contraction of the fibers, however, the pigment is spread over a greater portion of the surface and thus is manifest to the eye. Two sets of contrastingly coloured chromatophores, such as brown and green, expanding alternately, change the general hue of the animal from brown to green or the reverse as the case may be. The chromatophores, influenced by the eyes and skin through the pituitary gland, produce colour changes, as for instance in the African chamelons or the American lizard, *Anolis*, the false chamelon of the southern states. The effect of these chromatophores is greatly enhanced when there lies beneath them a reflecting layer formed of guanin, a substance allied to uric acid. This mechanism is particularly effective among fishes.

Physical Colours

Another change or play of colours is caused by surface structure, light falling upon which is reflected by finely incised parallel lines, often running in more than one direction, and thus undergoing dispersion into its component rays. The latter device is comparable to that used by physicists for spectrum analysis, and known as the Rowland grating. This instrument is generally made of spectrum metal which does not readily tarnish, upon which are ruled, with an engine of highest precision, some ten to twenty or even forty thousand parallel lines to the inch. A beam of light falling upon such a grating is broken up into its component colours, giving the rainbow effect known as the spectrum and comparable to that produced by the passage of the beam through a crystal prism. The sculptured surface of a beetle's wing or the scales upon that of a butterfly or the feathers of a humming-birds' throat produce the same result, except that instead of a series of colours such as those of the rainbow, but one may usually be seen from a given point, and this changes to another when the angle of vision is changed. The tropical butterfly *Morpho* ranges from blue to a greenish hue, while the ruby-throated humming-bird or the neck of some pigeons changes from a brilliant metallic red to a lustros green. The scales of *Morpho* when seen under the microscope exhibit two sets of striae perpendicular to each other, which accounts for the play of single colours rather than a spectrum, due to their mutual interference. Strangely enough, the actual colour of the scales as seen by transmitted rather than reflected light is neither green nor blue, but brown.

BIOLOGICAL SIGNIFICANCE OF COLOUR

Indifferent Colours

From the biological standpoint, the colours of animals may be considered under various heads. Of non-selection value but possibly of vital importance to certain ancestors of different environment are certain so-called "indifferent colours." Of such would be the brilliant scarlet of some of the filaments or fin-rays or body of certain deep-sea fishes. These colours are borne by heredity, and as they are not detrimental, being invisible in the Stygian darkness of the deep sea, they are not eliminated. Colours or markings such as the spots on the uniformly black fur of a melanic leopard, or those sometimes seen in the coat of a domestic horse are further illustrations.

Albinism is total absence of colour in hair, feathers, or skin, and even in the iris of the eye. The latter permits the colour of the blood to show, causing the pink eyes so characteristic of pure albinos. The hair and feathers are white because the tiny spaces which normally should be filled with pigment granules are full of air, which reflects all of the light rays just as froth or sea-foam does.

Melanism is the reverse of albinism, for instead of absence of pigment in the skin there is a profusion of black melanin, giving a totally black hue to the entire animal. In both albinism and melanism colour markings may be plainly visible, but in the same manner that the pattern shows in a piece of brocade or damask fabric. An albino peacock, for instance, whose feathers are entirely white, shows the eye-like markings so characteristic of the tail plumes of the normal bird, and the black leopard, as has been said, shows the spots in the same way.

Both albinos and melanos often arise as sports or saltations in a brood or litter of normally coloured individuals. That they are germinal variations is evident, for not only are they heritable mutations but they follow Mendel's law in the ratio of their appearance. Certain races of albino birds and mammals have been established among domestic forms, such as white mice, rabbits, chickens, and pigenons, while the silver fox is a melanic phase of the common red fox, *Vulpes fulvus,* which, though occurring sporadically among wild broods, is now a well-established and commercially profitable domestic variety, of which there are several distinct strains. In nature these sports should be distinguished from normally white or black species such as the various white species which inhabit the snowclad Arctic regions. Lack of pigment shows every degree of gradation from pure white through blotched or piebald individuals to those which show but faint traces of white.

Valuable colours

Valuable colours are such as evidently serve a direct physiological need. They have been classified under the following heads according to the uses to which they are put:

Sympathetic colours
Protective, of the hunted

Aggressive, of the hunter
Alluring colours
Warning colours
Mimetic colours
Signal and recognition marks
Confusing colours
Sexual colours

Sympathetic, Cryptic, or Concealing Colouration

It is that wherein the hue of the animal harmonizes with its surroundings in such a way that it blends into the background and loses its conspicuousness in order to escape from its enemies or to lie in wait for its prey, as the case may be. To the first, the name protective colouration is applied, as in the case of an Arctic hare, while the second group may be called aggressively coloured, the Arctic fox being an example. In the final analysis, however, both are protective, as it is just as essential to the fox that he be protected against starvation as it is to the hare that *he* be protected against slaughter.

The same species may vary in colour in two ways, known respectively as *local colours* and *seasonal colours*. In the first the species has a wide range over areas varying in general hue, so that the ground colour of the animal, if sympathetic, must also vary to harmonize. Several grasshoppers (Acrididae) whose hind wings are brilliantly coloured red or yellow have the fore wings (beneath which the hinder ones fold when at rest), as well as the remainder of the body, coloured to harmonize with the earth. Those found upon the area of the red shale in New Jersey, for instance, will have reddish brown fore wings, while the same species near the sea-shore will be light gray to harmonize with the prevailing sands. The gazelle, one of the most wonderful instances of desert adaptation, varies from white on the great sand plains to dark gray on the lava fields of volcanic districts. Among the hawk-moths the caterpillar of that found on the convolvulus *(Sphinx convolvuli)* when full grown is either green like its food-plant or brown like the ground beneath. It thus shows a double adaptation, each phase of which is apparently capable of protecting it to the same extent; as a matter of fact, however, the brown colour is more effective than the green, as we may learn from two facts. In the first place, the four young stages of the caterpillar are green, and it only becomes brown in the last stage, though sometimes even then it remains green. This suggests that the brown is a relatively modern adaptation, and probably would not have arisen had it not been better than the original green. In the second place, the green-coloured caterpillars are much less numerous at present than the brown ones, and this implies that the latter survive of tener in the struggle for existence (Weismann).

Another still more remarkable case is that of the aesop-prawn, *Hippolyte,* described by Gamble: "The wakeful hours of Hippolyte are hours of expansion. The red and yellow pigments flow out in myriads of stars or pigment cells [chromatophores]: and according to the nature of the background, so is the mixture of the pigments compounded to form a close reproduction both of its colour and its pattern: brown on brown weed, green on Ulva or seagrass, red on the red Algae, speckled on the filmy ones. A sweep of the shrimp net detaches a battalion of these sleeping prawns, and if we turn the motley into a dish and give a choice of sea-weed, each variety after its kind will select the one with which it agrees in colour, and vanish. At nightfall, *Hippolyte,* of whatever colour, changes, to a transparent azure blue: it stolidity gives place to a nervous restlessness; at the least tremor it leaps violently and often swims actively from one food-plant to another. This blue fit lasts till daybreak, and is then succeeded by the prawn's diurnal tint. Thus the colour of an animal may express a nervous rhythm."

A number of birds and mammals such as the ptarmigan, the Arctic fox, the varying hare, and the weasel which puts on royal ermine for its winter dress, show a change of colour from summer to winter, harmonizing with the browns of leafy soil or rock in summer and with the snow-covered ground in winter. The immediate stimulus to change on the part of the individual may well be increasing warmth or cold as the case may be, but temperature change is not believed to be the *direct* cause of the original assumption of the adaptation, for the common European hare, *Lepus timidus,* does not change its coat in spite of the cold. On the other hand, the varying hare, *Leppus variabilis,* also remains brown throughout the winter in southern Sweden, although the weather there may be exceedingly cold. In the higher Alps the same species remains white for six or seven months, in the south of Norway for nine months, and in northern Greenland it is always white, as the snow rarely melts, except in localized areas, even in summer. The lemmings also turn white in winter but experiments have shown that a captive lemming kept in a room in winter will not change colour until exposed to the cold, the cold acting as a stimulus which incites the skin to the production of white hairs.

Standard Faunal Colours

It has been found that each of the several different life condition under which animals are found is apt to make its impression upon its denizens in certain definite ways so that their habitat is usually readily inferable from their general appearance, and this is notably true of colour. For example, the desert animals are generally duns or grays such as the gazelle already referred to, the camel, and the lion. Plains-dwelling forms are wont to simulate the colour of dry grass, as in the case of the familiar "buckskin" horses which become invisible at distances at which black, bay, or white horses are readily seen. Jungle folk are often striped like the tiger or zebra, highly conspicuous forms when viewed in the menageries, but with colours which simulate the bars of sunlight and the lights and shadows among the tall jungle grasses. In the open, on the other hand, Roosevelt tells us that a little distance away the zebra's stripes become indistinct and he appears a uniform gray. Forest-dwelling forms are usually dappled, giving the effect of the splashes of sunshine caused by the pencils of light which fall through the interstices of the leaves. Instances are the leopard and jaguar, the fallow deer, the boa among serpents—although here the pattern is more definite but wonderfully harmonious. Some forms like the Virginia deer, the tapir, and the lion are spotted when young but more uniformly coloured when adult, which may well be of ancestral significance. Many forest insects are green, simulating the chlorophyl of the leaves very closely. Birds of temperate climates are rarely so, but in the perpetually green forests of the tropics green birds of many different and unrelated families are abundant. Green insects hibernate, generally in the egg state, during the sere months in northern lands, but green birds, unless they be migrants, would be highly conspicuous when the trees are leafless.

Arctic creatures are as a rule white, except the aquatic forms like the seals and walrus, and certain of these (*Phoca* spp.) are white when very young. But if a polar bear were brown or black he would inevitably starve, and on the other hand, a white animal away from the snow fields would be equally hard put to it to make a living.

Sea and air mark their inhabitants alike, that is, if they are aggressive or wandering forms. Many sea birds are steely gray or blue above and white beneath, which makes them harmonize more or less with the sea when viewed from above and with the sky when seen from below. Among the forms thus coloured are the gulls and terns. Many fishes such as the blue-fish and mackerel are similarly coloured and for like reasons, even though they are aquatic rather than aerial. One curious instance is cited of a pelagic snail, *Glaucus,* which floats belly upward on the surface of the sea. Here the colours—blue on the ventral ad silver-white on the dorsal side—are reversed with reference to the snail's anatomy, but normal from the standpoint of its life habits.

Nocturnal creatures also wear a proper uniform of mottled brown or gray such as one sees in the wild-cat or owl, which renders their prowling owner very difficult of discernment in the dim light, even to those with night-adapted vision.

Warning or Revealing Colours

These are the conspicuous reds and yellows such as one sees upon the bodies of poisonous or unpalatable animals like the hornet, coral snake, tiger salamander, Gila monster, and many caterpillars and butterflies. These creatures are practically immune from attack if they are recognized in time, so that the advertisement of their dangerous nature must be a very conspicuous one. It does not profit a nauseous butterfly if it is not eaten after it is killed, it is the killing that must be avoided, and a single stroke of a bird's beak might well be sufficient to render the butterfly *hors de combat.* Its character must be recognized at once before the chance of fatal injury occurs. It may be that a few fatalities on the part of the butterflies are necessary in order to impress each individual bird; on the other hand, the inherited instinct may warn the bird that conspicuously coloured animals are to be left alone, or tradition may take the place of instinct. At all events, most preying animals do recognize and avoid the warningly coloured forms within the scope of their natural environment, but may have to learn by bitter experience to avoid strange enemies. Experiments go to show that when hungry animals have been duped once or twice by having conspicuously coloured unpalatable caterpillars offered to them, they learn to discriminate and the colouring aids very largely in the attainment of this lesson. This is true even of the relatively unintelligent fishes.

While the warning colouration may well be the result of natural selection, it has been suggested that very abundant deposition of a waste-matter pigment may render an animal at once unpalatable *and* conspicuous. This, however, would not account for the mimetic colouring to be discussed later which renders a *palatable* insect conspicuous and therefore immune from attack, nor does it seem sufficient to account for the development of warning colouration in edible forms endowed with weapons of defense.

Warning colouration sometimes carries with it the assumption of an attitude which heightens the effect and may be designed to strike terror to the enemy's heart. Such for instance is the spreading of the "hood" of the spectacled cobra (cobra-de-capello) in which the flattening out of the ribs of the neck region displays the conspicuous markings upon its dorsal side to advantage. A small moth, *Smerinthus,* has large eye-like spots on the hinder wings which are concealed by the forward pair when the insect is at rest; when it is annoyed, however, the wings are raised in such a way that the great, staring, eye-like markings are displayed, an exhibition which must greatly impress the would-be assailant.

Mimetic Colouration

These will be discussed at greater length later in this chapter. In brief, it is a colour resemblance between an animal and any other object, animate or inanimate. It may be either such as to conceal or it may be a warning colouration, but in the latter instance it is merely a "bluff" though nevertheless of highly effective defensive value.

Signal Marks

They are apparently of very great importance among gregarious animals where mutual aid in time of danger is a characteristic. A herd of Virginia deer may be quietly grazing when one of the number becomes aware of danger. As he starts off, up goes the tail like a signal flag, showing the conspicuous white of the under side and the adjacent parts of the animal's body which the lowered tail had covered. This acts as a warning to the others and in an instant the herd is in full retreat. The American antelope (*Antilocapra*) has conspicuous areas of white hair on either rump which can be flashed in a similar manner through a spreading of the hair which reflects more light and forms a very effective danger signal. A still more familiar example is that of the cottontail rabbit, *Lepus sylvaticus,* whose signal flag gleams in the dusk and shows its young the way to safety. In each instance the signal is instantaneous, which may be of vital importance to the safety of the individuals.

Recognition Marks

These marks for use of other individuals of the same species, are such as the red and orange spots on the side of a brook trout (*Salvelinus fontinalis*) which render it so beautiful a fish. In general the trout obeys the law of sympathetic colouring, a lurking fish in a shadowy pool being almost invisible to the enemy above, as his dusky back is as harmonizing from that point of view as his light-coloured belly would be from below; but to creatures of his own kind and at his own level the characteristic speckledness is clearly discernible. These recognition marks are also borne by many insects and are often as in many butterflies and moths, visible only when the animal is in motion, for then no amount of protective colouration will generally avail its owner. Under such circumstances the sympathetic colours are on the exposed portions of the animal as it rests with wings folded, that is, on the upper surface of the fore wings in moths and the under surface of both pairs in butterflies, for in the former group the hind wings are folded beneath the fore, which lie roof-like against the sides of the body, while the butterflies bring the upper surfaces of the erectly held wings together, thus exposing their under surface.

Confusing Colouration

With many forms such as the moths and butterflies which we have just described, the recognition colouring may be very conspicuous, so that during flight, when no protective colouring would avail, the eye follows it readily. A sudden settling of the insect, the brilliant colour instantly disappears, and one loses sight of the insect completely, since its exposed colouring now harmonizes with the support on which it rests. Were the eye to follow the sympathetic colour during flight, the resting insect would be much more readily discerned, but the sudden disappearance of the conspicuous colour is highly confusing to a human observer, who looks in vain for the coloured object and overlooks it in its transformed state, and the same must be true of pursuing creatures as well. The Catocala moth, with red-banded hind wings is an instance in point, and this colouration is not at all confined to insects, as certain somber-coloured lizards have brilliant colour on the under side of the tail, which the eye follows and loses as the creature stops after one of its lightning runs and, lowering the tail, crouches in the sand, with which it now harmonizes to perfection.

Sexual Colouration

As we learned in the chapter on sexual selection the males of birds and other forms are often much more conspicuously coloured than their females, as for instance, the cardinal bird with his gorgeous red nuptial plumage as compared with his soberly a garbed mate. Other similar contrasts are seen in the oriole, bobolink, several ducks such as the wood duck and the eider, in the wild progenitor of domestic fowls, and in the related pheasants. Why the males are conspicuous is perhaps not so clear, but the protective colouration of the female must be of prime importance to the race, especially if she is nesting or a potential mother. The young males which have not yet reached sexual maturity are also protectively coloured, so that in many cases it is extremely difficult except through anatomical examination to distinguish the young male from the female bird, even though the colour contrast is so marked in the fully adult. The brilliant colour of the cardinal bird is gradually assumed, several years being required for its perfection.

The spotted coat of young animals, such as the deer, lion, and puma, is well known. The young Brazilian tapir is somewhat irregularly striped, which is also true of the young of the wild boar. The jungle fowl, the ancestral stock of our domestic chickens, are coloured in their adult life much like the game cock. The newly hatched chicks, on the other hand, instead of being plain yellow or dark-coloured as in most domestic breeds, bear two dark longitudinal stripes along the back. That all of these infantile colours are of superior protective value to those of the adult, especially under the conditions surrounding babyhood, is, in many instances, evident, but some may well be more of historic than of immediate significance.

In a higher insect such as a moth or butterfly, egg, larva, pupa, and adult may each have a protective colouration suited to the immediate environmental need and differing markedly in each successive stage. That these colours aid in increasing in percentage of survival in each of the several stages is evident.

BIOLOGICAL CAUSE OF COLOURATION

Some colouration is doubtless dependent upon the chemical material of which any organism is composed and the hue is due entirely to the absorptive powers of that substance; since oxide of iron is red, so is the blood of vertebrates, and since copper oxide is bluish green, so is the blood of the octopus, and so on. In many cases the colour of internal organs may be due entirely to the nature of the food consumed; thus the flesh of brook trout caught in central Massachusetts where there are few crustaceans available for food is pale, whereas trout from the Adirondack lakes where crawfish are abundant have flesh of a deep salmon pink. The bones of a gar-pike are green in colour through the deposition of green vivianite under certain physiological conditions. Thomson says that "the pigment substances are primarily waste-products, reserve-products, or by-products of the animal's metabolism, and in many instances the colours have no more significance for their possessors than the gorgeous autumnal tints of withering leaves have for the tree—that is to say, none!"

Pigment, aside from its protective or warning significance, may have other very real values. Such for instance are the definitely localized spots of colouring matter which are sometimes associated with end organs of the nervous system. This pigment absorb light, indirectly stimulating the nerve, and thus acts as a light-percipient organ. All eyes have this for their for their basic principle, but in some instances accessory organs which are not strictly speaking eyes and yet have a light-perceptive value are thus formed.

Other pigments found in fur or feathers may serve to absorb or reflect heat and thus be of value to the owner in addition to the function of sympathetic colouration, such for example as the pale grays of the denizens of the desert. Again, dark pigment such as that in the negro and other dark human races serves to cut off the ultraviolet rays of light which from their physiological action cause such discomfort to humanity. Hence a coloured man under a tropic sky is less liable to sunstroke than a fair one, and the development of human pigment, which is said to bear a direct ratio to the number of sunny days in a year in a given region, may well be due to the working of natural selection in eliminating the lighter coloured individuals rather than to the inheritance of the direct effect of the sun's rays in the production of pigment.

The structural features, striae and the like, which give rise to the physical play of iridescence, have been compared to ripple-marks or rhythms of growth such as the growth-lines on a shell, and the same may be true of the colour when it is laid down in concentric lines or cross-bars.

VALUE OF COLOURATION

The value of colouration to the animal is in many instances a very real thing and must play an important part in determining the creature's chances for survival. That it is not in every case operative, however, is undeniable, for, as Roosevelt says, speaking especially of the higher animals, some have so wide a range of activity, continually passing over such totally different landscapes and among such totally different surroundings, that the difficulty of devising a concealing colouration would be enormous (1911). Perhaps one of the most glaringly conspicuous of animals as seen in the zoological parks is the zebra. The question arises as to the value of this colouration, either as concealing colour or as recognition marks. Gregory has summarized the evidence as follows: "Pocock maintains with much force that the testimony of observers in the field has established the truth that the colouration of the coat renders a zebra invisible under three conditions, namely, at a distance in the open plain at midday, at close quarters in the dusk and on moonlight nights, and in the cover afforded by thickets."

On the other hand, the late E. C. Selous, a keen observer with a lifetime of field experience, writes in *Great and Small Game of Africa:* "Never in my life have I seen the sun shining on zebras in such a way as to cause them to become invisible or even in any way inconspicuous on an open plain, and I have seen thousands upon thousands of Burchell's zebras." He admitted the fact, that the bold black an white stripes on the coat of Burchell's zebras are not apparent at any great distance but stated that nevertheless the zebras, on account of their constant movements and by the violent switching of the tail, were allays conspicuous even in the moonlight. Moreover, the lions hunt the zebras chiefly by smell and at night, or lie in wait for them at the water holes. Colonel Roosevelt also vigorously supported the view that the striping was of no special use for concealment.

With reference to the value of the stripes as recognition marks, Selous through them of little service, but Ewart, in his famous Penycuik experiments in which a male zebra was mated with an Arab mare, had great difficulty in overcoming the reluctance of the zebra, for only after he had become accustomed to the strange appearance of the dark-coloured, unstriped mare could he be induced to mate with her at all. Zebras are such keen-eyed animals that they can hardly fail to recognize individual, much less specific, differences in their companions.

Nevertheless, it the colour was such as to protect the bird or animal at critical periods, not only from the standpoint of its own individual protection, but also from that of the race, as for instance during the bringing forth of the young or the incubation of the eggs, its value would be abundantly proved even though at other times the protection afforded was inadequate.

Whatever may be the initial cause of colour, its final perfection of adaptation for concealment, warning, or whatsoever service it may render may well be the result of the natural selection factor.

MIMICRY

Of all the devices adopted by nature for the protection of her children none is so marvellous in its perfection as those which are included under the head of mimicry—the resemblances which organisms bear to others and to inanimate objects in form, colour, attitude, and action, thereby either escaping observation or advertising an apparent harmfulness which is not at all real.

Just as colouration of animals was classed as protective and aggressive, so we may consider two aspects of mimicry, the protective mimicry being as before divisible into concealing and warning and the aggressive into concealing and alluring. Mimicry may be further divided into unconscious or passive, under which head by far the greater number of mimics fall, and in which form and colour make up the resemblance; and conscious or active, in which by its actions as well the mimic imitates the immune model.

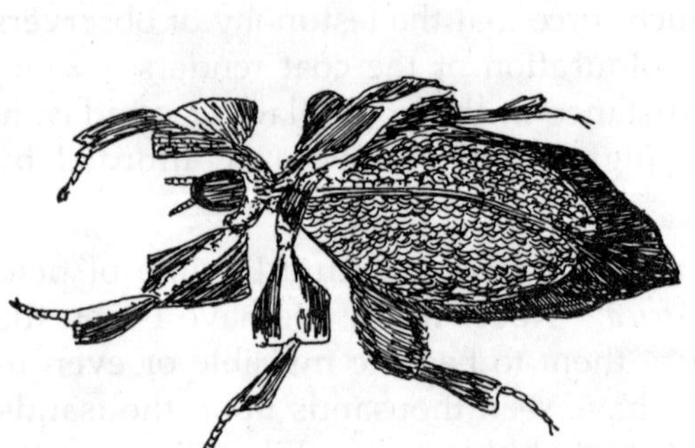

Fig. 10.1: *Phyllium*, "walking leaf," an example of insect mimicry.

Almost anything may be mimicked, from a wave-worn pebble to a twig of a tree on an active harmful insect. Thus all the kingdoms of nature, the inorganic, plant, and animal, furnish the host of models to be imitated where immunity is sought.

Protective Mimicry

As we have seen, has two aspects, concealment and warning, of which the first is by far the more common. An example is the crab *Cryptolithodes* whose smooth rounded from and texture and white colour harmonize so perfectly with the white quartz pebbles of the beach that one needs must turn the animal over, generally by accident, to see its real organic character as a living form. The immunity which it thus secures from enemies of its own humble station must be of real protection, though doubtless it is occasionally crushed by a larger form. Its mimicry is certainly unconscious, as action betrays its presence and stoical passivity is essential to safety. Another crab resembles wave-worn dead coral very closely; here, as the animal is carnivorous, the concealment may have a twofold purpose, protection against its enemies and aid in securing its prey.

The geometrid moths, whose caterpillars are the familiar measuring worms, are often not only protectively coloured but may mimic the tiwgs and smaller branches of various plants such as clematis, the birch tree, the pear tree, and many others. That on the birch, the caterpillar of *Selenia tetralunaria,* when it needs particularly to escape observation, grasps the branch with its two hinder pair of prop-legs and throws its body outward at an angle, in a rigid posture as though in a cataleptic state. Details of resemblance are now apparent, not only in colour and form but in excrescences which simulate the latent buds of a twig, while the somewhat pointed head and feet resemble terminal buds. So perfect is the resemblane and so long maintained the posture that even a trained observer often fails to see the creature or to be sure of what he sees until he actually touches it. Such mimicry is in the main unconscious, that is, in the colour and form, but as action precedes the inaction of the mimicking pose it is also in part a conscious one.

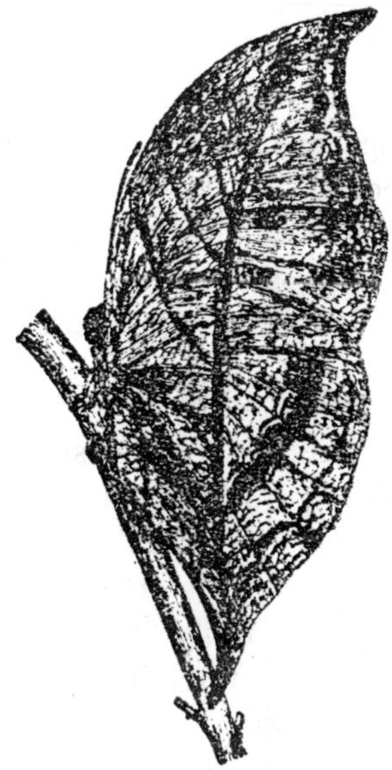

Fig. 10.2: Dead-leaf butterfly, *Kallima paralecta,* from India.

Yet another geometrid moth, *Schizura mucronis,* carries the mimicry still further, for not only does the rough-barked caterpillar mimic the twig to perfection, but the moth does also, although somewhat less effectively. The latter rests head downward against the bark of the actual limb with the stiffened body held out at an appropriate angle, the wings being folded around the abdomen in such a way as to heighten the resemblance.

Many moths mimic the rough bark in colour when in an ordinary resting posture upon it, but with them it is a matter of colour mimicry only.

Many of the walking-stock insects (Phasmidae) are also close mimics, with their slender body, attenuated limbs, sympathetic colouration, and slow movement, often stiffening into rigidity as do the geometrids. When moving, some of the walking-sticks have a peculiar tottering gait, but in one instance in Texas, while the shadow of the insect could be clearly seen upon the shadow of a bush because of this characteristic, only careful search revealed the creature in substance upon the bush itself. All of the American species of phasmids are wingless and there are comparatively few of them, but in the tropic and subtropic regions of the earth over six hundred species are known, many of which have other protective resemblances, some of them marvellous indeed. Perhaps the most perfect example is the leaf insect *Phyllium*. Here the wings and flattened and expanded body and limbs are all green except for irregular small yellowish spots which simulate the fungus or rust growths upon a leaf so that the total resemblance is very precise.

Many butterflies are also leaf-like in appearance, simulating not only the general hue of a dead or withered leaf but its petiole, midrib, venation, rust spots, and the clear places which sometimes occur in a diseased or injured leaf. One such butterfly is *Coenophlebia archidona* from Bolivia, in which the upper outer angle of the fore wing forms the apparent stem or petiole of the leaf; the most noteworthy instance, however, is that of *Kallima paralecta* from India in which the hinder wings are prolonged into the stem-like structure, and the other leaf-like markings are carried to an extreme of perfection. This butterfly is striking coloured above, blue-black with a reddish yellow or bluish white band, the recognition markings of the species. This colouring is seen in flight. On the instant of alighting, the insect practically disappears, for now the wings are folded together so that only their protectively coloured under surface is exposed. This is deceptively like a dead and sere leaf due to its colour, in which red and brown alternate with occasional spots bereft of scales which simulate dewdrops. In addition there is seen the midrib and often the lateral ribs of the leaf, and black and mouldy spots also occur.

Warning Mimicry

Some of the most remarkable mimicry of all is included under the caption of warning mimicry—mimicry of advertisingly coloured forms which are distasteful or even poisonous either when devoured or through the possession of poisonous fangs. Among reptiles there are certain brilliantly coloured poisonous snakes belonging to the same family (Elapidae) as the deadly cobras. These snakes are of the genus *Elaps* and are confined entirely to America. They are beautifully coloured, often in red and black alternating bands which in the coral snake are edged with yellow. While they possess a strong poison they are practically harmless to man because of the limited extent of the gape. Each of the several species of poisonous coral snakes is mimicked by other species of harmless snakes belonging to different genera, and while the markings are in no case exactly similar, the approximation is sufficiently close to render the imitators practically immune from attack except perhaps by mankind. Many other harmless snakes such as the hog-nose snake (*Heterdon*), while not coloured exactly like poisonous ones, will nevertheless flatten the head, rendering it triangular, and hiss and often strike to show how very dangerous they are. With lower animals this doubtless gives them a certain immunity from attack but on the other hand only serves to invite destruction at the hand of man. The mimicry in those species which imitate the corals snakes is unconscious of passive, that of *Heterdon* is conscious as it depends upon the actions of the snake to make it effective.

Among insects again there are a host of imitators, both active and passive; those which mimic their models in form, colour and action, such as the hairy, brilliantly coloured syrphus—or flowerflies which resemble the stinging bees and wasps; beetles which resemble wasps; clear-wing moths which mimic bees, and so on. Some of the most remarkable imitations, again, are among the butterflies, which resemble other distasteful and hence immune butterflies. The most familiar instance is that of the immune monarch butterfly, *Anosia plexippus* (= *Danais archippus* which is inedible, and its imitator, the viceroy; *Basilarchia archippus* (= *Limenitis disippus*), which would otherwise be destroyed, as it is palatable from the point of view of insectivorous birds.

Sometimes only the female of a butterfly will mimic an immune model, the male being coloured in a totally different way (dimorphic species). Again in widespread species more than one form of female will be associated with a given male, each female mimicking an immune species which happens to be locally abundant. In Africa the Danaid butterflies are unpalatable, while the Papilios are not. *Papilio merope* is a species which has scarcely varied at all in the male sex in the extent of its distribution through Africa, but in the female has almost everywhere lost the outward appearance of a *Papilio,* and assumed that of a Danaid, which is protected by being unpalatable; and not everywhere the appearance of the same species, but in each place that of the prevailing one, and sometimes of several in one region. As a result, these females now show a polymorphism consisting of four chief mimetic forms, plus the primitive form—that resembling the male. The latter has survived in Abyssinia alone, and even there it is not the only one, but occurs along with some of the mimetic forms (Weismann). Thus while in a general way the different types of female are locally separate, their areas of distribution often overlap, and, at the Cape for instance, one male and three forms of female have been reared from one set of eggs.

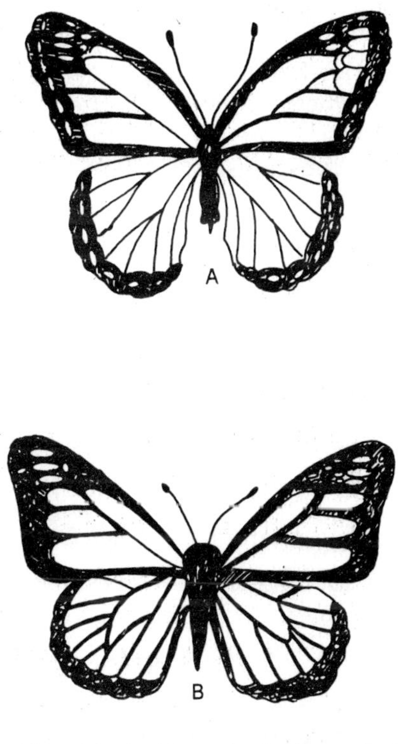

Fig. 10.3: A=Monarch, B=Viceoy butterflies.

Aggressive Mimicry

It is that shown by certain carnivorous forms such as the spiders found on golden-rod and other flowers, whose yellow bodies so harmonize with the flowers upon which they rest as to render them invisible to the visiting insects which from the spider's prey. Other spiders resemble oak galls or other vegetable growths, yet others the dropings of birds, all of which resemblances have the same ulterior design. These are all instances of concealing mimicry.

Another spider is described as resembling an orchid blossom more or less closely, both in colour and form. In this instance the resemblance is an alluring one and is advertising rather than sympathetic. It is doubtless a highly profitable adaptation to the spider. Another similar adaptation is that shown in a certain African lizard, protectively coloured except for a brilliant patch of colour at the corner of the mouth which acts as a lure for the unwary.

Simulation of Death

Fig. 10.4: A moth mimics a wasp.

But one aspect of mimicry remains to be discussed, again a conscious imitation, that of simulating death. This has been developed to such an extent in the American opossum (*Didelphis virginiana*) that the act is known in common speech as "playing possum." Whether it is an intentional performance on the part of the creature, or whether, as has been suggested, the animal faints away from fright, in which event it could hardly be called conscious mimicry, it certainly is again an adaptation of real value to its owner and may often save its life when attacked by an enemy that prefers to kill its prey. Many insects, especially hard-bodied beetles, which a fall will not injure, have a similar habit, as they drop like a pebble when one is about to seize them, lie inert where they fall, and are often searched for in vain among the leaves and grass beneath the bush upon which they were. This may be, as in the case of the opossum, an actual unconsciousness, but is even more efficacious, as the animals are so hard to find.

Wallace's Conditions for Protective Mimicry

Wallace long ago summed up the conditions which must be fulfilled whenever protective mimicry occurs. They are as follows:

1. Imitative species must occur in the same area and occupy the same station as the mimicked.
2. The imitators are always the more defenseless.
3. The imitators are always less numerous in individuals.
4. The imitators are differ from the bulk of their allies.
5. The imitation, however minute, is external and visible only, never extending to internal characters or to such as do not affect the external appearance.

BATESIAN AND MULLERIAN MIMICRY

Bates (1892), an English naturalist, proposed *Batesian mimicry*. In this mimicry, the animal is protected from the predators by an unpleasant taste or odour or possession of strings or irritating hairs.

The examples of Batesian mimicry are the butterflies of the family Helicondiae and Pieriedae. Heliconidae are unpleasant in taste and are not devoured by birds. Pieriedae are edible but they have the same colours a possessed by Helicondiae, therefore, they are confused by the birds.

Fritz Muller (1821-1897) a German Zoologist described this type of mimicry. Here two or more unpalatable species resemble each other. All of them have warning colouration.

Poluton compared Mullerian mimicry to the action of a group powerful firms, which became better known for a reduced cost by joint advertisement. *Beebe* and *Kennedy* (1957) gave an example of a moth which is unpalatable to birds, lizards and ants. This moth closely resembles unedible wasp. It is an example of *Mullerian mimicry*.

CAUSES OF MIMICRY

There are a number of views as regards to the origin of mimicry. Some of them are as follows:

1. **Natural Selection.** According to *Weismann*, the natural selection is the only factor in the production of mimicry. In opposition to this it has been argued that the mimicry, to be of any selection value, must be perfect at once and the minute variations which are supposed to be the exponents of natural selection, would be of no survival value in mimicry.
2. **Mutations** Another explanation of mimicry is that the mimetic form may have arisen as a large mutation. However, there is little doubt that whatever causes may be secondary operative in the production of colouration and mimicry, natural selection is the chief.
3. Due to the direct influence of the environment acting similarly on different species.
4. Due to a physiological response, such as *colour sensation*.
5. Due to sexual selection changed by the presence of other types of colour.

Objections to the Theory of Mimicry

(1) Birds and other predators do not eat butterflies and do not kill sufficient number of them so as to allow mimicry to evolve. *Keetswell* (1956) proved that melanic moths are eaten by birds. *Brower* (1958) described that birds do not eat unpalatable butterflies. *Poluton* and *Carpenter* proved that butterflie are eaten by birds. Sometimes, the marks of a bird's beak are also seen on the wings of the butterfly.

(2) *Uvarov* (1932) objected that many animals feed on ants and it is dangerous to mimic ants. Other animals do not eat ants. However, *inediblity* is a relative term which depends on the degree of hunger of insectivorous birds.

Significance of Mimicry

The main significance of mimicry is to protect the animals against enemies. This self defence is naturally of the survival value and of evolutionary significance. Thus, mimicry is a special type of adaptive colouration in animals. It is more specialized in spiders and insects. Genetic variability and natural selection work in balance to bring about the mimicry and it is of advantage to the species.

VARIATION

The basic prerequisite to evolution is undoubtedly variation, which together with heredity may be regarded as an undeniable fact. Therefore, as has been said, the cause of variation must be a contributory case of evolution itself. But not all variations are alike, and only certain ones, apparently, are heritable, hence it is necessary to understand quite clearly what sorts of variations exist, since those that can not be inherited can have no part in the evolution of a species, as they concern only the individual and not the race.

SORTS OF VARIATIONS

Variations may be grouped into three pairs of contrasting sorts—germinal or acquired, continuous or discontinuous, determinate or indeterminate—six in all. A given variation belongs in each of the three groups but must be either one or the other of the contrasting sorts within the group, as they are mutually exclusive. The first group is based upon the nature of cause of occurrence. It includes:

(1) Germinal Variations are Intrinsic

Such as arise in the germ-plasm itself due either to *recombinations,* which owe their origin to new groupings of the germinal factors or genes; or to *muttions* which result from relatively radical alterations in the gene complex. Changes due either to recombinations or to mutations are of course manifest only in the animal or plant body, the soma, and do not depend on external conditions for their origin. Such conditions may impede or arrest their development but cannot cause them. Often these mutations cannot appear until later in life, as, for example variations in adult size and proportions. They may appear at any period from the beginning of embryonic life to the death of the organism.

(2) Acquired or Somatic Variations or Modifications

These are imposed upon the organism during its lifetime and are due to *extrinsic* influence. Under which of these two heads a variation should fall is often almost impossible to decide except by experimental work, and sometimes not then, but the difference is of fundamental importance, for germinal variations only can be inherited by the offspring.

Examples of germinal variation are the occasional occurrence of supernumerary digits in man, the domestic cat, the horse, and other forms, for they can not be due to accident or to any external influence and are occasionally inherited through several generations. On the other hand, the loss of a digit in man through too close proximity to a circular saw would certainly be an acquired characteristic and long series of experiments have proved conclusively that such variations are without the pale of heredity.

Another example which occurs normally in nature would be the hive bees (*Apis mellifica*), in which the sex distinction between the workers and queens is an acquired variation, while that between drone and queens or worker is germinal. The queen is impregnated but once, the male elements being stored in the seminal receptacle, and used one at a time for the subsequent fertilization of the eggs. The workers build two sizes of cells in the brood combs, in the larger of which the queen lays an unimpregnated egg, while that laid in the smaller cell is fertilized at the time of laying. From the unfertilized egg there arises, invariably, a drone or male bee, while the fertilized egg produces a potential female. Hence the difference between male and female bee, depending as it does upon impregnation or not, is germinal. The ultimate fate of the impregnated egg is subject to the control of the workers, for while the vast majority of the eggs remain within the small worker-cells and, being fed upon a meager and comparatively in nutritious diet, develop into workers—the so-called neuters, which are in reality undeveloped females—a select few have their cells enlarged at the expense of the surrounding ones and are fed upon highly nutritious "royal pabulum" and ultimately develop into mature females or queens. It is said that even after three days of larval life, the humble worker, like the beggar maid of King Cophetua, may become a queen. Thus this variation, being imposed upon the animal during life, is an acquired one.

Indeterminate and Determinate

Variations may be further classified according to direction of occurrence, as (3) indeterminate, those fortuitous variations which are not subject to any law but occur in any conceivable direction of change; or (4) determinate, controlled by some unknown influence and confined to certain definite lines or directions of change, usually in an adaptive direction. Of the first of these two contrasting types there is no doubt, and they are the variations upon which the Darwinian factor of natural selection is supposed to operate. Illustrations are legion, as the comparison of any group of related organisms will show a variation of size which embraces every dimension of space and combination thereof.

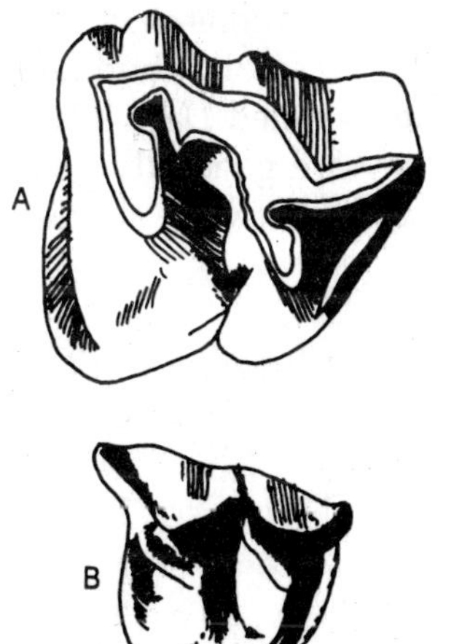

Fig. 10.5: Teeth of (A) rhinoceros (*Caenopus* sp.); and (B) coney (*Hyrax capensis*), much enlarged.

Of determinate variations we are not so sure; some claim that they have no real existence, that there is no such tendency to vary always in a given direction in successive generations, while others consider these variations, which they call orthogenetic (Gr. ὀρθός, straight, and γενευις, generation), *the* important ones in evolution.

What is apparently an example of determinate variations is that given by Kellogg, who described in 1906 the gradual but obvious change from one dominant type of colour-pattern to another in the leaf-eating beetle *Diabrotica soror.* It is apparently proved that such change is not explicable on a basis of intra-specific selection, nor can it be interpreted as a direct ontogenetic (pertaining to the individual) reaction in each succeeding generation to changing climatic conditions. It is believed to be an example of definitive orthogenetic variation.

Many paleontologists think they see repeated instances of orthogenetic variation, since they are, as a rule, so deeply impressed by the adaptive nature of the evolutionary process and the definitiveness of is direction, that they can not believe the restraining or selective action of the environment is enough to keep the breed true; for example, the sure way in which the cusps of the teeth in ungulate animals appear in the course of time, so that one can predict the ultimate tooth pattern at the end of an evolutionary series with fair assurance. The rhinoceroses, on the one hand, and the totally unrelated coney, *Hyrax,* on the other, have teeth so similar that Roosevelt calls attention to the fact in his *African Game Trails.* This is an instance of *convergence,* which is often the result of what is apparently orthogenetic variation. The overgrown antlers of the Irish deer or the tusks of the Jefferson mammoth which in each instance seem to be so far beyond the point to which natural selection would carry them, may be considered as further instances of this form of variation.

Continuous and Discontinuous

Variations may also be classed according to difference of degree, as (5) continuous, that is, small, abundant, and occurring in graded series. These, if also fortuitous, are the so-called Darwinian variations or fluctuations, to be acted upon by natural selection. They are generally quantitative rather than numerical variations, and the increment of change between successive generations is extremely slight.

Fig. 10.6: Eight-fingered hand, *Homo sapiens,* female.

In contrast to the continuous are the (6) discontinuous variations, which are mostly large and rare. They are also known as sports or salutations, although the distinction between them and the continuous is only one of degree, as the two sorts merge into each other.

The great mass of variations come under the head of continuous, the discontinuous ones being relatively rare, although many such have been recorded, as for example the numerous instances of variation from the standard number of digits in vertebrates. In mankind, six, seven, or even eight fingers to the hand have been observed while the doubling of the foot has brought the number of toes as high as nine on a side. Cattle and horses sometimes show multiple digits and, on the other hand, pigs with the two median toes united into one perfectly formed digit, the so-called, solid-hoofed pigs, occasionally occur. Sheep sometimes show four horns instead of two, but the peculiarity is that instead of having one pair behind the other as in the normal four-horned antelope (*Tetracerus quadricornis*) the horns always stand in a single transverse series across the skull. One family of goats on an isolated farm near Bozen has four horns which have been inherited for many generations (Bateson).

All of these are instances of meristic (Gr. *μερος*, part) or number variations and are likely to occur wherever any structure is repeated in numerical series: in the number of segments of worms and arthropods, or in the vertebral column, ribs, and muscles, and in the number of appendages. Radially symmetrical forms such as sea-anemones and starfish also exhibit meristic variation and it frequently occurs among plants, especially in the flowers or leaves, as in the four-leaf clover.

Other instances of discontinuous variations are not of this numerical sort but are mutations of colour, form, and size. Such, for example, are the famous new salutations of Lamarck's evening primrose (*Enothera lamarckiana*), which breed true and gave rise to De Vries' theory of the origin of new species by brusque salutations. Some of these vary from the original in size, one is a smooth-leaf form with a more beautiful foliage, and so on. Another example is that of the short-legged ram which suddenly appeared in Massachusetts in the year 1791. This ram proved to be a potent sire, capable of handling on his peculiarity to his progeny and thus was the founder of a breed of so-called Ancon sheep whose principal virtue was their inability to jump fences due to the brevity of their limbs; with the introduction of the more desirable merinos, breeders allowed the Ancon sheep to disappear.

SOURCES OF VARIATIONS

Sources of acquired variation are apparent. Any external condition, influencing the organism produces this type of variation. The external factors may be light, heat, cold, food etc. These variations, because they are somatic, may not be inherited. The cause of evolution, really speaking, comes from the germinal variation which is inheritable.

1. Light

When two animals or plants of the same species are grown under different conditions of light intensity, they show mark variations in their form. *Cuningham* proved experimentally that the flat fish, *Solea* developed pink colour on its ventral surface, which is normally white, when exposed to light. The upper surface, which is normally pink, turns white when the animals are kept in shed.

2. Temperature

A number of variations can be observed by lowering or increasing the temperature. *Tower* and *Agar* by their experiment an mice have shown, that mice reared in a warm room at about 21ºC were found to differ considerably from those reared in a cold room at 5ºC.

3. Water

It is also an important factor for normal growth. Plants growing in dry an arid soils undergo conserve water, while plants of moist areas show luxurious growth.

4. Nutrition

Nutrition, like water, is another very important factor for the growth of an organism.

5. Endocrine Glands

The hormonal secretions of endocrine glands, which influences the development and differentiation of various physical and mental characters are also responsible for causing variations. For example Axolotl larva of *Ambyostoma*. Due to thyroxine deficiency it remain as larva.

6. The use and Disuse of Organs

Use and disuse of organs also produce variations. The organ which is put to continuous use shows better development than less used. The constant disuse of an organ leads to atrophying of that organ.

7. Cytological Basis

The cytological or genetic sources of variations may be due to three reasons.

(i) Difference in the number of chromosomes

(ii) Change in the structure of individual chromosomes

(iii) Gene mutations.

The first two are collectively called *chromosomal aberrations.*

(i) Differences in the number of chromosomes. It is also called *heteroploidy.* It is of two types.

(A) Changes involving entire sets.

(a) Haploidy. Having are genome only (n)

(b) Polyploidy. Each genome is represented more than twice like triploidy (3n), tetraploidy (4n) and so on.

(B) Changes involving the number of chromosomes in one genome.

(a) *Monosomic.* Loss of one chromosome from diploid set (2n - 1)

(b) *Nullisomic.* Loss of both the homologous chromosomes from diploid set (2n - 2).

(c) *Trisomic.* Increase of one chromone in a diploid set (2n + 1)

(d) *Tetrasomic.* Increase in one pair of homologous chromosomes (2n + 2)

(ii) Change in the structure of chromosomes.

(a) *Delation*. Loss of one or more genes.

(b) *Dupliction*. Addition of one or more genes.

(c) *Inversion*. A chromosomal segment is inverted at 180°C.

(d) *Translocation*. Exchange of parts between non-homologous chromosomes.

CAUSES OF VARIATIONS

The causes of acquired variations are apparent, for every modification which the organism undergoes in its lifetime in response to any external condition whatsoever comes under this head. The influence of abundance or scarcity of food in giving the animal a great stature or one less great than that of its brethren, the influence of heat, cold, omnipresent enemies, or even the absence of exacting conditions: all these make themselves felt upon the individual in greater or less degree, but, as has been emphasized, this individual adaptation, however interesting it may be, is *not* evolution. Therefore in our inquiry into the subordinate factors of evolution, it is not the cause of *acquired*, but of *germinal* variations with which we are concerned.

Causes of Germinal Variations

As these variations are highly important raw material of evolution, it is at once apparent that the cause or causes which produce them are of fundamental importances prime movers of the whole process. Our ignorance of these causes is profound, nevertheless the following ideas concerning them are worthy of record.

First the inherent tendency to vary. This does not, however, state a cause but merely a fact and really begs the question. The reverse of the statement—the almost absolute impossibility of two organism being alike—is perhaps the better way to express it. When one thinks of the marvellous complexity of protoplasm, consisting as it may of sixteen or more elements with the atoms numbering thousands, held in an atomic structure so complex that the chemist cannot express it graphically, as he can simpler compounds such as water, he sees how remote is the possibility of any two particles of protoplasm being precisely alike. Furthermore the organism, no matter how simple, is made up of countless molecules, and these are in a continual state of chemical change due to energy traffic. How then can one possibly expect coincidence of detail in any two organisms, no matter how close their relationship? The amazing thing is not that they differ, but that they are so much alike!

Second, dual or biparental parentage (amphimixis) has been given as a principal cause of variation. That which is actually handed on to the next generation is the germ-plasm derived in part from each parent. This in itself is a highly complex material, consisting of chromosomes which, in man for instance, number 48. During the process of maturation of the germ cells, ova and spermatozoa, the number is reduced to 24.

These are combined in the process of fertilization so that the number of chromosomes in the initial cell of the embryo is once more 48. The process may be likened to two miniature packs of cards, each of which is elaborately shuffed, one-half the cards rejected, and the remaining halves combined into a new pack. This gives rise to an enormous number of possible re-combinations. If the packs contain 48 cards, instead of the normal 52, to correspond with the human chromosomes, the number of possible permutations will be 16,777,216 and for the average germ capacity during the sexual life of a pair of parents the total number of different combinations of fertilizable ova would be 300,000,000,000,000 (Dorsey from Thompson). No wonder the children of a family differ!

Third, actual observation shows that variation also occurs where there is but one parent, as in the Protozoa, where offspring are produced by the division of the mother cell into two daughter cells. Also in higher animals where the offspring come from a virgin mother without the intervention of the male (parthenogenesis), as in many insects—bees, lant lice, and the like; but here again there is the same process of halving the pack with the maturation of the ovum, giving opportunity, although not mathematically so great, for re-combinations in the offspring. In the protozoan *Paramecium,* Woodruff has observed a periodical reorganization of the nuclear apparatus (endomixis), which in some way seems to rejuvenate the race. Here again there may be an opportunity for new chromosome combinations to emerge.

The second and third might account for germinal variations which are the result of re-combinations of ancestral characters appearing in the offspring, but there are also the mutations, which are due to fundamental changes in the germinal constitution itself. To account for these our first statement might apply in part.

Another idea is that, in spite of the insulation of germ-plasm from extrinsic influence, nevertheless some of the more deeply saturating results of the ever varying environment may act directly on it and so bring about heritable change. This has been definitely shown by X-raying the germ-plasm.

A new line of research, which is yet far from complete, is the study of the endocrine or ductless glands, which are already known to have a profound influence upon individual development, both mental and physical, including that of the secondary sex characters. These glands do not form secretions in the sense that the salivary and other digestive glands do, but are connected only by the blood stream, into which are poured under proper stimulus the hormones or chemical messengers which they manufacture. These hormones are picked out of the blood stream by the appropriate organs and upon them exert a powerful influence. They pass through the reproductive glands as readily as elsewhere, and it is reasonable to suppose that certain of these hormones may react upon the germ cells as well as upon the body cells. This may prove to be a mechanism whereby external influence is exerted on the most intimate parts of the organism.

Chapter—12

Origin of Species

It has already been observed that populations evolve new characteristics largely as a result of changes in the organism-environment relationship, acting through the medium of natural selection. Since environments are very diverse and complex in both space and time, organism-environment relationships are always different for related populations living in different parts of the earth, and are perpetually changing as the earth's environments are altered in various ways. Changes in climate, in the distribution of land and sea, elevation and degradation of mountain systems, and particularly the increase, decrease, and extinction of different kinds of animals and plants all interact to produce these changes in organism-environment relationships.

We can start with a single relatively homogenous population, living in a particular area and possessing a reasonably large pool of genes, from which combinations adapting some of its individuals to new conditions can be sorted out by natural selection and established in new habitats. If this condition exists and if new habitats are available, the evolutionary line derived from the initial population will undergo ADAPTIVE RADIATION. Individuals entering new regions will found subpopulations which will differ in various adaptive characteristics from the ancestral population and from each other. In this way a POLYTYPIC SPECIES will evolve consisting of many races adapted to different conditions. The most easily recognized of these races, or groups of races, are usually called SUBSPECIES. From time to time, but by no means always, races can diverge from their ancestral populations in ways which decrease or prevent their ability to exchange genes with other races. This may come about either by reducing the frequency with which interracial hybrids are produced or by lowering the viability or fertility of the hybrids or their progeny.

When any of these barriers to gene exchange, known as ISOLATING MECHANISMS, have become strongly developed, the populations are said to be REPRODUCTIVELY ISOLATED from each other. They then rank as distinct SPECIES. If, thereafter, new environmental changes enable reproductively isolated populations to migrate into the same region, they can exist side by side, each population maintaining its own distinctive adaptive gene complex. Each of these newly evolved species can undergo its own distinctive cycle of adaptive radiation and speciation. In this way an adaptive radiation develops which is of a higher order than that which gives rise to a polytypic species. This results in the evolution of clusters of related species which form a GENUS.

Sometimes two differently adapted populations or subspecies of a poly-typic species can come together before reproductive isolation between them has been fully developed. In that case they can exchange genes to a limited degree, and progeny resulting from such HYBRIDIZATION may contain individuals adapted to environments different from those occupied by either of the parental populations, Although such hybrids are most often unsuccessful because of their weakness or sterility, some of them may give rise to new populations, well adapted to a new habitat which may be available. Under such exceptional circumstances, hybridization can greatly speed up the processes of evolution. In plants, hybrids may give rise to new species by doubling their chromosome number, a process known as POLYPLOIDY. In this way, a sterile hybrid can give rise immediately to fertile offspring, which will remain constant for the intermediate, hybrid condition.

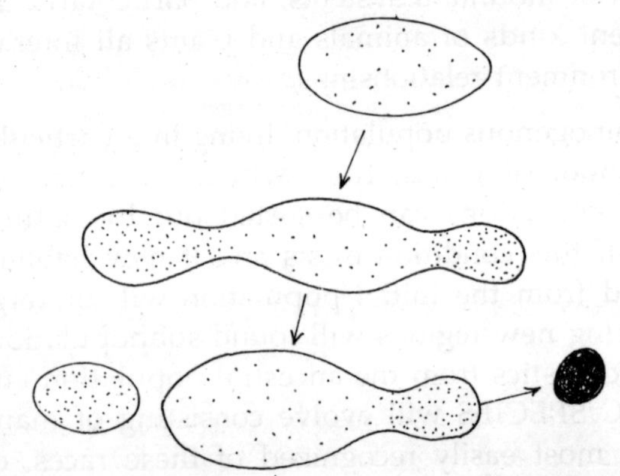

First stage.
A single population in a homogeneous environment.

Second stage.
Differentiation of environment, and migration to new environments produces racial differentiation of roces and subspecies (indicated by different kinds of shading).

Third stage.
Further differentiation and migration produces geographic isolation of some roces and subspecies.

Fourth stage.
Some of these isolated subspecies differentiate with respect to genic and chromosomal chonges which control reproductive isolating mechanisms.

Fifth stage.
Changes in the environment permit geographically isolated population to exist together again in the same region. They now remain distinct because of the reproductive isoloting barrirs which separate them, and can be recongnized as good species.

Fig. 12.1: Diagram showing the sequence of events which leads to the production of different races, subspecies, and species, starting with a homogeneous, similar group of populations.

This sequence of events is illustrated in Figure 12.1. The five stages shown in this diagram and explained in the legend, are somewhat over-simplified, but nevertheless give a general idea of the commonest way in which species can originate.

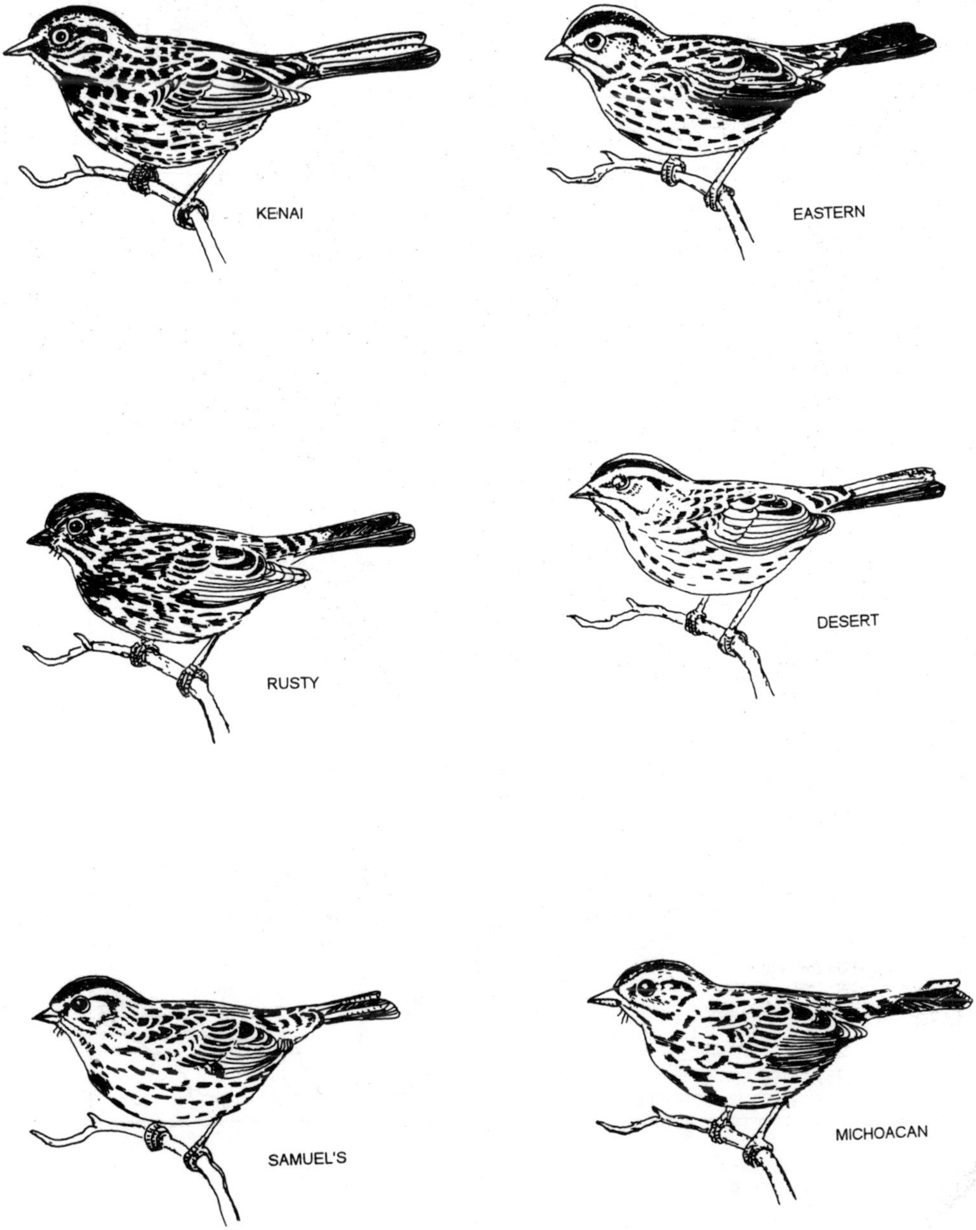

Fig. 12.2: Representative subspecies of the sang sparrow (*Melospiza melodia*) together with a map showing the distribution of the numerous subspecies in North America.

EVOLUTION BY SHELL

What is a Species?

The problem now before us in involved-almost stated-in the definition of species. Though taxonomists have long argued the nature of species and still differ in the meanings they attach to be word, elimination of some of the finer points at issue will give us a workable concept. A species is a group of individuals which have potential access to the same stock of genes. These genes have at least an opportunity of flowing freely through any line of descent within the species. Genes to which the members of a species do not have access belong to another species. Some sort of separation exists between the members of different species, such that the lines of gene descent do not intermingle. This means, of course, that they do not cross.

To bring about commerce between such groups of genes, individuals of different species would have to meet; in the higher animals they would have to copulate; eggs and sperm would have to commingle, then unite in fertilization; embryonic development would need to proceed normally, producing a functioning adult, which in turn would have to meet another individual, initiating a like chain of events. As between species, either the initial event does not occur, or some link in the chain is not forged. The stoppage occurs at different places for different pairs of species, but the effect is the same-the groups of genes are kept apart.

The general phenomenon of prevention of gene interflow is known as *isolation*. Some of the sources of isolation are imposed from without the organisms; some are within them. Examples of the more important ones will be reviewed.

Why should there be Species?

Living beings vary in size from microbes as small as 100-250 millimicrons in diameter to giants like blue whales and sequoia trees. They live on unlike foods, in unlike environments—in water, in humid jungles, and in dry deserts—and at temperatures as low as —23°C and as high as 80-85°C. Yet it is probable that the living world is monophyletic. If life arose only once, then all existing organisms are products of divergent evolution. The evolution was gradual; sudden radical transformations did not occur. Therefore, if all organisms that ever lived were preserved as fossils, one could find all transitions between all now existing forms of life. But only a minuscule fraction have been fossilized. Anyway, the living world is radically discontinuous. Not only does life present itself to our observation in the form of discontinuous quanta called individuals, but individuals usually occur in discontinuous arrays. Some examples of such arrays are mentioned above-mankind, the chimpanzee, gorilla, and orang. It is legitimate to ask, What is the meaning, the biological function, of the discontinuous arrays we call species?

It has been mentioned repeatedly in the foregoing chapters that the mechanism of Mendelian segregation and recombination generate a profusion of gene constellations. One parent heterozygous for n genes has a potentiality of producing 2^n genetically different gametes; two parents heterozygous for the same n genes have the potentiality of producing 3^n and for n different genes 4^n genetically diverse zygotes. Wright (1932) pointed out that these potentialities are far in excess of what can be realized. Even with over conservative estimates of the number of genes and gene alleles, the number of potentially possible genotypes in the living world is many orders of magnitude greater than the number of atoms and subatomic particles estimated by cosmologists in the universe. What exists is an almost infinitesimally small sample of what could exist.

Arrays of the gene combinations that exist have been winnowed by natural selection. Inviable or poorly viable combinations were not perpetuated, even if they ever appeared. This is not to say that all possible organisms that could live are living or have lived. Natural selection has no pre-established program to bring into being all that could live. It is useless to speculate what organisms might have emerged that did not emerge. The situation of the living world can conveniently by envisaged in the form proposed by Wright, a symbolic "map" of adaptive "peaks" and "valley". Such "maps" should have as many dimensions as there are genes with two, or several, or many alleles. With two dimensions, the two axes symbolize the allelic variants of two gene loci. The Darwinian fitness (adaptive value) of the combinations of these alleles is symbolized by contours, like in a topographic map.

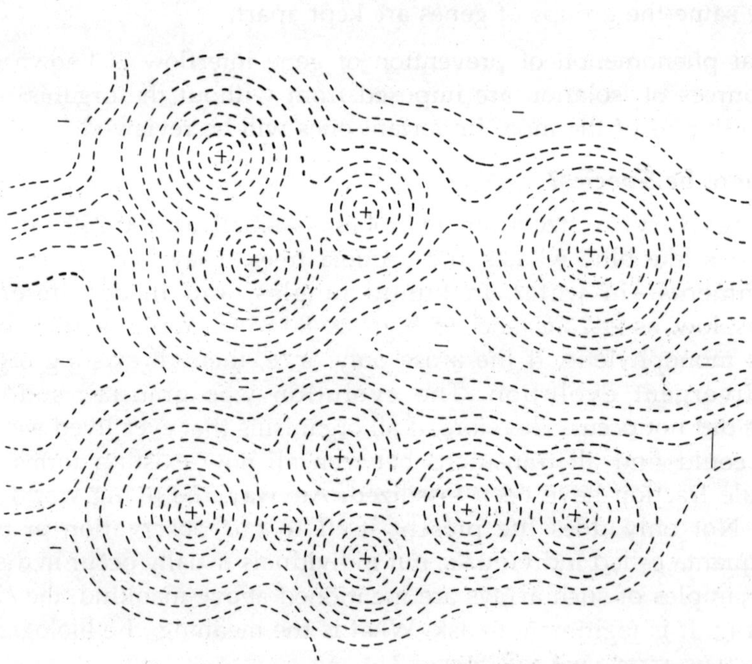

Fig. 12.3: Adaptive peaks and valleys in the field of gene combinations. The contour lines encompass genotypes with comparable adaptive values. Populations occupy adaptive peaks (+); the valleys (-) represent genotypic combinations with low adaptive values.

One can imagine a world where life exists in only one ecological niche, uses only one kind of food, and subsists in only one environment. A single species would then be enough. Perhaps a situation of this sort was realized when the first life arose from inanimate matter. But since then life has conquered and spread into many environments. Life has responded to the diversity of environments by evolving genotypes adapted to inhabit them. It has mastered many adaptive peaks. In the field of potentially possible gene combinations some combinations are adaptively coherent, while others are unfit to survive and reproduce. A species is an array of related gene combinations that occupy one of the adaptive peaks represented in Figure 12.3. The summit of the peak is held by the genetic "elite," the genotypes that are fittest in the environments in which the species lives. The carriers of the less-fit genotypes can be said to occupy the slopes of the adaptive peak. The least fit, e.g., lethals and other constituents of the genetic burden, perish in the "adaptive valleys." The question of why there should be species can thus be answered: because there are many adaptive peaks.

DISCONTINUITY OF ORGANIC VARIATION

Figure 12.3 shows several adaptive peaks, some of which are higher than others. High peaks symbolize species of high adaptedness, and lower ones are held by forms less well adapted but still able to survive. There can also be unoccupied peaks, i.e., opportunities for living that are nor exploited. Some unoccupied peaks may in time be settled if genotypes able to occupy them eventually evolve. Others may remain unoccupied if suitable gene combinations fail to emerge. How do we know that unoccupied peaks exist? There is good evidence that animals and plants introduced to new countries sometimes become highly successful pests or weeds, much more so than in their native countries. For example, the gypsy moth (*Porthetria dispar*) does relatively little harm to the vegetation in its native Europe, but in New England, where it was introduced, it has multiplied so much that it has denuded forests. There evidently was an unoccupied adaptive peak that the gypsy moth occupied quite successfully. One may speculate that man could well have lived in the early Tertiary or even Mesozoic periods; however, it was only in the late Pliocene and Pleistocene that the human species finally appeared. The human peak, this Everest of the adaptive landscape, was unoccupied for tens or even hundreds of millions of years. Evolution can be described as a series of occupations of previously vacant adaptive peaks.

The adaptive peaks in Figure 12.3 are shown grouped in clusters, separated by some shallow and some deep adaptive valleys. This symbolizes the two fundamental characteristics of organic diversity—its discontinuity and the hierarchial ordering. Instead of all possible gene combinations generated at random, one observes species between which, with exceptions to the discussed below, intermediates are absent. For example, no individual has ever been seen about which one could have the slightest hesitation as to whether is belongs to the species humankind or the species chimpanzee. However, the genetic and structural gaps between species of one genus are on the average smaller than between, genera, smaller between genera than between families, and so forth. Classifiers of animals and plant utilize discontinuities to achieve a natural and convenient system. The system. The system is "natural" in so far as the discontinuities are observable realities.

Any number of examples could be given to illustrate the discontinuous and hierarchical character of organic variation. We shall use the cat-like animals and other carnivores because the are familiar to most readers. All breeds or races of domestic cats belong to the species *Felis cattus*. They mate freely and produce fertile hybrid progeny. The European wildcat, *Felis sylvestris,* is a related but by now nearly extinct species. Hybrids of wildcats and domestic cats have been obtained in zoos, although their fertility has not been tested. The bocat (*Felis lynx*) is sometimes treated as a genus (*Lynx*) separate from *Felis;* crosses of bobcats with domestic and wildcats have been reported in zoos, but their viability and fertility are uncertain. The lion, tiger, leopard, and jaguar were formerly considered species of the genus *Felis,* but are now placed in a separate genus *Panthera (P. leo, P. tigris, P. pardus, P. onca)*. No hybrids between these species occur in nature, but they do mate and produce hybrids in zoos. There are reports of hybridization of lions and tigers; the hybrids ("tiglons") are viable and fertile, at least as females. The crosses lion X leopard, lion X jaguar, leopard X jaguar, and leopard X tiger have all been obtained, but the fertility of the hybrids is insufficiently studied. Crosses of species of *Panthera* with domestic cats are impossible on account of differences in size, but it would be interesting to try to achieve hybridization by artificial insemination. *Felis, Panthera,* and some other genera make up the cat family (Felidae). Together with the dog family (Canidae), bears (Ursidae), weasels (Mustellidae), and mongooses (Viverridae), these families constitute the order Carnivora.

Let it be reemphasized that discontinuity and a hierarchical ordering are universal in the living world. They are consequences of the fact that only a minority of the potentially possible gene combinations can give rise to viable organisms. Discontinuities and group hierarchies are observed among micro-organisms, even viruses, as well as among lower and higher animals and plants. The very universality of these phenomena, while it is a godsend to systematists, raises some knotty problems. Are species of, say, bacteria self-same or even analogous biological phenomena to species of grasses, trees, insects, birds, sexually reproducing and asexual, recent and fossil organisms,? It is merely a matter of convenience whether we place certain species in the same or in different genera? It is arbitrary to treat genetically distinct populations as races of the same species or as separate species? We have seen above, in the example of the carnivore mammals, that some species put in one genus can be placed in different genera.

Racial Differences within Species

Nearly every species is differentiated into races, and in many of them there exist races distinctive enough to be called subspecies. For example, a person familiar with the birds of the eastern United States can recognize the song sparrow (*Melospiza melodia*) by its size, colour pattern, and song. If he travels to the west coast of our continent, he will hear a song which reminds him of that sung by the eastern song sparrow but is not quite the same. When he sees the bird which is singing this song he will recognize still further differences. Furthermore, the song sparrows of the Pacific Coast are not all alike. Those of central California are recognizably different from the song sparrows of Oregon and Washington. Within the San Francisco Bay region, populations of song sparrows inhabiting hill slopes and marshes separated from each other by only a few miles are recognizably different.

We might ask at once, "Doesn't this mean that there are actually several different species which are all given the common name of song sparrow?" The biologist at once answers, "No, these different kinds of song sparrows are no more different species than are the various races of mankind." The reason for this answer is that wherever areas of geographic distribution of two different races or subspecies of song sparrow come close to each other, intermediate populations are found. One can never find two or more different races of this bird breeding in the same locality and remaining distinct from each other.

Species of plants also consist of many different populations. If we look at the pine trees which cover the slopes of a high mountain, we notice that those near the summit are much smaller than those near the base, and they are likely to be gnarled and spreading. The same is true of trees and shrubs which grow near the seashore and are exposed to constant winds. We might suspect that these differences result directly from the effects of the severe environment. This is true to a certain extent. Nevertheless, when seeds are taken from trees growing at various altitudes in the mountains, and their offspring are raised in a uniform nursery, those descended from trees growing at the lower altitudes grow faster and become larger than do the progeny of trees which were growing at higher altitudes. Evidently, therefore, natural selection has sorted out genotypes adapted to the conditions of their native environment. Such adaptive differentiation is not confined to size and rate of growth. For instance, many smaller plants or herbs survive the harsh winters of northern climates or high altitudes by dying back to their roots and becoming dormant during the winter months. If such plants are transported to a mild climate where little or not frost occurs, they will, nevertheless, become at least partly dormant during the winter. Related races of the same species which are native to the mild climate will grow actively during this same period.

In a number of instances, hybrids have been made between such differently adapted races or subspecies of plants. The hybrids are usually vigorous and intermediate in most respects between the parental races. When seeds are grown from such hybrids, they produce a large and extremely variable segregating progeny, no two individuals of which are genetically alike. Furthermore, when such progeny are grown in a variety of environments, individuals can be recognized to have adaptive properties different from both parents, as well as from the F_1 hybrid. The segregating progeny of a hybrid swarm, containing a greatly increased pool of genes controlling adaptive characteristics. If such hybridization takes place extensively in nature, the adaptive complexes of the parental populations can rarely be reconstituted in their original form.

Sometimes the differentiation of races within a species is not associated with marked differences in the climate or habitat which they occupy. For instance, the island of Celebes in Indonesia is inhabited by six different races of a species of carpenter bee (*Xylocopa nobilis*). These races inhabit very similar rain forests. The differences between them, though probably adaptive, are not related to any easily recognizable differences in their habitat. Quite possible, they represent different modes of adaptation to the same habitat, as explained at the end of the previous chapter.

Although most populations of a species inhabiting different regions are recognizably different from each other, this is not necessarily so. Several species of plants inhabiting two or more widely separated areas have populations so similar in the two areas that the difference between areas is probably less than the variation within a population or between neighbouring populations. An example is the skunk cabbage (*Symplocarpus' foetidus*), common in swamps throughout the eastern United States and found also in similar places in China. In these examples, geological and palaeontological evidence indicates that exchange of genes between the separated populations has not been possible for millions of years. Consequently we must conclude that separation of populations merely by distance is not enough to cause them to diverge and form different races. They must be subjected to differential selective pressures.

Some examples of mixing between races tell us, furthermore, that races formation is a reversible process. Races can disappear through hybridizing and combining their genes with those of other races. This is true even of races which have been separated and differentiated from each other for thousands or even millions of years. Among birds, this is shown by the wood-pecker-like birds known as flickers (*Colaptes auratus*) in the northern United States.

Among woody plants there are numerous examples of populations which have been separated from each other for millions of years and have become differentiated into entities generally recognized as species, but which are still able to exchange genes with each other. This condition has been found in pines, oaks, chestnuts, plane trees, and many others. For instance, the hybrid between the plane tree of the eastern United States (*plantanus occidentalis*) and its counterpart in southeastern Europe and southwestern Asia (*P. orientalis*) is often grown in parks and is fully fertile.

Separation by Space

Individuals belonging to different populations do not mate if they never meet. Sometimes mere distance would pre—clude intermating even if there were no other obstacle. Usually there are ecological factors which keep the species more or less apart even though they are near one another. The cormorant and the shag, two bird species belonging to the same genus, are close together on the coat of Europe, but the former feeds in shallow water, on fish that live near the bottom, while the latter feeds in deeper water, on free-swimming fish. Two subspecies of the mouse Peromyscus are kept apart because on lives on lake beaches, the other in woods. Two species of mosquito (genus Anopheles) in Africa are separated by salinity of the water. The larvae of both can live in fresh water, but one of them can endure brackish water and, perhaps to avoid competition, usually does live in such water.

Sometimes topographic barriers keep species apart. The Grand Canyon separates two species of ground squirrel, also one species are partially separated by the canyon, but meet either above or below it. Some of the steamer ducks (genus Tachyeres) are separated by water. These ducks propel themselves rapidly over the surface of the water by means of their wings, used as paddles. One species is limited to the region around the Straits of Magellan at the tip of South America; another species is limited to the Falkland Islands. These two species do not meet. A third species, with better locomotion, occupies both of these areas and crosses the water between. It has maintained itself a single species. The falls of the Nile River effectively separate the stream fishes, and certain genera found in the lower course of the river are not represented at all above the falls.

Isolation by Time

Two species of cockleburs produce their flowers at different times. In one of them the flowers do not open until after the seed capsules of the other are fully formed. Without any overlapping, they could not cross. Two species of Salvia are similarly separated by different times of flowering; but two other species of Salvia have overlapping flowering periods, and some hybrids are produced when the two kinds live together.

Specialized food habits separate two species of bees (genus Andrena) in Wisconsin. They visit different flowers and are active at different seasons corresponding with the times of blooming of these plants. Two moth species feed on different plants which are in best condition at different times, but the moths emerge at different times in the spring so that food is not, at least directly, the separating factor. In two grasshopper species the isolation is definitely produced by the time of emergence. One of these species lives over winter in an immature form and becomes adult early in the spring. The other develops from eggs laid in the spring and matures much later. Even with this difference, there is an overlap of about two weeks in the periods when adults are found.

Two forms of sockeye salmon in British Columbia ripen the germ cells at different times. One form spawns in August and September, the other from October to December, both in the same lake. The late-spawning kind has a degree of immunity to a parasitic copepod not possessed by the other, and there is some difference in colour. Such differences must arise because of lack of crossing between them. Taxonomically the two types are doubtfully considered subspecies of the same species.

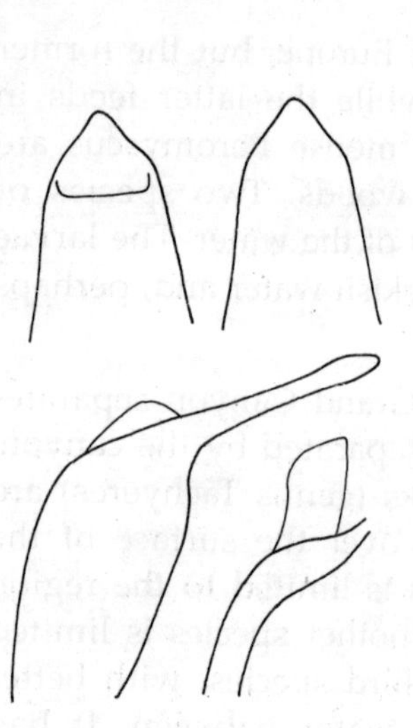

Fig. 12.4: Male genitalia of two species of lady beetle, *Hippodamia Aquinquesignata (left)* and *H. convergens*. These species have crossed and produced fertile offspring.

Diurnal cycles of activity almost separate two species of fish in the rivers of Indiana and adjoining states. The white crappie is active in the daytime, the black crappie at night. They have similar habits and food and live in the same streams. Here the isolation is not complete, for interspecific hybrids are occasionally found.

Bars to Mating

Members of different species which live together and are mature at the same time are kept from mating by various factors. There has long been, particularly among entomologists, a belief that the copulatory apparatus is so constructed that of a male would not fit a female on another species. This belief was named the "lock-and-key" hypothesis. It does not seem very well founded, however. Two species of lady beetles (Hippodamia) in which the male genitalia are strikingly different have been crossed, and the genitalia themselves shown to be just one of the inherited characters. Three pairs of cumulative genes would account for the specific differences in these structures.

Some species are kept from mating by psychological (Mayr prefers the world ethological) differences. These are of different sports. The courtship behaviour of animals is designed, not to influence choice of mates, but to arouse readiness for mating. The courtship of animals of different species is usually broken off in early stages because the behaviour patterns do not complement one another. A number of pairs of species of Drosophila have been tested, and there are few interspecific matings when members of the same species are available. The call notes of frogs and toads are found to keep species apart at breeding time. Patterns of display among fiddler crabs involve waving the chelipeds, raising the body, dancing, and other movements, and the details and combinations are different in different species.

The breaking off of courtship between species has sometimes been attributed to "species recognition," as if animals wanted their offspring to remain true to type. It seems more likely that the courtship simply did not arouse the desired reactions and was discontinued because it was ineffective.

Ethological barriers are often assumed when no others can be detected. Two species of mice (Peromyscus) breed freely together in the laboratory, and their offspring are viable. Yet in nature, though they meet over considerable areas, practically no hybrid are found. Two species of meadow larks are similarly inter fertile indoors, but even with broad overlap of their ranges and habitats there is seldom any natural crossing. The mallard and pintail ducks produce fertile hybrids in captivity, but do not cross in their very wide area of overlapping ranges.

Conjugation in *Paramecium aurelia* is barred or ineffective as between certain strains. This species-regarded as a single species by taxonomists—is broken up into eight physiological species (usually so called though besides their isolation they differ in size, rate of fission, frequency of autogamy, and time and temperature of conjugation). Each of these species comprises two mating types. Two individuals of the same mating type never conjugate, but an individual of any mating type may conjugate with a member of he other mating type of the same physiological species. This phenomenon of mating types is not isolation; rather it is like sex in other organisms. Between the physiological species, however, there is isolation. Neither mating type of species 2 will conjugate with either mating type of species 6. The isolation of these two is complete, and the point of isolation is the prevention of mating.

An interesting point concerning the isolation of physiological species of *Paramecium aurelia* is that it may rest on a single-gene difference. There has been much discussion of the simplicity or complexity of the genetic isolation of species among higher organisms, with most evolutionists taking the position that it must (or does) rest on a number of genes. In *P. aurelia,* however, a single mutation may give rise to an isolated group.

SPECIES AS DISTINCT SYMPATRIC POPULATIONS

The facts presented above tell us that related populations can be separated from each other for long periods of time, extending even to millions of years, and can be subjected to such different selection pressures that they come to differ markedly from each other while still separated. Nevertheless they can retain the ability to exchange genes freely and reunite into a single interbreeding population if they are brought together again. There exist, however, many groups of related populations which live together in the same area, but which remain distinct from each other without exchanging genes. These are the entities which we usually recognize as species. Anyone familiar with the animal life of the lakes and streams of the northeastern United States can recognize the leopard from (*Rana pipiens*), the wood from (*Rana sylvatica*), or the bullfrog (*Rana catesbiana*). Although these species occur commonly in the same area, they never intergrade. Some acquaintance with birds will enable one to distinguish between the song sparrow (*Melospiza melodia*) and the Lincoln's sparrow (*Melospiza lincolni*) (Figure 12.5). Populations of these species live together over

Fig. 12.5: Two related but sympatric and distinct species of North American birds, the Lincoln sparrow (left) and the song sparrow (right). From R. T. Peterson, *Field Guide to Western Birds* (Houghton Mifflin Company).

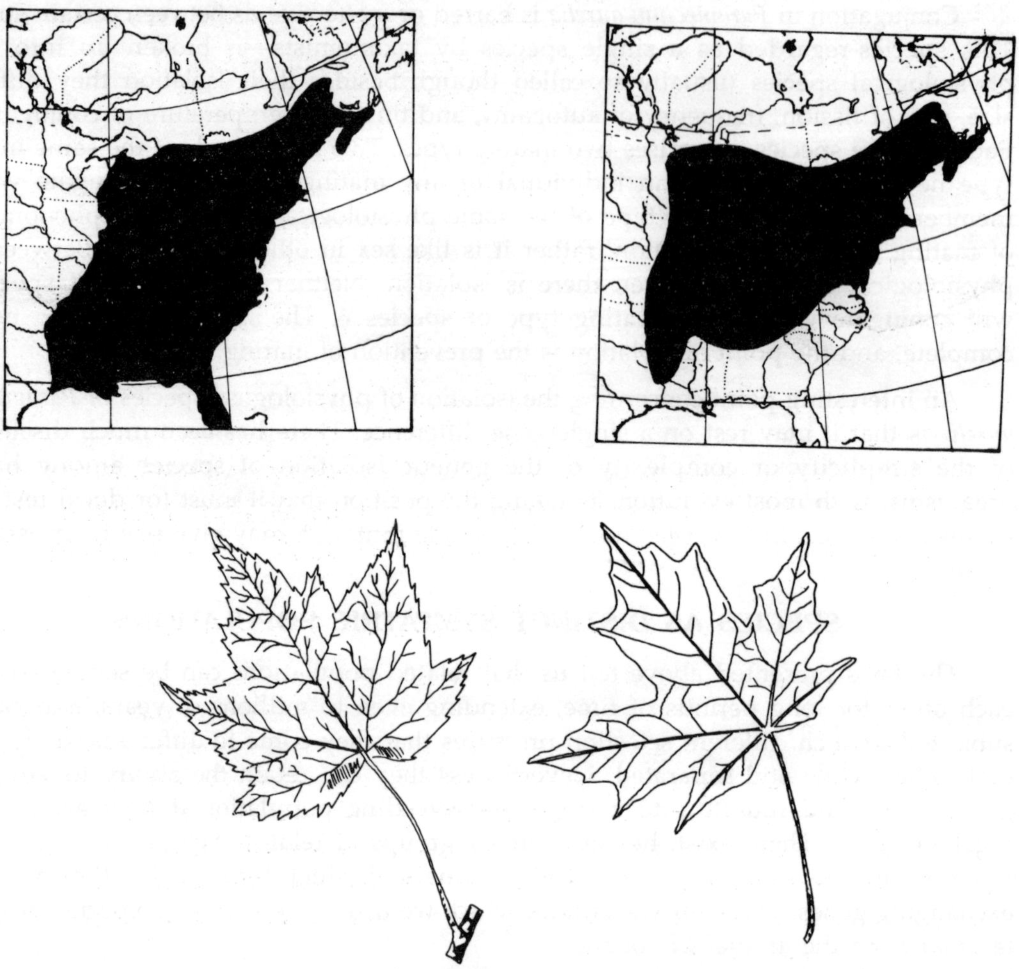

Fig. 12.6: Two related and sympatric species of trees, the red maple (left) and sugar maple (right). From Preston, *North American Trees* (Iowa State University Press.)

considerable areas of our country and regularly remain distinct from each other. They are apparently incapable of exchanging genes under natural conditions. The same is true of such easily recognized species of plants as the sugar maple (*Acer saccharum*) and red maple (*A. rubrum*) (Figure 12.6), long leaved pine (*Pinus palustris*) and scrub pine (*P. virginiana*), as well as ponderosa pone (*P. ponderosa*) and digger pine (*P. sabiniana*) in California.

At this point, the reader should become fully aware of what is meant by the phrase "incapable of exchanging genes under natural conditions." It does *not* mean that species can become distinct from each other only by evolving mechanisms which prevent them from hybridizing under any conditions, or which render the hybrids completely sterile. There are many examples of perfectly distinct and valid pairs of species which nevertheless can form fertile hybrids under artificial conditions. For instance, the common mallard duck (*Anas platyrynchos*) and the pintail duck (*anas acuta*) will hybridize readily and produce fertile offspring when kept in captivity, but due to their very different courtship and nesting habits, they do so very rarely in nature. Hence they can maintain themselves as distinct sympatric species even though they nest side by side in millions of ponds or creeks.

One of the basic questions asked by the evolutionist is: What factors in the evolutionary history of these related populations have made them diverge to such an

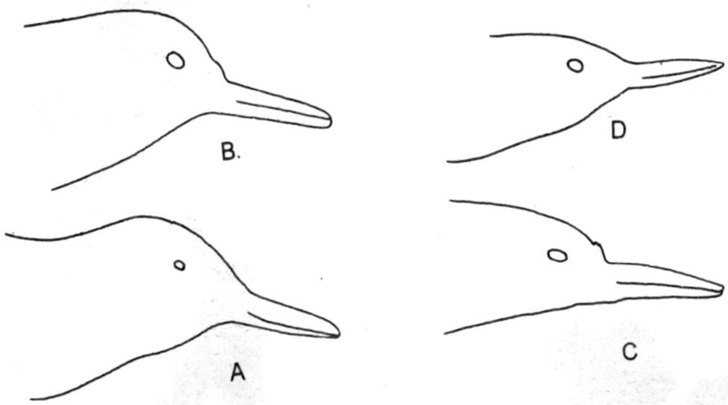

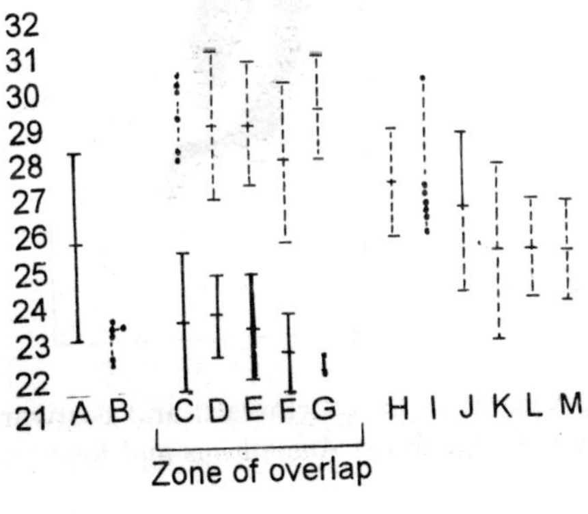

Fig. 12.7: Diagrams showing character displacement in respect to break size in two species of nuthatches (A-C and B-D) which inhabit southeastern Europe and southwestern Asia. In the region where the two species occur sympatrically (C, D) they differ from each other to a greater extent than in regions where only one species is found (A, B).

extent that they are no longer capable of exchanging genes? This question is so important that the remainder of the present chapter will be devoted chiefly to discussing possible answers to it. Its importance lies in the following fact. As long as populations retain their ability to exchange genes, this mixing will cause them to merge into a single population whenever they are brought together after separation, and will prevent them from diverging further. Consequently, they cannot become the progenitors of new and distinctive lines of evolution. If, on the other hand, populations have evolved distinctive characteristics of the sort which prevent them from exchanging genes with each other, they will remain distinct when brought together in the same region. Furthermore, natural selection will tend to reinforce this distinctness for the following reason. Each population can be assumed to have a gene pool of variability, which will include some variants more similar to and other more different from the individuals of the other species. Those more similar to the other species will be more

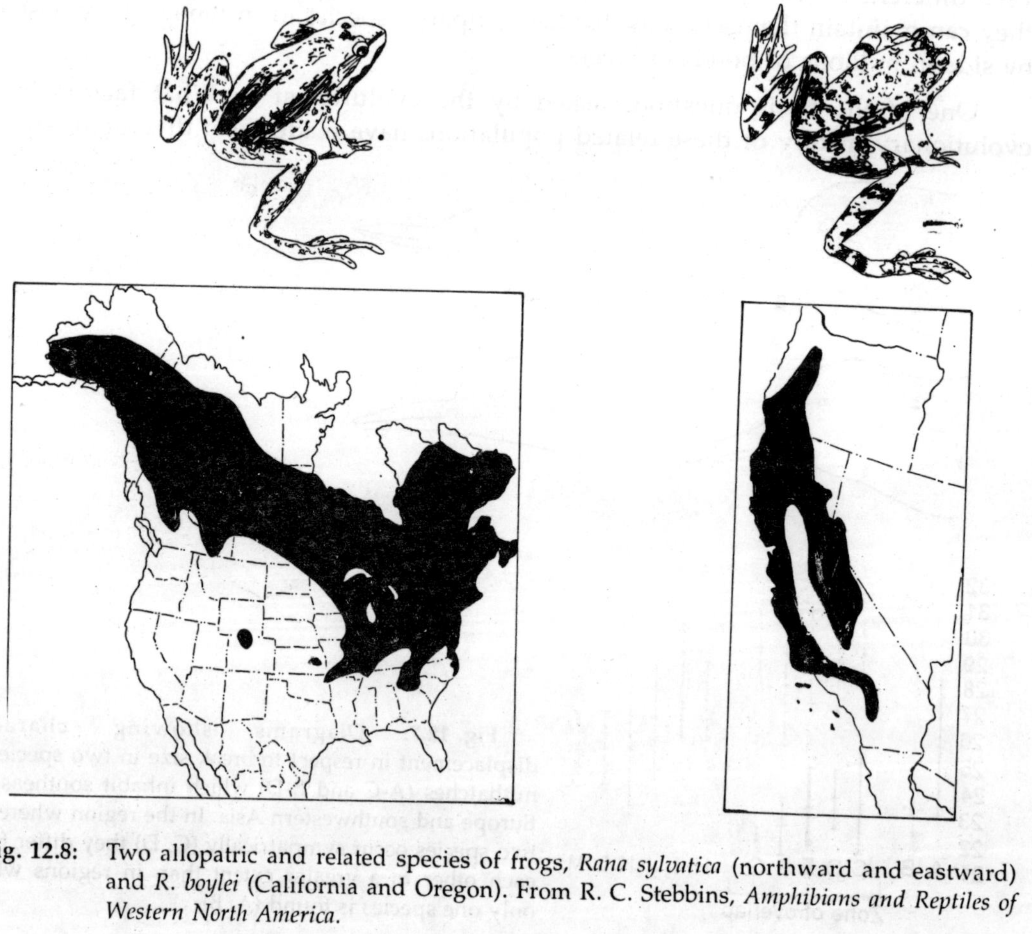

Fig. 12.8: Two allopatric and related species of frogs, *Rana sylvatica* (northward and eastward) and *R. boylei* (California and Oregon). From R. C. Stebbins, *Amphibians and Reptiles of Western North America*.

undergo its own distinctive cycle of adaptive radiation and speciation. In this way an adaptive radiation develops which is of a higher order than that which gives rise to a polytypic species. This results in the evolution of clusters of related species which form a GENUS.

Sometimes two differently adapted populations or subspecies of a poly-typic species can come together before reproductive isolation between them has been fully developed. In that case they can exchange genes to a limited degree, and progeny resulting from such HYBRIDIZATION may contain individuals adapted to environments different from those occupied by either of the parental populations, Although such hybrids are most often unsuccessful because of their weakness or sterility, some of them may give rise to new populations, well adapted to a new habitat which may be available. Under such exceptional circumstances, hybridization can greatly speed up the processes of evolution. In plants, hybrids may give rise to new species by doubling their chromosome number, a process known as POLYPLOIDY. In this way, a sterile hybrid can give rise immediately to fertile offspring, which will remain constant for the intermediate, hybrid condition.

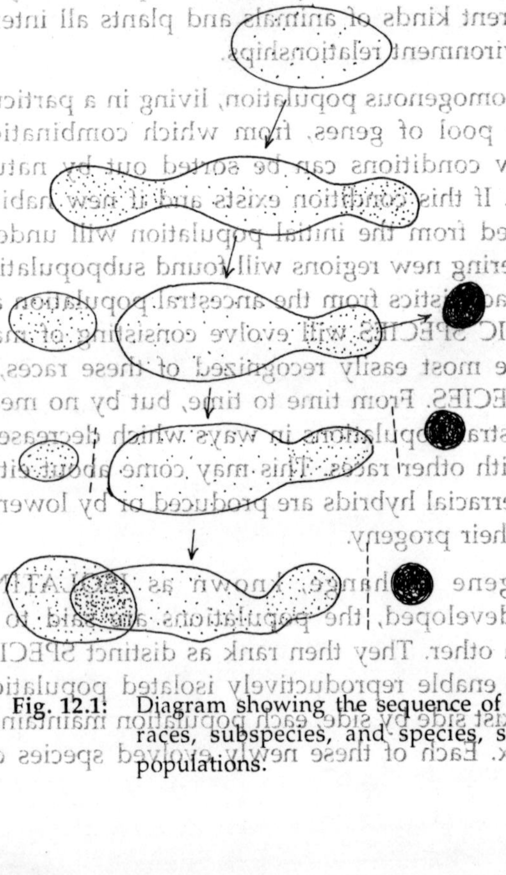

Fig. 12.1: Diagram showing the sequence of events which leads to the production of different races, subspecies, and species, starting with a homogeneous, similar group of populations.

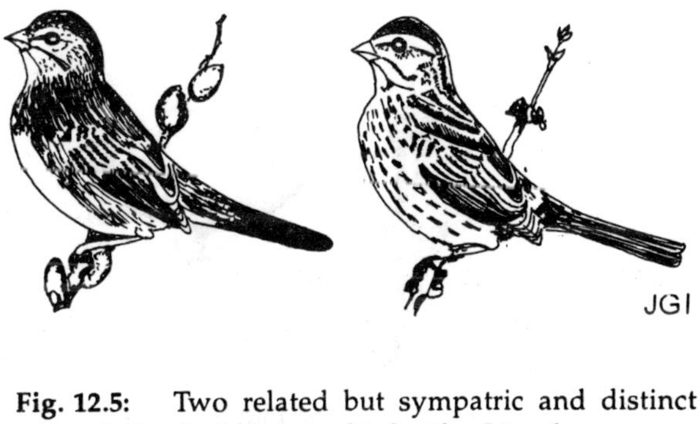

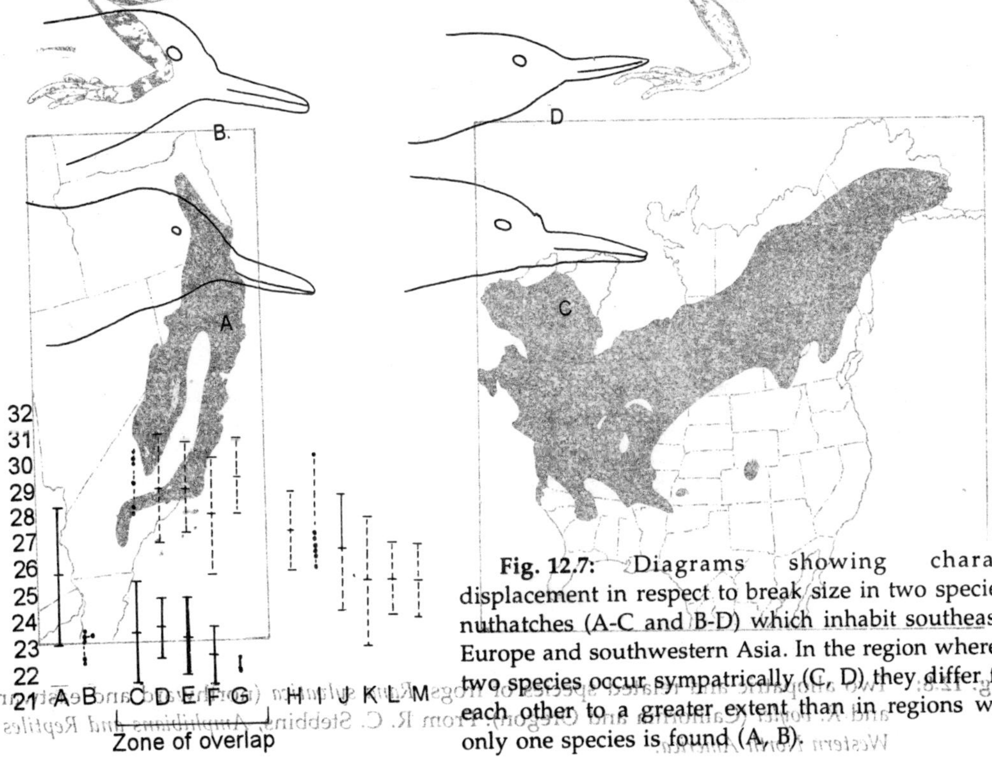

likely to come into direct competition with it for food and for places to hide from predators. They will, therefore, be at a reproductive disadvantage compared to those individuals of the population which are more different and are beginning to exploit their environment in a different way. Natural selection will, therefore, promote the divergence of two species which live together in the same areas but are unable to interbreed.

A good example of this effect is found in two species of birds known as rock nuthatches in southeastern Europe and western Asia (Figure 12.7).

The discussion in this chapter has revealed the need for distinguishing between populations or species which live together in the same area, and those which inhabit different areas. For this purpose, evolutionists have devised the term SYMPATRIC for the former and ALLOPATRIC for the latter. Thus, the eastern song sparrow and Lincoln

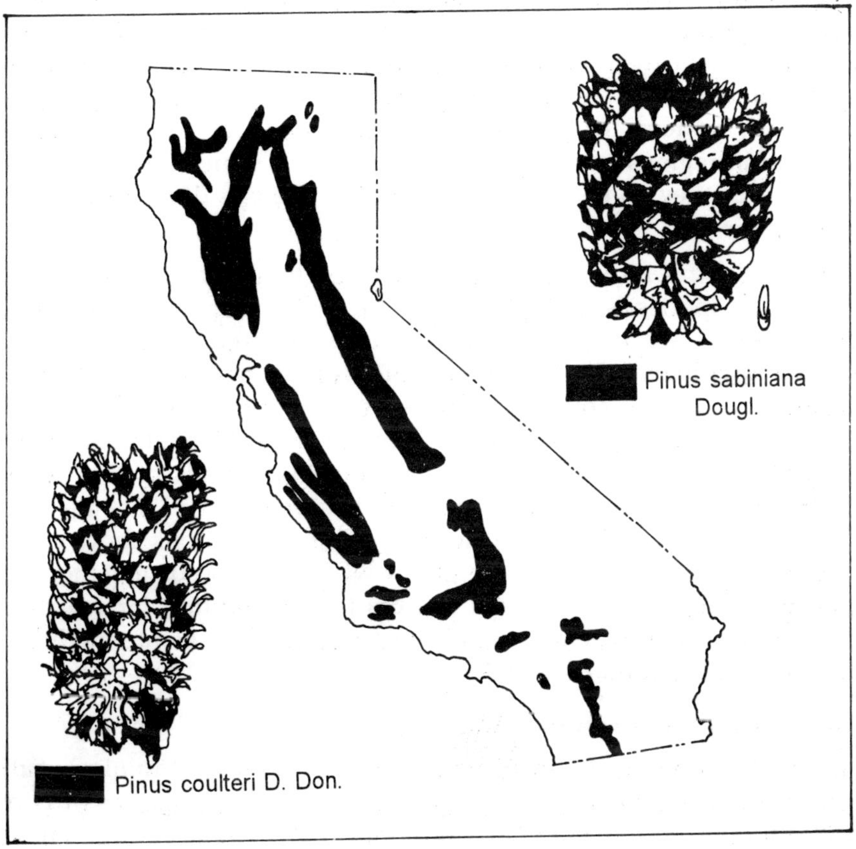

Fig. 12.9: Two related species of pine, the digger (*P. sabiniana*) and Coulter (*P. coulteri*), which are largely, allopatric but which are sympatric and remain completely distinct from each other in parts of central and southern California. Illustrations of cones from McMinn and Mainio, Pacific Coast Trees, University of California Press; map from Munns, Distributional maps of principal Forest Trees of the United States.

sparrow, as well as the red and sugar maples, are sympatric species. The northern wood frog (*Rana sylvatica*) and the western yellow-legged frog (*Rana boylei*), as well as the southeastern yellow pine (*Pinus echinata*) are allopatric species (Figure 12.8). You must recognize that the difference between sympatry and allopatry is not absolute, but one of degree. There are many examples of related species which are allopatric in some parts of their distributional ranges, but sympatric (i.e. overlapping) in others. A typical example is that of the two large coned pines, digger (*P. sabiniana)* and Coulter (*P. coulteri*) in California.

Because of its importance in promoting further divergent evolution and because the diversity of the flora and funna inhabiting a region could not have evolved without it, the ability of two populations to exist sympatrically without losing their identity through hybridization and gene exchange is regarded by evolutionary biologists as being of prime importance in recognizing them as distinct species. When the related populations of an area are considered, those which "pass the test of sympatry," that is, those which retain their identity even though they live close enough together so that cross fertilization between them is possible, are regarded as separate species. On the other hand, systems of populations which may have evolved very different allopatric races in regions remote from each other, but which intergrade and lose their identity whenever they come into contact, are regarded as different race or subspecies of the same species.

Unfortunately, this criterion cannot be applied to all populations about which the status as races or species is in doubt. There are many allopatric, completely separated populations which are obviously related to each other, but between which no intermediate individuals exist. These populations have never had a chance to take the "test of sympatry," and so could be either species or subspecies. In such examples we can only guess at what would happen if they did occur sympatrically. Sometimes our guess can be made more intelligent by bringing individuals of the two populations together and attempting artificial hybridization between them. There are other situations in which sympatric or partly sympatric systems of populations show varying degrees of distinctness or intergradation in the places where they occur together. If we assume that the origin of species is a gradual process (an assumption for which, as we shall see, there is much evidence) we would expect to find such intermediate degrees of distinctness between populations.

Prevention of Cross-fertilization

Even when mating occurs there frequently are obstacles to union of the gametes. The obstacles are usually of a physiological nature. In many species of Drosophila the mucous membrane of the vagina swells markedly after mating. The swelling is stimulated by the spermatic fluid, even without any sperm in it. If the mating has been with a male of the same species, the vagina returns to normal size within a few hours. In an interspecific mating, however, it may remain swollen for days. A female which reacts thus strongly may not mate again. Eggs are sometimes destroyed in a swollen vagina, and in any case their fertilization is hindered. Certain pairs of species have this insemination reaction, as it is called, sufficiently reduced to permit the production of some interspecific hybrids.

The amount of sperm introduced by a male of a different species may be less than in an intraspecific mating. This is true in *Drosophila pseudoobscura* and *D. persimilis*, as determined by examination of the spermatheca of the female after mating. Foreign sperm are also less viable in the passages of the female. These two phenomena should help reduce the number of interspecific fertilizations.

Often the reason for lack of fertilization is not known. The lady beetle *Hippodamia tredecimpunctata* copulates with another species of the same genus, but no hybrid offspring an no development of the eggs have ever been obtained from either of the reciprocal matings. Whether spermatozoa enter the eggs in uncertain.

In plants it is very common for the pollen tube to grow more slowly through the style of a flower of another species. If it arrives at the egg after the flower withers, there is not fertilization. Also, the style of another species may be longer than the one the pollen has to traverse in its own species. A cross between a short-styled species used as a female and a long-styled species as male succeeds better in Datura than does the reciprocal cross. Sometimes the tube bursts in a foreign style, and the sperm nuclei travel no farther. Sometimes pollen does not grow at all in a foreign style.

In certain marine animals, bringing germs cells together does not depend on behaviour, growth, or reactions; the eggs and sperm are simply poured out into the sea water at the same time. While there may be some ecological separation of closely related species, this cannot be depended on to keep eggs and sperm apart when water is free to move. Eggs of sea urchins have been experimentally expose to their "own" and foreign sperm and have experienced cross-fertilization less than 2 per cent as often as intraspecific fertilization. There is some block to fertilization by foreign sperm, some incompatibility between the cells of different species. This hindrance may be lessened by changing the alkalinity of the sea water, but that scarcely explains the physiological nature of the protoplasmic barrier.

Inviability of Species Hybrids

If gametes of different species have succeeded in uniting, there still may be complete isolation of the contributing species. Some of the physiological species of *Paramecium aurelia* referred to in an earlier section are isolated in this way. Species 4 and 8 mate almost half as readily as do the two mating types of either of them, but none of the F_1 (exconjugants) survive. Species 1 and 5 mate about 40 per cent as readily as do different mating types of one species, and the F_1 are viable; the F_2, however, are almost completely inviable.

Fertilized eggs from attempted species crosses of fish have shown abnormal cleavage, elimination of chromosomes into the cytoplasm, and early death. The lady beetles *Coccinella novemnotata* and *C. transversoguttata* have rarely been crossed, and the fertilized eggs produced rather normal-looking embryos in the shell; but the hybrid larvae never hatched.

In certain species crosses of the Jimson weed Datura the pollen tubes reach the embryo sac, and the egg is fertilized, but no seeds develop. In one Datura cross the fertilized eggs cleaves three times, to produced eight cells, then stops. The endosperm may develop farther than the embryo, and to such disharmony is attributed the failure of certain species crosses in wheat. In some crosses of flax species (genus Linum) a weak hybrid may be helped to grow by removing the seed coats. The growing embryo appears unable to break through these without aid. In other flax crosses removing the seed coats is not enough; the embryo has to be dissected out and put into a nutrient sugar solution. With that assistance it may develop into a self-maintaining plant in soil.

In crosses of certain moth species, only male hybrids survive. The females die in the pupal stage. In many species crosses there is a disparity of the sexes produced. Irregularities of the chromosome mechanism of sex determination is doubtless involved in all these instances.

Hybrid Sterility

One of the best-known methods of preventing the mixing of species waits to the very end of the hybridizing process before it interposes its veto. Hybrids may be formed, be vigorous, and have every appearance of being normal organisms, yet be sterile. Even if they are only partially sterile, they are at so great a disadvantage as compared with members of either contributing species that the transitional group is never effective. Interflow of genes between species is stopped or restricted to negligible proportions by this reduced fertility. One of the earliest-known species hybrids, the mule, offspring of a female horse and a male ass, is a very vigorous organism, superior in many respects to both of the parent types. Yet seldom are the reproductive organs sufficiently normal to act as such. Fertility has no necessary connection with general vigor.

The best-understood cause of sterility of hybrids is the irregularity of the germ cells they produce. A clear case of intersterility due to chromosome irregularity and nothing else is that existing between related plant species differing in their ploidy. If a tetraploid species has arisen and achieved a fair degree of stability through mutations that give it a functional diploidy but with the tetraploid number of chromosomes, the tetraploid species is at once more or less isolated from its parent diploid species. Any hybrid of theirs would have three chromosomes of each kind, and distribution of chromosomes at meiosis would necessarily be irregular. The troubles arising from such irregular distribution were pointed out in the earlier discussion of polyploidy; those troubles pertain equally to hybrids of species with unequal ploidy.

Hybrids of species with the same number of chromosomes may be partially sterile if the parental chromosomes are not strictly homologous, that is, if their genes are different or are arranged differently. The chromosomes of such a hybrid do not pair, or pair only loosely. Without such pairing, they are not distributed regularly in meiosis. Few of the germ cells receive a single chromosome of every kind as they do within species. Germ cells are of many different kinds, and fertilized eggs are mostly not viable. The few normal cells they produce could, if lucky, produce surviving individuals; but they provide only a very narrow bridge for the transfer of genes between species.

Hybrid sterility resulting merely from lack of homology of the chromosomes from the parent species is practically limited to plants. While irregular distribution of chromosomes at meiosis may occur in hybrids of animal species, something else, more fundamental, is almost always involved. And in plant hybrids these more fundamental things are quite commonly involved as well.

Sometimes there are genes producing intersterility, and these genes act in a variety of ways. Sterility genes acting *within* a species are well known, such as the "asynaptic" gene in corn. This gene prevents chromosomes from pairing properly, even though these chromosomes are strictly homologous. The result is irregularity of chromosome combinations in the germ cells, and few viable offspring are produced. Many plant species have recessive genes which impose sterility on individuals homozygous for them. Now, such genes could be imagined to create intersterility of species if the recessive sterility genes became dominant in the presence of a foreign genotype, that is, in a hybrid. Sterility gene *a*, harmless in its own species so long as it existed only in *Aa* individuals, would on crossing be in *Aa* individuals but then be supposedly dominant. Such species hybrids would, on the hypothesis, be sterile. Yet to produce any considerable proportion of sterile species hybrids, gene *a* would have to be fairly common in its own species. Being common, it would occur in many *aa* individuals, rendering them sterile. That would produce so great a handicap to its won species that natural selection would work strongly against the system. No example of this scheme is known.

A somewhat different one-gene isolating scheme has been proposed by Castle. It involves sensitization of a gene by its allele, a sort of anaphylasix. Castle assumes that a new dominant gene arises in the inert part of a chromosome. This creates an allele in the opposite chromosome, likewise in the inert part, by sensitization. Having arisen by this method, these alleles will kill an individual which subsequently receives both of them. To act as a barrier to species crosses, each of these alleles would have to become incorporated in a separate population. How they would do this, when both arose in the same individual, is difficult to imagine; yet Castle points to crosses in chiccry and peas which seem to him to support his view. To most geneticists the origin of a new gene out of inert material will probably seem more difficult than the separation, into different populations, of genes that are produced in the same place.

A more workable scheme is two sets of dominant interacting genes. Each set, by itself in one species, is harmless; but when the two sets are brought together in a species hybrid, the dominant genes interact to damage the reproductive function. Theoretically one dominant gene in each of two species would suffice to make such a scheme work. Actually, in the best-understood sterile hybrids, there are more of these genes. *Drosophila pseudoobscura* and *d. persimilis* when crossed produce completely sterile male hybrids and partially sterile females. While these species differ in four chromosomes inversions, that is not the reason for the sterility of their hybrids. The trouble with the hybrids is not merely that their chromosomes enter irregularly into the germ cells; their whole maturation process is badly deranged. In one of the two reciprocal crosses, the testes of the males are small, indicating disorganization before the stage of chromosome pairing is reached. Genes must be responsible for this situation. Appropriate genetic tests show that the disturbing genes are in most of the chromosomes; hence there must be a number of them. In back crosses of the hybrid females to the parent species, sometimes there is only one maturation division, and spermatozoa are not produced. There is thus little chance for genes to flow from species to species even through the limited fertility of hybrids of one sex. Other Drosophila species produce similarly sterile hybrids whose gonads show degeneration prior to the time of pairing of chromosomes. This is true of *D. melanogaster* crossed with *D. simulants;* and when *D. psecudoobscura* is crossed with *D. miranda,* the male offspring have small testes and the females lay many eggs which seldom hatch. This evidence that sterility comes from genes, rather than from unmatched chromosomes, seems to be fairly general. Indeed, as stated earlier, hybrid sterility arising solely from chromosomal features is practically limited to plants.

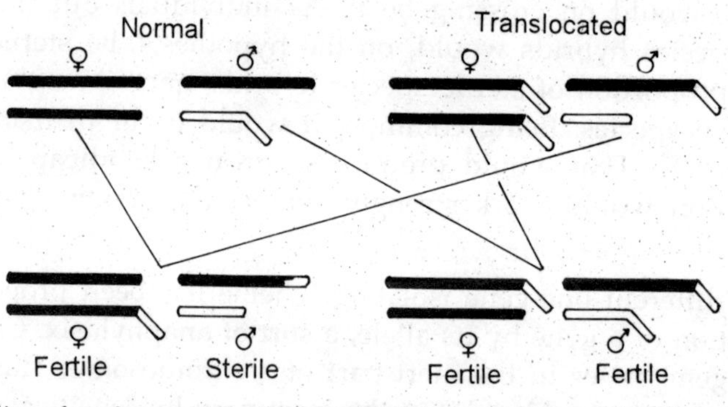

Fig. 12.10: Sterility of males produced by a racial cross involving a translocated segment of the Y chromosome in Drosophila.

Stern has used the fertility function of the Y chromosome in Drosophila to explain hybrid sterility of one sex in certain crosses where the other sex is fertile. From irregular distributions of the chromosomes it was early shown that a male *D. melanogaster* without a Y chromosome is sterile. Two segments of this chromosome are involved, and both must be present to produce fertility. If one of these segments were translocated to an X chromosome, a race could be established whose X chromosomes both carry the one Y segment, while the other segment remains as a Y. If this race were crossed with the normal flies, one cross would produce fertile offspring of both sexes; the other reciprocal would produce fertile females but sterile males. Two sub-species of *D. macrospina* shown this relation of sterility and sex in their two reciprocal crosses, but no translocations between Y and X chromosomes have been found in them.

Intersexes

When the normal mechanism of sex determination is disturbed, individuals combining in some degree the characteristics of both sexes may be produced. Such individuals are called *intersexes,* and they are usually sterile. Since intersexes may be the result of hybridization, they are of interest in relation to the isolation of species.

The best-known intersexes are not species hybrids, but only racial hybrids. They were produced in the gypsy moth by Goldschmidt in a long series of experiments. Moths were obtained from various parts of Europe and Asia and were found to have different sex potencies in different regions.

In moths the male has two Z chromosomes (many geneticists call them X just as in mammals, and flies), the female only one, with perhaps a W chromosome (also called Y). The Z chromosomes carry pre-ponderantly male genes, while femaleness is represented mostly in the cytoplasm. Two male-favouring Z chromosomes in female-favouring cytoplasm produce a male, while it there is only one Z (with or without a W), it is overridden by the cytoplasm and a female results.

Now, Goldschmidt found that races of gypsy moths differ in the "strength" of the male genes in the Z chromosome and of the female influence of the cytoplasm. In some races both Z chromosomes and cytoplasm are weak; in other races both Z chromosomes and cytoplasm are strong. So long as matings occur only within one race be it weak or strong, the balance of two Z's or of one Z against the cytoplasm produces clear-cur distinctions between the two sexes. If, however, a strong female is crossed with a weak male, a strong and a weak Z pitted against strong cytoplasm may not produce a male. An intersex may result. And if a strong male is crossed with a weak female, even a single strong Z might overbalance the weak female influence of the cytoplasm, and not a female but an intersex be produced. The sexes differ in wing pattern, size of body, and business of antennae, and of course in their reproductive organs. The intersexes show various intermediacies and combinations, while their reproductive organs are reduced.

In general, the races of gypsy moths which lived in neighbouring areas were not sufficiently different in the strength of their sex factors to produce intersexes on crossing. Races from regions far apart, however, often yielded intersexes on crossing. There was one exception to this rule-two neighbouring Japanese islands are occupied by races thus different, but these islands are separated by a wide strait. In natural situations, therefore, there are not many intersexes. Also, these populations are not different species, merely races. Nevertheless, the sex factors could very well be of different strengths in different species, and this means of isolation is open to them. Sterile intersexes are known in few other animals, some ducky, body lice, and guinea pigs among them.

Elimination of Hybrids without Sterility

Sometimes species hybrids disappear without, so far as is known, being sterile. What ecologists have regarded as an instance of such elimination is found in the hybrids of the blue-winged and golden-winged warblers. The blue-wing breeds in bushy low pastures in a northern range, the golden-wing in uplands and oak woods to the south. They overlap in river valleys of Wisconsin, New Jersey, New York, and Connecticut. The species differ in appearance chiefly in the colour of the under parts, whitish and yellowish, and the colour of the throat, plain or black. These colour characters segregate so cleanly that they are regarded as simple genetic units, with white under parts and plain throat dominant. The bluewing has one of the dominants (plain throat), the goldenwing the other dominant (whitish under parts). The throat colour may just possibly not be quite so simple; it could be the product of two genes.

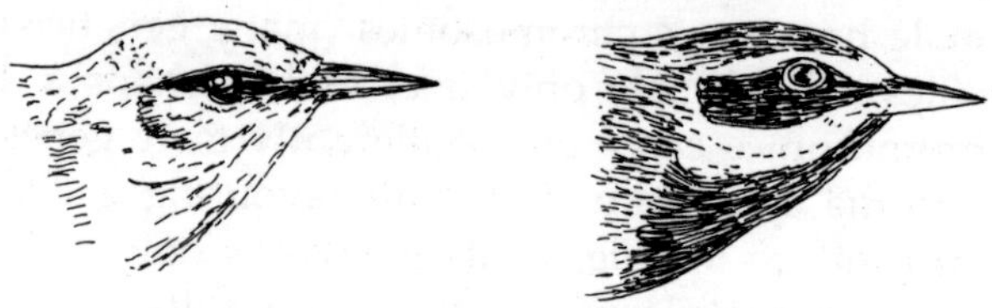

Fig. 12.11: Blue-winged (*left*) and golden-winged warbler. (*From Champman, handbook of Birds, Appleton-Century-Crofts, Inc.*)

In the regions of overlap there are birds with both dominant characters. They are regarded as hybrids and are called Brewster's warbler. They must breed together, for occasionally birds with the two recessive characters are found; these are known as Lawrence's warlber. The double recessive is understandably much rarer than the double dominant. Backcrosses to the parent types are also known.

Neither of these hybrid types is very common. Certainly they have not produced any general invasion of either the blue-winged or the golden-winged species by the genes of the other. The fact that the hybrids must breed together and with the parent species indicates they are not sterile hybrids, though there are grades of sterility in many known hybrids of other animals. Ecologists have generally held that in some way the environment has selected against these hybrid birds. Each species is fitted to its own kind of environment. Even if the hybrids were intermediate in their ecological requirements, so the argument goes, there would not be enough transitional habitat to maintain the intermediate type.

Speciation

It will be noted that, among the agencies which keep species genetically distinct, some are environmental; others are within the organisms. There has been room here for argument as to just when, or whether, a given population has attained the status of a separate species. Dobzhansky has proposed that speciation shall mean only the origin of reproductive isolation. Geneticists seem inclined to adopt that definition. Ecologists, on the contrary, would prefer to let speciation refer to genetic differentiation of two populations that in fact do not interbreed, irrespective of whether they could interbreed if brought together. Two populations in slightly different ecological situations, kept apart because of some physiological difference which fits them for their respective areas, may well develop genetic differences without losing their ability to interbreed. They have lost interbreeding by ecological separation, but have not lost their *capacity* to interbreed. By the ecologists' definition, the two groups may be regarded as separate species; speciation has already occurred. By the reproductive-isolation definition they are still one species. The latter definition has the advantage or relying on one criterion, though grades of reproductive isolation will still lead to disagreement as to whether, in specific instances, speciation has yet occurred. The word speciation, as a name for reproductive isolation, is somewhat unfortunate in containing within itself so little indication of what it is made to mean.

Choice of a definition here should probably depend on what one wants definition to do. If the prime requirement be ease and sureness of application, the single criterion would probably be best. Linnaeus used arbitrary criteria, such as numbers of stamens, to distinguish taxonomic groups, but in doing so brought together some plants which by modern standard are not closely related. A nonbiological example may also be useful. A bungalow without a roof would hardly be called a house; it might be called a court or, if it had no floor, perhaps a patio. Possibly a roof, supported by posts but no walls, would also be called a house. If so, the sharp definition of "house" would be "a roof". Yet if walls were subsequently built under the post-supported roof, the name "house" would doubtless be applied to the whole structure, including the walls. So it could be argued that, though speciation has not been completed until there is reproductive isolation, any genetic differentiation between the isolated groups would still be part of the speciation process. Such differentiation could happen either before or after the reproductive isolation arose. If speciation is to include the genetic differentiation arising before reproductive isolation, it should probably include any that arises after reproductive isolation. By this definition, speciation would be more or less continuous. Usage will decide what meaning it to be attached to the word; but if what happens is understood, names will not be too important.

How long two separated populations could continue to be, and to become, genetically different while still retaining the capacity to interbreed is a matter of some importance. The time was when intersterility was supposed to result from accumulation of genetic differences of any kind whatever. Even now it is sometimes said, though rarely, that two populations become different in so many respect that they can no longer interbreed—the implication being that any mutation in one which does not occur in the other contributes its mite to reproductive isolation. If the only source of reproductive isolation were the failure of chromosomes to pair in the mature species hybrid, the presence of different forms of genes in the parental chromosomes might favour such a failure to pair. But as we have seen, most hybrid sterility in animals is genic, not chromosomal; and there are many earlier points where the interflow of genes may be stopped. There is a fundamental genetic concept which might be held to support the idea that all kinds of mutation contribute to isolation: it is that no gene produces a character by itself; it always cooperates with other genes; indeed it may interact with the whole genotype. Despite this truth, each mutation affects certain features of an organism more than it affects others. In this sense there are still genes for eye colour, genes for growth, genes for this or that specific physiological activity. It seems unlikely that all these genes relate in an important way to reproduction.

Many of the visible mutations of Drosophila have, by suitable crosses, been collected into one stock of flies. This stock is enough different from the wild type to the considered a different species if only it were reproductively isolated from the wild type. But it is not so isolated; it is perfectly fertile in crosses with normal flies. The collected mutations have not contributed to isolation.

In nature, however, Drosophila species are reproductively isolated while exhibiting very slight visible differences. Species' distinctions are notoriously small in these flies. Many times have two species been regarded as one until, in breeding experiments, their reproductive separation has been demonstrated. Reproductive isolation is being produced at a high rate, relative to detectable changes of other kinds. It may possibly be said that we do not known all the physiological mutations which are occurring, and that they all may be contributing, in crosses, to reproductive isolation. This suggestion need not be rejected as an attempt to rescue the preconceived idea that all variation helps to create reproductive isolation; it is answered by the fact that Drosophila's visible mutations, which like all characters are also physiological, demonstrably do not contribute to isolation.

It seems necessary to conclude, then, that some mutations, in combination, help to produce reproductive isolation, and others do not. Drosophila appears to be producing the former kind rather rapidly. Reproductive isolation may occur before many other types of differentiation have arisen. Theoretically it could arise before *any* other differentiation has occurred. Reproductive isolation *could* be the *only* distinction between species.

Spatial Separation before Reproductive Isolation

There are strong theoretical reasons for supposing that genes which, on interacting, produce reproductive disturbance must arise far apart if they are to effect isolation. They must arise in separate populations. There need be no barrier separating these populations; they could be parts of the same species and genetically alike, but at distant parts of the range.

To illustrate the argument, it may be assumed that only two dominant complementary genes are involved. They are of the sort which, if one existed harmlessly in one population, the other harmlessly in another population, would if brought together in a hybrid interact to derange the reproductive process. Assume that one of these already exists and has already spread through an entire breeding population. If, now, the second complementary gene arises in the same population, the individual bearing both dominant genes would be at once sterile. To ease this handicap, suppose that the first dominant gene has not spread through the entire population but that it is common. If the second dominant gene then arises in an individual not carrying the first, the heterozygous mutant would be fertile; and if it mates with another individual not possessing either of the dominants, it will have fertile offspring. The half of its offspring which carried the new dominant would be able to breed with others like themselves or with individuals not carrying either dominant. Random matings would of course lead to unions of both dominants, and some sterility would follow. This loss of individuals would be a handicap to both dominant genes; selection would work against the success of both in the same population. If the effects of the two dominant genes on reproduction were less drastic, so that sterility would be incomplete, both genes *could* become established, but natural selection would still operate strongly against their success.

It is assumed, therefore, that such isolating genes or sets of genes arise separately. At distant points in the range the genetic make-up of the population may be somewhat different, even if only because of genetic drift. If one population occupies a slightly different ecological niche, there is more reason for genetic difference. Thus there may already be some differentiation of characters having no relation to reproduction before the reproductive isolation occurs. The complementary isolating genes are not any more likely to arise *because* there has been other differentiation first; it is merely that populations at a distance from one another can hardly, as a random matter, be identical.

Advantage of Isolation

So widespread is the reproductive isolation of species—fully interfertile species are uncommon—that most evolutionists have concluded that there must be some advantage in the system. They hold that it must have developed under the guidance of natural selection. This would mean that populations cut off from the supply of one another's genes must have left more descendants than did populations which intermingled and interbred freely. Since ultimately it is individuals which produce descendants, those which could breed with other individuals like themselves, but not with others differing from themselves in some group of characters, left more descendants than did those which could breed with different kinds. Wherein lies the advantage of this reproductive isolation?

The answer is that there is need of a certain amount of conservatism in evolution. Gains once made must be held. Recombination, even with the brakes that chromosomes put upon it, could supply organisms with far more readjustment than their slowly changing environment demands. Not only does environment demand no more than slow change; it even penalizes too rapid change. An organism that produces too many kinds of offspring sees most of them rejected. Instead of the dozen types of environment which those offspring would be fitted to enter, there is usually only one environment. The offspring that were ready to try something new are lost. In the next generation, the ones which conservatively imitated their parents, and so survived, are a larger fraction of the total population. The place of their experimenting fellow is taken by the unchanged or less changed progeny of other parents—parents which did not provide them the means of making many new combinations.

One way to reduce the variability of progeny is to limit the mating of parents to individuals more or less like themselves. If such limitation is desirable, as pictured above, any individual which has developed an isolation mechanism preventing its breeding with distinctly unlike organisms will furnish the unchanged type of offspring which the unchanged environment can accommodate. These offspring will survive, carrying with them their isolation mechanism. If the advantage is real, the selection may be expected to continue until the isolation device is a characteristic of the whole group. Artificial elimination of hybrids of *drosophila pseudoobscura* and *D. persimilis*, in experiments by Koopman, was followed by reduced frequency of hybrids, which is interpreted as supported for the view that, once a degree of reproductive isolation arises, it may be increased by selection.

Such isolation could be carried too far. Sometimes a change of environment will call for a new type of organism, and enough variation must be available to make the adjustment. Isolation must no, then, set aside a population which is wholly uniform; any specific step toward isolation of such a group would presumably be nullified, perhaps long after, by destruction of the population. Nothing is designing these steps; they are all trial and error. What seems an advantageous procedure now may prove fifty generations hence to have been mistake. By that time the better alternative, which could have been "chosen" fifty generations earlier, may not longer be available. Some mistakes are doubtless irrevocable. That is the way natural selection must work.

This adaptive picture of reproductive isolation is the one that most evolutionists now paint. Some evolutionists base it one the thesis that environments are rather sharply defined, that habitats do not merge gradually into one another. If species did cross, a population possessing an intermediate group of characteristics would presumably result, and an intermediate environment would be needed to accommodate it. If there never had been any isolation, the variation of organisms would spread over the entire range which their various genes could provide. There would be no gaps separating one moderately variable group from another moderately variable group. The environment, to fit such organisms, would have to display all grades of its various components in all combinations. This, these evolutionists say, environment does not do.

To what extent environment actually falls into distinct types, unconnected by gradations, is of some importance in connection with the supposed origin of isolation. With respect to temperature, environment *does* supply intermediate situations. No area of one temperature, or range of temperatures, is sharply marked off from another area having another temperature or range. There are always intergradations, and these are found in areas large enough to accommodate large populations of organisms. Atmospheric moisture likewise graded. While it is subject to great temporal variation, so that only averages may be compared, regions of dry air are not sharply separated from moist situations. At the earth's surface, bodies of water do give over suddenly, at the water's edge, to dry land. Yet in Michigan, north of the terminal moraine, there are abundant swamps and marshes furnishing almost any gradation of the water-land contrast. These water gradations are much less extensive, however, than the temperature gradations. Organisms requiring merely damp situations would have smaller choice. No adequate study has been made of all the components of environment, to determine how sharp the areal separations are. Wnat ecologists call "niches" have not usually been demonstrated to be the sharp-angled recesses which the appellation seems to imply. The could often be only "undulations," or at best "scallops," so far as actual ecological measurements show. Perhaps we think of them as niches because we see definite animal or plant types occupying them. If we *should* be reasoning from organisms to habitats, the adaptiveness of reproductive isolation might prove to be an illusion.

It is thus possible to argue on both sides of the question whether, or to what extent, reproductive isolation has been produced through selection. Muller is of the opinion that isolation is a by-product of other characters that have been established through selection, though it is not necessary to make it a by-product of *every* character thus evolved. Huxley likewise regards separation of species as being super-imposed upon the other evolutionary processes and as having little bearing on them. Isolation, he holds, is an inevitable consequence of evolution, but irrelevant to its main theme. It would be thinkable that in general isolation is useful, but that it has been overdone by living things. Though the parallel is not too close, one may call attention to the part that isolation plays in Wright's computation of opportunities for evolution. Partial isolation of small populations in a large group is estimated to afford the best opportunity for such change. Wright is postulating the best chance for change by recombination, without considering whether the change so provided might be too rapid for the change of the environment. Also, the isolation he postulates is not reproductive; it is merely spatial. Were the isolation complete, it would nullify any advantage of having the small differentiated populations. The suggestion has been made that complete reproductive isolation of species might not be the most adaptive arrangement; such isolation might restrict variation too much. Perhaps a small amount of species hybridization would be an advantage.

Anyone who is convinced that every common feature of life on earth has been made what it is by natural selection can point to the occasional species crossing in animals and the more frequent species crossing in plants, and say here is just the right amount of leakage in the isolation of species. An alternative conclusion would be that some things happen that are not particularly beneficial. A result that can be arrived at in many different ways is more likely to be common, and at the same time accidental, than is a consequence of only one kind of event. Reproductive isolation certainly has arisen in many ways. Is its prevalence a consequence of these multiple origins, at least in part, or *must* it be advantageous in order to be so common?

SIBLING SPECIES

In other instances we find groups of species which are so much alike that they are almost impossible to distinguish from each other by their outward appearance, but which nevertheless are separated from each other by a number of well-developed isolating mechanisms. These are known as SIBLING SPECIES. A good example is found in the malaria carrying mosquito of Europe (*Anopheles maculipennis*) and its relatives. This is a group of six species which are so much alike that they cannot be told apart on the basis of the adult mosquito. Their larvae or "wrigglers" are, however, somewhat different from each other, and their eggs are so different that officers of the U.S. Army Sanitary Corps, who were in Italy during World War II, found that they could teach an untrained GI to distinguish the different species by their eggs. This recognition is of great practical importance, since some of the species carry malaria and some do not. Consequently the adults do differ from each other in respect to invisible physiological characteristics.

These mosquito species are isolated partly by differences in mating behaviour, and partly by hybrid sterility, which is complete in F_1 males of any combination. Consequently, they easily pass the test of sympatry. Populations of two or more species may exist together over very wide areas without the slightest indication of gene exchange between them.

Sibling species are very common in other flies, such as *Drosophila*. Indeed, they occur in all groups of animals. Well-known examples among birds are the tyrant flycatchers, and among fishes the whitefishes in northern Eurasia and North America. Others have been studied in butterflies, crickets, "water fleas" (*Daphnia*), limpets, and sponges. They are apparently less common in plants, but have been found in the tarweeds (*Holocarpha*) and Gilias of California, and in some genera of grasses (*Elymus, Panicum*).

From the genetic and evolutionary point of view, sibling species are just as valid as are those separated by conspicuous visible characteristics. Their existence tells us, however, that the genetic differences which build up isolating mechanisms can occur independently of those which produce clearly recognizable differences between species. To understand the origin of species, therefore, we must ask ourselves what kinds of evolutionary processes can be particularly responsible for the origin of isolating mechanisms.

THE EVOLUTION OF ISOLATING MECHANISMS

Based upon the large body of evidence which was reviewed in the first four chapters of this book, we now realize that two conditions must be fulfilled before populations can diverge with respect to any characteristic which is determined by many separate genes. First, divergent pressures of natural selection must be acting upon them. Second, they must be well enough isolated from each other so that the initial genetic divergence will not be "swamped" by gene exchange. These conditions must be fulfilled for reproductive isolating mechanisms as much as for any other kind of difference. Consequently, the processes which lead to the origin of species are not of a special kind. They are not qualitatively different from the processes which give rise to divergent races and subspecies. They consist of the divergent action of natural selection upon those particular kinds of differences in genes and chromosomes which can form isolating mechanisms.

The origin of most of the prezygotic isolating mechanisms can be the direct result of selection for divergent adaptations. Habitat isolation can clearly come about through adaptive radiation of populations into the different habitats found in any region. Seasonal isolation can be acquired while populations occur in separate regions with different climates. For instance, among the pines of California there are several examples of pairs of sympatric species which differ markedly in time of pollen shedding. One member of each pair sheds its pollen and fertilizes its cones in February or March, and the other in April or May. In each example, the early flowering species occurs by itself in regions which are farther south, lower in altitude, or nearer the coast than the area of sympatry, and in which the winter climate is relatively mild and dry. The later flowering species, on the other hand, occurs by itself in more northerly regions, or at higher altitudes, where conditions for cross pollination and fertilization are very unfavourable in February and March. Consequently, the most likely evolutionary history of these species pairs is that they were originally allopatric. The later flowering species evolved its distinctive seasonal rhythm in climates which were not favourable for pollination until late spring. When the species later became sympatric in a region having a mild climate, gene combinations for late flowering had been so firmly established that they persisted, even though selection pressures favouring them were not nearly so strong.

In plants, natural selection for flowers adapted to pollination by particular species of insects or other animal vectors could take place where the vectors concerned were unusually abundant, and other vectors were relatively scarce. For instance, most species of larkspur (*Delphinium*) have blue or purple flowers, and are pollinated by bees, whole vision is most acute in the violet and ultraviolet range of the spectrum. In northern California there exists a species with scarlet flowers (*D. nudicaule*). It is pollinated by hummingbirds whose vision is most acute in the red and infrared range of the spectrum. This red larkspur occurs in most, shady canyons where bees are relatively uncommon and hummingbirds are more abundant. Another flower which is pollinated by hummingbirds, the "California fuchsia" (*Zauschneria*), flowers in late summer and autumn when relatively few native bees are flying. There are, however, many examples of related species having different pollinators which cannot easily be explained on this basis.

The origin by natural selection of ethological isolations is very poorly understood. Weakly developed barriers of this sort are occasionally found between allopatric races of species and have presumably originated as by-products of selection for adaptation to different environments. Artificial selection in populations of *Drosophila* for differences in bristle number and other characteristics have sometimes led to some degree of ethological isolation between the selected populations. There is, therefore, some reason for believing that ethological isolating mechanisms can be built up as a by-product of divergent selection for adaptive difference, but much more research needs to be done before we can be sure of this.

Genic disharmony in the development of hybrids, which includes both hybrid inviability and developmental hybrid sterility, is most likely to evolve as a by-product of selection for different rates and temperature optima for various processes of cellular metabolism. The beginning of this process is seen in the divergence of the races of the leopard frog, mentioned earlier in this chapter. It has continued to the stage of isolating closely related species in the case of certain Japanese fishes, known as loaches (*Cobitis taenia* et aff.).

The origin of segregational hybrid sterility is best explained by the fact that the various genes which constitute a particular adaptive complex are often grouped together on a small segment of a chromosome. As populations invade new territories, we might expect these adaptive clusters to become modified by adding or subtracting genes. This could be accomplished most easily by means of small chromosomal re-arrangements. Evidence that racial differences of this sort have evolved in plants is provided by detailed studies of chromosome structure in different races of plant related to the lily family (*Trillium kamchaticum*) in northern Japan. This plant has very large chromosomes which, when exposed to cold temperatures at the time of mitosis, reveal an intricate pattern of differences in staining capacity. Within most populations of Japan, individual differences with respect to some or all of the chromosomes are found, and heterozygosity for different chromosomal types in the rule. Moreover, the different chromosomal types have distinctive geographic distributions, associated with corresponding differences in the outward appearance of the plants. So far as is known, these chromosomal differences are too small to give rise to any lowering of fertility in hybrids between individuals carrying different chromosomes. Nevertheless, chromosomal differences between. *T. kamchaticum* and other species of *Trillium*, which are great enough to make their hybrids sterile, are greater in quantity but are otherwise similar to the interracial differences within *T. kamchaticum*.

Chromosomal differences of this sort may in some instances arise in rapid succession, in response to rapid, drastic alterations in the environment. Among the species of *Clarkia*, a genus of annual flowering plants found in California, there are several species which are found only in a very restricted area, and which are closely related to more widespread species occurring in nearby regions. When crosses were made between these localized species and their more widespread neighbours, the F_1 hybrid in every case proved to be heterozygous for many chromosomal differences and were consequently sterile. The places where these species occur are known to have been subjected to drastic changes in climate in recent times. Consequently, a likely explanation of their origin is that they were derived from their neighbours viz a rapid succession of chromosomal changes, which brought together groups of genes adapting them to their new environments.

It not hard for us to see that the accumulation of numerous genetic differences constituting the system of isolating mechanisms that separate any two related species could not arise if gene exchange were occurring even at a low rate between the populations concerned. Such exchange would act directly to break down the isolation, while the selective pressure toward maintaining and increasing it would be indirect in nature. Consequently, those isolating mechanisms which enable two populations eventually to exist side by side and pass the test of sympatry must arise while they are separated from each other to such a degree that gene exchange between them is impossible. The degree of separation needed is still not clear, and probably differs according to the organism involved. Organisms which easily travel for long distances, such as birds and the larger mammals, probably need to be separated in diffrent geographic regions before isolating mechanisms between them can be built up. The same is true of the trees of the temperate zone, which have pollen and seeds that can be borne for long distances either by the wind or by strong flying birds. For small, sedentary organisms, and particularly for specialized parasites, the amount of spatial isolation needed in much less.

If we assume that geographic separation is usually needed before reproductive isolation and the origin of distinct species can take place, then we might expect to find and characteristic distributional pattern in groups containing species which are actively evolving. Populations between which isolating mechanisms are partly developed to varying degrees should be separated geographically in regions either close to each other or having somewhat similar environmental conditions. Distinct, sympatric species should be more strongly isolated from and more different from each other than are the related allopatric populations.

Just such a pattern has been found in plants of the daisy family known as "tidy tips" (*Layia*) in central California. These populations have been carefully analyzed in respect to the morphological differences between them, their geographical distribution and habitat preferences, and the degree of sterility and of chromosomal differentiation found in hybrids between them. On the basis of geographic distribution they fall into three groups. One group consists of a pair of species, one in the region about San Francisco Bay, and the other in the foothills of the Sierra Nevada directly east of there. The second group contains three species with occur about 200 miles south of San Franscisco, one on the coast, one about 50 miles inland, and the third about 100 miles inland. The sixth occurs by itself in southern California, but extends northward so as to be partly sympatric with members of both of the two other groups.

When these species are artificially intercrossed, vigorous hybrids can be formed between all of them, indicating that they are all fairly closely related to each other. Hybrids between the two species within the first group and the three within the second group have about 25 to 50 per cent of normal fertility. This indicates that these populations are on the borderline between subspecies and species. One could not be sure what would happen if they should occur naturally in the same region; they might pass the test of sympatry or they might become merged through gene exchange. Hybrids between populations belonging to the first group and those belonging to the second are much more sterile, and their chromosomes do not pair normally at meiosis.

These two groups have certainly reached the stage of separate species. Finally the single species in the third group forms almost completely sterile hybrids with all of the others, and its chromosomes are well differentiated from theirs. Where it occurs sympatrically with two of the other species it occupies completely different habitats, and there is absolutely no sign of gene exchange between them. Here we can see in diagrammatic form three stages in the process termed by evolutionists GEOGRAPHIC SPECIATION.

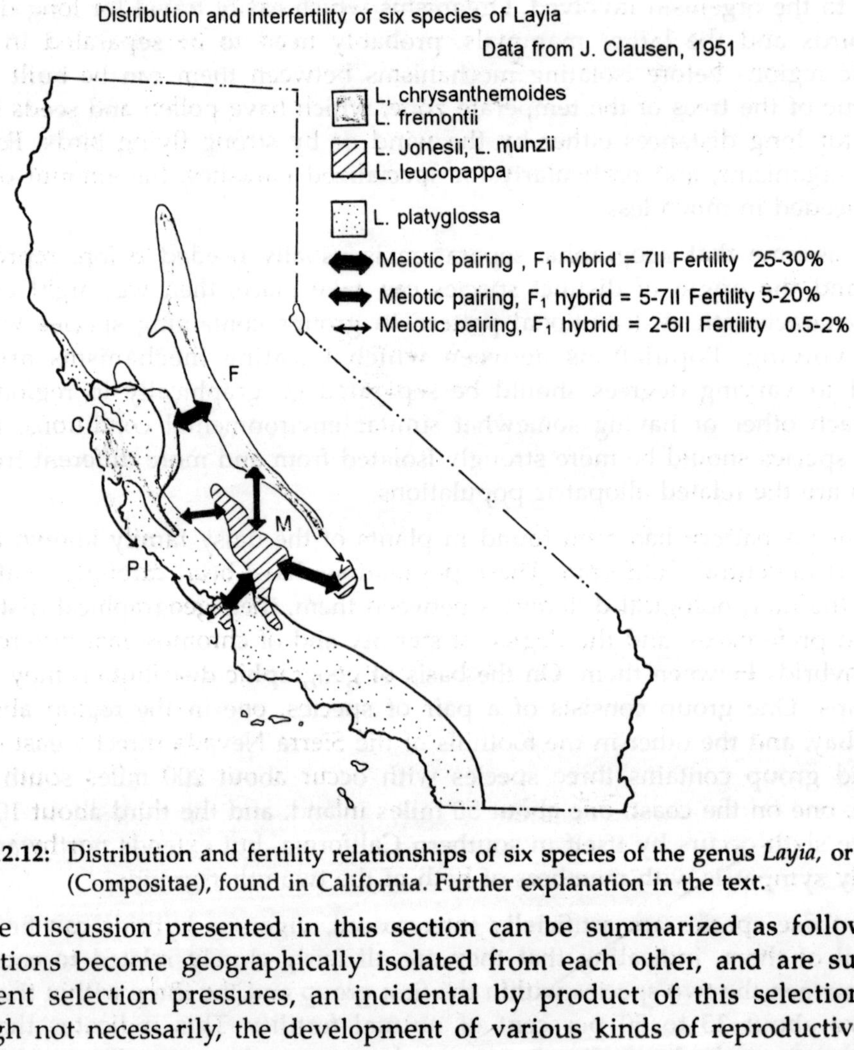

Fig. 12.12: Distribution and fertility relationships of six species of the genus *Layia*, or "tidy tips" (Compositae), found in California. Further explanation in the text.

The discussion presented in this section can be summarized as follows. If two populations become geographically isolated from each other, and are subjected to divergent selection pressures, an incidental by product of this selection may be, although not necessarily, the development of various kinds of reproductive isolating mechanisms. Once these have become well developed, the two pupulations rank as species, and can become sympatric again without losing their identity.

Under these conditions of secondary sympatry, reproductive isolating mechanisms can be reinforced by the *direct* action of natural selection on the barriers themselves.

This is accomplished by the direct action of natural selection in the following manner. Individuals which tend to mate with members of a foreign population will produce fewer vigorous, fertile offspring than will those which mate exclusively with other individuals of their own species. Consequently, mutations or gene combinations which favour mating between individuals of the same species will have a positive selective value, and will spread through the populations. The effectiveness of such selection has been shown by an experiment in which populations of two species of *Drosophila* separated by a partly developed barrier of ethological isolation were placed together in a cage. The experimenter then removed systematically all of the F_1 hybrids formed, thus penalizing reproductively those flies which mated with the other species. After sixteen generations flies of the two species were found to be almost completely isolated from each other.

These results show that ethological isolation can be reinforced by natural selection when two sympatric species often come in contact with each other. We might expect natural selection to reinforce similarly any isolating mechanism which tends to prevent hybridization at relatively early stages of the process. For instance, if two sympatric species often form sterile hybrids which automatically reduce the reproductive potential of both of the parental species, genetic changes which in any way prevent the hybrid zygotes from being formed will raise the reproductive potential and hence increase the evolutionary success of these species. In animals this can be done by reinforcing ethological isolation, and by various mechanisms which prevent fertilization even though copulation has taken place. In plants, natural selection might be expected to favour increasing divergence in flower structure between sympatric species. It would favour genes which prevent pollen of a foreign species from germinating on the stigma of a species, or would cause foreign pollen tubes to grow more slowly than those of the same species. In several groups of flowering plants, such as the milkweeds (*Asclepias*), Gilias, and ground-cherries (*Physalis*), barriers which block the initial stages of hybridization are strongly developed between sympatric species.

Based upon the concept of geographic speciation followed by reinforcement of isolating mechanisms, we should expect to find different relationships between species inhabiting unstable regions as compared to those inhabiting regions with more stable environments. In regions where the climate is changing relatively rapidly in terms of the geological or evolutionary time scale, or in which mountain building movements and volcanic activity are taking place, areas having a particular set of environmental conditions are changing their position rapidly, and the organisms inhabiting these areas must change their geographical distributions accordingly. In such regions, we should therefore expect to find many examples of allopatric populations on the borderline between races and species, as well as many species which have only recently become sympatric and are separated by relatively poorly developed isolating mechanisms. On the other hand, regions which have been climatically and geologically stable for long periods of time possess environmental niches which have remained the same for corresponding lengths of time. In such regions, we would expect to find fewer examples of species which have recently altered their geographic ranges. The opportunities for geographic speciation would be less. We should, however, expect to find more examples of species which have existed side by side in the same region for very long periods of time, and which have consequently undergone a long period of reinforcement of reproductive isolation by natural selection. These species should, therefore, be sharply distinct and well isolated from each other.

A comprehensive review of hybridization in fishes has revealed exactly this situation. In regions like the streams flowing into the desert from the mountains of Southern California, which vary greatly from season to season according to the amount of rainfall, and the relatively new, unstable lakes in the glaciated regions of northern Europe, the fish fauna is poor in species, and the species are poorly defined. Hybridization between species is common. The same is true of the lakes and streams of Scandinavia and Canada, which were formed after the retreat of the glaciers a few thousand years ago. On the other hand, the coral reefs of tropical regions are an old habitat which has remained stable for a very long period of time. There the fish fauna is much richer, and includes many groups of related species which tare sympatric over wide areas, but are very where very distinct and strongly isolated from each other. This situation is further evidence for the theory that rapid evolution is associated with habitat disturbance, and evolutionary stabilization is the result of long continued stability of the environment.

Chapter—13

Adaptive Radiation

Adaptation to environment is one of the most obvious and at the same time remarkable qualities of living organisms; in fact, it sums up nearly the whole result of evolution, and the real cause of it all lies in the germs-plasm itself, although it is manifest only in the somatoplasm of the animal. Operating the dissimilar or unrelated groups whose life conditions are comparable, it has given rise either to parallelism or to convergence, so that animals of remote stocks have come to look and act alike in so remarkable a way that one sometimes imagines relationships where none actually exists. Such creatures have been called homoplastic (Gr. *όμός*, same + *πλάsμα*, anything molded, *e.g.*, by environment) in contradistinction to homogenetic (Gr. *γέvos*, race) forms. Convergence is especially manifest in features concerned with locomotion, food-getting, offense, and defense.

The converse of this is known as divergence, where creatures of the same or closely related stocks have gone their several ways in their search for food and safety, giving rise to such varied adaptation that the several extremes of evolutionary lines depart markedly, not only from the norm or stem form, but also from each other. This idea of divergence under new and strange conditions is not new but was recognized by Lamarck, who called it "enbranchement" and by Darwin, who used the term "divergence" itself to express the idea. Finally Osborn clearly stated it in the form of a law which he first called adaptive radiation. This law may be re-stated in his own words:

Law of Adaptive Radiation

"Each isolated region, if large and sufficiently varied in its topography, soil, climate, and vegetation, will give rise to a diversified . . . fauna. The larger the region and the more diverse the conditions, the greater the variety of mammals which will result. From a primitive stem form radii go out in four diverse directions, the adaptations being mainly those of limbs and feet—also of teeth, but that of the teeth and feet do not necessarily parallel."

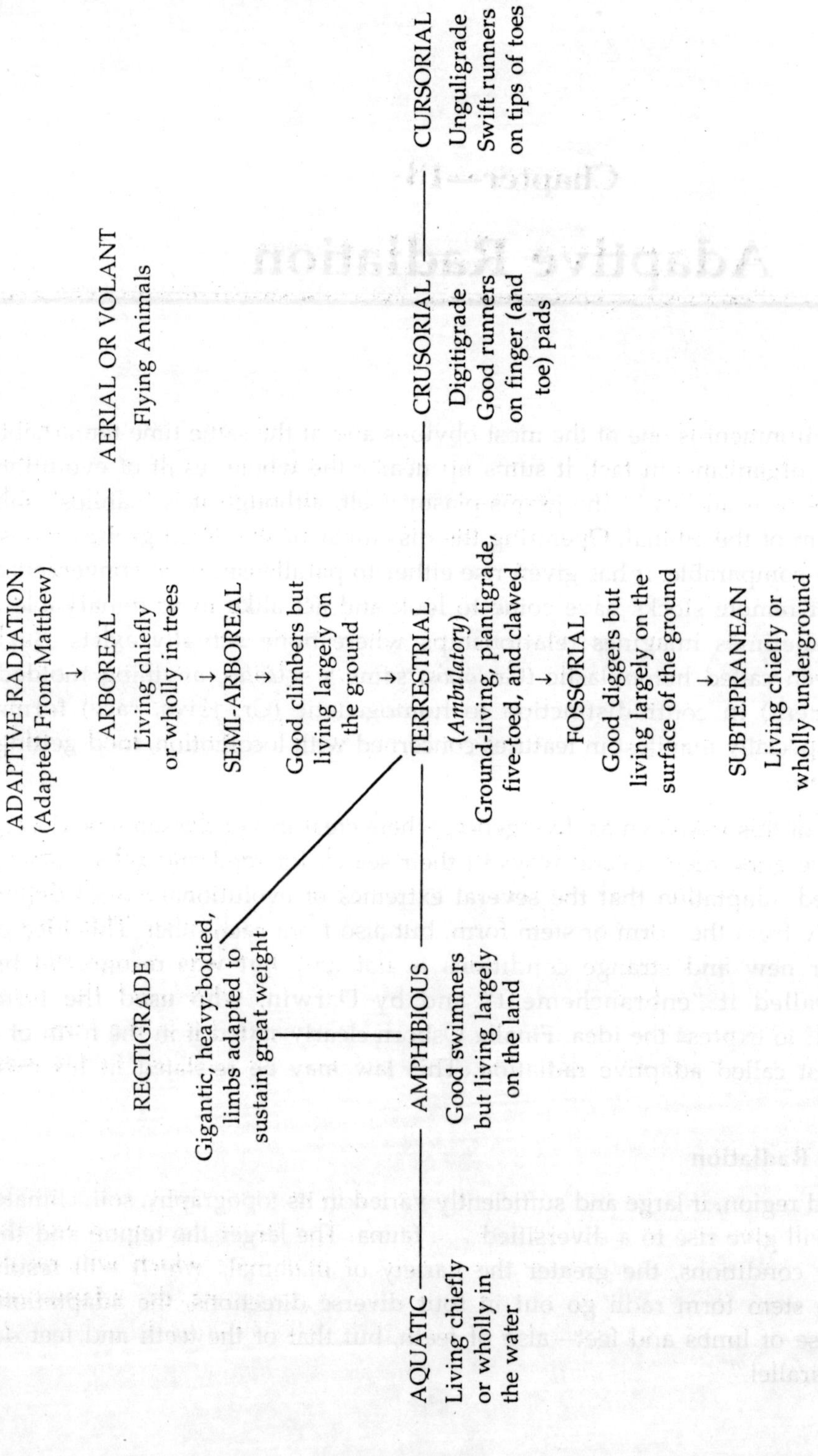
ADAPTIVE RADIATION
(Adapted From Matthew)
AERIAL OR VOLANT
Flying Animals
ARBOREAL
Living chiefly or wholly in trees
SEMI-ARBOREAL
Good climbers but living largely on the ground
RECTIGRADE
Gigantic, heavy-bodied, limbs adapted to sustain great weight
AQUATIC
Living chiefly or wholly in the water
AMPHIBIOUS
Good swimmers but living largely on the land
TERRESTRIAL
(Ambulatory)
Ground-living, plantigrade, five-toed, and clawed
CRUSORIAL
Digitigrade
Good runners on finger (and toe) pads
CURSORIAL
Unguligrade
Swift runners on tips of toes
FOSSORIAL
Good diggers but living largely on the surface of the ground
SUBTEPRANEAN
Living chiefly or wholly underground

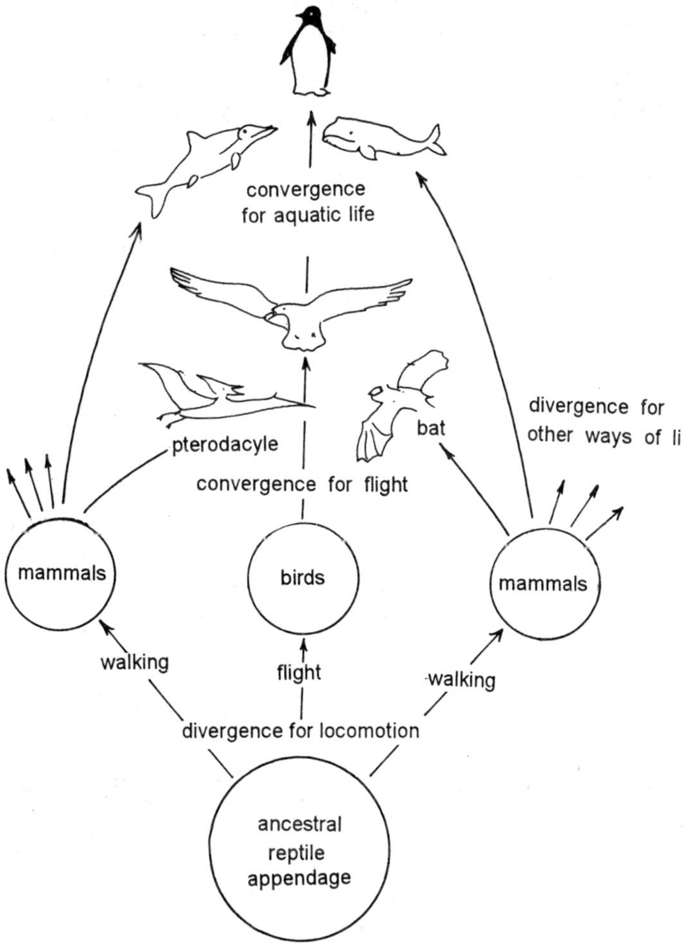

Fig. 13.1: Divergent and convergent Adaptation

While the term adaptive radiation fairly described the process as it occurs when the vertebrate classes are considered, yet *adaptive branching* more accurately expresses what takes place in the orders and families, which do not radiate in all directions, nor always in straight lines, but in branches that often turn and change their direction.

Osborn applied his law mainly to the mammals, but it is equally true of other groups such as the reptiles, especially during the Mesozoic, before the mammals came to their own. The central or focal type is conceived as a short-limbed, ambulatory form, with five clawed digits upon the approximately equally developed feet. It was small and generalized in its feeding habits, insectivorous (that is, devouring insects, worms or such creatures) or omnivorous and therefore with simple short-crowned teeth.

Adaptive Radiation (Teeth)
(Adapted from osborn)

Carnivorous *(Fish)* *(Flesh)* *(Carrion)* *(Omnivorous)*

Herbivorous *Grazing* *Browsing* *Frugivorous (Fruits, nuts, etc.)*

Insectivorous stem

Omnivorous

Myrmecophagous *(Ant-eating)* (dentition reduced)

The adaptations appearing under each of these radii will be discussed in detail in this and subsequent chapters.

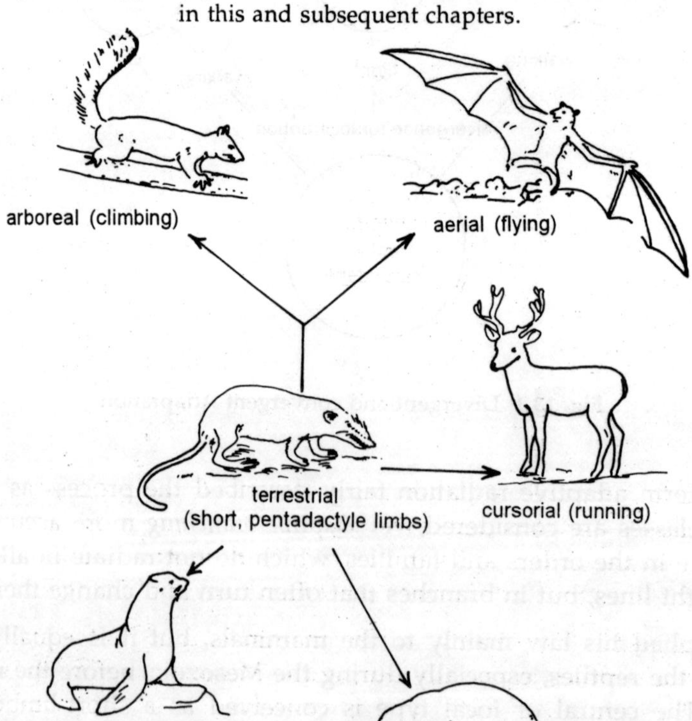

Fig. 13.2: Adaptive radiation in limb structure of mammals.

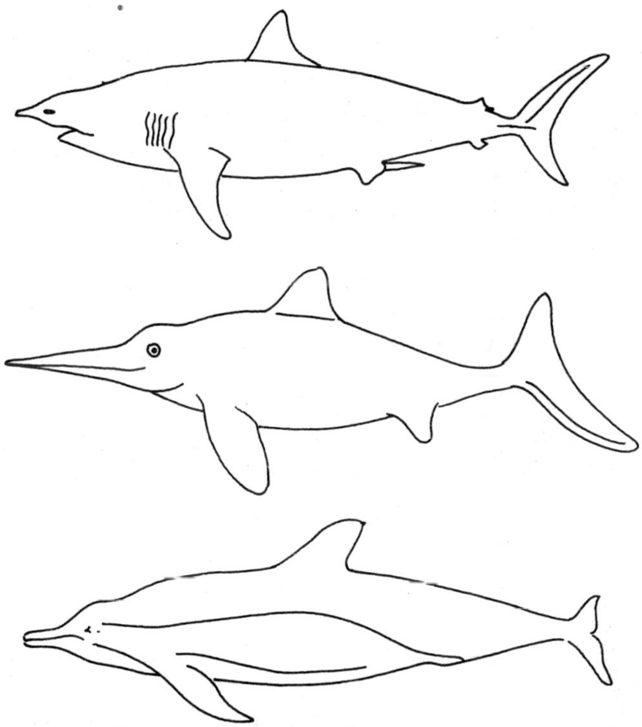

Fig. 13.3: Convergent evolution exhibited by, a fish, reptile (*Ichthyosours*) and a mammal (whale).

The impelling causes of adaptive radiation are the need of food and the need of safety, and while the minor roads along which evolution is possible are many and devious, they all lead in four general directions: over the earth's surface where speed would become the great desideratum (cursorial), beneath the surface to the subterranean realm (fossorial), above the surface into the trees (scansorial), or finally into the air (volant) and into the water to become denizens, wholly or in part, of the aquatic realm. Sub-terranean life results in degenerative specialization, so that a return of fossorial creatures to the terrestrial habitat is rare, while the aquatic realm modifies its inhabitants so profoundly that there is no known case of a return on the part of a lung-breathing form which has been adapted to aquatic life. These two roads—to the waters and the subterranean fastnesses—are one-way trails along which many go but none return.

Primary and Secondary Acquired Adaptations

Adaptations are direct or primary if they lie in the original direction of adaptation, outward along any of the four radii, as from terrestrial through scansorial to volant; or they may be reversed or secondary, when the return direction is indicated. While from the nature of its flying mechanisms, which involves the hind limbs, it impossible to conceive of a secondarily flightless mammal such as the bat, nor are any such recorded; nevertheless there are many instances of flightless birds, which have reverted to their original environment. They were probably ground or sea birds, however, before flight was lost. There is thus a reversal of the *direction* of evolution, but this does not actually contradict the law of irreversibility of evolution, which says that an organ once lost can never be regained and that a specialised form can never again become generalized. A further example will make this clear. The marsupials are either all arboreal to-day or give evidence in their anatomy of arboreal descent. One striking arboreal feature is a grasping great toe or hallux on the hind feet. This, being offset, opposes usually the fourth digit, thus forming an admirable prehensile organ whose grasp of a branch it is difficult to loosen. The terrestrial kangaroos, on the other hand, have become adapted in a wonderful way for speed over the earth's surface, and among other things have entirely lost this grasping hallux, as reduction of the lateral digits, much or little, invariably accompanies evolution for speed. There is, however, a tree kangaroo, *Dendrolagus,* which is very much at home in its arboreal retreat; having no grasping great toes, it has to rely upon its claws and broadened soles for security and they are well suited to the task, although the hallux would undoubtedly be better. *Dendrolagus,* while arboreal, shows in the loss of this useful members a terrestrial ancestry, and back of that, in common with all marsupials, an arboreal one. Hence for it the course of adaptation has been reversed, but the reversal is one of function and not one of structure, for however useful the grasping hallux might be in the present environment, its loss during the terrestrial career of the ancestors is irrevocable. While the atrophied organ therefore is lost forever, its old-time function may be secondarily acquired by other organs should the need arise.

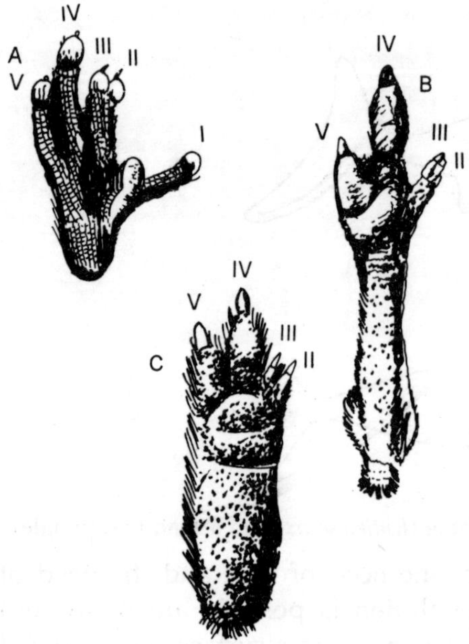

Fig. 13.4: Hind feet of marsupials. A, opossum, *Marmosa pusilla,* with grasping hallux, arboreal; B, kangaroo, *Macropus dorsalis,* without hallux, digits II and III syndactyl, cursorial; C, tree kangaroo, *Dendrolagus ursinus,* without hallux, arboreal (secondarily).

Other instances would be those of marine reptiles like the ichthyosaurs, or whales among mammals, both of which were derived indirectly from a fish ancestry, but which during the subsequent terrestrial life of their later progenitors lost the water-respiring gills. Perhaps the greatest menace to the safety of the modern whale is the necessity of rising to the surface to breathe, but the loss of gill-breathing is irrevocable and no organ apparently has been able to take its place. Primarily these forms were terrestrial animals, secondarily they have in many marvellous ways been adapted to the new environment, but owing to the law of irreversibility their adaptation apparently may never be perfected. Rodents are normally gnawing, herbivorous, terrestrial animals, certain ones being aquatic and fish-eating, the muskrat devouring both fish and fresh-water clams.

Local Adaptive Branching

The law of adaptive radiation which has been cited refers to evolution on a broad scale and includes the entire mammalian or reptilian fauna of a continent. Local adaptive branching, on the other hand, is restricted to the differences arising within a group of closely related forms whose life habits are in the main similar. Africa supports to-day two very distinct types of rhinoceros—the square-mouth or white and the pointed-lip or black rhino (see Fig. 13.5). With the square lip go long-crowned or hypsodont (Gr. *υψι*, on high, and *ὀδους*, tooth) teeth, while the pointed lip is correlated with short-crowned or brachyodont (Gr. *βραχυς*, short) grinders. The differentiation is one therefore of feeding habits, the first largely grazing, the second browsing, and the local distribution of the two types is in harmony with this distinction in that the square-mouth rhino inhabits the open country, especially the broad grassy valleys between the tracts of brush south of the Zambesi River, while the pointed-lip species, on the other hand, is found in the wooded and watered districts from Abyssinia to Cape Colony. Geographically therefore the rang of the black rhinoceros includes that of the white, but is more extensive, as the species is far more common; locally the two habitats are distinct and have given rise to feeding and consequently structural differentiation. Similar instances are repeatedly met with, especially among prehistoric forms such as the horses. Differences, slight at first, will in the course of time be accentuated so that the ultimate descendants are very unlike each other. A specific instance, comparable to the African rhinoceros, is that of the American Miocene horses *Hypohippus* and *Merychippus*. The first had short-crowned, uncemented, browsing teeth and well developed lateral hoofs, forming a foot fitted for yielding ground like that of the forested areas. The second had complex, long-crowned, grazing teeth adapted to grasses, and feet fitted for the hard prairie soil. The distinction between these two races may well have had its inception during late Oligocene time when differentiation of habitat was already established, but we cannot with our present evidence clearly separate the phyla until Lower Miocene at least. Local adaptive branching therefore gives rise to different phyla of the same group of animals in a relatively small areas.

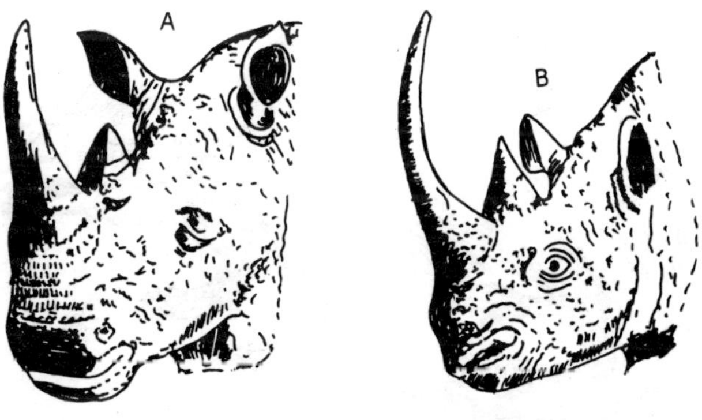

Fig. 13.5: African rhinoceroses. A, square-mouth or white rhinocerous, *Rhinoceros simus;* B, pointed-mouth or black rhinoceros, *R. bicornis*.

Continental Adaptive Radiation

On the other hand, is on a broad scale and, as has said, embraces the entire fauna of a given class (*i.e.*, Mammalia). Such radiations have appeared more than once, sometimes as repetitions of evolution within the same area, or again the radiations may be contemporaneous, but within the limits of distinct isolated continental masses.

Contemporaneous Radiations

As we learned in the chapter on geographical distribution, the world's surface may be divided into three coequal but not coextensive realms: Arctogæa, Neogæa, and Notogæa, of which the first includes all of the northern hemisphere, and Africa; the second, South America; and the third, Australia. These three great realms have been the centers respectively of three remarkable adaptive radiations of mammals during Tertiary time, when all three were isolated each from the other in such a way as to prohibit absolutely for a long period all faunal interchange. This isolation is true of Australia to this day. Hence while the South American mammals have practically ceased to exist, (Australia's fauna, except for certain overseas migrants introduced largely through human intervention, remains practically as it has been, composed of mammals of Mesozoic rather than of Tertiary type. Between Australia and Arctogæa there some very remarkable parallelisms, for, upon a more limited scale, nature has nevertheless repeated herself, though not exactly, in the lesser continent. A comparison of the two radiations may be made).

In *Arctogæa* the central or focal type is represented in a general way by the Malayan insectivore, *Gymnura*. This creature belongs to the hedgehog family, Erinaceidæ, but the more familiar European hedgehog, *Erinaceus*, is somewhat too specialized in its quill-like hairs and burrowing habits and consequent modification to serve as the focal types.

Along the cursorial radius we have in the dogs, especially the swift fox and the coyote, creatures of amazing speed although digitigrade, that is, running upon the toes. Unguligrade or hoofed runners are best represented by the deer, the African and Mongolian antelope, but especially by the horse-like animal, the ghorkhar (*Equus anager*), which inhabits the Persian desert, an animal so swift that adults in good condition can neither be ridden down, unless by relays of horsemen, nor taken with greyhounds (Lydekker).

Fig. 13.6: The most primitive living representative of the Insectivora, *Gymnura*.

Arboreal forms are represented by the squirrels, some of which—the flying squirrels—have learned to launch themselves into the air and take long soaring leaps, supported by folds of skin between the limbs and the sides of the body, and hence have become semi-aërial. Truly volant or aërial forms are represented among mammals by the bats, which, however, belong to a different order insectivorous stock, although no annectant forms are now alive. One remarkable insectivore, *Galeopithecus*, which will be described more in detail later, is a wonderful glider, and while not in the line of bat descent shows up quite clearly how such an adaptation could have arisen.

Aquatic-amphibious forms are represented by the *Potomogale* among insectivores, the muskrats and beaver among rodents, and the otters, but especially the seals, among carnivores; while the truly aquatic types are such as the porpoises and whales, which are apparently modified ancient carnivores, and the sea-cows (manatee, and dugong), which were derived from the ungulate stock.

Subterranean forms are the insectivorous shrews and moles and the rodent woodchucks, gophers, and mole-rates (*Bathyergus*). Of these the true moles and the golden moles (Chrysochloridæ) of the Cape of Good Hope, two entirely independent evolutions of mole-like forms, are the most extreme.

Notogæa

The Australian mammals, as we have learned, are practically all marsupials or pouch-bearing mammals, whose young are brought forth at an extremely early stage of development, transferred to the pouch-covered nipples, and there nurished as 'larvæ" until mature enough to stand the vicissitudes of ordinary animal childhood, when they are virtually reborn. There is no placnta, except for one or two rare vestiges (*phascolarctos, Perameles*), and the young, being inadequately nourished compared with placental mammals, cannot stand competition with the latter even in their adult condition. Such was the stock with which Australia was inoculated, in the Mesozoic age and out of that stock, due to the immenseperid of isolation which has elpsed, there has arisen an adaptive radiation paralleling that of Arctogæa in many remarkable ways.

Perhaps the creature nearest the focal form among marsupials is the opossum. The opossum is now an American form and therefore in its existing condition cannot represent the actual ancestor, but it gives a good impression of what the Australian stem form was like. Opossums are, however, extremely old, persistently primitive forms, of which near relatives are found in Cretaceous rocks. They have survived cotemporaneously with their diversified descendants. The opossum is arboreal, but it is quite probable that the marsupials had all passed through a tree-inhabiting stage before their great radiation began.

Cursorial adaptation in the marsupials is well shown in the Tasmanian wolf, *Thylacynus*, which is very dog-like in contour, development of limbs, and digital reduction. It is perhaps a little less extreme in its cursorial adaptation that the jackal (*Canis aureus*) for instance, as the distal segments of the limbs, whose relative lengthening is generally a speed requirement, are still proportionally short. *Thylacynus* clearly occupies the station of the wolves and dogs of Arctogæa, a fact which has caused its extinction when brought into competition with true canids (*i.e., Canis dingo*).

The herbivorous bandicoots and kangaroos are highly speedy and represent the ungulates of Arctogæa, the bandicoots (*Chœropus*) being quadrupeds whose hind foot has but a single dominant digit, although other vestigial ones are present. The kangaroos, on the other hand, are bipedal when running, with feet which are remarkably effective.

Arboreal marsupials are still numerous, the so-called "sugar squirrels" or phalangers being very similar to our squirrels in appearance, while the Arctogæan flying squirrels are represented with great fidelity by the Notogæan flying phalanger (*Petaurus sciureus*). A further extreme of arboreal adaptation is shown in the koala (*phascolarctos cinereus*), slow-moving, tailless, with splendidly adapted hands and feet, the former having two fingers offset against the other three, and the latter the hallux against the other four, especially the fourth and fifth. In many ways the koala resembles the lemurs or half-apes of Arctogæa, especially such forms as the loris, *Nycticebus.*

No truly volant forms comparable to the bats exist in the Australian marsupial fauna, although placental bats are present, nor is there record of their having existed there during geologic time.

Aquatic types among marsupials are restricted to amphibious forms, truly aquatic types comparable to the Cetacea being unknown and improbable, for such would lack the isolation and security from competition so necessary to marsupial evolution. Strangely enough, however, there is not recorded instance of even an amphibious marsupial in the existing Australian fauna, although South America possesses one, the water opossum (*Chironectes*), whose hind feet are webbed and whose general appearance reminds one of the muskrats, though of course, unlike the latter, it is exclusively carnivorous in its diet. It is highly improbable, however, that Australia has never possessed an amphibious marsupial.

Notogæa is not without its amphibious type, for the duck-bill or duck-mole (*Ornithorhynchus*) is a remarkable combination of burrowing and swimming animal, the projecting swimming membrane of the fore foot being folded back against the palm, exposing the powerful claws, when the creature wishes to dig. It its life habits the duck-bill is not unlike the muskrat, but the food differs, as it feeds upon various small water-animals, such as crustaceans, insects and their larvæ, snails and worms, which are dug out of the soft mud by the tender muzzle. *Echidna*, an ally of *Ornithorhynchus,* while having a pouch for the protection of the newly hatched young, is not a member of the order Marsupialia but of the yet more primitive Monotremata—the egg-laying mammals—which while strictly Notogæan in their existing distribution, are nevertheless not of the present Australian adaptive radiation, as that implies a common, *i.e.,* marsupial, ancestry, but are survivors of an earlier evolution.

Fossorial Notogæan animals are admirably represented by the marsupial "mole," *Notoryctes typhlops,* a recently discovered type, which bears a remarkably close resemblance in its molar teeth, external form, and mode of life to the South African golden moles (Chrysochloridæ) of which we have spoken. But they cannot be related, for *Chrysochloris* is a placental insectivore and *Notoryctes* a marsupial. They therefore represent one of the most interesting cases of convergent evolution of which we have knowledge.

Thus it will be seen that the Australian adaptive radiation practically coincides with that of Arctogæa, almost every animal in the former assemblage having some approximate equivalent in the latter. The reverse, on the other hand, does not hold true, for the Arctogæan realm is relatively so vast and its range of habitat conditions so much greater, that its fauna is proportionately more extensive and varied.

The ancient fauna Notogæa has but a dubious future, for the artificial introduction of several placental mammals has already made serious inroads into the integrity of the marsupial ranks, the dingo having driven the Tasmanian wolf from the Australian mainland while the rabbits and sheep, through their destruction of the food supply, have reduced the marsupial herbivorous population by cutting off their means of subsistence. To this must be added the destruction of marsupials by mankind, for the hides, both for furs and leather, are commercially known as opossum and kangaroo respectively, It is quite probable that, with few exceptions, this ancient radiation is doomed to speedy extinction.

Successive Radiations in Time

There have been three successive radiations among mammals of which the first, however, that during the *Mesozoic,* was comparatively unimportant because of the terrestrial dominance of the dinosaurs and other reptiles. During this radiation the known mammals were all small, although differentiated into three or four phyla from the standpoint of their dentition and implied feeding habits. Thus some were insectivorous, some perhaps more distinctly carnivorous, and some herbivorous, possibly fruit or nut-eating (seed-eating?).

It was not until the great reptilian extinctions at the close of the Cretaceous that the mammals really had their opportunity, at least in the known fossil-bearing regions; and with the vacating of the terrestrial realm by the huge cold-blooded brutes, the higher creation began its development in the known areas. Palaeontologists soon find records of carnivorous forms (creodonts) of various sorts, swift cursorial ungulates (condylarths), and slow-moving, grotesquely armed types (amblypods). There were also sloth-like forms (tæniodonts) and probably ancestral marsupials. Of all of these so-called archaic mammals, but few survived the Eocene; some were sooner to later rendered extinct through competition with invaders of the third or Tertiary radiation, others may have evolved into certain of the higher types, yet others, driven southward into Neogæa and Notogæa, may have formed the stock of part at least of their subsequent faunas when isolation closed the door upon later invaders.

The *Tertiary* radiation was the great Arctogæan radiation of modernized mammals, so called because their descendants still exist; at all events on higher group has appeared, and the story of mammalian evolution in this world, owing to the present dominance of man, is virtually over. Whence the progenitors of the modernized mammals came we have no positive knowledge, but circumstantial evidence points to boreal Holarctica—the circumpolar lands of the northern hemisphere—as their pristine home. For in the Cretaceous and early Eocene indications point to a warm climate and abundant forestation in the north, ideal conditions for the well-being of the nodal type and its immediate derivatives. Increasing cold in the northland drove forests and their fauna south down the three great continental axes so that we find creatures of much the same sort, such as the ancestral horses, appearing synchronously in the Old World and in the New. Early in the Eocene their deployment was well under way, the Oligocene saw certain lines already approaching racial death and others wondrously varied, while during the Miocene the culmination was reached and never before nor since was there so great a mammalian florescence. By the Pliocene they had begun to wane. The mammals of to-day, including man himself, are the remnant of this Tertiary radiation.

Tooth Radiation

The mammals are, with few exceptions (Cetacea, anteaters, etc.), characterized by a heterodont (ετερος, different, and όδους, tooth) dentition, that is, the teeth, in contrast to those of most reptiles, for instance, are differentiated into several sorts with, as a rule, very distinct functions. Those in the front of the mouth, the incisors, are mainly prehensile, the canines grasping, tearing, or for defense or offense, the premolars sometimes shearing, or, like the molars, grinding teeth. The least mentioned molars and premolars show the greatest structural modification to meet their owner's requirements, the incisors and canines being more conservative, as their use shows less variation in the different orders.

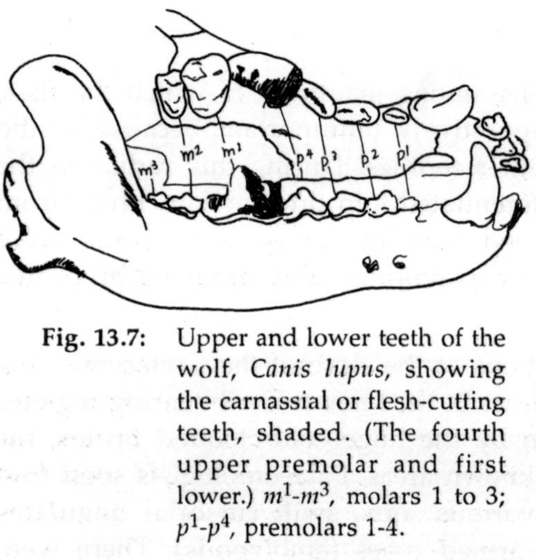

Fig. 13.7: Upper and lower teeth of the wolf, *Canis lupus*, showing the carnassial or flesh-cutting teeth, shaded. (The fourth upper premolar and first lower.) m^1-m^3, molars 1 to 3; p^1-p^4, premolars 1-4.

The *insectivorous type,* those of the stem form, are low-crowned, simple, with few cusps, generally sharp-pointed and suitable for crushing feebly prey. With the *carnivores,* the cheek teeth become high-crowned, trenchant, shearing structures, and the jaws have little or no lateral play. This reaches its highest specialization in the cats wherein the true grinding teeth are almost entirely wanting. In the dogs, with a less exclusive diet, more of the grinders are retained, although the shearing teeth are well developed, while in the bears, carnivores with a strong tendency toward an herbivorous diet (omnivorous), the cheek teeth are broad-crowned, tuberculate grinding organs, and shearing teeth are wanting. Carrion-feeding forms have, as a rule, blunter teeth, while in fish-eating creatures like the seals the teeth have become secondarily simplified and are all prehensile. The toothed whales (Odontoceti) have lost their tooth differentiation and the teeth are practically all alike—

simple, slightly recurved, grasping cones. The teeth are sometimes absent in the upper jaw (sperm whale, *Physter,* in which, while germs of upper teeth are present, they are apparently functionless), and in the whalebone whales (Mystaceti) they are entirely absent, except for functionless vestiges in the unborn young, and their place is taken by the curious baleen or whalebone which hangs from the palate and by means of which small molluscs and crustaceans are strained out of the water to serve for food.

Herbivorous forms have the incisors fitted for seizing and cutting the vegetation. In the ruminants, however, they are absent in the upper jaw, the lower incisors biting against a horny pad. Canine teeth are of little importance to the herbivore unless they are used for defense (musk deer), or as in the swine for uprooting the vegetation upon which they feed. The grinders vary amazingly in the pattern of the cusps and crests borne upon the crown but in general may be divided into short-crowned, brachyodont teeth fitted for succelent leaves and twigs; or long-crowned, hypsodont teeth whose great length and complex structure are adapted to meet the wear imposed by the harsh grasses and other vegetation which form the staple of a grazing animal. Perhaps the greatest degree of perfection has been reached by the modern horse (*Equus*), along one line of descent and the elephant, more especially the extinct woolly mammoth (*Mammonteus primigenius,* compare) along another.

As we have seen, a carnivore like the bear may become adapted to an all-round diet and hence be *omnivorous,* but the term is best applied to a direct line of evolution from the insectivorous stock, such as the primates, and especially mankind. Here the incisors and canines show little distinction and the cheek teeth are very simple and primitive, with bicuspid premolars and molars with few (four or five) low cusps.

The *myrmecophagous type* represents the height of specialization, for in its extreme development the teeth have utterly disappeared, the jaws are reduced, the mouth is incapable of opening except at the extreme anterior end and had become tubular, with a highly extensile and prehensile tongue. This organ is provided with an adhesive substance and when thrust into a ant-hill and withdrawn brings away great numbers of the unfortunate insects, which are swallowed without mastication. As a special adaptation to such rather insurgent prey, the interior nostrils are carried far backward so as to be in direct communication with the windpipe to prevent any of the creatures from wandering the wrong way. An analogous adaptation is seen in the marsupials to avoid suffocation when the milk is forced into the mouth of the feeble young by the mother's muscles, and in the whales and again in the crocodiles, both of which devour their prey under the eater and would run the risk of strangulation were it not for this device.

Chapter—14

Cursorial and Fossorial Adaptation

CURSORIAL ADAPTATION

Animals living on land need bodily structures enabling them to run away quickly. Thus these adaptations are also called *speed adaptations*. These adaptations can be seen in reptiles like snakes, lizards, birds like *Emu, Ostrich, Rhea, Moa* (all are flighless birds) etc. and more perfectly in mammals. They are characterised by following changes.

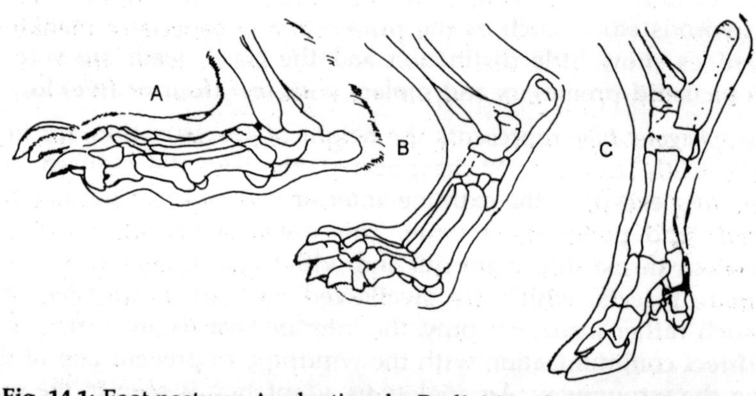

Fig. 14.1: Foot postures A=plantigrade, B=digitigrade, C=unguligrade

Speed adaptation has been developed in a very wonderful way in terrestrial (cursorial) types, in aquatic, and finally in aërial forms. Of these we will first discuss the terrestrial.

Vertebrates Showing Speed Adaptation

The following is a list of vertebrates which show this adaptation for speed:

CLASS REPTILIA. REPTILES

Order Squamata. Lizards and snakes.

Especially certain of the Agamidæ. *Chlamydosaurus*, the frilled lizard of

Australia. Numerous others, especially desert-inhabiting forms.

Dinosaurs.

Order Saurischia. Carnivorous and amphibious dinosaurs.

Most of the carnivorous dinosaurs, especially the smaller, more agile forms: *Podokesaurus, Hallopus, Cœlurus, Compsognathus, Ornithomimus* and *Ornitholestes.*

Order Ornithischia. Herbivorous beaked dinosaurs.

Nanosaurus and the Camptosauridaæ.

CLASS AVES. BIRDS

Division Ratitæ. Flightless cursorial birds.

Especially emus, cassowaries, ostriches, and rheas, and probably the extinct Dinornithes, the moas of New Zealand.

Division Carinatæ. Flying birds.

Of these few are adapted for speed except perhaps the rails, one of which (*Aptornis*) is flightless.

Certain of the Limicolæ, plovers, oyster-catchers, etc., are also speedy for short distances.

CLASS MAMMALIA. MAMMALS

Order Marsupialia.

Dasyuridæ, especially *Thylacynus,* the dog-like Tasmanian "wolf."

Peramelidæ, the bandicoots.

Macropodidæ, the kangaroos.

Order Rodentia.

Lepordæ. Hares and rabbits.

Caviidæ. Cavies.

Dipodidæ. Jerboas.

Order Carnivora.

Canidæ. Dogs.

Felidæ. Cats. Especially *Acinonyx,* the cheetah or hunting leopard.

Cohort Ungulata.

Order Perissodactyla

Hyracodontidæ. Extinct cursorial rhinoceroses.

Equidæ. Horses, living and extinct.

Order Artiodactyla.

Protoceratidæ. Extinct.

Antilocapridæ. American prong-horn antelope, *Antilocapra.*

Cervidæ. Deer.

Bovidæ. Especially the African antelopes.

Giraffidæ. Giraffe and okapi.

Tylopoda. Camels.

Order Litopterna. South American ungulates.

Especially the Proterotheriidæ, the extinct "pseudo-horses."

Body Contour

All speedy animals, whether terrestrial, aquatic, or aërial, have the body moulded externally in such a way as to offer the least resistance to the medium through which they pass. Owing to the greater resistance of water, this is especially true of the aquatic, nevertheless a speedy cursorial type also shows it, though not always as well when at rest as it does when in action. A race-horse with head and neck extended, ears thrown back, and every tense muscle of its wonderful body working with machine-like precision, sows the beautiful contour of a perfectly adapted mechanism. The body spindle-shape, of stream-line contour, the lines extended into the neck and head without a break in their even curves—all are calculated for swift passage through the air with a minimum of resistance.

MECHANISM

Loss of General Utility

The propelling organs in cursorial forms are the limbs exclusively, so that, aside from the resistance-lessening contour, this adaptation concerns itself chiefly with their modification, of which the first is the loss of general utility. This is especially true of the hind limbs, because they are the more efficient drivers and therefore, especially in forms whose adaptation has not gone very far, are likely to be somewhat in advance of the fore limbs in the degree of their evolution. For this several reasons may be assigned: (1) The extended hands pull the body forward in running, (2) the fore part of the body is usually heavier than the hind part and requires larger limbs to support and propel it; (3) running is a sort of leaping on all fours, and the hands are larger and wider to take the impact when the animal falls forward; finally (4) the fore limbs, being nearer the mouth and hence perhaps somewhat concerned in food-getting, are the last to lose their general utility. Two notable instances of this accelerated evolution of the hind limbs over the fore are the early four-toed horses of the Eocene (Hyracotheres), in which, while the hand still retains its four digits, the foot has but three. It is not until Oligocene time that the additional finger is lost and the evolution of both limbs becomes parallel. In another Oligocene form, *Protoceras,* a curious artiodactyl with remarkable excrescences upon the skull and dagger-like canine teeth, the hand is four-toed while the foot has but two.

Change in Foot Posture

The primitive terrestrial foot is plantigrade (Lat. *planta,* sole, and *gradi,* to walk), which means that the entire palm or sole rests on the ground, neither wrist nor ankle being raised. Almost the first step in speed adaptation is the lengthening of the limb and this may be accomplished without the actual elongation of a bone, merely by rising upon the toes. While the bear, raccoon, and the primates such as the baboons and man are plantigrade, probably secondarily so, a large proportion of modernized mammals have become digitigrade (Lat. *digitus,* finger, toe), walking or running upon the digits themselves, with the bones of the wrist (carpal) and ankle (tarsal), the upper ends of the palm (metacarpal) and the sole bones (metatarsal) clear of the ground. Some of the speediest of animals—dinosaurs, birds, dogs, all mammals in fact but the ungulates—have merely perfected the digitigrade gait, developing special sole-pads for the absorption of the shock of impact, and have never gone beyond it. The ungulates or hoofed animals, on the other hand, walk on the modified nail or hoof (unguligrade, Lat. *ungula,* hoof), the distal toe-bones (ungulas) being depressed or flattened and not, with rare exceptions, compressed or claw-like. The hoof has reached the highest degree of perfection in the horses; in other related but non-cursorial types like the rhinoceros the hoofs bear little of the weight, as a broad cushion-like pad serves instead.

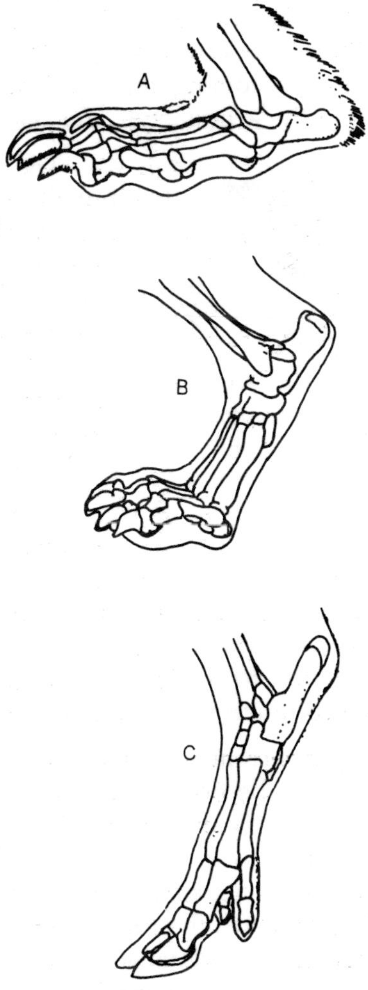

Fig. 14.2: Foot postures. A, plantigrade, bear; B, digitigrade, hyæna; C, unguligrade, pig.

Ungulate animals show all gradations from the semi-plantigrade condition of the slower forms to the high, stilted hoofs of certain of the African antelope, notably the klipsringer (*Oreotragus saltator*). A truly unguligrade condition has never been attained among reptiles, although certain beaked dinosaurs had depressed instead of claw-like unguals. *Triceratops* and *Stegosaurus* were certainly far from speedy; hadrosaurs, on the other hand; which also bore this type of ungual, were bipedal and doubtless possessed a fair measure of speed when well under way, though little celerity of movement is indicated.

Certain ungulates like the modern camels have become secondarily digitigrade, the foot having retrogressed as an adaptation to the yielding desert sands. This is not accompanied, however, by any material loss of speed, as the camels are among the most remarkable travelers of all terrestrial forms.

Loss of Digits

Plantigrade animals are generally five-toed; there are of course exceptions, but the elevation of the heel or wrist generally carries with it digital reduction, digitigrade animals becoming four-toed, unguligrade four-, three-, two-, or even one- toed, two toes in the artiodactyls and one in the perissodactyls being the irreducible minimum.

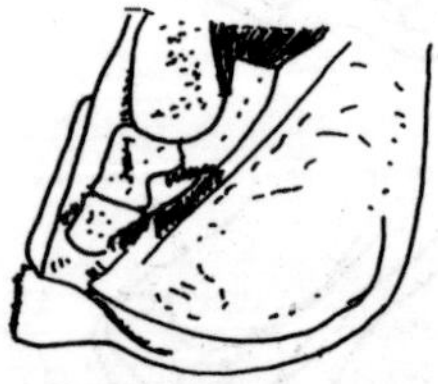

Fig. 14.3: Rhinoceros foot sectioned to show the supporting pad.

The frilled lizard of Australia, *Chlamydosaurus*, is five-toed but the lateral toes are shorter than the median ones, which is almost universally true except in aquatic types such as the seal and otter. Hence when *Chlamydosaurus* runs on its hind feet, as it does when startled, the outer and inner toes are raised off the ground and the animal makes a three-toed track. If this were the habitual gait of the creature, the lateral digits would be rendered practically useless and would follow the course of all useless organs and become reduced, whatever the philosophical explanation of the mean whereby this is accomplished.

Environment as well as speed adaptation has its influence in determining digital reduction, for it will be accelerated if the ground is hard, as in the prairie-evolved horse with but one remaining digit, or the prong-horn antelope of similar environment with two. On the other hand, the Miocene forest horse, *Hypohippus*, retained the lateral toes as functional organs just as the reindeer and caribou (*Rangifer*) have to-day, as an adaptation to a yielding footing, while contemporary relatives had in each instance evolved much farther along the line of digital reduction.

Concurrent with the loss of digits, especially if the foot be lengthening after the manner to be described below, comes a compacting of the bones of the palm and sole (metapodials) and often this is carried so far as to give rise to actual fusion of these elements into a "cannon-bone." The dinosaurs, with one doubtful exception, never attained an actual fusion, although in many respects, especially in *Ornithomimus* of the Cretaceous, the foot is very bird-like. The birds, on the other hand, always show a fusion of the metatarsals. Among mammals the carnivores do not form a cannon-bone nor does the marsupial wolf; but all of the speed-adapted ungulates do. Ancient ungulates, however, had the metatarsals separate, and we can often witness the fusion in fossil series (camels, etc.) when the proper degree of speed adaptation has been reached. Among rodents, the jerboa, a tree-toed bipedal form, has a foot and metatarsus so wonderfully bird-like that one almost has to count the phalanges of the digits to be sure he has a mammal before him.

Reduction of Fibula and Ulna

The fore arm and shin, that is, the second segment from the body, have each typically two bones: in the arm, the radius and ulna, and in the leg, the tibia and fibula. These are both developed in slow-moving forms or where the fore limb still has considerable general utility, especially if the rotation of the hand on the arm is retained. On the other hand, cursorial forms, especially if the limbs are exclusively locomotor, tend to reduce the ulna of the arm, the proximal end only being present in extreme cases to form the elbow joint. They also lose the fibula of the leg, which may be reduced to the merest vestige.

Loss of Universal Movement

The entire motion of the limbs becomes pendulum-like, that is, restricted to movement in but one plane, the exception being at the hip and shoulder, where universal movement is still retained through the development of a ball-and-socket articulation. The necessity for this is apparent, first to avoid interference between fore and hind feet when running, since a dog, for instance, at top speed brings his hind feet well in advance and outside of the fore. A second need is that of lying down and rising again, which would be practically impossible were the movement at hip and shoulder restricted to the fore-and-aft plane. With other articulations, those of ankle and wrist, knee and elbow, and between the digits, the tendency is toward rigid limitation of movement in unguligrade, less so in digitigrade forms. This is accomplished by the development of tongue and groove joints such as ere discussed under kinetogenesis. These are very perfectly sown in the hind limb of a modern horse, as well as at the elbow joint, forming in each instance an articulation permitting movement through a wide arc in one plane of space and none whatever in any other. These joints, while they may be broken, cannot be dislocated.

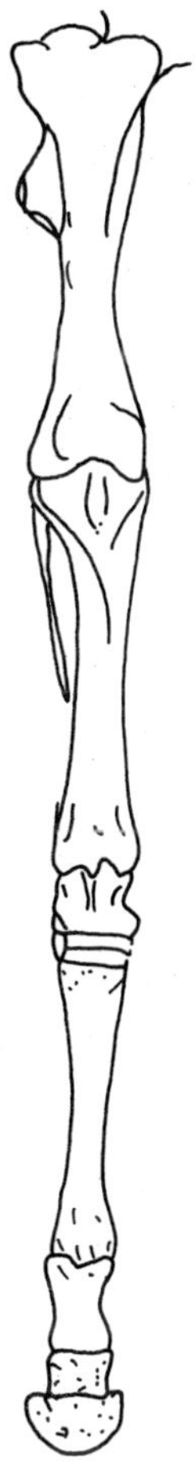

Fig. 14.4: Hind limb of horse, *Equus caballus,* to show pulley-like joints for the restriction of movement to one plane.

The limbs are compound levers, for not only is there motion of the limb as a whole but also between its component parts. The lengths of each of the several segments bear definite relations to the speeds developed and also to the loads they have to carry. Those forms which, like the elephant, are mighty of frame, have a type of limb which is in marked contrast to that of a horse. To the former type has been applied the term graviportal (*i.e.,* weight-carrying), to the latter cursorial, although both are adapted to increase their owner's traveling power. In the graviportal type such as the mastodon, the foot is short and the thigh and shin relatively long, whereas in a cursorial form the foot elongates and the thigh is conservative. This may be shown by a comparison of ratios between the middle metatarsal and the femur, thus:

Ratio: $\dfrac{\textit{Length of metatarsal III}}{\textit{Length of femur}}$

Graviportal		*Subcursorial*		*Cursorial*	
Uintatherium	.10	Eohippus	.50	Equus	78
Mastodon	.11	Tragulus	.56	Antelope	1.00
Elephas	.13	Mesohippus	.57	Odocoileus	1.00

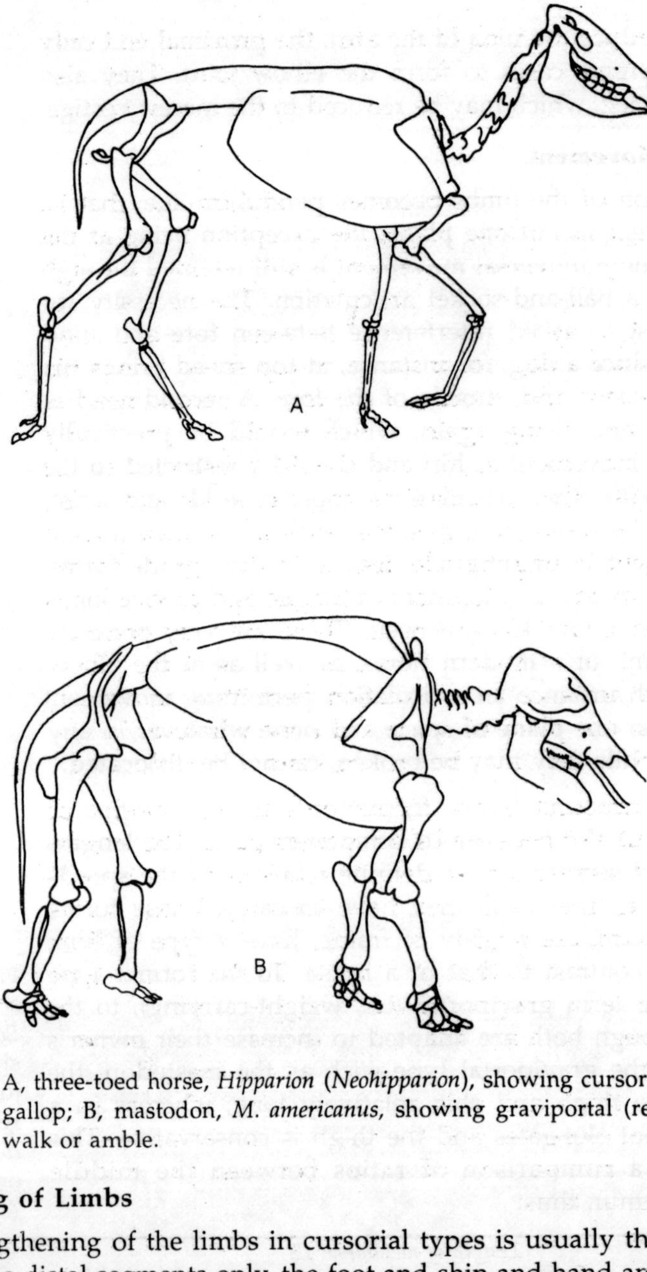

Fig. 14.5: A, three-toed horse, *Hipparion* (*Neohipparion*), showing cursorial adaptation for run or gallop; B, mastodon, *M. americanus*, showing graviportal (rectigrade) adaptation for walk or amble.

Lengthening of Limbs

The lengthening of the limbs in cursorial types is usually thus accomplished by a growth of the distal segments only, the foot and shin and hand and fore arm increasing, but rarely the thigh or upper arm. This increase is therefore both actual and relative. This is well shown in comparing the limb of a horse with that of a man, for while the human thigh may actually exceed the length of that of the horse, the horse's foot, which is measured from the end of the hoof to the hock, the equivalent of the human heel, may be two and a half times that of the man.

This lengthening of the distal segments, which is for the purpose of increasing the length of stride, would be of little avail were the muscles not concentrated at the proximal end of the limb, their power being transmitted by long slender tendons to the lower leg and foot. One effect of this may be best understood by comparing the limb with a pendulum. The length of the pendulum determines the scope of swing, but the position of its center of gravity controls its speed or rate of beat. To accelerate the beat, therefore, the bob is moved upward, to retard it, it is moved downward, the arc of pendular motion remaining constant. This concentration of the muscles at the proximal end of the limbs has the same effect as rasing the pendulum bob, and by this device—long slender limb and high but powerful muscles—the maximum length of stride and speed of movement are obtained. While this may well be an important reason for the concentration of muscle, the parallel with the pendulum is not exact for the opposing muscles not only serve to initiate the swing but also to damp it. Perhaps the chief reason, therefore, its that by raising the insertion points of the thigh-muscles focussing around the knee, the *angles of insertion* of many muscles are increased and this gives higher propulsive components, *across* the shaft of the femur; at the same time the muscles are shortened and made thicker, which increases their power and speed of contraction. It can readily be seen that a limit may be reached beyond which bone will not stand the strain to which it would be subjected, although bone is a wonderfully efficient material. Hence one would expect to find the greatest speed developed on the part of creatures of small to moderate size—the antelope of Africa or horses like the wild ass of Persia (*Equus onager*) the speed of which has been mentioned and which reaches a stature of but 11½ hands. The Mongolian wild ass or kiang (*Equus kiang*), according to Roy Chapman Andrews, who chased one for 29 miles in a motor car, averaged 30 miles an hour for the first 16 miles and then when it began to slow down, still ran four more miles at a speed of 20 miles an hour. The modern race-horse is relatively small compared with some other breeds, and the limit of weight, size, and speed consistent with safety seems to have been approximately reached.

Ratios

Lengthening of limbs also implies, at any rate in a quadruped, the concurrent lengthening of neck and skull in order that the animal may readily reach the ground for food and drink. Hence the various parts of an animal's frame bear definite ratios to one another and this may also extend to individual bone proportions, a definite "speed index" being recognizable. This makes it possible through the law of correlation to gain some insight into the habits of extinct and little-known forms through the study of comparatively fragmentary remains.

BIPEDALITY

A two-footed mode of progression as an adaptation to speed has been repeatedly evolved among vertebrates, as follows:

Reptiles:

Lizards, several occasionally bipedal.

Dinosaurs, two evolutions.

Birds:

One evolution.

Mammals:

Marsupials, one evolution.

Rodents, three evolutions.

In all, eight or more times.

The erect posture of man was probably not originally a speed adaptation, nevertheless speed has always been a vital factor in human evolution, in all offensive and defensive operations. The human foot, which was originally a climbing structure, has been readapted for bipedal walking and running. The long thigh and shin of modernized man increase the stride materially in contrast to those of the gorilla and chimpanzee. The Neanderthal man had short stocky limbs as compared with the existing species, but doubtless could outrun any of the anthropoid apes.

Reduction of Fore Limbs

In bipedal creatures the hind limbs form the exclusive means of *rapid* propulsion, hence all of the cursorial adaptations which have been mentioned apply solely to them. The fore limbs, on the other hand, retain their generalized character, as their sole use is for resting (kangaroo, herbivorous dinosaur) or for slow locomotion while feeding. They often serve as very efficient organs of prehension. Because of the marked division of labor between the limbs, disparity of size soon arises and, as cursorial adaptation is perfected, the fore limbs become smaller and smaller in proportion until at length, as in the ultimate carnivorous dinosaurs, they are so absurdly reduced that it is difficult to conjecture their use.

Counterpoise

Some sort of counterpoise is always necessary in a semi-erect biped and the tail usually assumes this function. In the kangaroo and in the dinosaurs it is a powerful organ and serves as a prop, like a third limb, when the creature rests without coming down on all fours. The tail may be comparatively short and heavy in larger forms or extremely long and slender in more lightly built creatures, on the principle that an ounce at the end of a sixteen-inch lever is as effective as a pound on one but an inch in length. Many dinosaurs and bipedal lizards have a long, attenuated tail; this is especially true of the dinosaur *Podokesaurus*, a Triassic form from the Connecticut valley, and of the Australian frilled lizard *Chlamydosaurus*. Among mammals the kangaroos have a relatively short, heavy tail; the jerboa on the other hand has a very long one terminating in a tuft of hair, which through its resistance to the air adds effect to the counterpoise.

No existing birds have a long tail, that is, as regards the actual tail itself, although the feathers may be long. These, as in the pheasants, may subserve a balancing function. The true cursorial birds, Ratitæ, are practically tailless, but maintain their balance with ease, the head and neck sufficing. The ostrich raises its wings as an aid in running, but with the practically wingless cassowary or the emu the head and neck alone must serve.

Shortening of Neck

In bipedal mammals there is a tendency toward reduction in the length of the neck, especially in the rodents such as the jerboa, in which cursorial adaptation is extreme and there is a remarkable cervical reduction associated with the shortened skull. There is of course no diminution in the number of neck vertebræ, for the number, seven, is with two or three exceptions (sloths and manatee) constant among mammals; but the vertebræ themselves are shortened and tend to coalesce into a rigid mass of bone. Thus in the rodent *Pedetes* cervicals 2 and 3 are so closely articulated as to eliminate motion, in *Perodipus* the axis (2d cervical) and next two vertebræ are fused, while in *Dipus* (jerboa) all of the cervicals except the atlas (1st cervical) are coössified as in the whales. As we shall see, the shortening of the neck may be also an aquatic adaptation, since it occurs in the whales and sirenians (sea-cows).

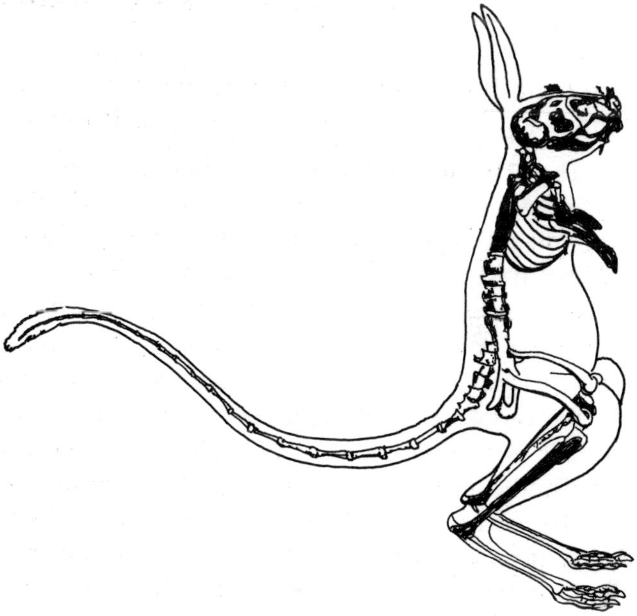

Fig. 14.6: Skeleton of jerbos.

MENTAL PRECOCITY

Animals which depend upon speed for safety, as do the ungulates or the whales, cannot have helpless young. Such must either be brought forth in some secluded den or carried about by the mother. Carnivores and rodents have very feeble young, but they are kept hidden until about to shift for themselves. The kangaroo, on the other hand, must carry her offspring with her and this undoubtedly proves a very heavy handicap to the race when competition with higher forms prevails, for the destruction of the mother means that of the young as well. With all other forms which depend upon speed for speedy, the young animal must be able to keep up with the herd almost at once. Hence there is no period of helpless infancy, but the new-born deer or horse or camel, with its grotesquely long limbs, has the relative mental alertness of a very much older dog or rat, although the ultimate mental attainments of the ungulate may not be so great. As an illustration of precocity, Andrews describes as experience with a ten-day-old baby antelope. For four miles he seldom went slower than 25 miles per hour and for five more miles he averaged 15 miles per hour. He circled too quickly for the car in fact, which had to average about 40 miles to overtake him.

SIGNIFICANCE OF CURSORIAL ADAPTATION

Not only does speed adaptation give rise to some of nature's most beautiful and perfect machines, but it seems to have a much deeper meaning which has been summarized by Broom. He is speaking of Permian reptiles:

"The African, or more preferably the South Atlantic type, is chiefly remarkable for the great development of the limbs. . . . What may have been the cause we can not at present tell, but it was a most fortunate thing for the world. It was the lengthened limb that gave the start to the mammals. When the Therapsidan [mammal-like reptile] took to walking with its feet underneath and the body off the ground it first became possible for it to become a warm-blooded animal. All the characters that distinguish a mammal from a reptile are the result of increased activity—the soft flexible skin with hair, the more freely movable jaws, the perfect four-chambered heart, and the warm blood. It is further singularly interesting to note that the only other warm-blooded animals, the birds, arose in a similar fashion from a different reptilian group. A primitive sort of dinosaur took to walking on its hind legs, and the greatly increased activity possible resulted in the development of birds. Birds were reptiles that became active on their hind legs, mammals are reptiles that acquired activity through the development of all four."

Back of all this lay the impelling natural cause. The earliest known mammals are late Triassic, the first recorded bird Middle Jurassic; the inference that both stocks arose in Permian time is justifiable from the degree of evolution which each class had attained by the time the actual record of their existence begins. Schuchert tells us that early in the Permian the climate of the lands seems everywhere to have been arid or semi-arid and that this condition lasted into Jurassic time. One characteristic of desert animals of to-day—the lizards, gazelle, Persian ass—is *speed,* for the creature must fare widely for food and drink if he would fare well. Again we are told that during the Permian there was a period of extensive glaciation with a severity of climate, especially in the southern land masses, as great as, if not greater than, the polar one of Quaternary time, although, like the latter, the Permian glacial period had warmer interglacial intervals as well. The incentive for speed already given, rendering the development of warm blood possible, the devastating cold would soon place a premium upon such as did develop it and eliminate those which did not. From this fortunate relation of cause and effect might well have arisen on the one hand the primal mammal, making human evolution possible, and on the other hand the ancestral bird.

FOSSORIAL ADAPTATION

Classification

Bionomic

Fossorial types may be classified bionomically in a manner which denotes the degree of their adaptation to subterranean life. Thus there are, first, those whose habitation is above ground but which dig for their food. There are several such, like the swine and the elephant, but aside from the mere digging mechanism—snout, tusks, etc.—there is but little fossorial adaptation to be noted. Nevertheless, such as it is, it may have a far-reaching secondary effect upon correlated organs. Thus the entire modification of the skull, jaws, and trunk, besides the tusks, of the elephant—structures which sum up almost all of the recorded evolutionary change of the most remarkable of beasts—are either directly or indirectly the result of fossorial habit.

Secondarily, there are those creatures which dig for retreat and which show still greater modification, especially in the body and limbs; but those whose food is above ground are not so profoundly modified as those whose food also is found below. The latter, the wholly fossorial, exhibit the extreme of adaptation which is to be discussed.

Zoölogic

Zoölogically, the partially and wholly fossorial animals are the lowly, more primitive and defenseless or unambitious members of their respective phyla. They include representatives of every great vertebrate class except the fishes, but the appended approximate summary is based almost entirely upon existing forms.

CLASS AMPHIBIA

Cæcilians.

CLASS REPTILIA

Sphenodon

Uromastix.

Legless lizards.

Desert snakes.

CLASS AVES

Burrowing owl.

"Starlet" so., of Mingan Islands.

Cliff swallow.

CLASS MAMMALLA

(All wholly fossorial mammals are primitive, small, plantigrade, pentadactyl, with moderate to large claws, and relatively defenseless.)

Order Monotremata. Entire order.

Order Marsupialia.

Wombat (*Phascolomys*).
Dasyure (*Dasyurus*).
Kangaroo-rat (*Bettongia*).
Pig-footed bandicoot (*Chœropus*).
Marsupial mole (*Notoryctes*).

Order Edentata.

Armadillos.
Aardvark (*Orycteropus*).

Order Insectivora.

Common moles (*Talpa, Condylura*).
Golden mole (*Chrysochloris*).
Shrew-mole (*Scalopus*).
Water-shrew (*Crossopus*).
Desman (*Mygale*).
Hedgehog (*Erinaceus*).
Oryzoryctes.

Order Rodentia.

Hares (*Lepus*).
Ground-squirrels (*Spermophilus*).
Prairie-dog (*Cynomys*).
Woodchuck (*Arctomys*).
Pocket-gopher (Geomyidæ).
Mole-rats and bamboo-rat (Spalacidæ).
Octodontidæ.
Paca (*Cælogenys*).
Viscacha (*Lagostomus*).
Bathyergidæ.
Siphneinæ.

Order Carnivora (retreat only).

Otter (*Lutra*).
Ratel (*Mellivora*).
Javanese skunk (*Mydaus*).
American badger (*Taxidea*).

Cohort Ungulata.

Few dig for food: swine, elephants, mastodons.

MODIFICATIONS

Body Contour

The density of the medium through which the burrowing form must penetrate necessitates a spindle or fusiform shape, not so beautifully modelled as in swift aquatic types, but such as will nevertheless offer but little resistance to subterranean passage. The greatest diameter in most fossorial forms lies near the shoulder, due in part no doubt to the necessary strengthening of the shoulder girdle. This is true of the mole; in other forms, such as the echidna and platypus, the breadth of the stout body brings the greatest diameter further back, but in neither of the latter instances is the fossorial adaptation so extreme as in the former. Legless forms, snakes, lizards, and cæcilians, have a cylindrical body.

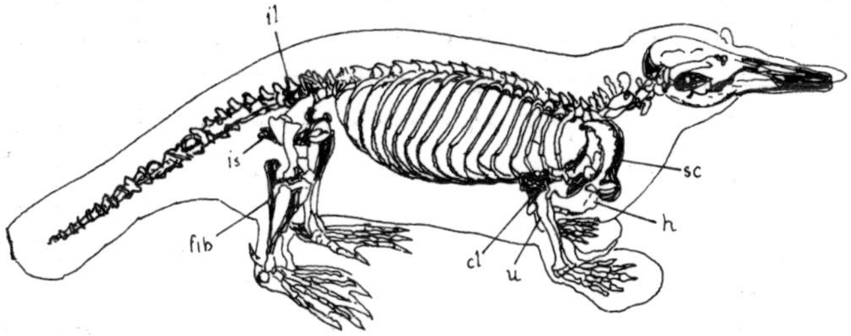

Fig. 14.7: Skeleton of platypus, *Ornithorhynchus*; *jib*, fibula; *h*, humerus; *il*, ilium; *is*, ischium; *ol*, olecranon; *sc*, scapula; *u*, ulna.

Forward of the shoulder the head tapers rapidly so that the contained skull is subconical and lacks the widely expanded zygomatic or cheek arches of many surface types. This last feature, however, is correlated also with the feeble jaw musculature, as almost all wholly fossorial types subsist upon feeble prey such as insects and worms, which require little strength of jaw or tooth.

Tail

The tail, even in semi-fossorial animals, is usually short, for not only would a long tail seriously incommode a subterranean creature, but its chief uses in the mammal seems to be in connection with a cursorial habit, hence there is manifestly little need of its development in forms in which no premium whatever is placed upon speed. Thus in the hedgehog, ratel, and woodchuck the tail is short, while in the wombat and the several moles it is vestigial. The tail of subterranean forms may have its uses, for it is said to serve as a valuable tactile organ.

Eyes and Ears

In truly fossorial types the eye, through disuse, tend to become vestigial. They would also be a source of injury in burrowing forms were they well developed. Two factors determine in general the degree of visual reduction, the duration and completeness of subterranean life. Thus "in the pocket-gophers (Geomyidæ) and Bathyergidæ the eyes are small; in [the mole-rat] *Spalax typhlus* they are mere black specks among the muscle (although retaining a relatively complete structure); in the marsupial mole (*Notorycts typhlops*) they are imperfectly developed and functionless; in *Talpa* they are vestigial; in the Cape golden mole (*Chrysochloris*) the eyes are covered with skin" (Shimer).

The external ears also tend to disappears as in aquatic types, for not only is their position such that they would be a decided obstruction to burrowing, but they would tend to gather the soil within the ear and thus obstruction rather than aid in the collecting of sound waves, and further, their true function, just mentioned, can only be effectively rendered in the open air. Hence in the Geomyidæ and the otter, which is also aquatic, the external ears are small, in the ratel (*Mellivora*) they are very minute, in the Bathyergidæ (Cape mole-rat, etc.) they are reduced to a slight fringe of skin around the aural aperture, while in the wholly fossorial moles all three sorts are bereft of external ears entirely, which is equally true of the Monotremata. The last-mentioned group, however, are so very primitive that they may never have had them.

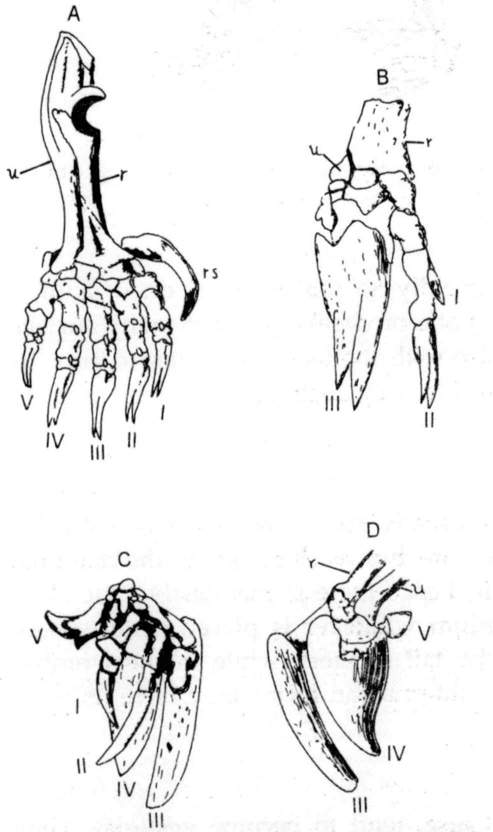

Fig. 14.8: Hands of moles. A, common mole, *Talpa europæa*, B, golden mole, *Chrysochloris*, C, palmar aspect, and D, dorsal aspect of marsupial mole, *Notoryctes*, I-V, digits; *r*, radius; *rs*, radial sesamoid (os falciforme); *u*, ulna.

Digging Mechanism

The *snout* forms an important organ for digging in the swine and in the hog-nosed snake (*Heterodon*), in each of which the organ is truncated and upturned at the tip. In the swine and in the mole *Talpa* there is a prenasal bone developed at the tip of the cartilage of the nose. This bone doubtless serves the same purpose in each instance, that of reinforcing the snout as an aid in digging.

The *incisormeth* particularly, and sometimes the canines (swine), form very adequate fossorial organs. If the incisors, they are usually procumbent or forwardly directed as in the pocket-gophers and thus protrude in such a way as to add to their effectiveness. The tusks of the elephants, as has been mentioned, and the procumbent lower tusks of the shovel-tusked mastodons in particular, were very effective digging instruments if one may judge from the manifest wear to which they have been subjected.

The *fore limbs* are by far the most important fossorial organs of all, hence their development and modifications have reached an extreme. All of the limbs are very short and stout in wholly fossorial forms, for not only is the speed which length of limb implies unnecessary, but long limbs would be a positive detriment in a form working in such close quarters as the average burrow. Length would also increase the mechanical disadvantage which always accompanies a lever whose weight arm is long in proportion to the power arm. In some semifossorial types, on the other hand, such as the hares and the pig-footed bandicoot, there is need of speed sufficient to offset the advantages afforded by short limbs in their partial subterranean life.

The hand particularly is broad and stout, with long claws, and it differs materially from the foot, as the two members have undergone divergent specialization, the hand loosening the earth while the foot not only throws it further backward but also serves to drive the creature ahead and resists the occasional backward thrust received from the hands. In the common mole, *Condylura,* the hand is as broad as the entire body is high, so that a single sweep of this very efficient organ will clear a space wide enough for entrance, hence the digging is very rapidly accomplished. This broadening of the hand in the mole and in the echidna is effected not only by having all of the original five fingers fully developed, but by the further addition of a bone (*os falciforme*) which increases the breadth of the palm materially. In the golden mole (*Chrysochloris*), there are but four fingers, of which the two middle ones are greatly enlarged and bear powerful claws, so that while the hand is not broadened it is still very effective.

The bones of the fore limb are always very strong, with prominent tuberosities for muscular attachment. This is especially true of the ridges at the proximal end of the humerus, which serve for the insertion of the powerful shoulder muscle, and those which aid in rotating the hand. The olecranon process, which is the extension of the ulna beyond the elbow-joint, is notably large in fossorial forms, as it is the insertion of the powerful triceps muscle which serves to straighten the arm, hence the greater the olecranon the more effective the leverage. One peculiarity of wholly fossorial forms lies in the narrow shoulders, for the fore limbs must not dig too broad a burrow and yet must be sufficiently developed to have adequate leverage. As a consequence the shoulder sockets must be as near each other as possible, and this is effected in one of two ways: generally by a shifting forward of the entire shoulder girdle and upon the neck, as in the monotremes and northern moles; or, as in the golden mole (*Chrysochloris*), by a hollowing out of the walls of the chest, the ribs and sternum (breast-bone) being convex inward instead of outward. The shifting forward of the shoulder in the true mole is accomplished through the elongation of the first sternal segment (*manubrium sterni*), carrying with it the remarkably short clavicles or collar-bones. A powerful shoulder is essential to resist the great muscular strain, hence in digging types, as in those which climb and fly, the clavicles are retained and are fully developed. In the monotremes, not only are the clavicles present, but they are reinforced by a large T-shaped episternum and a pair of powerful coracoids which extend from the shoulder socket to the presternum. In no other mammals do these latter reptilian elements reach the breast-bone, but they are generally represented by vestigial processes. Emphasis has been laid upon the retention of many primitive features by the monotremes and much of this may be due, as in the present instance, to their fossorial mode of life. It is sometimes difficult, however, clearly to distinguish cause and effect.

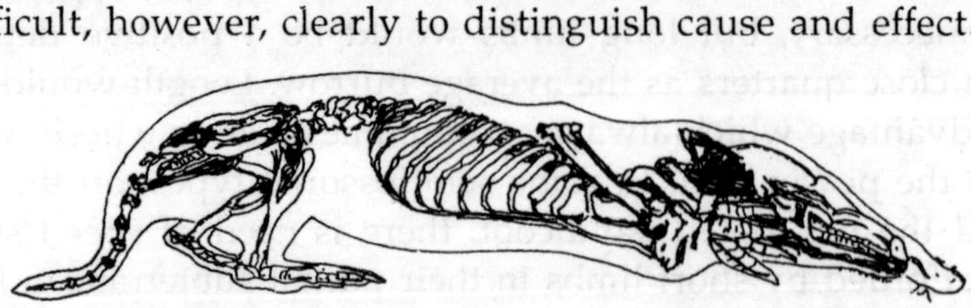

Fig. 14.9: Skeleton of common mole, *Talpa europæa.*

The divergence of function which has been emphasized between fore and hind limb has resulted in a decided difference in their relative power. In the *hind limb* the femur is by no means as robust as the humerus, nor are its tuberosities for muscular attachment so excessively developed as in the latter bone. Greater strength is given to the lower leg by a partial fusion of the tibia and fibula, and the calcaneum or heel bone is very prominent, as the increased leverage thus gained aids very largely in pushing the animal forward by extending the foot more powerfully. In the mole *Talpa* there is a large sesamoid bone at the side of the tibia corresponding to the *os falciforme* of the hand, but otherwise the foot exhibits none of the great modification which the hand has undergone. In the hind limb of the monotremes there is an unusual extension of the proximal end of the fibula beyond the articulation at the knee, comparable to the olecranon process at the elbow and doubtless subserving the same mechanical need.

As the stress is largely transmitted from the hind limbs forward in the direction of the axis of the body in subterranean creatures, instead of vertically in opposition to gravity as in most terrestrial forms, the *ilium* and *ischium* have become elongated in the fore-and-aft direction and lie parallel to the vertebral column instead of forming an angle therewith as usual. The ilium is also fused, usually throughout its entire length, with the sacrum, which in some forms, notably the armadillos, consists of a large number of coössified vertebræ.

In addition to the firm fusion of the sacral *vertebræ,* those of the loin and neck also tend to coössify to give greater strength and firmness in pushing the animal through the earth. It is also possible that the peculiar intercentral ossicles between the lower portions of the lumbar vertebræ of the mole *Talpa* and the hedgehogs (Erinaceidæ) may be of use in strengthening the vertebral column. Such structures are comparatively common among reptiles, but are extremely rare among mammals except in the tail. The neck vertebræ vary, for in the wombat (*Phascolomys*) and the armadillos the cervicals are wide and depressed, and in the latter several of them are commonly fused together or anchylosed as in the whales. In the mole the fourth, fifth, and sixth cervicals are much elongated and overlap each other. This region of the neck is that covered by the forward-shifted shoulder girdle, which undoubtedly accounts for this modification. The transverse processes of the lumbar vertebræ and the muscles which are attached to them exhibit comparatively little development, as the strains transmitted are largely longitudinal with little or no lateral movement.

Hibernation

Winter sleep is a necessity on the part of subterranean animals living beyond the limits of the tropis, because of the absence of green vegetation which forms either directly, or indirectly by supporting insects and worms, the mainstay of their diet. Digging through frozen soil would be so insuperable a task that it is also prohibitive of active underground life. An exception to this rule would be the lemming (*Myodes lemmus*) whose "food is entirely vegetable, especially grass-roots and stalks, shoots of the dwarf birch, reindeer-lichens, and mosses, in search of which they form, in winter, long galleries through the turf or under the snow" (Flower and Lydekker).

Chapter—15

Aquatic Adaptation

Aquatic realm is one of the most important habitat. Where a large number of animals live. It is specially important because it covers a large area of the universe. Adaptations to water habitat are of two types: primary aquatic adaptations which includes primitive gill-breathing vertebrates (fishes); secondary aquatic adaptations which are acquired as in reptiles and mammals.

PRIMARILY AQUATIC VERTEBRATES

By primarily aquatic forms is meant the fishes, which have served has a terrestrial ancestry, but have evolved from more primitive aquatic progenitors. As a consequence their adaptation to a watery medium is perfect and they do not suffer as those secondarily adapted do through their inability to breathe water. They are, therefore, the primitive gill-breathing vertebrates, sometimes with accessory air-breathing organs, it is true, but retaining the gills as the chief or only respiratory organs throughout life. But this is not all, for the dense medium in which they live has exerted so profound an influence upon them that, with some exceptions which only serve to emphasize the rule, they are stamped in a common mould, giving them so great a similarity of form that no one fails to recognize a fish.

Contour in an aquatic vertebrate is all-important, for nature as a marine architect conforms to rules of mathematical precision, the study of which has aided very largely in ship designing. The head, body, and tail are compressed into a beautifully curved stream-lined form, the entire surface of which is accurately rounded, with no protuberances which would retard the swift passage of the animal through the water. The head is sub-conical, the edges of the jaws and gill-covers fit precisely, and even the eyes conform accurately to the curvature of the head. Viewed from the front, the outline of a typical swiftly swimming fish like the Spanish mackerel (*Scombermorus maculatus,*) is a perfect ellipse, and the fine which are so prominent when viewed from the side can scarcely be seen, as they are reduced to thin keel-like lines. The greatest circumference of the fish is approximately 36 per cent of the length, measured from the snout, the "entering angles" being very similar for a great many different fishes.

From the point of greatest cross-section the lines of the "run." As it is called, are also very similar—smooth, hollow curves which freely permit the return of the displaced water.

Locomotion is effected by lateral undulations of the flexible body, aided by the fins. The latter are not, therefore, primary but accessory locomotive organs. The body is thrown into a series of alternate curves which begin at the head, pass along the body, and disappear at the tail. The static water enclosed in the incurved places is pressed upon, causing the fish's forward movement. Of course in an ordinary fish there is some lost motion or "slip," for the water is not sufficiently resistant to oppose the thrust of the body, but in the larval eel, which is a thin ribbon-like creature whose great relative depth presents a large lateral surface, the swimming is so precise that if a pencil is held in one of the hollows the fish's body passes without touching it at all. The other extreme is seen in the swimming efforts of an ordinary snake which, while it does progress, loses a great deal of motion because its lateral is not sufficiently great.

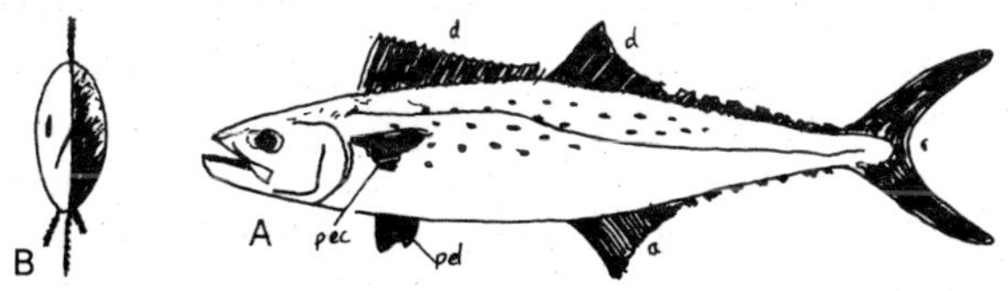

Fig. 15.1: Type of swift-swimming fish, Spanish mackerel, *Scomberomorus maculatus*. A, side view: *a*, anal fin; *c*, caudal; *d*, dorsal; *pec*, pectoral; *pel* pelvic. B, front view.

The increase of resistant surface is obtained in the fish through the development of the unpaired fin-folds of the skin, stiffened by fin-rays of elastic bone or cartilage. These unpaired fins may be more or less continuous from the head along the mid-line of the back, around the tail and forward along the under side as far as the vent. Usually, as in the Spanish mackerel, this continuous fin-fold is broken up into a number of distinct fins, developed where stresses arise, and disappearing where there are none. The fins along the back are known as dorsal, that around the tail as caudal, and that beneath the body as anal. Of these the caudal is by far the most important as a propelling organ. The mackerel also shows horizontal keel-like ridges on either side of the tail.

The fish also has lateral or paired fins corresponding to the fore and hind limbs of the terrestrial vertebrate. Of these the pectoral fins lie just behind the gill apertures at the shoulder, while the pelvics are more variable in position, though normally they should lie on either side of the vent. The paired fins when held out from the body may be simple keels like the unpaired fins; the pectorals especially, however, may have additional functions, as they serve as stabilizers and also to check the animal's way. Perhaps the primitive function of the paired fins was to maintain a horizontal position, as the removal of the pectorals tends to make the fish dive, and of the pelvics, to rise to the surface.

Swim-Bladder

As a further adaptation to aquatic life all fishes above the sharks may have a structure known as the swim-bladder—a hollow organ filled with air or gas, the present function of which is largely hydrostatic, in that it maintains the fish at a certain depth of floatation. If the creature wishes to sink, the body is slightly compressed through muscular contraction, and with it the swim-bladder. This lessens the bulk, and the weight remaining constant, the specific gravity is increased, with a resultant loss of buoyancy. Relaxation of the muscles has the contrary effect, the compressed gas in the bladder expands, increasing the fish's size and thereby decreasing the specific gravity, and the creature rises. The principle is comparable to the method whereby the depth of flotation of a submarine vessel is controlled, although the mechanism differs, for here water is a admitted to the ballast tanks to be driven out again by compressed air, and thus the ship sinks or rises as the case may be.

The swim-bladder is an outgrowth of the alimentary canal and is a very important organ in the evolution of higher forms, for it is the homologue of the lung of the terrestrial animal, and there can be no doubt that the air-bladder of certain ancient fishes, the primal function of which may well have been respiratory rather than hydrostatic, actually evolved into a lung when the drying up of the waters left them stranded and they were forced to become air-breathing. Thus arose the progenitors of all terrestrial vertebrates.

SECONDARY AQUATIC ADAPTATION

Secondarily aquatic forms are the lung-breathers which, through stress of circumstances—inhospitable lands where food was scarce or competition severe—were forced to return once more to the primal home of their remote ancestors. There is always, as has been said, the handicap of lung-breathing, but otherwise the adaptation of certain of the more extreme forms is little short of marvellous.

As an offset to this handicap it should be remembered that lung-breathing made possible or at least accompanied a tremendous advance in other organs and their functions, such as a higher brain with a consequent psychological advancement. When the lung-breathers turned back to the waters, they readopted only the externals of fish life, but kept in varying degrees the higher brain and the more efficient methods of aëration of the blood and of locomotion. And again and again they easily won a place on the lower level against the most highly specialized of the inferior grades.

Amphibious vertebrates, as the name implies (Gr. *ἀμφί*, double, and *βίος*, life), spend part of their time on land and part in the water. They are really terrestrial forms, showing partial aquatic adaptation only, which rarely extends beyond the possession of webbed feet and a laterally compressed swimming-tail, which may bear fin-like expansion along its upper margin, and sometimes a lack of ossification of certain of the wrist and ankle bones. The class Amphibia, which embraces the historically transitional forms in the original landward migration, modified representatives of which still exist, breathe typically by means of gills during their youth, and sometimes throughout their life. Other abandon their gill-breathing at the approach of maturity and become as essentially terrestrial as a reptile.

Instances of amphibious life among forms above the class Amphibia are numerous, but one or two instances will suffice. The Galapagos lizards (Fig. 15.2) are instances of forms terrestrial from choice but aquatic from necessity, for, it will be remembered, they live on certain rocky islands of the Galapagos group, swimming out to the surf line and diving for the sea-weed upon which they feed. There is evidently no menace to their safety ashore but their aquatic excursions seem to be made in actual dread of bodily injury, as their behaviour indicates. Here a flattened swimming-tail and slightly compressed body are the extent of their aquatic adaptation.

Another instance wherein the habits are inferred rather than the result of observation is that of the late Cretaceous hardrosaurs a group of dinosaurs of which our knowledge is very complete, for not only have all parts of the skeleton and dentition been preserved to us, but fossilized mummies as well creatures which died in the open and simply dried up so that bone and sinew, hide and even portions of the flesh have been preserved in great detail. These creatures had a wonderful battery of teeth, an adaptation to a very harsh herbage, presumably the Equisetales or horsetail rushes which are found preserved with them. There is evidence of adaptation to submerged, underwater feeding on the part of the hadrosaurian dinosaurs, which, together with the splendid swimming-tail, webbed hands and feet, imply an aquatic or at least amphibious type. Add to this the fact that the creature was defenseless, utterly devoid either of weapons or armor, and that its arch-enemy *Tyrannosaurus* was also terrestrial, and we have evidence which points to a reversal of the life conditions of the Galapagos lizards, for with the dinosaur food was on land, and safety and retreat in the waters. One would therefore expect the greater relative degree of aquatic adaptation which they show.

Fig. 15.2: Galapagos sea-lizard, *Amblyrhynchus cristatus.*

The marine turtles have gone a long step further in that here both food and safety are fund at sea and only one need, which does not concern all, but only the females,

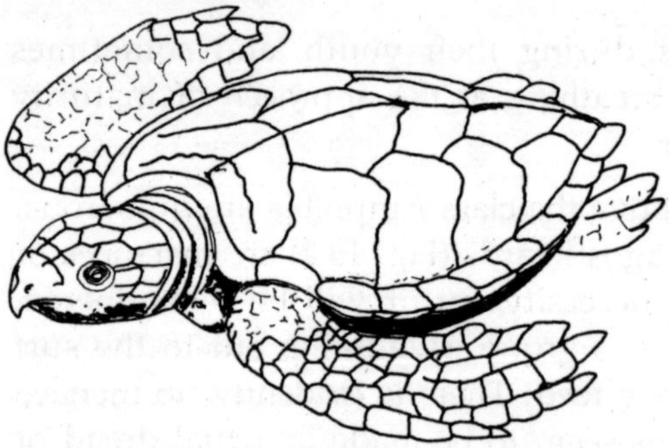

Fig. 15.3: Marine turtle, "hawksbill", *Chelone imbricaia.*

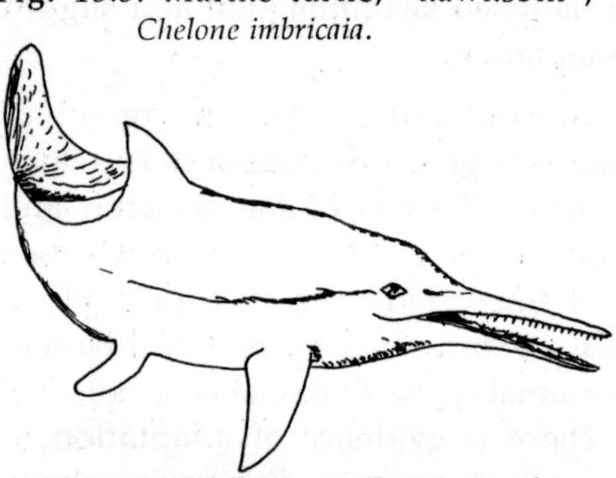

Fig. 15.4: Ichthyosaur. Restoration by Knight.

that of egg-laying, brings them to the land at all, and this annual shoreward migration is fraught with dire peril which nothing but the most urgent summons would cause them to face. There is reason to believe, moreover, that in some species the males never come ashore.

The final stage in reptilian adaptation was shown by he ichthyosaurs (Fig. 15.4) of the Mesozoic whose perfection for their life conditions equalled that of the modern whales, even to the utter abandonment of shore going, for there is abundant evidence from the contained embryos in several known specimens that these medieval high-seas corsairs brought forth their young alive and therefore did not need to go ashore for egg-laying. This is a necessary part of ultimate aquatic adaptation, for no egg laid by a truly air-breathing vertebrate (*i.e.* allantoic egg), can be hatched in the water, as the enclosed embryo would drown as certainly as would a submerged adult.

A list of secondarily aquatic vertebrates follows:

CLASS REPTILIA

	Amphibious	*Aquatic*
Order Chelonia	Swamp and river turtles	Sea-turtles
Order Sauropterygia	Nothosaurs	Plesiosaurs
Order Proganosauria		All
Order Ichthyosauria		All
Order Thalattosauria		All
Order Parasuchia	Phytosaurs	
Order Crocodilia	Crocodiles and alligators	Extinct sea-crocodiles
Order Squamata	Marine iguanas Mosasaurs	
Order Dinosauria	Amphibious sauropods, hadrosaurs	
Eight orders		

	CLASS AVES	
Order odontocæ		*Hesperornis*
Order Ichthyornithes	*Ichthyornis*	
Order Pygopodes	Loons and grebes	
Order Impennes		Penguins
Order Tubinares	Petrels, albatross, puffins, etc.	
Order Steganopodes	Gannet, cormorant, frigate bird, pelican	
Order Herodiones	Flamingo	
Order Anseres	Ducks and geese	
Order Gaviæ	Gulls, terns, auks	Great auk
Nine orders		
	CLASS MAMMALIA	
Order Monotremata	*Ornithorhynchus*	
Order Marsupialia	Water opossum (*Chironectes*)	
Order Insectivora	Aquatic shrews, desman, etc.	
Order Rodentia	Muskrat, water rat, beaver, capybara, etc.	
Order Carnivora		
Suborder Fissipedia	Sea otter, otter, mink, etc.	
Suborder pinnipedia		All
Order Cetacea		All
Order Ungulata	Hippopotamus	
Order Sirenia		All
Eight orders		

Thus there are twenty-seven orders of secondarily adapted water-inhabiting vertebrates all told, some of which are exclusively aquatic. Many of these, of course, are now extinct, so that the whole number is far in excess of those of any one time. Of aquatic lung-breathers, the honour falls to the reptilian order Proganosauria, which are the first in point of time, as their remains are found entombed in rocks of Permian age not very long, relatively, after the reptiles were established.

Body Contour

As in primarily adapted forms, the body contour becomes stream-lined, the neck constriction disappears, the tail enlarges, and the same "numerical lines" prevail. This assumption of the fish-like form is best seen in the fully aquatic orders—Ichthyosauria, Cetaces, Sirenia, Pinnipedia—and to a lesser extent in several other groups. The minor factors contributing to this general effect are, first, *skull modification.* This includes a shortening of the cranium or brain-case, which becomes proportionately higher and wider, with a consequent effect upon the proportions of the brain, which is likewise short and wide. The facial portion of the skull, on the contrary, tends to elongate so that many forms, especially those which subsist upon active prey, as the porpoises and ichthyosaurs did, have an elongated slender snout or rostrum. The zygomatic or temporal arch in the Cetacea is also reduced almost to a vestige.

The *neck* shortens very materially and there is a loss of mobility in the swifter, tail-driven types. In those forms like the plesiosaurs, whose paddle propulsion was necessarily slower, an elongated, supple darting neck was necessary to overtake the swiftly moving prey. In the whales, while the number of cervical vertebræ is the standard mammalian seven, they may be fused into a solid, compressed mass of bone, whereas the neck of the manatee among Sirenia, with but six cervicals, forms one of the three exceptions to the standard number, the other two, as have seen, belonging to the sloths.

In old-fashioned reptiles derived from primitive stocks which early became adapted to aquatic life, such as the ichthyosaurs, the *vertebraæ* retain their ancient simplicity, having simple biconcave bodies or central like those of fishes. In higher forms the vertebraæ tend to simplify secondarily, due to the fact that the body, being water-borne, is equally supported throughout, hence the vertical stresses which are the result of gravity inland animals are practically eliminated, and the thrust or driving force is transmitted longitudinally through the column. This simplification includes the centra or bodies of the vertebræ and especially the secondary articulations or zygapophyses. The several processes may become reduced in the trunk region but elongated in the tail to provide greater area for muscular attachment.

The *sacrum*, that portion of the vertebral column which articulates with the pelvis and which, therefore, in land animals has to withstand and transmit the supporting impact of the hind limbs, is more or less reduced indirect ratio to the loss of supporting or propelling function on the part of these appendages. Thus, in the ichthyosaurs, cetaceans, and sirenians the sacrum cannot be distinguished from the other vertebræ and while its approximate locality cannot be far from the vent, the actual identity of the former sacral vertebræ is lost.

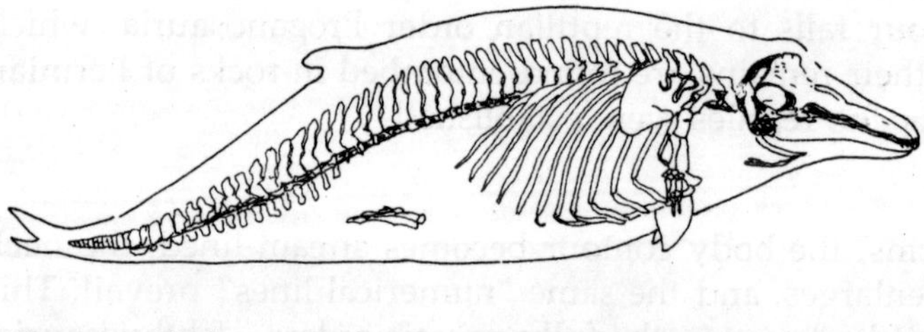

Fig. 15.5: Skeleton of porpoise. The vestigial pelvic bones are shown embedded in the flesh.

The *chest* of the truly aquatic type tends to become cylindrical; land forms, on the other hand, generally show lateral compression, especially if they are quadrupedal in their mode of progression, with the body of the ground. The chest in aquatic types is also modified in such a way as to bring the internal cavity higher toward the back. This insures greater stability of floatation and increased lung capacity and is accomplished by the ribs tending to become highly arched dorsally and then to move upward from their point of attachment on the centra to the transverse processes. A gradation in this last feature may be demonstrated by taking first the ichthyosaurs, wherein the rib articulation is entirely central, both rib facets—that for the capitulum or head and that for the tuberculum—being on the body of the vertebra, a unique feature which makes an ichthyosaur centrum unmistakable. In the Pinnipedia the position is transitional, whereas in the whalebone whales the articulation of ribs and vertebræ is extremely loose, probably to allow great mobility of the chest for the rapid respiration necessary after prolonged submergence. In the Sirenia and Cetacea the diaphragm has become horizontal in position instead of being practically vertical as in most quadrupedal mammals.

The *bones* of swimming forms tend to become light and spongy, the interstices in those of the whales being filled with oil. Exceptions to this are found in forms which, like the walrus and Sirenia, derive their sustenance from the bottom, the walrus feeding on molluscs, the sea-cows subsisting on submarine vegetation just as their bovine namesakes graze in terrestrial meadows. This necessitates a blasting which is secured by increased weight of bones, like the great, wide, massive ribs of the manatee.

Externally, the secondarily adapted sea-vertebrate is characterized, like the fish, by the elimination of retarding excrescences, hence in the course of their evolution aquatic mammals have lost all trace of *external ears*. This not only renders the contour of the head smoother, but removes a practically useless appendage, for the pinna of the ear has for its especial use the collection of aërial sound waves, a function which is valueless in a submerged form. Thus the ears are reduced in amphibious mammals, and are totally lost in the whales and true seals and walrus, thought retained in reduced condition in the sea-lions (Otariidæ) which spend much of their time ashore). The occasional atavistic occurrence of external ears in the porpoise has been noted.

The *external nostrils* or nares tend to forsake their old terminal position at the end of the snout and move toward the apex of the head as in most of the whales, ichthyosaurs, phytosaurs, and mosasaurs, mainly in forms with reduced mobility of neck in which the vertex of the head first appears above the surface of the water. This one feature is invariably indicative of aquatic life. The nostrils are often capable of being closed, as is also seen in desert-adapted forms like the camel as a protection against drifting sand.

The *eyes* in amphibious types also tend to shift higher on the face as in the hippopotamus, whose nostrils, eyes, and ears can appear above the surface of the water while all the rest of the head and body remains submerged. The advantage of this is obvious. In truly aquatic types, on the other hand, the eye does not shift its position, as aërial vision has largely lost its importance, but instead of this the eyes become adapted to aquatic vision which, because of the denser medium, requires a different curvature of the lenses. They adapt their vision to the air by contracting the pupil to a narrow slit, vertical in the seals and horizontal in the whales, or by an actual shifting of the position of the crystalline lens. Penguins which pursue their prey submerged are curiously limited in their aërial vision.

Locomotive Mechanism

Fleshy, fin-like expansions of the body-wall occur in the whales and ichthyosaurs and also probably in the extinct thalattosaurs and sea-crocodiles (thalattosuchians) (*Geosaurus*). These *fins*, as in fishes, may be dorsal and caudal. The dorsal fin is a triangular structure, essentially equilateral in the ichthyosaur, very and narrow in the killer whales, especially *Orcinus rectiponna*. On the other hand, some whales such as *Delphinapterus,* the white whale, and *Balæna,* the right whale, lack the dorsal find entirely; while in the sulphur-bottom, *Badaenoptera musculus,* the fin is small and situated well aft upon the tail. The development of this fin must be entirely a response to mechanical needs and correlated with a certain bodily form and peculiarity of locomotion, just as the deep fin-keel of a sailing yacht would be superfluous upon a motor-driven craft.

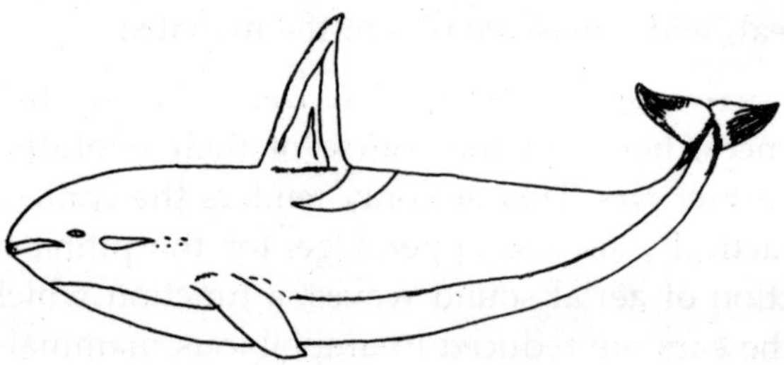

Fig. 15.6: Killer whale, *Orcinus rectipinna*.

Secondarily aquatic types very often go back to first principles and readopt the old wriggling or undulatory motion of their pristine ancestors, to which fins, etc., are only subsidiary. This also accompanies elongation of body, multiplication of segments, and loss of limbs (as in the zeuglodonts and many long-tailed aquatic types).

Thus two methods of propulsion are seen among aquatic types, even in those whose adaptation has passed the amphibious stage. Williston has called them "oar propulsion" and "tail propulsion". In the former, exemplified by the turtles and plesiosaurs the limbs are more nearly equivalent, in size and propulsive power than in tail-propelled types. In the latter, which is included in the undulatory form of locomotion, that seen in the ichthyosaurs, whales and sirenians, the hind limbs tend to disappear until finally not external vestige is discernible, though slender bones, representing the pelvis, thigh, and sometimes the tibia, may be found deep buried within the flesh.

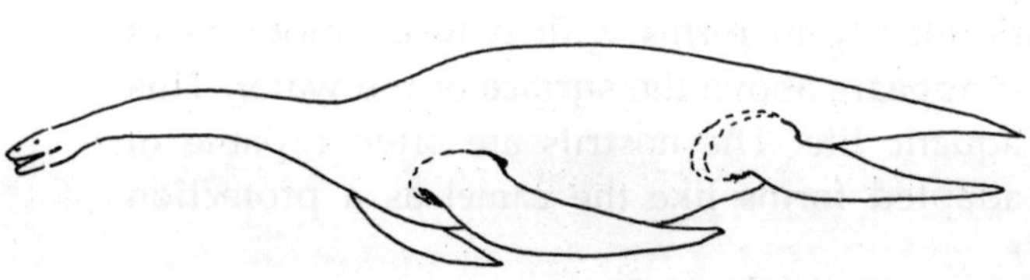

Fig. 15.7: Plesiosaur, *cryptocleidus*.

It is characteristic of secondarily aquatic vertebrates that where unpaired fins are developed they are never supported by the skeletal elements, known as fin rays, as they are in fishes; they may, however, be strengthened by masses of dense connective tissue.

The ichthyosaurs were extinct forms ranging in time from the Triassic to the late Cretaceous. The earlier species moved largely by means of the limbs, the later ones almost exclusively by the tail. In the former the hind limbs were nearly as large as the front ones, while in the later ichthyosaurs, as the tail developed the hinder paddles were reduced in size until they were often very much smaller than those in front. More than eighty years ago Sir Richard Owen, the great English anatomist, noticing a curious downward dislocation of the till at its mid-length in many articulated skeletons, came to the conclusion that the ichthyosaurian tail must have borne a terminal fleshy fin quite like that of the whales and sirenians, but it was not until forty years later

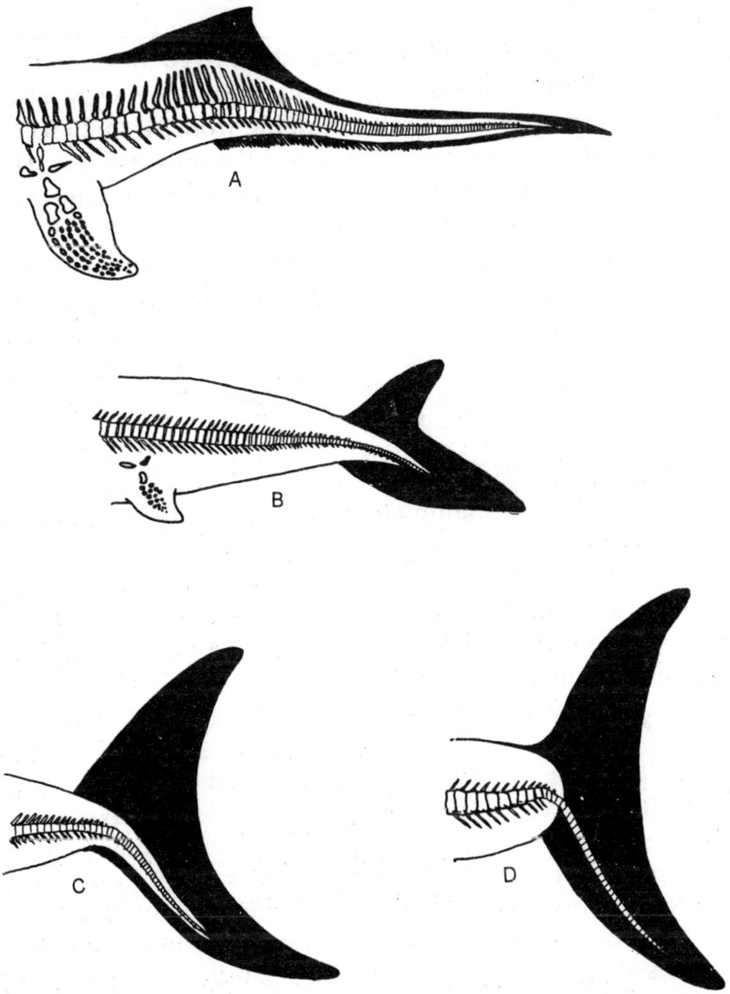

Fig. 15.8: Propelling tails, caudals, of ichthyosaurs, showing the development of the organ. A, *Mixosaurus nordenskjöldi,* Trias, Spitzbergen; B, *Icthyosaurus quadriscissus,* young, Upper Lias, Württemberg; C, same species, adult; d, *I. trigonus* var. *posthumus,* upper Jurassic, Solenhofen, Bavaria.**Fig.**

that specimens showing the actual outline of the body and fins were found and Owen's conjecture verified. His only error lay in the supposition that the caudal fin was horizontal like those of the mammals, whereas it is vertical like that of a shark. The ichthyosaur tail is diametrically opposed to that of the shark, however, for in the latter the backbone is deflected upward into the superior lobe of the fin, whereas in the former it extends along the front margin of the lower one. This may be due to the fact that the caudal arose as a low fin-like expansion along the upper side of the tail, not an uncommon thing among air-breathing vertebrates, and ultimately developed into the widely expanded fish-like fin of the later forms. The diagram of the ichthyosaur caudals will make this clear.

The Thalattosuchia or sea-crocodiles were a short-lived race, their remains being confined to the rocks of Upper Jurassic age in Europe. They were also of moderate size compared with other marine vertebrates, ranging from 10 to 20 feet in length. But few forms are known, of which *Geosaurus* is perhaps the most typical. This type shows a sharp downward bend toward the end of the tail as in the ichthyosaurs and the inference the like them the creature bore a well developed caudal fin is undoubtedly correct. Strangely enough, however, the hind limbs were much larger than those in front, probably a character inherited from its shore-dwelling ancestry.

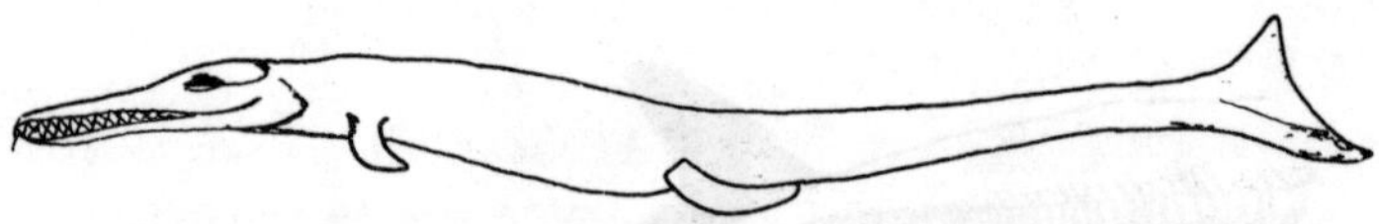

Fig. 15.9: Marine crocodile, thalattosuchian, *Geosaurus*.

The caudal fin of the marine mammals differs markedly from that of the raptiles in being horizontal instead of vertical and in the symmetry of its two halves or flukes, the bone dividing the tail into two equal parts rather than running into one lobe to the exclusion of the other. In the Sirenia the same is true, the manatee having a rounded tail, while in the dugong (*Halicore*) and the exterminated Steller's sea-cow (*Rhytina*) it was notched like that of a whale.

Limbs

Webbed feet are the first natatorial adaptation, and the degree of webbing is a very variable thing. It may consist simply of lobe-like lateral expansions of skin on either side of the toes, as in the coot, or of actual connections from digit to digit. Extreme aquatic adaptation, on the other hand, implies the development of a paddle in which there is a loss of mobility of the various joints, the entire skeleton of the limb being enclosed by the skin in a single mass showing no external division into fingers and toes. The result is the production of a flexible paddle of great aquatic utility, but ill adapted for coming ashore. As a further modification the individual phalangeal bones increase in number (hyperphalangy) and, in some instances, one or more additional rows are seen, as though extra toes over the normal five had been added (hyperdactyly). There is also a change in relative proportions of the various segments as in cursorial adaptation, the humerus or femur diminishing in length while the other individual bones may shorten much more, although from their increase in numbers the area represented by the digits may become very much extended. In some instances, as an especial modification, for example, in the round-headed dolphin, *Globicephalus*, while the length of two or three of the digits may be very great, the others may be much reduced.

The reduction of the hind limbs with the development of the tail has been mentioned, together with their total loss externally in the aquatic mammals. Not only are their vestiges discernible in the from of the bones hidden in the flesh, but as Kükenthal has shown, external traces (*Anlagen*) of them are visible in fœtal whales (*Megaptera*, the humpback whale) on either side of the vent.

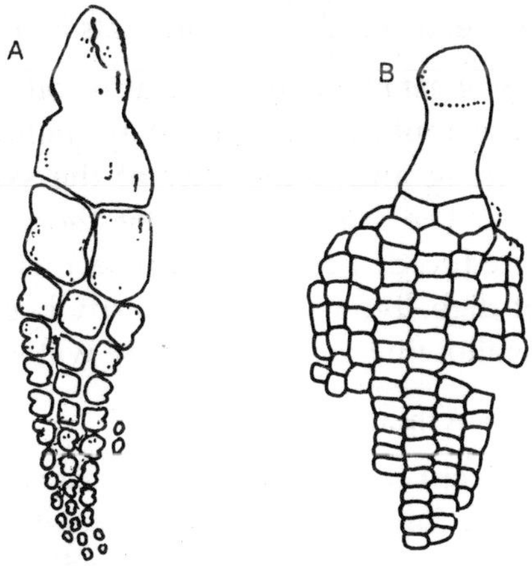

Fig. 15.10: Ichthyosaur paddles. A, left anterior paddle of *Merriamia zitteli*, Trias, California. B, same, of *Ichthyosaurus platydactylus*, Lower Cretaceous (Aptian), Germany.

Integument

The modification undergone by the skin in aquatic animals is in the line of reduction of armoring, of hair, of skin glands, muscles, and nerves. *Loss of armor* has taken place in the ichthyosaurs, for reptiles are primitively armored and in the wonderfully preserved specimens from Holzmaden, Bavaria, the only remaining traces lie on the anterior edge or cutwater of the fins, and, in one specimen at least, along the mid-dorsal line from the head to and on either side of the dorsal fin.

The *loss of hair* is characteristic of marine mammals, and the reason seems to be that whereas hair is a wonderfully good non-conductor of heat when dry, in that it retains a blanket of still air around the body, it is of little value in the water. Hence the fur seals (Otariidæ) which retain more terrestrial characters than do the hair seals (Phocidæ), such as the external ear and the ability to use the hind limbs for progression on land, still wear a thick under coat of fur—the sealskin of commerce. The hair seals with their more perfect aquatic adaptation come ashore more rarely and have lost the furry under coat entirely, retaining simply the so-called contour hairs, so that their hides are valueless as fur although used for other purposes. The whales and sirenians have lost all traces of hair except perhaps a few bristles around the mouth, but most of them are well covered with fœtal hair before birth, which points to an ancestrally hairy condition. In a few instances, such as the white whale or Beluga and the narwhal, even the fœtus has lost its hair, thus showing the extreme of specialization.

As a compensation for the loss of hair, the mammals have developed a *layer of fat* in the connective tissue beneath the skin (subcutaneous tissue) and this serves admirably for the retention of the bodily heat which would otherwise radiate out very rapidly into the surrounding water. Even with this device much heat apparently is lost, for one of the greatest impediments experienced in shipping to New York porpoises caught on the coast of the Carolinas was the difficulty of keeping the water in the shipping tanks, and therefore the animals contained in them, sufficiently cool for health. This shows that the porpoises develop so great an excess of heat that a more efficient heat-retaining integument is not necessary; the heat excess being correlated with the high speed which these animals can attain.

Skin glands such as sweat or oil glands have their value impaired if their secretions are continually washed away, hence their reduction in aquatic animals; while the thickening and immobility of the skin has resulted in the atrophy of its muscles and nerves. The *salivary glands* of those forms which devour their food under water are also reduced, possibly because their secretions would be too largely diluted to have great digestive value, and partly because the mechanical function of lubrication to aid in swallowing is subserved by water taken in with the food.

Mouth Armament

Except in the sea-cows and walrus the jaws, being no longer used for mastication, but only for the prehension of relatively feeble prey, have very largely lost their power, so that the coronoid process and other areas for muscle attachment are reduced. The teeth also suffer modification, usually in the direction of simplification, the increase of numbers, or on the other hand total loss from one jaw (sperm whale) or both (baleen whale). In the later instance the function of the teeth is taken by the remarkable been or whalebone to be described later. The marine reptiles have simple prehensile teeth fitted for the retention of slippery prey, except the sea-turtles whose teeth had disappeared long before their aquatic adaptation began.

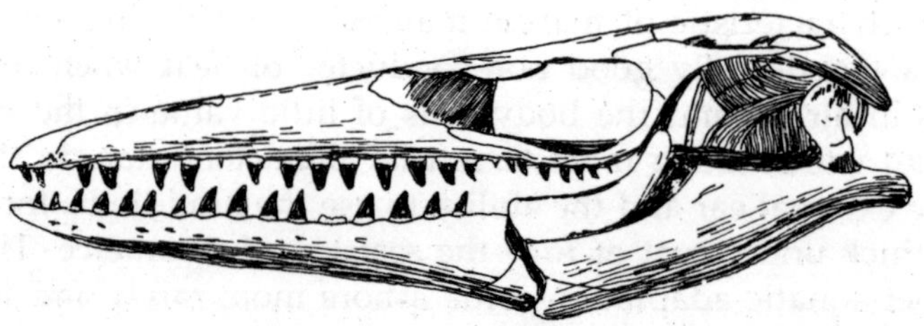

Fig. 15.11: Skull of a mosasaur, Clidastes, Cretaceous (Niobrara), Kansas.

The Cretaceous marine lizards or mossasaurs show a remarkable adaptation for the prehension of prey, for not only were the rims of the jaws armed with slightly recurved teeth, but others were borne upon the roof of the mouth as well. The lower jaws instead of being firmly united together at the symphysis (chin) were connected by an elastic ligament and each jaw was doubly articulated with the skull through a very movable intervening bone, the quadrate. The most remarkable thing, however, was an extra joint in the mid-length of each jaw bone so that it could be bowed but into a decided curve and thus, as in snakes, effect the swallowing of prey larger than the normal gape of the mouth.

The whales show a very interesting tooth gradation. Ancient whale-like forms (zeuglodonts) still had the tooth differentiation characteristic of mammals, the cheek or molar teeth bearing many cusps arranged in longitudinal series (see Fig. 15.12). In the modern whales, in which the teeth are retained, they are generally simple cones and are as a rule much more numerous than the normal number for a placental mammal, which is forty-four, eleven in each half of each jaw. In *Globicephalus,* the ca'ing whale, the number may be over one hundred, and the dolphins, *Delphinus* and *Inia,* may have twice that number. As we have seen the sperm whale, *Physeter,* is devoid of functional teeth in the upper jaw, in spite of the fact that it feeds upon active cephalopods (giant squid). *Monodon,* the narwhal, has but a single tooth, although its mate may be vestigial, in the form of a long, spirally twisted tusk or "horn," really a modified incisor. In the female the horn is reduced. In the curious toothed whales, *Ziphius* and *Hyperoödon,* there is but a single tooth in each mandible. The latter genus has small teeth in the upper jaw, which are also present but functionless in the former.

Fig. 15.12: Zeuglodon skull, *Prozeuglodon atrox,* Upper Eocene (Birket-el-Qurun), Egypt.

In the baleen whales, which may possibly represent a distinct whale-like evolution from that of the toothed forms, the teeth are entirely absent except for vestiges in the embryo, which, however, never cut the gum. In their place there has been developed the remarkable baleen or whalebone. This is a horny outgrowth from the epithelium lining the mouth and may be compared to an exaggeration of the transverse ridges on the palate characteristic of many mammals. Each piece of whalebone is triangular, attached by its base of the roof of the mouth, with the free inner margin frayed out into numerous threads which form the straining apparatus. When the creature feeds it rushes open-mouthed through swarms of whale food or "brit" (pelagic organisms which are largely pteropod molluscs); then the mouth is partly closed and the water is forced out between the plates of whalebone by the tongue; thus the organisms are strained out and subsequently swallowed.

The ichthyosaurs, while as a rule well toothed, also reduced the dentition so that in the American genus *Raptanodon* and the European *Opthathalmosaurus,* which were related if not identical, the teeth had become vestigial and functionless, although the germs were still present in the jaws. The convergence of the older ichthyosaurs toward the modern whales is thereby rendered all the more complete and certainly the diet of fish and cephalopods in the toothed forms, as shown by the fossil waste voided from the alimentary canal (coprolites) in the ichthyosaurs, was the same in both instances. But the food of the toothless ichthyosaurs is harder to conjecture; there is certainly no evidence of the development of anything comparable to baleen, as has been suggested.

Precocity

Mental precocity is as necessary in the gregarious aquatic animals are among the cursorial, and they soon show ability to keep up with the mother. The new-born young of whales are from one-quarter to one-sixth the length of the parent and a porpoise about half the length of its mother has been seen maintaining its position as readily as she at the bow of a 15-knot ship.

Speed

Record of speed are difficult to obtain, but porpoises are known to keep pace with a 39-knot destroyer, sheering off ahead of the craft with the utmost ease, and this speed is maintained with, as a rule, an almost inappreciable vibration of the highly efficient propelling tail, as seen from above.

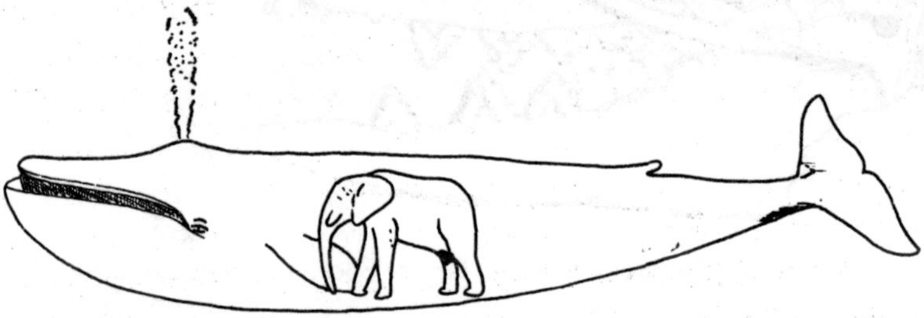

Fig. 15.13: Sulphur-bottom whale, *Balœnoptera musculus*, with the African elephant "Jumbo," *Laxödonta africana*, drawn to scale.

Size

Water-borne animals exceed all others in size, for the energy exhausted by terrestrial creatures in overcoming gravity may here be turned into growth force. The largest recorded terrestrial animals which live are the elephants, of which "Jumbo," an African specimen, had a height of 11½ feet and a weight of six and one-half tons. African elephants grow to 13 feet and the great imperial elephant of the American Pleistocene may have exceeded this by a foot. In comparison with a sulphur-bottom whale, however, with a length of 87 feet and an equivalent weight in tons, the elephant becomes insignificant. The largest strictly terrestrial dinosaur, *Tyrannosaurus*, reached a length of 47 feet and a standing height of 18 to 20 feet. The bulk of body, tail, and hind limbs was also great, but the amphibious exceeded it, although the disparity of size was not comparable with that of whale and elephant. *Brontosaurus,* one of the most ponderous dinosaurs, had a length of 67 feet and an estimated weight when alive of 38 tons. *Diplodocus,* while more slenderly proportioned, was at least 87 feet long; and the remains of *Gigantosaurus* from East Africa (Tendaguru) indicate an animal nearly equalling *Dialodocus* in length and havier than *Brontosaurus.*

Chapter—16

Desert Adaptation

The popular conception of a desert is of a vast level stretch of drifting sand, devoid of vegetation except in the rare oases, and peopled by camels and the more venturous of humanity—nomadic Arabs and the like. While certain desert areas do conform more or less closely to this description, it by no means applies to all, nor is it even characteristic of the greater portion of the arid stretches of the earth's surface. The desert does not imply sterility, but simply the possession of a rainless, or nearly rainless climate, where drought may be said to be almost continuous; for in the true desert there is no regularly recurring rainy season, and such rains as may fall are intermittent and uncertain in their character. Some years, even torrential rains may fall for a few days, as the existence of considerable numbers of dry water courses, dry saline lake beds, and the marks of ancient torrents conclusively proves. During other years, practically no rain may fall; indeed, in many of the principal deserts, according to the reports of residents inhabiting the various oases, it is stated that sometimes for several years in succession not a drop of rain has been known to occur. In tropical countries desert conditions prevail if the average rainfall is under 15 inches a year, in temperate under 10 inches, and only the most utter deserts are absolutely rainless, as, for example, the desert of Atacama in South America.

Extend of the Desert

The combined area of all of the deserts of the world, that is, the portion where evaporation of water is in excess of precipitation, and where the water courses have no outlets to the sea, has been estimated at eleven and a half millions of square miles, or over one-fifth of the entire land area of the globe. Thus it will be seen that, while neither desert plants nor animals can exist in any marked proportion to those of humid areas, they must, nevertheless, form a large and peculiarly adapted flora and fauna, thus constituting a problem of considerable importance.

DESERT LIFE

The three great needs of every organic group are that its members may find food and safety and be able to continue their kind in future generations. This is of course true of desert forms, but the conditions are such that two necessities stand out in marked contrast with the others. These are moisture-getting and moisture-conservation, and defense against both the physical and organic environments. The struggle for existence is exceedingly severe and the weeding out of the unfit ruthless.

MOISTURE-GETTING

The securing of an adequate supply of moisture is the prime requisite of desert forms, for, as we have seen, its scarcity is the one unvarying characteristic of the desert. Water therefore must not only be obtained by diverse means, but also effectively conserved.

Plants

Even where the surface sands are dry and burning there may be available ground waters, therefore many plants send down deep exploring roots which seeks and find the subterranean water supply. Among such plants is the date palm (*Phœnix dactylifera*) which flourishes only where its roots have access to water. Along the banks of the Nile where the eroding action of the river has bared the roots of the palms, they may be seen going to a very great depth until they reach the water-bearing strata. Thus the presence of the date palm implies either subsoil water or that it is watered artificially from a well, as rain may seriously injure it. "In the land where rain seldom falls and where an incandescent sun daily pours down its fiery beams, out of a sky in which a cloud is rarely seen, the date tree finds its most congenial home" (Madden).

In the Sonoran desert of our Southwest, the most characteristic tree is the mesquite (*Prosopis juliflora*) and here again the roots are of great length, sometimes penetrating to water-bearing strata at a depth of 50 to 75 feet, although only under exceptional conditions does the tree itself reach 20 to 40 feet in height.

In addition to these deep penetrating tap roots there is often another series which spreads horizontally through the superficial soil, and by means of which the plant avails itself of such occasional rainfall as does not penetrate the deeper layers. In younger plants of *Cornegiea gigantea,* for instance, the main root system is that associated with the tap root, which has the dual function of anchorage and absorption. As the plant grows, however, the lateral system develops, widespread from the original root, and takes over the function of absorbing water, leaving that of support entirely to the tap root.

Moisture suspended in the air and which, while it may form, is not actually precipitated in the form of rain, must be rendered available, and this may be brought about by one of two ways, possibly by the deliquescence of the salty encrustation so characteristic of desert plants or by the mechanical collection of dew by the hairs which bedeck their surface. In either case, the water instead of evaporating back into the atmosphere may be largely absorbed by the plant. The saline coat and the pubescence are therefore hygroscopic (having the property of readily imbibing moisture from the atmosphere) in their effect. Whether or not the plant can utilize the moisture thus collected is, however open to question.

Animals

With the animals almost the sole source lies in the juices of the food, whether plant sap or the blood of prey. Many creatures have learned to depend on such a source entirely and have abandoned the ordinary means of quenching the thirst by drinking water even when, as occasionally happens, it becomes available, although the collected dew may be absorbed with the plant food. To what extent the skin may have the hygroscopic power is not recorded, but Willey discovered that the moloch (*Moloch horridus*), a highly spinescent lizard which lives in the sandy deserts of Australia, absorbed water like blotting paper when a live specimen was put into it.

MOISTURE CONSERVATION

Plants

The first method of water conservation is by means of reservoirs. Of these Van Dyke says: "The storage reservoir below ground is not an unusual method of supplying water to the plant. Many of the desert growths have it. Perhaps the most notable example of it is the wild gourd. This is little more than an enormous tap root that spreads out turnip-shaped and is in size often as large around as a man's body. It holds water in its pulpy mass for months at a time, and while almost everything above ground is parched and dying, the vines and leaves of the gourd, fed from the reservoir below, will go on growing and the flowers continue blooming with unruffled serenity. In the Sonora desert there is a cactus or a bush growing from a root that looks almost like a hornet's nest. This root is ... again a water reservoir like the root of the gourd."

The cacti contain large masses of watery pulp in their stems, the water being renewed during each humid season. Some of these, like the giant saguaro (*Carnegiea gigantea*) and the still more gigantic saguesa (*C. pringleii*), consist of huge, fluted columns, the first from a foot to a yard in diameter and 10 to 60 feet in height, sometimes single, again branching like a candelabrum. The latter consists of from three to ten such columns rising from a single root. These have woody skeletons like vertical ribs corresponding to the flutings of the column and supporting the pulpy mass.

The barrel cactus or bisnaga (*Echinocactus emoryi*) affords an easily obtained supply of drinkable water for the thirsty Indians of the Sonoran desert. The water is secured by cutting off the top of the cactus and then mashing the pulp with a stick. It may then be squeezed with the hands, the plant tissue thrown away, and the moisture remains behind in the cavity thus formed. From the agave or "century plant" (maguey) of Mexico the sap is obtained by tapping the plant in the center of the crown of leaves This is allowed to ferment and forms the Mexican national beverage, "pulque."

A thickened bark to hold water is found on many trees like the sangre—de-dragon and torote (*Jatropha* spp.) which are also protected from loss by evaporation by a glazed or varnished surface. If a tree is wounded the sap flows, thickens to a varnish, and heals the cut.

Animals

Water storage in animals has not been extensively investigated, except perhaps in the camel, in which the rumen or first division of the stomach has developed in its walls what are known as water cells. These are cavities with a narrow mouth which can be closed by a sphincter muscle, acting like the drawstring of a bag. When the beast drinks, not only is the stomach filled, but these water cells as well. The water contained in the stomach is of course at once absorbed into the circulation and distributed, but the stored water is later doled out to the stomach and thence to the blood through the relaxation of the sphincter muscles. Hence the camel can travel from oasis to oasis with comparative ease, but Lord Wolseley's *Soldier's Pocket Book* tells us that when worked for three days without water, a camel's condition runs down rapidly, and after long journeys of five or six days without water, it requires several days to regain its strength, while if it is allowed to go down below a certain point, it will take weeks, perhaps months, to pick up again, and if it is old it may never recover (Madden).

Avoidance of Evaporation

Plants avoid loss of water principally by leaflessness. No desert shrub or tree is large-leaved, and with most the leaves are either very small or, as in the cacti, entirely absent, the thickened stems serving such respiratory function as is absolutely necessary. Where leaves occur they either fall off at once with the coming of drought or fold themselves so as to present the smallest possible surface for evaporation, or they may hang edgewise to the sun.

The form of the plant is also such as to offer the least surface in proportion to the bulk. To be ideal this form should be spherical, which is of course hardly possible, nevertheless a robust cylindrical shape with a hemispherical termination is common and some cacti are reduced to the hemisphere alone, which, with its armament of spines, looks not unlike a well stocked pincushion. The cylindrical form intercepts more sunlight when the sun is low and the light proportionately weak; as the sun rises and gains in strength, however, less and less is received by the cylindrical sides until a condition of total reflection is reached. The relatively small hemispherical summit will now give the plant a very material advantage over the cactus whose bulk is entirely contained within the hemisphere. The fluting of the surface which often occurs is of course opposed to the principle of surface reduction, but is possibly a mechanical necessity for stiffening a tall column with little internal strength, also for allowing a bellows-like expansion with temporary increase of stored water.

The hairiness which has already been mentioned as a mechanical aid to moisture-getting is also of conservation value, as it resists transpiration by day as well as collecting dew by night; and the varnishing or waxiness so characteristic of desert plants has the same economic function of conserving moisture. It also reflects sunlight, giving total reflection when rays are slanting.

Animals conserve water by being non-perspiring, for while perspiration may have a physiologic value, not only to aid in ridding the body of some effete waste, but also, through its evaporation, in keeping the body cool, these benefits are offset by the ill effect of the loss of moisture from the system where water is scarce and precious. Practically non-perspiring mammals are thus a response to arid adaptation.

DEFENSE

Plants

The first defensive need of the plants is *anchorage* against the almost continually blowing winds, and the same tap roots which serve so admirably for water-getting also aid in securing the plant in the place where it would grow. The so-called Russian thistle (*Salsola tragus*), which is a widespread and trouble-some weed in many parts of the central and western states, is firmly anchored until its growth is accomplished and it has ripened its seeds, then, loosening its hold, it is driven before the wind and so scatters its seed far and wide. The seeds may have been brought into Canada by the Russian immigrants with their wheat, but now it own migratory power have spread it far beyond its older limitations.

Spinescence, which is a very prevalent characteristic of desert plants, doubtless serves very adequately the purpose of defense, although it may also admit of another explanation, for spines tend to diminish as humidity increases and increase where it diminishes. But the numbers of grazing animals against whose depredations the spines are thought to be effective diminish in direct proportion to the increase in aridity and hence where spinescence is most highly developed. This would seem to weight against the belief that such structures are the result of adaptation for defense. Nevertheless, as Van Dyke says:

"The sahuaro, the bisnaga, the cholla, and the pan-cake lobed prickly pear would have a short life and not a merry one if they were left to the mercy of the desert prowler. As it is they are sometimes sadly worried about their roots by rabbits and in their lobes by the deer. It seems almost incredible but is not the less a fact, that deer and desert cattle will eat the cholla—fruit, stem, and trunk—though it bristles with spines that will draw blood from the human hand at the slightest touch."

The jack-rabbit will also gnaw the flat cactus, *Opuntia,* avoiding the spines, thus sadly mutilating it but never quite destroying the plant.

"Nature knows very well that the attack will come and so she provides her plants with various different defenses. The most common weapon which she gives them is the spine or thorn. Almost everything that grows has it and its different forms are many. They are all of them sharp as a needle and some of them have saw edges that rip anything with which they come in contact. The grasses, and those plants skin to them like the yucca and the maguey, are often both saw-edged and spine-pointed. All cacti have thorns, some straight, some barbed like a harpoon, some curved like a hook. There are chollas that have a sheath covering the thorn—a scabbard to the sword—

and when anything pushes against it the sheath is left sticking in the wound. The different forms of the bisnaga are little more than vegetable porcupines. They bristle with quills or have hook-shaped thorns that catch and hold the intruder. The sahuaro has not so many spines, but they are so arranged that you can hardly strike the cylinder without striking the thorns."

Of the *chemical characteristics* which the plants of defense, the first to assail one's senses is the strong aromatic character of desert vegetation. Both odors and tastes are very pronounced and may prevent the plant from being eaten just as the spinescence does. A notable instance is the very characteristic sage brush of the western plains which no mammal will eat except the jack-rabbit and no bird except the sage hen, and the flesh of each is so thoroughly impregnated by the sage that no human being will eat either of them if any other food is available Neither the rabbit nor the chicken will eat the sage during their first summer and then the flesh of each is most palatable. The greasewood is another disagreeable tasting plant with sticky varnished leaves which nothing will eat, and the sangre-de-dragon has a blood-red sap which is so powerfully astringent that the Indians use it to cauterize bullet wounds.

Thus, aside from being disagreeable to the taste and smell, many plants are actually poisonous, some cathartic, others emetic in their effects, while one, the loco weed (*Astragalus hornii*) produces, aside from bodily alterations, symptoms in domestic stock comparable to insanity, hence the term "locoed" as applied to one mentally deranged (Spain. *loco,* insane). Horses and cattle do not touch the loco plant for the first time voluntarily, but as with many other forms of narcotic, the habit accidentally acquired is almost impossible to break.

It would seem again as though these chemical characteristics were of great defensive value, and doubtless they are; but whether that is a sufficient *cause* for their being what they are, or whether it is due to desert conditions regardless of the presence of devouring animals is not quite so clear. The aromatic character is so prevalent among desert plants, whether used as food or not, that it leads one to question whether or no defense is a prime cause in the evolution of this characteristic. The gazelle, one of the most admirably adapted of desert animals, is confined to tufts of wiry grass and a few dwarf and strongly scented aromatic herbs for food. That these agree with it is attested by the fact that the number of gazelles in some parts of the Arabian desert is extraordinary and vast herds have occasionally been met with.

Another conception of the origin of the chemical characteristics of desert plants is that they are merely a response to aridity. It is well known that saline lakes are characteristic of arid climates, and to the geologist the presence of beds of salt or gypsum or other alkaline substances is indicative of desert conditions in bygone times. These salt lakes are thus described by Pirsson: Part of the river's burden "consists of various salts in solution and such salts are carried by all streams, even if, in a given volume, the water appears so fresh that they can only be detected by chemical means. In ordinary rivers these salts are discharged into the sea, but in interior drainages [*i.e.* such as end in evaporation[since they cannot be dissipated by evaporation like the water, they must constantly accumulate at the point where the drainage ends.

"If the river ends by dwindling, its lower part finishes in a stretch covered with salt deposits, sometimes in wet seasons converted into a salt marsh or shallow salt lake, and known as a salina. Examples of these are found in the Tarim River which ends in the desert of Gobi in central Asia, in the Desaguedero River which carries the drainage from Lake Titicaca in Bolivia, and in many other places. But if the end of the drainage system is a lake the latter is bound in time to become salt through the concentration of these substances, and such salt lakes are features of arid or desert regions in all the continents." Examples are the Dead Sea in Palestine, Lakes Baikal and Balkash and the Aral Sea in Siberia, and in North America the Great Salt Lake in Utah, Pyramid Lake and others in Nevads, and Mono Lake in California.

The idea has been expressed that the chemicals which impregnate desert plants may have an analogous source—water more or less laden with these salts is drawn from below by the roots and, being dissipated ultimately into the air by evaporation, leaves the substances behind as a residue which may in turn be elaborated by the plant into the characteristic aromatic or saline chemical. If this be true, it would account for these characteristics without involving the need of defense, but one would expect to find, as in thorny growths, a lessening of the chemical attributes as one passes from the arid to a more humid region.

Animals

Against physical conditions the animal must find means of protection and the first of these conditions is that of the *temperature extremes* so characteristic of arid climates, for while moisture in the atmosphere is transparent to light, it is opaque to heat. Hence where the air is laden with water—either diffused or in the form of clouds—not only does less solar heat penetrate to the ground during the day but less is radiated into space during the night. Therefore some of the highest temperatures recorded are not under the equator where it is humid but in northern Africa and in Asia Minor where dryness prevails. Solimos in his *Desert Life* informs us that the "sand temperature one day marked 146° Fahr.—temperatures of 170° Fahr. were also found near Jaffa, and Duveyrier found the Sahara one day over 182° Fahrenheit.' But these would probably be sun temperatures. On other occasions he tells us 'Matches were lit by touching the sand with them. Blankets drawn over one another or even slightly shifted, blazed up like sheet lightning. The heat was 109° Fahr. among clothes in a trunk, and 111° Fahr. in the wind: combs, vulcanite, bone or horn, became brittle and useless." Again, the German traveller, Doctor Barth, after describing the great heat of the desert in southern Tripoli, mentioned that the guide of the Arab caravan begged him to beware of the cold during the night, which he represented as very intense, and it is a fact well known to all travellers that though the heat in the Sahara is often very extreme by day, the nights are uniformly cold, when the hot wind does not blow (Madden).

Thus the desert animal, unprovided as he is with additional garments, must seek out "the shadow of a great rock in a weary land" by day or if none be available he must burrow and thus escape the extremes of heat and cold, which might prove fatal.

Defense against the ubiquitous *sand* is a prime need, for sand assails there vulnerable portions of the creature's economy—the eyes, ears and nostrils. Desert reptiles, among other marked adaptations, have these vital organs protected as follows: "In the digging species the nostrils are directed upwards instead of forwards; in most of the snakes they are protected by complicated valves, or they are reduced to small pinholes. The eyes of *Typhlops* [a burrowing desert snake] are overhung by the head-shields. In *Agama* and *Phrynocephalus* [lizards] the margins of the lids are broadened into plates and are furnished with peculiar scales. In *Teratoscrincus* the upper lid is enlarged. The lizard *Mabuia* has the lower lid enlarged, with a transparent window in it, so that the eye can be closed without impeding sight, an arrangement carried to the extreme in *Ablepharus* [whose lower eyelid is transformed into a transparent cover, which is fused with the rim of the reduced upper lid, exactly as in the Lacertine genus *Ophiops*]. The ear-opening is either small, or protected by fringes of scales, or it is abolished, *e.g.*, in *Phrynocephalus*" (Gadow).

The camel again shows its desert adaptation in the large eyes, necessary for an animal which relies so much upon vision for security, but they are guarded by long, abundant eyelashes, and the level carriage of the head, seen also in the desert ostrich, brings the eyes as far as possible above the sand—nine feet in the camel, seven in the bird. This is also a response to the reflected heat of the blistering sand and rocks. The camel's nostrils are capable of being closed like eyelids, and the apertures of the ears are protected by hair.

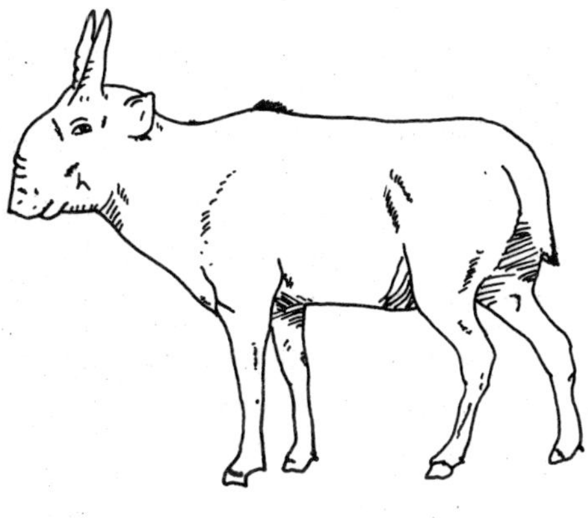

Fig.16.1: Saiga antelope, *Saiga tartarica*.

The saiga antelope (*Saiga tartarica)*, is a typical desert form found fossil (Pleistocene) in England but now confined to eastern Europe and western Asia. This peculiar beast has a large and much inflated nose, the two nostrils being widely separated and very short and the narial apertures of the skull placed very far back like those of certain extinct types (such as the horse *Hivpidion,*). This whole device is for the exclusion of sand from the breathing apparatus while the animal is grazing.

Coloration is the first and perhaps the most conspicuous of the adaptations which furnish protection *against other animals*. No greens are seen among desert creatures, but they are generally sympathetically tinted—gray, brown, or red, harmonizing with the colour of the sand or rock. Sometimes the same species will show double colour adaptation to suit local conditions. Thus the gazelle, Madden tells us, "furnishes a conspicuous and very beautiful example, so closely do their coats resemble the general colouring of the landscape [white on sand, dark grey on volcanic rock] that in the distance, and when at rest, it is scarcely possible to distinguish them from the surrounding sand and stones." Van Dyke says of the coyote: "He is cunning enough to know... that you cannot see him on a desert background as long as he does not move; so he sits still at ties for many minutes, watching you from some little knoll. As long as he is motionless your eyes pass over him as a patch of sand or a weathered rock."

Warning coloration is often seen in the desert, and as a rule it is not merely the *mimicry* of a dangerous form but the actual livery of one which should not be molested if the assailant would escape unscathed, for venomous spiders and insects and reptiles are common. Here for instance is found the conspicuous yellow and black Gila monster, the only lizard with poison fangs; and conspicuously marked wasps, while not, perhaps, actually dangerous to humanity, are to the creatures with which they compete.

Hard surface and spinescence are as characteristic of desert insects and some reptiles as they are of desert plants. The moloch with the thirsty skin which has been described is covered with thorn-like scales, and the horned toad is a conspicuous American instance of spinescence. Here, as with the plants, the spinescence loses its intensity of development as one passes into the more humid regions, for while in the horned toads of the Southwest the horns at the back of the head are long and prominent, in those of Wyoming and Nebraska they are much less conspicuous.

Venom has already been referred to as a desert attribute. Van Dyke again sums the matter up most admirably, although he speaks of more than the reptiles with which he introduces his remark. He says of them:

"They are given the most deadly weapon of defense of all—poison. Almost all of the reptiles have poison about them in fang or sting. We are accustomed to label them 'poisonous' or 'not poisonous,' as they kill or do not kill a human being; but that is not the proper criterion by which to judge. The bite of the trap-door spider will not seriously affect a man, but it will kill a lizard in a few minutes. In proportion to his size the common red ant of the desert is more poisonous than the rattlesnake. It is reiterated with much positiveness that a swarm of these ants have been known to kill men. There is, however, only one reptile on the desert that humanity need greatly fear on account of his poison and that is the rattlesnake... The rattle is indescribable, but a person will know it the first time he hears it. It is something between a buzz and a burr, and can cause a cold perspiration in a minute fraction of time... Every animal on the desert knows just how venomous is that poison. Even your dog knows it by instinct. He may shake and kill garter-snakes, but he will not touch the rattlesnake.

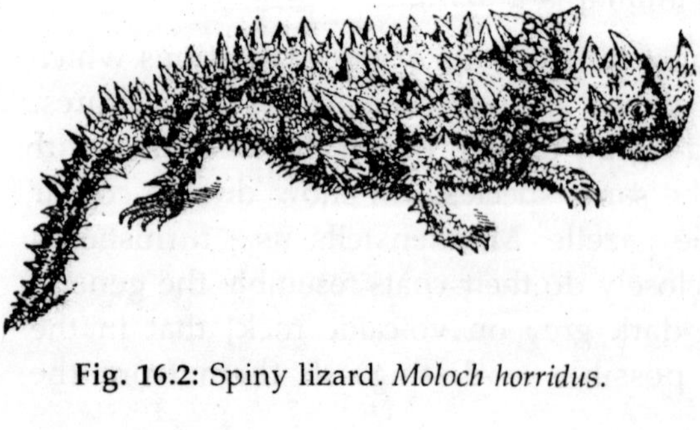

Fig. 16.2: Spiny lizard. *Moloch horridus.*

"All of the spider family are poisonous and you can find almost every one of them on the desert. The most sharp-witted of the family is the trap-door spider—the name coming from the door which he hinges and fastens over the entrance of his hole in the ground. The tarantula is simply an overgrown spider, very heavy in weight, and inclined to be slow and stupid in action. He is a ferocious-looking wretch and has a ferocious bite. It makes an ugly wound and is deadly enough to small animals. The scorpion has the reputation of being very venomous; but his sting on the hand amounts to little more than that of an ordinary wasp. Nor is the long-bodied, many-legged, rather graceful centipede so great a poison-carrier as has been alleged. They are all of them poisonous, but in varying degrees."

The skunk of the Southwest is greatly feared by the natives as its bite is supposed to produce skunk hydrophobia. In reality, however, it is blood-poisoning that results from the bite, for the skunk feeds on carrion and its teeth, defiled with the remains of its last, meal, inject some of the putrescent material into the wound caused by its bite. It is probable that in many of the desert creatures the dangerous reputation is altogether undeserved. The native cook of a certain Yale expedition to the Southwest had the entire personnel in terror of its life from the only arachnid which does *not* possess poison fangs!

Speed, is a characteristic of the dwellers in arid climates, for they have to travel far and wide for food and drink or to pursue or evade pursuit as the ease may be. There is so little cover and the creature in motion can be seen so far that speed becomes a prime requisite in many unrelated types. Many of the lizards are so swift that the eye loses sight of them as they dart from rock to rock, and some run on the hind limbs for greater rapidity. Some birds, as the ostrich or the road runner or chaparral cock of the Southwest, are noted for their speed. The gazelle, wild ass, and camel again are great travellers, while a desert jack-rabbit finds his match only in a greyhound—an ordinary wolf or coyote will not attempt to chase him, for they realize the hopelessness of it. The wester antelope, *Antilocapra,* even with a limb shot from under him, has been known to outdistance a horse, and with all four limbs the run is so easy and so deceptive that a gunner almost always underestimates the speed and places his bullet behind the fleeting game. Even the domestic cattle on the western ranges have an case and speed in running which is a revelation to an easterner.

This speed shows itself in the slender form, long limbs, and often in the sand-adapted feet of the desert wayfarer. Those of the camel, with the broad, yielding pad and secondary retrogression from unguligrade to digitigrade, are splendidly fitted to the yielding sands. Of the desert lizards, *Eremias* has very large crural (shank) shields; *Scapteira* has the digits broadened out into shovels; others *e.g., Phrynocephalus* and *Teratoscincus,* have long lateral fringes on the digits, a very rare arrangement among geckoes, occurring elsewhere among them only in *Ptenopus* and *Stenodactylus,* which are likewise inhabitants of the desert (Gadow).

The **senses** of sight, hearing, and smell are highly developed as contrasted with those in forest-dwelling forms, for instance, whose vision is notoriously dull even though hearing and smell may be acute enough. Sometimes about all one can see of a desert rabbit, frozen into immobility through fright, are the large, clear, watchful eyes which never seem to close.

Intelligence is also a desert desideratum, especially among defenseless, that is, non-venomous forms, for the Gila monster seems to be superlatively stupid while the others must match their wits as well as length of limb if they would survive.

As we shall see in later chapters, the widespread desert and semi-desert conditions which have occurred from time to time in the geologic past have been of prime importance as evolutionary stimuli, and have given rise to some of the most momentous changes in organic life.

COLONIAL LIFE

The plant life of the desert is not evenly spread as in humid regions, but is in scattered, discontinuous colonies with long stretches of verdureless sand in between. Moreover, these communities are composed of several species, not only of one, for while the inter-specific struggle in the desert is intense, that against a relentless physical environment is more so, so that coöperation becomes of greater moment than exclusive self-advancement. McGee says:

"The various plants of the [Seri] district, including those of the distinctive types, are communal or commensal, both among themselves and with animals, to a remarkable degree; for their common strife against the hard physical environment has forced them into coöperation for mutual support. The tufts or clusters in which the vegetation is arranged express the solidarity of life in the province; commonly each cluster is a vital colony, made up of plants of various genera and orders; and forming a home for animal life also of different genera and orders; and, although measurably inimical, these various organisms are so far interdependent that none could survive without the coöperation of the others."

The origin of the communities has been well summed up by Van Dyke: "Mesquite springs up on the plain, birds nest in the branches, drop seeds of cacti some of which like vines cannot stand alone, armature, of cacti and mesquite combine for protection. Windblown grass seeds lodge about the roots and grasses grow and seed beneath the shelter of the branches. Small mammals seek some protection and dig holes among the roots, making channels for rain and fertilizing the spot with rejectamenta. Annual and semi-annual plants take root in the sheltered and fertilized soil beneath the cacti and mesquite and in season the place becomes a miniature garden of foliage and bloomage. Then come certain ants for seeds, flies and wasps for nectar, and birds nest in the branches. Such communities dot the vast plains, intermediate stretches are practically lifeless."

ORIGIN OF DESERT LIFE

The desert floras are all local adaptations of migrants from more humid regions, except the cacti and torotes, which seem to be the products of aridity, since neither is represented in the flora of humid regions. Of the animals, all are forms whose range is greater than the desert limitations; nevertheless some of them, like the camels, are so perfect and ancient an adaptation to desert life that when, during the Miocene, we find the first indications of retrogression of the feet we can safely infer the beginning of aridity if we had no other evidence. A few animals are desert-adapted, others may be merely temporary migrants desertward, returning to more salubrious places from time to time. Many of these latter creatures go the desert during its periods of fertility and the numbers of species which have been noticed as occasional visitors is very large; but as the waters dry up and the vegetation begins to wither, these take their departure in common with the desert nomads and their flocks; of whom a regular exodus then begins (Madden).

Chapter—17

Cave and Deep—Sea Adaptations

Origin of Caves

Those caves which are available as animal habitats are invariably the abandoned channels of underground rivers which prevail in limestone regions. This limestone is soluble in water so that instead of being confined to the surface of the earth the waters sink under ground enlarging the crevices through which they descend until the level of ground-water is reached when the movement becomes lateral and subterranean channels of considerable magnitude are formed. If the level of ground-water becomes depressed the waters then tend to flow at lower levels and the process may be repeated the older channels being abandoned except perhaps in time of flood just as the ancient surface courses were. This abandonment may be due to a gradual lowering by erosion of the river valleys which serve as outlets for the region but is more generally due to a gradual uplift of the land to a new level for manifestly where the ultimate outlet of the river system is the ocean the subterranean streams cannot descend below sea-level.

Subsequent subaërial agencies—rain and wind and frost—denude the surface laying bare portions of the caverns so as to afford ingress for the cave-dwelling plants and animals and subterranean life begins. Still further surface erosion brings more and more of the cavern to light until it may ultimately be destroyed. The far-famed natural bridge of Virginia is a relic of what was once the roof of an extensive cave or series of caves.

These caverns form discontinuous units of environment as there may be no means of communication from one to another. The creatures within them therefore if not longer suited to epigean (Gr. *ἐπί*, on and *γη*, earth) life are virtually confined and must in extreme cases be the product of a very local evolution. In certain of the larger cavern regions the individual caves do communicate or at any rate are part of the same subterranean drainage system. Hence the aquatic types which they contain may intermingle within the limits of the system but they may be nevertheless confined entirely within such limits.

Distribution of Caves

In North America there is a large cave area extending through parts of Indiana Kentucky Tennessee and Missouri embracing the so-called Mammoth Cave region. Mammoth Cave itself is ten miles in length and its tortuous channels cover a distance three times as great. In Edmonson Country Kentucky alone there are said to be at least four thousand sink-holes or pipes through which surface waters pass under ground and five hundred known caves. Wyandotte Cave in Indiana Luray Cavern in Virginia and the Carlsbad Caverns are noted. As one would naturally infer the region north of the Ohio river constitutes a separate system or bionomic unit from that to the south.

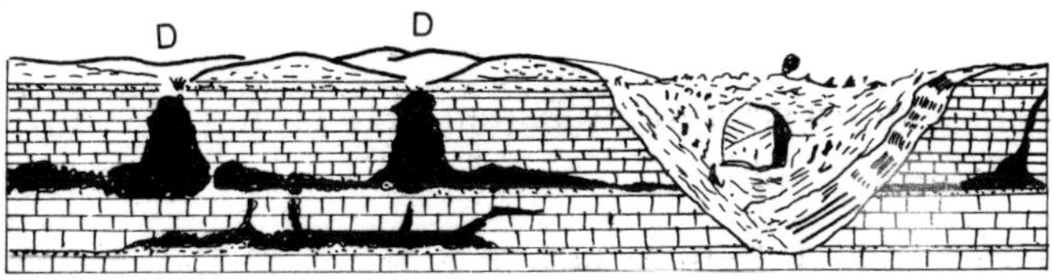

Fig. 17.1: Diagram showing the formation of eaverns in limestones. AA clay bed; BB limestones The arch is the remnant of a former cave forming a natural bridge. DD sink holes leading to domes below.

In Cuba there are extensive caves in communication with the sea and in Europe there are others notably in Carniola and other portions of Austria and Illyria and in the Basses Pyrénées in France and Spain. The last together with those of the Dordogne region in western France and the Cantabrian Mountains in Spain, are of great human interest in that they formed the abode of man in late Paleolithic time, and contain on their walls the remarkable art of the Stone Age.

Few caverns exist in the glaciated regions of Europe or North America, almost all of them lying south of the great terminal moraine.

Geological Age of Caverns

The older caves like the Port Kennedy Cave Schuylkill River, Pennsylvania antedate the last glacial invasion and contain the remains of a Pleistocene fauna such as the tapir, peccary, and the ground sloths *Megalonyx* and *Mylodon*, in all about forty species, of which all but twelve are now extinct. These remains were swept into the cave by floods and not drawn into it by animals as in the case of many European caves (Osborn). The forms represented are thus all epigean and do not represent a cave fauna such as we are to discuss.

The caverns in which living creatures are now found are all Pleistocene and although certain older caves may have been reëxcavated the fauna is probably not older than the Glacial period. The cave fauna as such is therefore, geologically speaking, of relatively recent date, which accounts for the fact, to be emphasized later, that the adaptive changes which the creatures have undergone are almost all the result of degenerative specialization, no new structures having arisen owing to the brevity of time.

NATURE OF THE ENVIRONMENT

Two physical characteristics stand out most strikingly: absence of light, and nearly uniform temperature, the first especially being of decided and far-reaching effect upon cave-dwelling forms. In connection with these features the caves may be divided into three regions as follows:

Cave Regions

1. *Twilight (dysphotic) or transitional region.* This lies just within the mouth of the cave and is bounded by the distance to which light penetrates from without. Conditions vary therefore from those of the outer world to those of the real cave.
2. *Region of fluctuating temperatures,* wherein may be felt the diurnal or seasonal variations of heat and cold.
3. *Inner cave region,* where light is absent and where the temperature remains comparatively constant regardless of the varying epigean conditions. For example, the Shawnee Cave at Mitchell, Indiana, where the extreme of thermometric variation the year rounds is only 2.2° Centigrade (about 4° Fahrenheit). This cave is open at both ends and the air flows through it freely.

These regions vary greatly in different caves. Some Cuban caves are entirely of twilight character, due very largely to their being sea caverns, the water at their mouth diffusing the light and thus causing it to penetrate much more deeply into the recesses than it otherwise would. These caves usually contain an abundant fauna, consisting largely of regular or accidental visitors from without. In certain ice caves which conserve the winter temperatures, that is, such as can be entered by the prevailing winter winds but not by those of summer, the entire portion beyond the twilight area may belong to the second section, while tropical caves the second section may not exist at all.

Moisture

The degree of moisture in the cave conditions, while practically uniform in some parts of a cave, may fluctuate more than any other element in other portions. The great increase is of course found in the streams and pools, especially after a rain, for then the River Styx of Mammoth Cave may rise 60 feet above the low-water level, while in Shawnee Cave certain portions are filled to the ceiling. The water fluctuates, therefore, in amount, clearness, and swiftness with outside meteoric conditions. On the other hand, in the upper parts of Mammoth and Wyandotte caves the dust may lie undisturbed for years, for in the former there are still shown the tracks of oxen made in 1860, while in the latter are to be seen impressions of the moccasined feet of Indians.

Barometric Pressure

Barometric pressure varies as without and there is also variation in the air as well as in the water currents. The air currents are caused by (1) flow of water, (2) epigean breezes, (3) changes in atmospheric pressure, and (4) differences in temperature.

On the whole the biological environment is comparatively simple and the adaptations are largely a response to lack of light, scarcity of food, and uniform conditions.

FLORA

The absence of light and the dryness are inimical to plants, especially to green plants, which, as we have seen, form the ultimate food supply of all animal creation. Hence the source of food of cave creatures is a very important problem. The plants which do exist are such as dwell in darkness and there have been recorded the following:

Mammoth Cave:

1. Common fungus, identified as *Ozonium auricomum* Link (=*Byssus aurantiaca*), which is also found upon wood in caves of Great Britain, Germany, Michigan, and elsewhere.
2. Reddish, button-shaped fungus.
3. Green mold.
4. Long white mold.
5. Fungus growing on the body a cave cricket (*Hadenæcus subterraneus*).
6. *Agaricus* sp., forming when observed in 1881 a natural bed of mushrooms.

Weyer's Cave:

1. *Peziza* sp.
2. A colourless Agaric.

Luray Cavern:

Two species of fungi described by Hovey:

1. Long white mold hanging in festoons.
2. *Mucor stalactitis*, a supposed new species.

Bacteria seem to be absent if one may judge from the following evidence (Hubbard):

"Washington's Hall [Mammoth Cave] is a chamber of the largest size, and for many years the lunching place of tourists. . . . `The larger masses of gypsum afford convenient seats and tables for picnickers, and are strewn about with chicken bones and bits of food. The accumulation of such rejectamenta is very great, to be reckoned perhaps by the cartload; yet, notwithstanding the presence of so much offal, kept perpetually moist by contact with the gypsum sand, not the slightest taint is perceptible in the air of the chamber; only at close quarters the recently deposited morsels give off a peculiarly rancid odor. As before, in the Rotunda, I was struck with the conviction that decay in the cave is an exceedingly slow process, accomplished mainly through the agency of a few fungi. . . . But if bacteria are absent, other scavengers in abundance attack this food material. I found it swarming with the larvæ of *Adelops* and the maggots of a small fly (*Phora*). The imagos of the beetle and puparia of the fly were also present in countless numbers.'"

On the other hand, the decay of mine timbers is generally rapid, dependent upon warmth and moisture. This may be far from the entrance. Here, however, the bacteria may be brought in with the timber or by the constant passing in and out of the miners, or they may be blown in with the enforced ventilation. The conditions are hardly the same as in a natural cave.

FAUNA

Temporary Fauna

The temporary visitants to the caves need hardly be discussed as they exhibit little or no special adaptation to cavern life. The bats come the nearest among the temporary forms, and their adaptation, which is truly marvellous, is largely a nocturnal response which serves incidentally for cave-dwelling. The brown bats (*Myotis lucifugus*) of the Wyandotte Cave are described (*Brooklyn Museum News*, February, 1913) as returning daily to the cave for shelter. They cling in masses or small groups or singly to the vertical walls, the ceiling or any dry parts of the various caverns, some relatively near the entrance, others far within the caves at distances and through galleries puzzling enough to man even when supplied with brilliant lights. It is still a great problem to known what the special sense is, or what may be its degree of reflex or possibly conscious reaction on the bats themselves, by which they conduct their flight through such galleries and among such numerous stalactites and stalagmites. Sometimes the masses of bats are so extensive that with a stick one may scrape from the wall a quantity sufficient at one sweep to fill a bushel basket.

"Marvel how the bat discerns
Some pitch-black cavern's fifty turns,
Led by a finer tact, a gift
He boasts, which other birds must shift
Without, and group as best they can".

Browning

Permanent Fauna

Only one *mammal* aside from the bats has been recorded, and that is the white-footed mouse (*Peromyscus leucopus*), which seems to be in a sort of transition between an epigean and true cave-dwelling form. This creature is characteristic of temperate woodlands. Where there are no rocks, these mice burrow and form extensive underground galleries; they even persist where the woods have long been cut away in open fields and meadows. In rocky woodlands they seek crevices among the rocks. When they collect in the caves, such as the Wyandotte, there is no evidence of their burrowing or of using any choice in the selection of nesting places other than in breaks or against overhanging walls. They were taken 200 yards from the entrance of the cave in totally dark galleries. These mice are not blind as yet, but have bulging eyes and long tectile whiskers and ears, all of which may be looked upon as partial adaptations to darkness, whether cave or not.

No *birds* are permanent cave-dwelling forms, as there are on blind birds. Some nocturnal birds such as owls use the caverns for daily refuge, and there dwells in the mountainous regions of northern South America and the island of Trinidad the "oil-bird," *Steatornis caripensis,* related to the owls and goat-suckers, which nests exclusively in caves. This creature was made known to science more than a century ago by Alexander von Humboldt, but was re-described by Roosevelt, and figured quite extensively in the daily press as the Colonel's "bewhiskered devil-bird".

No *reptiles* are permanent or even temporary cave dwellers. Eigenmann found one turtle a little way in, probably an accidental ingress, and a copperhead snake at the mouth of a cave. There are many blind *burrowing* lizards and snakes, but, as we shall see, the cave fauna has not been recruited from fossorial types, the only apparent exception being the mouse mentioned above.

As for the *Amphibia,* there are on blind frogs or toads, hence no cave-dwelling ones are included in the permanent fauna, for jumping in the dark is directly at variance with the ancient proverb which advises observation in advance of saltation as the only safeguard. Blind crickets inhabit caves, it is true, but they are armored creatures whose light weight and general organization render concussion of the brain a very remote possibility.

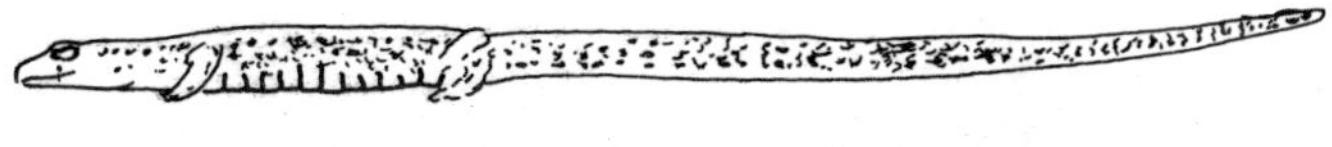

Fig. 17.2: Epigean salamander, *Spelerpes longicauda,* with eyes and colour.

Of cave salamanders, on the other hand, there are several in North America, some with normal eyes in the twilight zone and others with degenerate eyes, adapted to cave life, as there are no blind epigean salamanders. The normal-eyed forms belong to the genus *Spelerpes* of which *S. maculicauda* is red and black while *S. stejnegeri* is yellow and brown. The former is found from Indiana through Kentucky to Missouri, the latter southwest of Missouri. Neither has progressed appreciably in the direction of cave degeneracy. This is not true, however, of the remaining two salamanders, for *Typhlotrition spelæus,* which is light brown and lives in southwestern Missouri, has normal eyes in its larval state, with marked degeneracy of these organs upon becoming adult. The eyelids fuse and the rods and cones of the retina all disappear.

Typhlomolge rathbuni is a perennibranch salamander, that is, one with permanent gills, and has extremely degenerate eyes which have no muscles and are covered with skin and therefore functionless. *Typhlomolge* is bereft of colour and has very long, slender limbs. It gives an impression of extreme emaciation. This salamander inhabits the subterranean rivers of Texas, all the specimens thus far known having come from the water of an artesian well 188 feet deep, near San Marcos (Gadow).

Proteus anguinus is a peculiar form restricted to the underground waters of Carniola, Carinthia, and Dalmatia in Europe. Near Trieste are the huge caverns of Adelsberg which are especially noted for the occurrence of the "Olm," as the Germans call it. Here it dwells in absolute darkness, in an almost constant temperatures of 50° Fahrenheit. "Their total length is scarcely one foot. The whole body is white, occasionally suffused with a slight fleshy, rosy tinge, while the three pairs of gill-bunches are carmine-red. They are easily kept in captivity, and live for many years, provided three conditions are strictly adhered to, viz., fresh and clean water, an equable

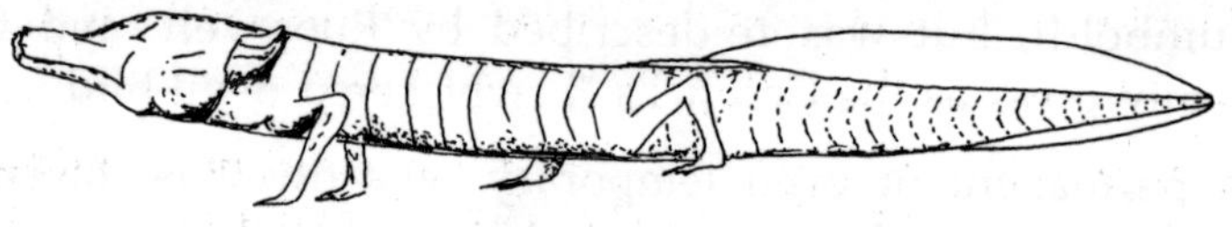

Fig. 17.3: Degenerate cave salamander, *Typhlomolge rathbuni,* without eyes or colour.

low temperature of about 50° F°.=10°C., and darkness. The question of food is not so very important, since specimens are known to have existed for years, although they refused to take any nourishment. How far darkness is an absolute necessity is not known. Anyhow, the white skin is almost as susceptible to light as is a photographic plate. If light is not absolutely excluded the white skin becomes in time cloudy, with grey patches, and if kept exposed to stronger light, the whole animal turns ultimately jet black" (Gadow).

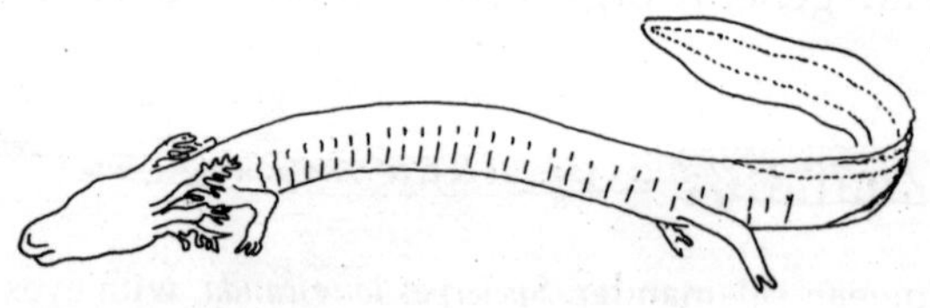

Fig. 17.4: Blind *proteus anguinus,* from the underground waters of Carinthia, Carniola, and Dalmatia.

While totally blind, these creatures are nevertheless annoyed by the approach of a candle flame, showing that the body is sensitive to light. They are guided to their prey mainly by the vibrations which its movements give rise to, possible also by the sense of smell.

The *fishes* include the largest number of cave-dwellers. The cave catfish, the species known as *Gronias nigrilabris,* is a partially blind form found in the caves about Conestoga Creek, in Lancaster Country, Pennsylvania, and is evidently derived from local epigean species. The eyes of many catfishes are not highly developed and are of little use in detecting food, so that their further reduction as an adaptation to cave life is not remarkable.

The great majority of North American cave fishes belong to the family Amblyopsidæ, and all of the species, whether cryptozoic (Gr. *κρυπτός*, hidden, and ζωή, life) or epigean, show signs of degeneracy, eight having degenerate eyes, five mere vestiges, six dwelling permanently in caves, one in a spring, the others in open streams. This family contains several genera, of which *Chologaster cornutus*, the Dismal Swamp fish, represents the epigean ancestral type. In this fish the eyes, though small, are normally developed and the colours are also preserved. Other species of *Chologaster* possessing eyes and colour, but also provided with the tactile papillæ which are so highly developed in true cave forms, are found in cave springs in Tennessee and southern Illinois. The small cave fish *Typhlichthys subterraneus*, from the Mammoth Cave, is supposed to be a direct descendant of *Chologaster*. Here the eyes are present in the young but become defective and useless in the adult, and are partly overgrown with other tissues. *Amblyopsis spelæus* of Mammoth Cave is one of the most interesting of cavern fishes. Eigenmann describes the senses of this creature at some length. The animal is totally blind, but has what he calls"terminal buds," interpreted in part as gustatory (taste) organs, scattered over the entire head, especially on the lips and snout. Experiments indicate that if they are present on other parts of the body they are few in number. While these fishes are no doubt able to taste by means of the buds on the lips and snout, practically all of the food is found by means of the tactile sense. The young up to 40 mm. in length do not have terminal buds developed, and so their food is located entirely by touch, for the sense of smell plays but a minor part, if any.

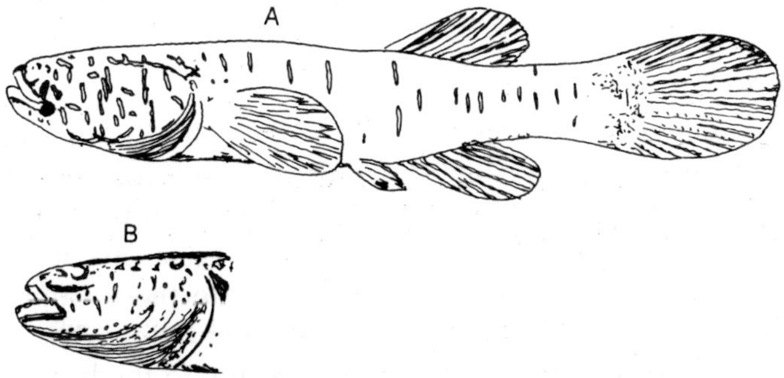

Fig. 17.5: A, Cave fish, *Amblyopsis spelæus*, eyeless; B, head of *Chologaster papilliferus*, possessing eyes but blind. Special tactile organs highly developed in both forms.

The tactile sense enables the fish to detect vibrations in the water originating at least 18 inches away, and it turns and swims toward the point of agitation with unerring accuracy. Agitations behind the animal are not so quickly detected as those in front or on either side. As the fish lives in quiet waters this sense becomes of the utmost importance to it, far exceeding in value that of smell or taste.

Although blind, *amblyopsis* avoids even the diffused light of a room remaining in the dark compartment of the aquarium until night, when the specimens all collect in the light compartment. Bright sunlight is irritating, and even a lighted match held near the aquarium causes very general and active movement. The direction of light does not influence the action of these fishes, but their behaviour is due to a perception of difference in the intensity. In *Chologaster* young fish an inch in length react more strongly to light than older ones, even if their eyes are destroyed, and one part of the body is as sensitive as another to a beam of strong light. *Amblyopsis* also shows a strong positive reaction toward violet rays and a strong negative one toward pink and blue (heliotropism).

The ear of *Amblyopsis* is anatomically normal, but no ordinary noises such as screaming or striking together metal objects over the fish have the slightest effect upon its actions. It seems therefore to be totally insensible to sound waves, although highly sensitive to other vibrations.

Amblyopsis feeds on occasional young fish of its own or some other species; it also eats the blind crayfish and *Crangonyx* and perhaps *Cæcidotea,* though the latter lives concealed under stones. The crayfish feeds upon the two lesser Crustacea, so that sooner or later their substance may form a part of the *Amblyopsis* even if they do not contribute directly to its menu.

The invertebrate fauna shows in general the same modifications among its members as are found among the vertebrates.

Summary of Modifications

The adaptations which cave animals have undergone are all in the nature of retrogressions or modifications of preëxisting structures. The principal modifications, therefore, fall under the four following groups, of which the first is:

1. Loss of Pigmentation

Loss of pigmentation or bleaching, in response to lack of light. The Cuban cave fish, *Stygicola,* a modified marine fish, is in a transitional condition, as the individuals range from pale violet to steel blue. The former should be found in the deeper recesses of the caves but as a matter of fact both light and dark fishes are found side by side. As has been said, the diffused light penetrates deeply into these Cuban sea caves, which may account for the apparent slowness of depigmentation. The rosy colour of the *Proteus,* especially the carmine of its gills, is simply due to the hue of its blood showing through the pigment-less skin. Proof of this, if such were needed, would be the alternate paling and reddening of the gills with the varying flow of the blood during respiration.

2. Reduction of the Eyes

These may degenerate in the manner shown above, the individual parts of the eye becoming imperfect and then functionless, or the entire eye maybe still further reduced, covered by other tissues, or even become entirely obsolete. Blind forms, as has been shown, may nevertheless be sensitive to light and even heliotropic. As a partial recompense for the loss of vision, tactile organs, including antennæ, or possibly the sense of smell, may be increased in efficiency. Cave fishes and salamanders are highly sensitive to ordinary vibrations, although in some known instances deaf to sound waves.

3. The organs of Digestion

The *organs of digestion* may be considerably modified, due no doubt to the scarcity of food, which not only implies long fasts and irregular digestive activity, but a highly efficient utilization of whatever of nutrient value the food may contain.

4. Slender Bodies

Slender bodies and long attenuated appendages are also characteristic of cave dwellers. The term "depauperated," which means falling short or the natural size from being impoverished or starved, expresses the general appearance quite clearly, as almost all of the effects, other than those produced diretly by the absence of light, are due to scarcity of food. Quiescent conditions seem conducive to length of appendages, for this characteristic is seen in other localities, as on the bottom of the sea.

Convergences

As one would infer, convergences exist because of the simple and uniform character of the environment, which offers little opportunity for variety of habitat or habits, hence of necessity all cave creatures tend to look alike.

Food Supply

The scarcity of food has been emphasized, but the ultimate source of food is yet to be discussed, for in its final terms animal food is derived from green plants, which in their turn depend upon sunlight. Hence nature's great epigean laboratory must be the ultimate source of the food, and its transference into the cave is largely a matter of chance. Some is doubtless brought in by temporary visitants whose rejectamenta form a merger supply for the busy scavengers which in turn perish that others may be fed. Some organic food is blown or carried in by the subterranean streams, but if may readily be seen that the amount of food available is in inverse proportion to the size of the cave and directly in proportion to that of the entrance.

Theories to Account for Cave Modifications

Several theories have been advanced to account for cave modifications, among which are the following:

1. Cessation of selection or panmixia would probably account for the reduction of certain structures, although whether it would cause them entirely to disappear is a mooted question. The reversal of selection may account for the loss of such structures as the eyes, for it is conceivable that in the deeper recesses of the great caves the presence of such easily injured organs might become a menace rather than a mere matter of indifference.

2. Most explanations of the origin of the retrograde characters of cave animals involve the inherited effects of disuse, in other words, the inheritance of acquired characters, of which so many good authorities deny the possibility.
3. In view of the difficulties in the way of this, it seems that the theory of the cumulative effect of *determinate* variations may explain the adaptation for cave life which makes possible the origin of cave inhabitants and their further modification after becoming cave forms (Banta).
4. The forms were adapted first and changed their environment to suit.

Perhaps a compromise explanation embodying portions of those listed above may come nearer the truth than any one exclusively. It may be that creatures already weakened and tending to degenerate entered the caves for retreat from the strife of epigean existence for which they were unfitted, and that starvation and absence of light increased the individual degeneracy and finally has its effect on the race. Nevertheless it may well be that certain of the modifications shown by cave-dwelling forms are merely ontogenetic, until heritable coincident mutations arise.

Origin of Cave Faunas

Four theories have been advanced by Eigenmann to account for the origin of cryptozoic faunas. These are:

1. Accidental carrying in. This probably plays little or no part in cavern-peopling, for frogs and fishes placed within the caves have speedily departed, either to the epigean region or to oblivion, and such attempts by man have never succeeded in establishing a single new cave form.
2. Step by step colonization, the species penetrating further and further into the deeper recesses in successive, generations.
3. The animals may have developed along with the caves *pari passu*, a theory which is rather unlikely.
4. The animals elsewhere adjusted to do without light may have gathered voluntarily into the caves. This, as has been emphasized, does not mean fossorial forms but those which lived in crevices or under bank and rocks where feebler creatures would be likely to take refuge. The white footed mouse has such a habitat in rocky woodland, and it is only when it lives in woodlands where no rocks exist that it burrows and thus proves the exception to the rule; this burrowing, furthermore, may well be secondarily acquired habit.

This last theory of cavern-populating conforms most nearly with that which seems most plausible for the origin of cave modifications.

DEEP-SEA LIFE

Environment

Extent

While the oceans cover 64.5 per cent of the earth's surface, the deep sea is more limited in extent; nevertheless it embraces about one-half of the entire globe. Most geologists have become convinced of the relative permanency of the oceans, there having been no essential shifting of their areas, at least since the close of Proterozoic time.

Physical Characteristics

The most remarkable characteristics of the sea in their effect upon animal life, are found in number, of which the first is *cold*. Below about 3000 feet the temperature is 37°F. or less, all diurnal and seasonal fluctuations of temperature cease, and the waters are equable in their darkness and degree of hear the world over, whether they lie beneath the pole or under the vertical ray of a tropical sun. In the great "deeps" the water is ice-cold, averaging about 32° Fahrenheit. Ocean currents from the polar regions, sinking ever deeper, flow toward the equator with a deliberation which renders their rate of progress immeasurably slow; nevertheless they bring with them to the greatest depth the polar cold, and oxygen; food, on the contrary, sinks directly from the sun-lit upper strata into the depths.

This intense cold is believed to date from the Glacial period, for with the melting of the ice the cold waters, pouring into the oceans, sank into the depths, so that the Pleistocene was a critical time for all forms of life, not only for the plants and animals of the glaciated lands and the shallow-water life of the northern and southern oceans, but evidently for the scarcer denizens of the abyss as well.

Below the limits of disturbance caused by the greatest of storm waves, about 600 feet, and below the average depth of tidal action, all *movement is exceedingly slow,* and there are only the imperceptibly deliberate diffusion streams which mingle the waters. However, about the oceanic islands and in places along the continental slopes down to about 3500 feet the bottoms are without ooze accumulation.

The distance to which *light* penetrates varies with the angle of the sun's rays and the clarity of the water; but on the average 200 fathoms is the limit of its penetration. This, as in the case of cave-dwelling forms, has its effects both direct and indirect on the organisms of the deep.

While atmospheric *pressure* at sea-level, depending as it does on the weight of the overlying column of air, is, on the average, nearly fifteen pounds to the square inch, when one goes downward below the surface of the water the pressure increases enormously owing to the greater relative weight of water. This rate of increase, which is in direct proportion to the depth, is, in round numbers about one ton to the square inch for every thousand fathoms. Even at sea-level the burden of the air pressure upon our bodies would be unbearably great were it not equalised within and without, and so it is with marine forms. The submarine diver in his inflated dress is strictly limited as to depth, and so is the ship and is not transmitted throughout its mass as it is in a deep-water animal. Hence the submersible is crushed and destroyed at a relatively slight depth compared with that beyond which pressure adapted creatures can penetrate with security.

Summary of Bionomic Features

Briefly, these are the four factors just discussed—cold, quiescence, darkness, pressure; add to these the total absence of living green plants beyond the light zone and the summary of conditions is complete.

Sea Floor

Because of its influence of marine life the character of the ocean bottom must be determined. Away from the continental margins the ocean floor is vast undulating plain with occasional volcanic and other elevations which may or may not reach the surface, and here and there profounder abysses—the so-called deeps. This floor is carpeted with a mass of material, of terrigenous (born of the land) origin near the shores, and organic in nature further out until the profounder depths are reached, where the organic materials cease. Thus the continental shelf from tide limits to the depth of 100 fathoms is covered with terrigenous materials, mainly sands with occasional areas of mud washed into the sea by rivers, the debris of subaërial and stream erosion, and also that made by the waves in cutting back the land. Out to 200 miles from land, more or less, and therefore at depths up to and over 1000 fathoms, both terrigenous and organic materials prevail, but beyond these limits the former cease except for water-logged volcanic material-pumice stone and ash—to which should be added the extra-terrestrial bodies called mateorites.

The organic remains are known collectively as oozes, and consist of the shells and partially decomposed remains of tiny organisms, Protozoa, Mollusca, and plants, which dwell in the upper portions of the sea and whose bodies fall as veritable rain into the depths when their allotted span is measured. The oozes are:

Pteropod ooze, consisting of a certain percentage of the limy shells of petropods or sea-butterflies (Mollusca) mingled with other organic remains. This ooze is not very extensive, comparatively speaking, and is found at a limited distance from land. In depth it ranges from 500 to 1500 fathoms.

The second kind of ooze, known as foraminiferal or *Globigerina ooze,* contains a high percentage of the remains of Foraminifera, small marine Protozoa whose shells are also composed of carbonate of lime. Of these the genus *Globigerina* is so characteristic and abundant as to give its name to the mass. This is the most extensive ooze of all, covering vast areas of the Atlantic and Antarctic oceans. Its vertical range is from 1500 to 2500 fathoms. The lower limit is caused by the increase of carbon dioxide dissolved in the water, which in turn is determined by increasing pressure.

The presence of this carbon dioxide adds to the solvent power of the water, hence limy shells cannot exist beyond 2500 fathoms, and, if the water's depth exceeds that limit, they are dissolved before being buried by later accumulations.

Radiolaria are also Protozoa which secrete delicate glassy tests or skeletons of silica. These, while fragile and easily broken, resist solution far beyond the disappearing point of the calcareous shells. *Radiolarian oozes* are not widespread, but occur in the tropical regions of the Pacific and Indian oceans. They cover but 3.4 per cent of the ocean floor.

Diatom ooze is formed of the siliceous shells of minute plants, of which a single species is particularly abundant in Antarctic waters, where the Diatom oozes cover an area estimated at 10,000,000 square miles or 6.5 per cent of the entire sea bottom. The average depth is about 2000 fathoms.

Beyond the oozes comes the *red clay,* the most widely distributed of oceanic deposits, as it covers 36 per cent of the oceans. In depth it ranges from 2500 fathoms downward and really includes the limits of the Radiolarian ooze, for Murray says that the latter may be regarded as a variety of red clay containing many radiolarian skeletons. This red clay is derived chiefly from the disintegration of the pumice and other volcanic materials which have long been acted upon by sea-water. In the greatest depths there is of course practically no limy material present, but in lesser depths the clay merges either into Globigernia or Radiolarian ooze. The rate of deposition of the red clay is very slow, as there have been found in its grains of meteoric dust, the undissolved ear bones of whales, and the teeth of huge sharks (*Carcharodon megalodon*). The latter are not known to have survived the Pleistocene period, and yet the teeth lie so near the surface of the deposit as to be picked up in an ordinary dredge, occasionally in great numbers, one dredge load from the central pacific containing 50 ear bones of whales and 1500 sharks' teeth.

Food Supply

The significance of the oozes from our present viewpoint is that they constitute the importers of food into the depths. Many of the creatures live in the great pelagic zone where life-giving heat and light are prevalent and where green plants (algæ) flourish. These algæ form the food of the ooze animals and the latter, dying, descend through the refrigerating cold to the depths below. Although certain bacteria may be present, disintegration must be very slow, so that the animal substance remains available as food for a considerable time. The ooze is in a state of suspension, probably a foot or more in depth, and to this come fishes with feeble masticatory organs to feed upon the ooze creatures, and in their turn form the food of the more aggressive, predaceous sort. From the nature of things the animals are largely carnivorous.

DEEP-SEA FAUNA

All phyla of animals are represented in the deep-sea fauna, but they are of course only water-breathing organisms, hence the higher Arthropod (insects, etc.) and all vertebrates above the fishes are debarred.

Origin

None of the deep-sea forms is the product of a locally evolved race, but they are all simply adapted migrants from the lesser depths, either the pelagic or the littoral realms. Thus their modifications are merely a response to the physical conditions of the deep sea which, while it gives rise to convergences as does the cave life, does not entirely mask relationships with the shoal-water forms. It was thought that because of its changelessness the deep sea would prove the last refuge of certain very ancient forms of marine life, elsewhere extinct, such for instance as the Paleozoic trilobites. But extensive submarine exploration, notably that by the English ship "Challenger," the American "Albatross," and the Norwegian ship "Michael Sars," has revealed many abyssal organisms, none of which is old, geologically speaking, or if ancient, without living relatives in shallow waters. In other worlds, no Paleozoic survivals are exclusively deep-sea in their present distribution, in fact, there are here no Paleozoic stocks at all.

It may safely be said that the deep-sea invertebrates do not antedate the Triassic and that the majority have affinities with Jurassic and Cretaceous types. The vertebrates, on the other hand, are yet more recent, for, with the exception of the sharks and chimæriods, none is older than the Lower Cretaceous and the exceptions mentioned may well be relatively recent downward migrants, although they may be representatives of creatures which have survived elsewhere at least from the Carboniferous.

Adaptive Characters

The adaptive characters of deep-sea animals are summarized by Günther as follows:

1. The animals are *frail and weak,* with very little earthy matter in their bones. The flesh is thin and flabby in consistency.
2. *Simplified colours* are the rule, some being red or brown, others either pearly gray or black, although some fishes have scarlet fins. None, however, has the diversified colours of the light zone. The scarlet fins may be an inheritance from a colourful ancestor, or, as in the blind *Proteus* of the Austrian caves, may be due to the colour of blood showing through a pigment-less skin. The pick eyes of an albino are analogous. One fish, *Macrurus filicauda,* reverses the standard faunal colouring of sea fishes in that it has a black belly and is silvery on top. This may be explained as a response to the reversal of the source of illumination, for such as there is, is due entirely to the luminescence emanating from the animal life, which is largely benthonic. Hence the enemy looking down on the fish from above sees only the silvery back which blends with the luminous environment, while to a foe looking up from below the black belly would blend with the overlying darkness and thus become practically invisible.

3. Some deep-sea fishes are *blind,* others have *telescopic eyes,* yet others have *eyes like concave mirrors,* either of the two latter types serving to absorb the greatest possible volume of light rays.
4. There are, as a compensation for the loss of vision, certain *tactile organs,* long feelers and slender attenuations of the fins such as may only be used effectively in quiescent waters.
5. Almost all are *luminescent.* The light-giving function may be either diffused over the entire body or on the belly or localized in highly modified organs. These may take the form of round, shining, mother-of-pearl-coloured bodies, embedded in the skin, either larger bodies of an oval or irregularly elliptical shape, placed on the head in the vicinity of the eye, or smaller round globular bodies arranged symmetrically in series along the side of the body and tail, especially near the abdominal profile, sometimes along the back. Teeth and mouth may be luminous.

This luminescence seems, in part at least, to have been acquired *before* the deep-sea migration occurred, necessitated by the nocturnal pelagic habits mentioned above. Such forms are diurnal vertical migrants, sinking below the light zone during the daytime and coming to the surface at night. This habit may have aided in their adjustment to pressure and cold and thus made easy their abyssal journey, accomplished probably by step-by-step migration extending over many generations. Luminescence is absent in all fresh-water fishes, for the only places wherein they could find perpetual darkness—the profounder depths of certain lakes and inland seas—are of too brief duration, geologically speaking for the evolution of new structures such as these. The chemical character of sea-water as an aid to their establishment in marine fishes has also been mentioned as a reason for their absence in fresh waters. But many insects are luminescent and none such are marine.

6. Many deep-sea fishes live on decaying ooze, and hence have lost their *powers of mastication.* Others are rapacious, with powerful jaws.
7. Many show wonderful contrivances for the *care of young;* others produce young in enormous numbers.

Vertical Distribution

Life exists at all depths, specimens having been taken even at so profound a distance downward as 8000 metres or 26,00 feet, nearly five miles, but the numbers of individuals decrease as the dredge descends to greater and greater depths. The abyss is, however, comparatively rich in species though poor in individuals. Thus at 500 fathoms the dredge may contain a hundred examples of the same animal, whereas in greater depths there may be two each of ten different species.

Sir John Murray has determined the number of bottom-dwelling forms at the various levels to be:

Down to	200	meters,	about	4,200	species
" "	2000	"	"	600	"
" "	4000	"	"	400	"
Over	5000	"	"	150	"

It should be remembered, however, that the operation of dredging in very deep waters is one involving much time and expense, so that a very small proportion of the abyssal floor has been actually explored. Then, too, the exploration is done blindly and the element of chance must enter very largely into the result, especially if the creatures to be secured are at all alert and quick of movement.

Summary

The absence of green plant life must be emphasized, but, with this exception, all the phyla of organisms found in shallow water are represented. Among the *invertebrates*, the sponges form an important element in the deep-sea fauna, and cœlenterates such as the corals, hydroids and their allies, while not so numerous as the sponges, are also well represented. Echinoderms, among which are the brittle-stars and stalked crinoids, the latter the survivors of a formerly extensive group, are present, and holothurians or sea-cucumbers are also found. The modern stalked crinoids are rarely found beyond a depth of 200 fathoms, although the free-swimming species, *Bathymetra obyssicola,* has been obtained at a depth of 2900 fathoms. One is, however, generally suspicious of the recorded depth of pelagic forms, for while the depth to which the dredge actually descended may be beyond dispute, the contained creatures, except of course benthonic forms, may possibly have been caught in transit and at a much higher level. Starfishes, while living mainly at moderate depths, do descend to 2000 fathoms or more. The sea-urchins have been dredged from depths of 2000 to 3000 fathoms but those from the greater depths haver very light thin shells comparable to the bones of the abyssal fishes. Of the annelid worms, while the majority are undoubtedly littoral, deep-sea forms are also common, characteristic specimens having been dredged by the "Challenger" from a depth of 400 fathoms off Teneriffe. These were tube-building worms, but other species also occur.

Bryozoa range to 3000 fathoms. Brachipods have been dredged from 2900 and probably go much deeper, a single species having a vertical range of nearly 800 fathoms. Of the Mollusca, the Plecypoda have a very great vertical distribution, that of the mussel *Mytilus phascolinus* being from the shore to 3000 fathoms. In the greater depths the shells of bivalves are exceedingly delicate, being sometimes quite transparent; in the littoral some, cn the other hand, thick-shelled pelecypods are the rule. The majority of gastropods are shallow-water forms, although a number of them are fund at depths of from 1000 to 2000 fathoms. As in the pelecypods, the shells and their ornamentation are more delicate than in their shoal-water relatives, and their colours are faint, "though often this is counterbalanced by the brilliancy and beauty of the iridescence, and even the non-iridescent abyssal species give out a sort of shèen, which is wanting in their shallow-water allies" (Grubau).

Cephalopods inhabit waters of moderate depths only, but because of their prevailingly great swimming powers only the benthonic or sluggish sorts are taken in the dredge, our knowledge of the more aggressive sort being derived largely as a result of accident or as a by-product of the sperm-whale fishery, for they constitute the bulk of the dietary of this huge cetacean.

Of the arthropods, we have barnacles from 3000 fathoms, ostracods from 2000, decapods such as crabs, shrimps, and lobsters ranging down to 2500, though true crabs are largely shallow-water forms, few occurring below 400 fathoms.

Vertebrates

Of the elasmobranchs, the true sharks have been taken at depths of from 345 to 400 fathoms and the rays down to 608 fathoms. None show deep-sea characteristics, however, except perhaps the luminous shark, *Spinax niger*, which ranges from 500 to more than 1500 fathoms. The chimæroids or silver sharks on the other hand, often with huge eyes and a long attenuated body and tail, are distinctly abyssal in their appearance of these, *Chimæra affinis* ranges from 200 to 1300 fathoms, *Callorhynchus* down to 600, *Harriotta* from 700 to 1000, and *Chimæra monstrosa* down to 1000.

The Esociformes (pike-like) group of teleost fishes includes the cave forms, Amblyopsidæ, described above, in addition to two families with deep-see members, such as *Cetomimus*, with a huge mouth, small teeth, no scales, and very small eyes. Another, *Ipnops*, is paradoxical in that while it has no eyes it has nevertheless two large, luminescent organs on the head! These may be alluring, to aid in securing prey. Of the Clupeiformes or herring-like fishes, the deep-sea family Stomiatidæ is characterized by delicate scales or none and by well-developed luminescent organs. Anguilliformes are eel-like, and of these the larva of the true eel, known as *Leptocephalus*, is to be included among the abysso-pelagic forms. Another, *Gastrostomus*, with very long, slender body and enormous jaws, is a highly characteristic abyssal type. Gadiformes, cold-like forms, are not old geologically, ranging from Oligocene and Miocene to Recent. The deep-sea representatives have the mouth reduced in size, dentition reduced, eyes very large, trunk shortened and tail tapering to a filament. Thus, while cod-like in general aspect, they are nevertheless distinctly abyssal, but being relatively recent downward migrants their adaptations are not yet profound.

The Pediculati (anglers) include a remarkable group of fishes whose paired fins have become adapted for bottom crawling, and the anterior fin-rays of the dorsal fin function as a lure, unwary fishes being brought within the influence of the horrible jaws and incontinently engulfed. The shoal-water allies may be protectively embellished with filaments, like fronds of sea-weed, to permit them to lie in wait in a bed of algæ without detection. The deep-sea forms, on the other hand, are bereft of such devices, which would be manifestly out of place where no sea-weeds exist, but they are peculiarly adapted none the less of their habitat. As they are not benthonic, the pelvic fins are absent, and so are the scales; the colour is black and instead of a seaweed - simulating lure, the fin-rays are tipped with luminous organs to attract the prey (*Linophryne*). One, *Oneirodes,* is itself blind, but here the value of luminescent organs is at once apparent

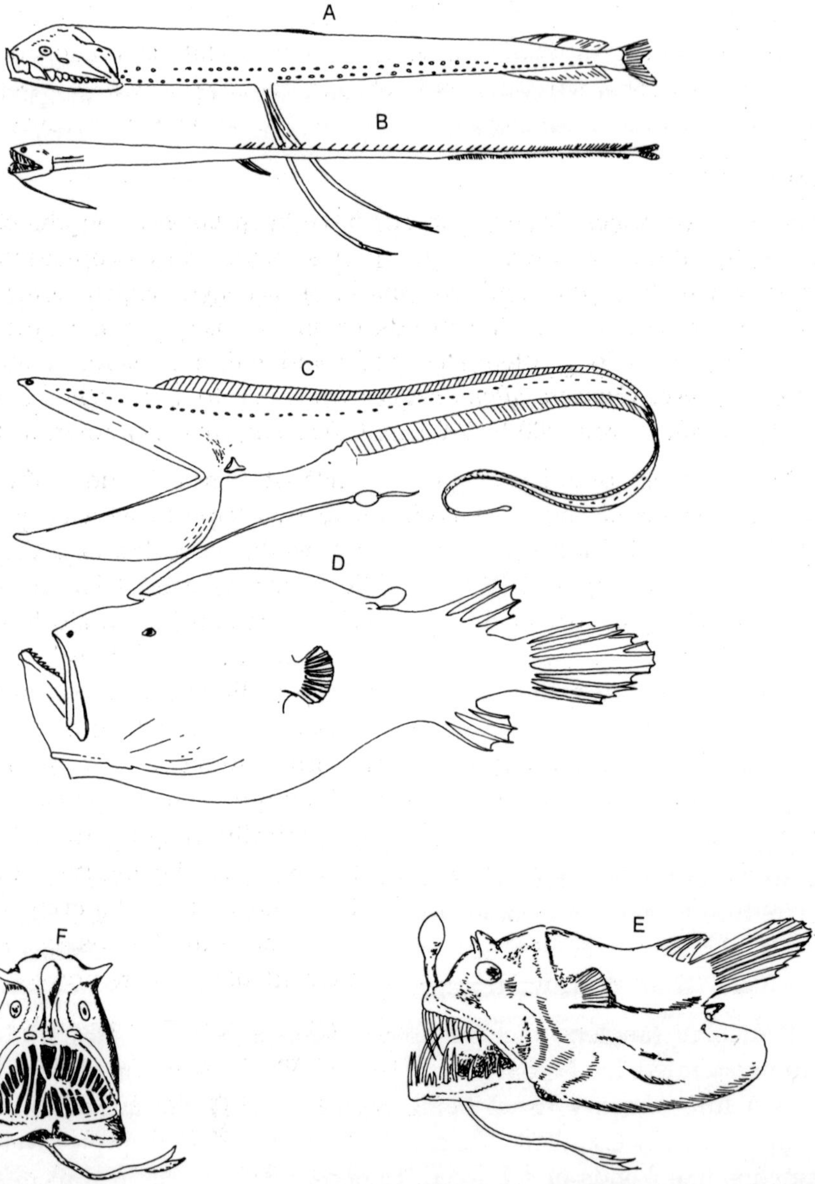

Fig. 17.6: Deep-sea fishes. A, *Photostomias guernei,* length 1.5 inches, taken at 3500 feet; B, *Idiacanthus ferox,* 8 inches, 16,500 feet; c, *Gastrostomus bairdii,* 18 inches, 2300-8800 feet; D, *Cryptopsaras couesii,* 2.25 inches, 10,000 feet; E, F, *Linophryne lucifer,* 2 inches.

The deep-sea fishes have been spoken of as weak and flabby of flesh, but that is when released from the terrific pressure to which they are adapted. The relatively great jaws and cruel mouth armament which many possess imply creatures of marked physical prowess, decidedly out of harmony with their apparent frailty when within the scope of human observation.

Another characteristic of deep-sea fishes which has not been mentioned is their very small size, for whereas the dimensions of shallower water forms may often be measured in feet, a length of several inches is distinctive of many of these grotesque "monsters" of the abyss.

Perhaps the most striking feature of the deep sea and of the cave as well is their changelessness. There is nothing to mark the flight of time—no day or night, no sequence of seasons, and the sameness of conditions year in and year out must have a marked influence on the inhabitants. The stream of individual life as well as of evolution is quickened by environmental changes and is lulled into repose where change is not. Thus these forms, when once they are adapted to the peculiar conditions which, as incoming migrants, they have to meet, practically cease their evolution as they diminish their individual activity and consequent metabolism. Life for them is the antithesis of the feverish activity of surface forms, and will be until the next great geologic revolution puts a period of their existence.

Chapter—18

Volant Adaptation

Next to water as a moulding environment comes the air, for the creature which inhabits either is surrounded on all sides by a homogenous medium so that it becomes uniformly and beautifully modified to offer the least possible resistance to the attainment of speed. Water-inhabiting forms, however, whether primarily or secondarily adapted, may become so thoroughly at home that in the latter group no evidence of their former habitat is outwardly visible. Aërial creatures, on the contrary, are never exclusively such and must return to the trees or earth or sea when they wish to rest. Hence their adaptation is always a double one and as a consequence cannot reach the extreme of specialization of water-borne types.

CLASSIFICATION OF FLIGHT

Passive or Gliding Flight

Flight is of two sorts: first, passive or gliding flight, in which, with the exception of the fishes, the creature merely takes an initial leap from a high point and, held up by certain sustaining organs and impelled by gravity, glides to a lower level, sometimes covering a horizontal distance of many yards. Aside from the initial impetus there is no locomotive force other than gravity, so that the flight is comparable to a gliding airplane without engine power.

1. Adaptation for Gliding

The gliding flights are performed by fishes (*Exocoetus*) amphibians (*Rhacophorus*), reptiles (*Draco*) birds (Ostriches) and mammals (lemurs).

The following adaptations are noted:

(i) **Development of Patagia.** Except in fishes, the patagium is a fold or series of folds of skin supported by ribs. In *Draco*, there is a pair of large wing-like membranes, which extends outwards from depressed body and each supported by 5 or 6 ribs. In flying lemur (*Galeopithecus*), the patagium extends from the back of the neck to the tip of tail even including digits.

(ii) Enlargement of Pectoral Fins. In the flying in (*Exocoetus*) large pectoral fins are present, which serves as organ of flight. Pelvics are smaller while the lower lobe of tail is longer. *Exocoetus* flies upto 200-300 meters. Another examples are *Pantodon.* (African flying fish). *Gigantopterus* (Extinet flying fish) etc.

(iii) Webbing of Feet. The tree frog, *Rhacophorus* is the only amphibian in which volant adaptations is found. It has webbed feet by which it has long prolonged leaps. The digits terminate in adhesive pads. Rudiments of patagia infront and behind the arms are also present.

True Flight

True flight, on the contrary, implies power, so that there is sustained movement through the air, whether the flight be brief like that of a domestic hen or supported on the tire-less pinions of an albatross. True flight has evolved three time among vertebrates: in the reptilian pterodactyls, the birds, and the bats. Whether flying fishes should be included is a much disputed question. True flyers may move the wings with varying degrees of rapidity from the extreme speed of a humming-bird's wing to the measured cadence of a winging crow. Many birds (and doubtless some of the ancient pterosaurs) also sail or soar on apparently motionless wings for hours at a time after having gained their altitude by a flapping rise. They are in reality gradually descending in a great spiral, although by taking advantage of the shifting currents of air they may retain their elevation with little apparent expenditure of energy.

An extreme adaptation of this last method is that seen in the albatross, whose majestic flight is thus described by Hutton: "With outstretched, motionless wings he sails over the surface of the sea, now rising high in the air, now with a bold sweep, and wings inclined at an angle with the horizon, descending until the tip of the lower one all but touches the crests of the waves as he skims over them. Suddenly he sees something floating on the water and prepares to alight; but how changed he now is from the noble bird but a moment before all grace and symmetry. He raises his wings, his head goes back, and his back goes in; down drop two enormous webbed feet straddled out to their full extent, and with a hoarse croak, between the cry of a raven and that a sheep, he falls `souse' into the water. Here he is at home again, breasting the waves like a cork. Presently he siretches out his neck, and with great exertion of his wings runs along the top of the water for seventy or eighty yards, until, at last, having got sufficient impetus, he tucks up his legs and is once more fairly launched in the air".

The flight of the albatross seems to be sustained on motionless wings and yet it will follow the wake of a ship with all of the apparent ease with which a school of porpoises precedes her bow. In the latter, when viewed from above, there are no *visible* propelling movements except at long intervals when a few rather vigorous dorso-ventral undulations of the tail are seen which, however, do not seem to accelerate the creature's speed appreciably. A closer view, especially if one be more nearly on a level with the water, shows the tail to be in rapid but minute vibration all the time, and this intense movement is sufficient to keep the creature ahead of a 39-knot destroyer (the record) and even this does not seem to be the limit of its speed. The progress of the albatross is apparently analogous, for, as Moseley says, "I believe that albatrosses move their wings much oftener than is suspected. They often have the appearance of soaring for long periods after a ship without flapping their wings at all, but if they be very closely watched, very short but extremely quick motions of the wings may be detected. The appearance is rather as if the body of the bird dropped a very short distance and rose again. The movements cannot be seen at all unless the bird is exactly on a level with the eye. A very quick stroke, carried even through a very short arc, can of course supply a large store of fresh momentum".

Doubtless the albatross takes advantage of every shift in the breeze, which is made up of a complex of varying air currents, tilting this or that wing to gain whatever lifting power the air can give. That this "jockeying" of the air currents is a very great aid is attested by the fact that on a calm day the albatross cannot sail, but must flap heavily to sustain itself in flight. It is analogous to man's flight in an engineless glider.

MODIFICATIONS

Body contour in volant animals has been emphasized and is second only to that of the purely aquatic forms in its degree of perfection for the lessening or resistance.

Sustaining Surface

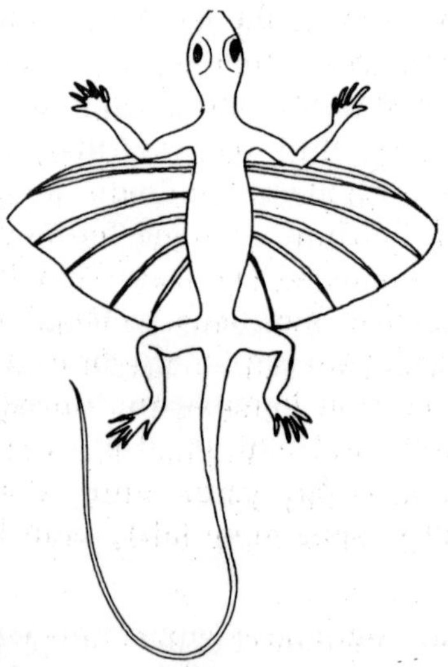

Fig. 18.1: "Flying dragon," *Draco valans.* (After Lull).

The sustaining surface is primitively, except in the fishes, a fold or series of folds of the skin known as the *patagium* (Lat. *patagium,* an edge or border). This may be supported in various ways, but with one exception, in the little lizards (*Draco* spp.) which inhabit the Indo-Malayan region, the limbs form the chief supporting agents. In the "flying dragons," *Draco,* just mentioned, the body is depressed and the sides extend outward into a pair of large, wing-like membranes, supported by five or six elongated ribs. The entire device can be folded like a fan against the sides of the body when not in use. The soaring powers are not very great but when resting among the luxuriant foliage of their habitat the animals are said to resemble butterflies in their habit of opening and closing the wings.

Most soaring mammals have the patagium supported between the fore and hind limbs, and sometimes the skin-fold extends in front of the fore limb to the neck and again between the hind limbs and the tail. Perhaps the extreme of development may be seen in *Galeopithecus*, the so-called "flying lemur," for here the patgium extends from the sides of the neck to the tip of the tail, even including the digits, which are webbed as though for aquatic life.

Where the patagium is supported mainly by the elongated fore limbs and their digits, true flight ensues as in the pterosaure, wherein the enormously elongated outer finger sustains over half the membrane, and in the bats, whose second, third, fourth, and fifth digits perform a like function, the first alone being free. In both group the membrane extends from the arm to the sides of the body and also to the front of the hind limb. An interfrmoral membrane, which, may have existed, has not been demonstrated in the pterosaurs but is variably present in bats.

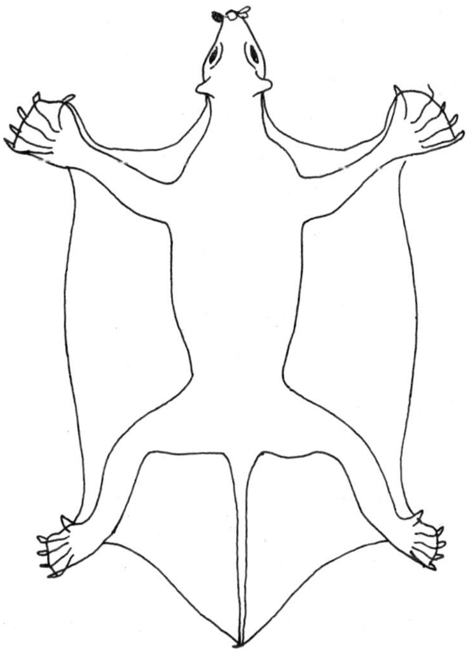

Fig. 18.2: *Galeopithecus valans*.

Feathers are structures which are absolutely diagnostic of birds, since no other group of animals had developed them, and indeed their complexity is such that there is little likelihood of nature's repeating herself in their evolution as she has done many times in that of simpler structures. Birds have traces of patagia in front of and behind the arm which may have had a very adequate supporting function before the feathers usurped their place; but in all known birds the main buoyancy is provided by the broad vanes of the remiges (supporting feathers) of the wing and the rectrices (steering feathers) of the tail which collectively form the most perfect device imaginable, except, perhaps, the insect's wing.

The feather has been called "nature's masterpiece," and while simply & modified reptilian scale has reached a complexity so great that its component parts may be counted by the hundreds of thousands. It is thus described by Evans: The feather consists of a quill or calamus and the rhachis or shaft. On the rchachis a double series of barbs are developed, carrying a similar double series of barbules, the barbules again giving rise to barbicels, which in the distal rows usually terminate in hooklets. These catch in the folded margins of the next proximal row, thus producing a firm surface. Each flight feather, therefore, forms a membrane-like supporting device, the several feathers of the wings being collectively sufficient to maintain the bird in the air even when a few at a time are lost as during the moulting season.

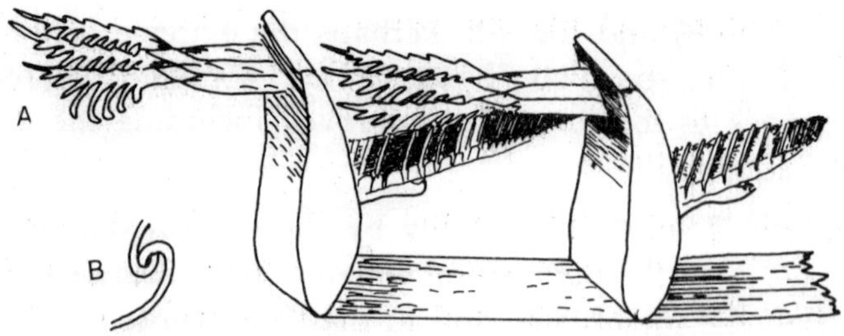

Fig 18.3: Structure of a feather. A, small portion of feather, with pieces of two barbs, each having to the left three distal barbules, and to the right a number of proximal barbules, many of them belonging to adjacent barbs; B, hooklet of distal barbule, interlocking with flange of proximal barbule.

Wing

The wing, as we have seen has been three times evolved, twice with patagia and once with feather. A comparison of the three types is of interest, beginning with the *bat wing* which is the latest in time and hence naturally the least modified. Here the humerus is well developed, the radius long and curved, and the ulna, from loss of general utility, vestigial as in cursorial types. The pollex or thumb is always free and clawed for crawling and climbing. In the smaller bats, Microcheiroptera the second finer, although distinct, is not free from the third but is attached to it distally, the two combining to support the anterior margin of the wing. The fourth and fifth digits are well developed. In the Magacheiropters (Fig. 18.3), or fruit bats, the second digit is independent of the third and bears a claw like the first.

In the *pterodactyl wing* the radius and ulna are more nearly equal, the former being somewhat smaller. The next segment consists of a very heavy fourth metacarpal, bearing the great wing-finger, and three extremely slender metacarpals supporting the first, second, and third digits, which are small but free and clawed. There is also the bone known as the "pteroid" which lies in front of the fore arm and in directed inward toward the shoulder. It is supposed to have supported the anterior margin of a small prepatagium which lay in front of the arm from the wrist to the neck (see Fig. 18.11). The single wing-finger, presumably the fourth, is huge and formed the entire anterior support of the patagium beyond the wrist. One curious feature of the pterodactyl wing lies in the position of the principal joint, the wing being flexed between metacarpal and proximal phalanx, rather than at the wrist as in birds and bats.

The *bird wing* is the most specialized of all, for here not only are the digits reduced to three, but these are more or less fused together so that, with rare exceptions, their sole function is that of flight. In all modern birds, therefore, there are three unequally developed metacarpals which are firmly coössified. The digits are represented by one or rarely two thumb phalanges which support the so-called bastard quills, while the second digit, which in much the largest, bears two phalanges, and the third, one. Claws are sometimes borne on the first and second digits of modern birds, while *Archæornis* a reptilian bird of the Jurassic, had a claw on each of the three free fingers. The alar or wing expanse is provided by the feathers, since the patagium, as we have seen, is vestigial. These feathers, known as remiges, are borne upon the hand (primaries), and on the arm (secondaries). Overlying their basal portion are several rows of coverts, protective feathers known as major, median, minor, and marginal. The importance of the wing coverts lies in the fact that they close the interstices between the quills of the flight feathers and give the wing a continuous area to oppose the buoyancy of the air. Birds have an advantage over both bat and pterodactyl in that lost or injured feathers are renewed, while serious injury to the patagium may impair its owner's powers of flight for life.

Fig. 18.4: Wing of pheasant, showing the two "bastard quills" borne on the first digit, the ten primaries on the hand, and the sixteen secondaries on the fore arm.

Pneumatic Bones

Hollow, air-filled bones are found in the birds and pterodactyls and in many ways they show a remarkable community of design. For instance, there is in the humerus of both pterodactyl and bird a foramen for communication between the respiratory organs and the cavity of the bone, but that is not so remarkable as the fact that in each instance the foramina correspond in position, form, and size, and that they are not one large hole, but in each case a reticulation of small perforations, one beyond the other. So far as our knowledge goes, pneumatics seems to have been universal among pterosaurs, but there are no degenerate or flightless pterosaurs known. On the other hand, birds do not all possess it in equal degree for, as one would expect, it is absent from the Ratitæ, nor is it developed equally in all birds with flight. Coupled with the pneumaticity in birds is a remarkable development of air sacs, principally in the abdomen, but in other portions of the body as well. These serve not only to lighten the specific cavity of the bird, but also to aid in respiration, for the lungs of birds are inelastic and do not hang freely in the body cavity as they do in mammals, but, by means of the abdominal sacs, air is drawn *through* them, not merely into them. Hence there is no unused portion of the lung containing residual air, as in mammals, and respiration is much more effectively accomplished. This is necessary, for with the rapid flight there is a high expenditure of energy, and the respiratory and nutritive organs and those of circulation need to be ample and efficient to meet the demand upon them.

Sternum and Shoulder Girdle

Not only is the sternum or breast-bone well developed in creatures with true flight, but in bears a median keel or carina for the origin of the pectoral muscles which wield the wings. To resist the contractile force of these muscles the shoulder girdle is made very rigid by the development of the clavicles and, in the birds, of the heavy pillar-like coracoids as well. Coracoids are lacking in the bats, and clavicles in the pterosaurs, but in birds both elements are present. In the pre-Cretaceous pterosaurs the scapula is saber-shaped and united to the coracoids at an angle of less than 90°, exactly as in carinate birds. The Cretaceous pterodactyls differ, however, in that the scapulæ, while articulating at right angles with the coracoids, are directed toward the vertebræ, uniting with their neural arches. In the great pterosaure *Pteranodon* (Fig. 18.11) the scapula articulates with the coalesced spines of several coössifed vertebræ which constitute the so-called notarium, the whole mechanisms being comparable to the pelvic arch although on a much larger scale.

In the flying (carinate) birds the scapula and coracoid are united by an articulation; in the flighless Ratitæ, on the other hand, they are firmly coössified, and form an angle with one another greater than a right angle. In the Ratitæ, as the name implies (Lat. *ratis,* raft) the sternum is raft-like, and bears no keel.

Brain and Sense Organs

True flight implies a good brain and the perfection of certain sense organs which, as well have seen; are developed in direct ratio with the locomotive powers of any animal. The brains of bird and pterosaur are curiously alike in that each has broad, well developed hemispheres (cerebrum) which touch the cerebellum behind, and the optic lobes are much enlarged.

Both birds and pterodactyls were well endowed with visual organs, the eyes in each case being large and having the so-called sclerotic plates, structures which are rarely elsewhere present except in certain reptiles (dinosaurs, ichthyosaurs, mosasaurs, etc.). Their function may be to resist variable pressure and also to aid in the rapid focussing which is of vital necessity in bird and pterodactyl. The bats, while reputedly blind, often have very good vision, especially during twilight. They also have a most marvellously developed tactile sense, which does not seem to need actual contact for the discernment of approaching objects. The remarkable ears and facial appendages of certain bats are the principal seat of this sense and the patagia and interfemoral membrane are also highly sensitive.

FLYING VERTEBRATES

Fishes

There are enumerated several genera of flying fishes, each of which represents a separate volant adaptation. Of these the first to be mentioned are the several species of the genus *Exocætus,* allied to the skippers and garfish, which live in all tropical and subtropical seas where they fly in shoals in their efforts to escape the relentless tunny and albicore These flying fishes are trim-build creatures with large pectoral fins, which are the main organs of flight, and variably developed but much smaller pelvics. The loser lobe of the tail is invariably the longer and aids in giving the final impetus to the fish as it leaves the water and also in accelerating its speed if in the course of its flight it comes near enough to the surface of the sea. The length of flight is said to vary up to 200 or 300 yards and it is sometimes sufficiently high to strand the creature on the deck of an ocean-going craft. Whether the flight of *Exocœtus* is true flight or merely a soar is a much disputed question and the evidence, briefly summarized, is as follows:

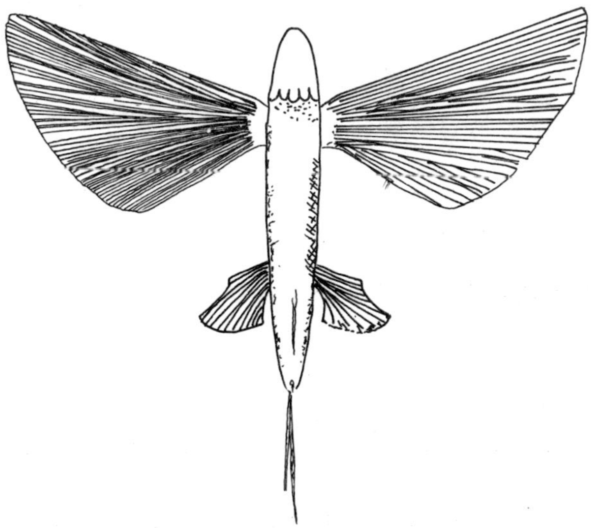

Fig. 18.5: Flying fish, *Exocœtus spilopterus.*

The wing expanse is hardly sufficient for such extended soaring. The wings (pectoral fins) do vibrate, but whether due to muscular effort or to friction, as a flag is flapped in the wind, is not clear. The muscular development seems insufficient for true flight, but on the other hand it is more highly developed than in allied non-flying fishes. It may well be that while true flight as such does not exist among fishes, rapid wing vibration insufficient in itself to support or drive the animal may aid in maintaining or prolonging a soaring flight of which the main propulsive effort is acquired by the tail before leaving the water. At all events, their flight is remarkable and the creatures are one of the most interesting features of the storied tropical seas.

The various species of *Dactylopterus* are known as the flying gurnets and while the flight is by no means as sustained as in *Exocœtus* of the former fishes Moseley rites: " I have distinctly seen species flying gurnets move their wings rapidly during their flight ... especially in the case of a small species of *Dactylopterus* with beautifully coloured wings, which inhabits the Sargasso Sea." Moseley likens the flight of the gurnets to that grasshoppers.

There is an African flying fish found in the Congo and Niger rivers—*Pantodon,* a form but three or four inches long-which leaps out of water and flutters through the air of some distance. Still another flying type is *Gastropelecus,* a small, compressed fish with long and curved but not particularly large pectoral fins, which occurs in the rivers of British Guina. It skims along the surface of the water for 40 feet or more, beating the water with its pectoral fins. Then it leaves the water for a distance of five to ten feet and when exhausted falls sideways into the water again (Eigenmann). *Pegasus volitans,* a little fish found along the coasts of Japan, China, India, and Australia, skims a short distance above the surface of the water by means of its broad pectoral fins.

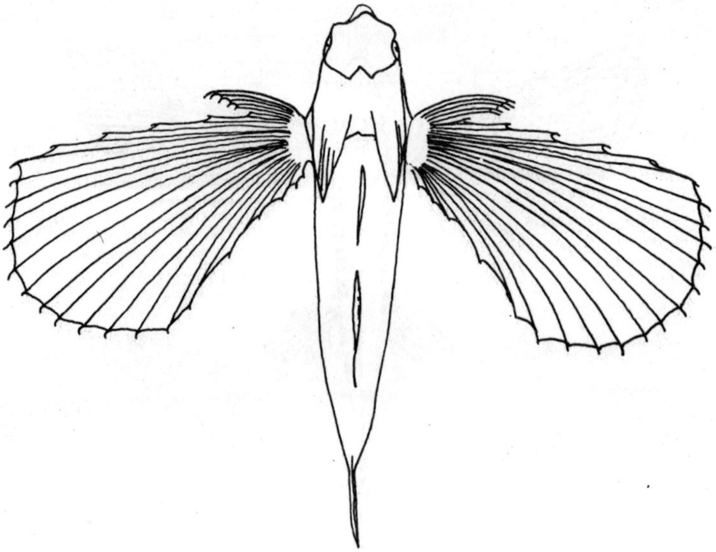

Fig. 18.6: Flying fish, gurnet, *Dactylopterus volitans.*

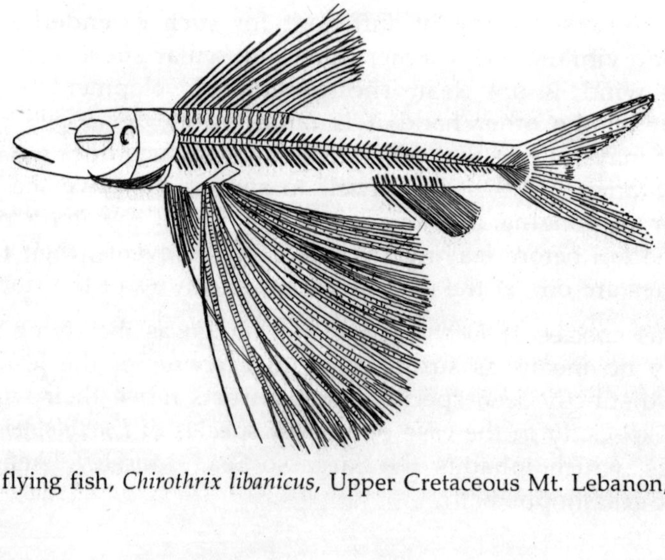

Fig. 18.7: Fossil flying fish, *Chirothrix libanicus,* Upper Cretaceous Mt. Lebanon, Syria.

Several extinct forms have been described as "flying fishes." These are: *Dollopterus* from the Middle Trias (Upper Muschelkalk) of Jena, *Thoracopterus* and *Gigantopterus* from the Upper Trias of Austria, *Exocœtoides* and *Chirothrix* of the Upper Cretaceous of Mt. Lebanon, Syria. The last mentioned is of particular interest on account of the huge size of the pectoral fins, which seem to imply powers of flight fully equal to those of the living *Exocœlus* and *Dactylopterus*. We have therefore among fishes no fewer than ten separate adaptations to aërial conditions, one or two, possibly three of which approach very near to, if they have not attained, true flight.

Fig. 18.8: Flying frog, *Rhacophorus reinhardtii*.

Amphibia

The only volant adaptation among amphibia is that of the tree-frog, *Rhacophorus* whose webbed feet sustain it in the prolonged leaps to which it is addicted. This genus includes a large number of species in the Oriental realm, especially in Borneo. The digits terminate in adhesive pads, in common with those of other tree-frogs, and are connected by web-like expansions of the skin. There are also rudiments of patagia in front of and behind the arms. In *Rhacophorus pardalis* the total alar expanse is about three square inches, which would imply rather feeble gliding powers.

Fig. 18.9: Lizard, *Ptychozoön homalocephalum*.

Reptilia

Lizards include at least two genera and several species of gliding forms, of which the most remarkable is the flying dragon, *Draco*, already referred to in which the patagium is supported by a number of extended ribs. They occur principally in the Malay Peninsula and Archipelago and average some eight to ten inches in length.

Ptychozoön is the flying or fringed gecko of the Malay countries, which is bedecked with lateral expansions of skin along the sides of the neck, body, tail, and limbs, and between the toes. While these may aid in breaking the creature's fall, they may also, coupled with the colour, serve a cryptic function and render the animal less conspicuous against the bark of the tree upon which it rests.

Several so-called flying snakes are recorded, such as *Chrysopelea*, the flying snake of Borneo, which descends obliquely through the air, its body rigid, and the ventral side concave to sustain the creature in its fall.

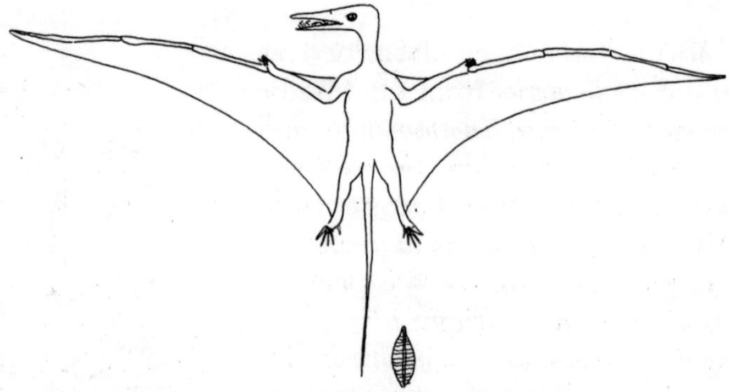

Fig. 18.10: Pterodactyl, *Rhamphorhynchus phyllurus*. At lower right, vertical rudder-like expansion at end of tail.

The pterodactyls or flying dragons of the Mesozoic were a very remarkable group of reptiles whose first recorded appearance is in rocks of the Rhætic or uppermost Triassic period. They range through the Jurassic and on into the Upper-Cretaceous, when they become extinct through racial death. They were undoubtedly akin to the birds, but that simply means, in all probability, derivation from a common, possibly Permian ancestry; nevertheless the two groups show a number of highly comparable, homoplastic characters, some of which have already been referred to. The remarkable thing is that, like the turtles, they first appear fully developed and characteristic of their order, with no record thus far discovered of their antecedent evolution, and the subsequent changes are mainly increase in size, perfection of the shoulder girdle articulation and loss of tail and of teeth. In size they range from that of a sparrow to the mightiest of nature's airplanes for the replica of the late Cretaceous *Pteranodon* mounted at Yale measures 13 feet 6 inches in alar expanse and Eaton is authority for the statement that at least one individual, judging from the relative proportions of the bones which have been preserved, had an estimated breadth of 26 feet, 9 inches from tip to tip. The pterodactyls posessed true flight, which in those from the Kansas chalk must have been sustained, as their remains are found in association with marine reptiles, fishes, and invertebrates, apparently far from the ancient shore. Three of the principal horizons whence these pterodactyls come, the Lias of Lyme Regis of England, the lithographic limestone (Upper Jurassic) of Bavaria, and the Kanas chalk, are all marine in origin, which is also true of the source of the Mesozoic birds. It is highly probable therefore that in each instance we have not as yet knowledge of the great bulk of the group, but only of a few of aberrant habits and adaptation.

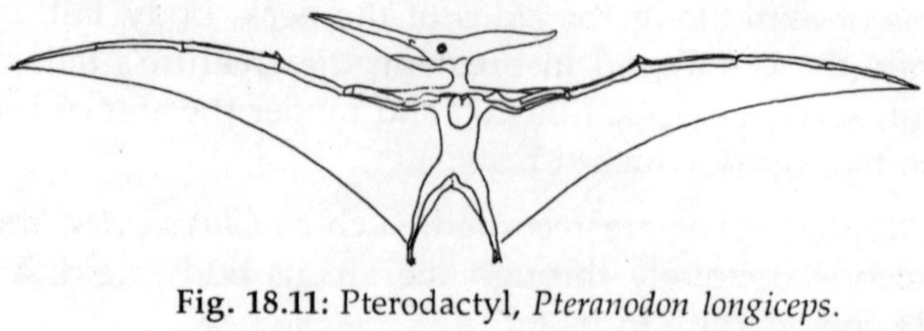

Fig. 18.11: Pterodactyl, *Pteranodon longiceps*.

Birds

The birds in all probability include but a single evolution to aërial life, although certain excellent authorities "believe more or less firmly that possibly birds had not one, but two points of origin, and feel that if we could follow back their lines of descent we should find that the ostriches came from one and the birds of flight from another" (Lucas). If this be true, the ratite or ostrich group as a whole, with a single exception (tinamou), have degenerated and lost the power of flight, although an examination of the skull and skeleton shows them to have been descended from flying normal birds; whereas in the carinate or flying birds loss of flight, while it has occurred (flightless rail, penguin, dodo, etc.), is relatively extremely rare. Flightless pterosaurs and bats, on the other hand, are inconceivable, as their flight mechanism involves the hind limbs which have, as a consequence, largely lost their terrestrial locomotion function; whereas birds, being double adapted can lose their flying powers and still progress easily on the ground or in the water, as their legs are not thus involved.

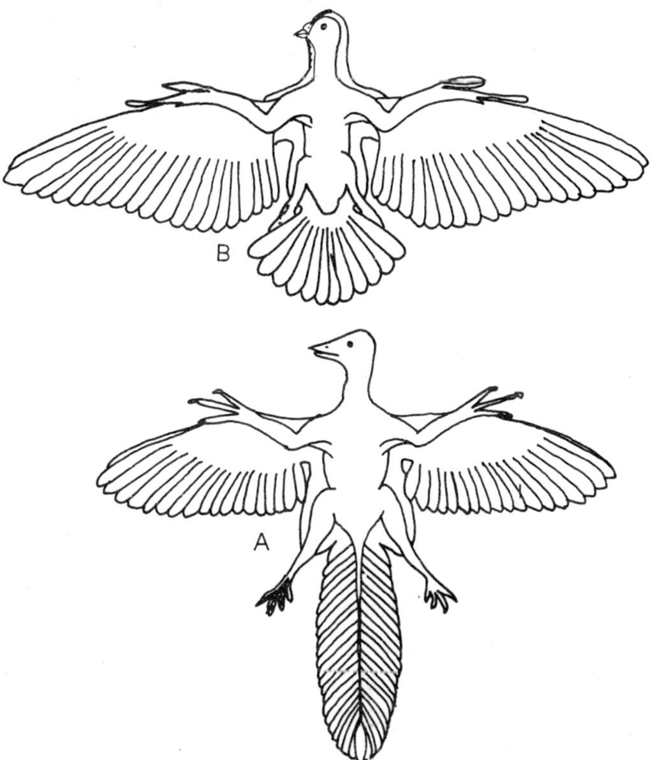

Fig. 18.12: Reptilian bird, *Archæopteryx* (A), compared with pigeon, *Columbalivia* (B).

Birds first appear in time in the Upper Jurassic (Solenhofen limestone) long after the initial record of the pterodactyls. These first birds, of which but two or three specimens have been recovered, are known as *Archæpteryx* and *Archæornis* and are so reptile-like that were it not for the preserved feathers it is doubtful whether they could be surely proved to have been birds. The reptilian traits are teeth, free clawed fingers in the hand, feeble breast-bone, abdominal ribs, etc.

The perfection of aërial adaptation among birds is superlative, as records will show compared with those attained by the man evolved airplane. With regard to speed, the record is held by a house-swallow (*Chelidon urbica*) which flew from Ghent to Antwerp, a distance of 32 miles, in 12½ minutes, an average speed of 153 miles an hour. Some bird speeds as compared with those of airplanes are given as follows, the figures being in miles per hour: white storks at 4200 feet altitude, 48;

mallard, 50; rook, 45; gannet, 48; geese, 55; lapwing, 40 to 45; golden plover pursued, 60. Swifts (*Apus apus*) at 6000 feet above Mosul easily passed and circled an airplane doing 68 miles, and a lammergeier made 110 while nose-dividing to escape a plane. For distance, the record is held by an albatross in the Brown University Museum, which flew 3150 miles in 12 days—probably more, as it rarely flies in a straight line. The weight of this bird was 18 pounds, wing spread 11 feet 6 inches, area 7 square feet. As regards height, the great vulture rises from 7000 to 15,000 feet and Humboldt, a very accurate observer, saw a condor hovering above Mt. Chimborazo, whose summit soars 20,498, feet into the blue.

Mammals

Among the mammals there are upward of thirteen separate volant adaptations, one of which, that of the bats, attained the power of true flight. Among the flying forms the first to be mentioned are the *marsupials,* of which the flying phalangers include several unrelated species: *Petaurus* spp., *Petauroides volans,* and *Acrobates pygmæus.* These are characterized alike by having a well developed skin fold along the sides of the body between fore and hind limbs, and a feebly developed one in front of the fore leg. In each genus the flying form is especially related to a separate type of non-flying phalanger.

In *Petauroides* the flying membrane extends from wrist to ankle, but is very narrow along the distal segment of each limb. The tail is very bushy except for its prehensile tip, which is naked on the under side. The tail, together with the long fur of the body, must supplement to a considerable extent the buoyancy of the patagium. This genus, with it single species, includes the so-called Taguan flying phalanger found in Australian from Queensland to Victoria.

Petaurus has a much broader patagium and there is a naked prepatagium as well. The tail is very large and bushy, but lacks the naked prehensile tip of the preceding form. There are three species of this genus, ranging over New Guinea and part of Australia.

The genus *Acrobates* includes two small species of flying phalangers which have narrow patagia extending from the elbow to the knee along the flank. The long fringing hairs borne by the patagium, together with those on either side of the tail, aid materially in flight. *Acrobates pygmæus* is found in New South Wales, Queensland, and Victoria, while a second species, *A. pulchellus,* is a native of Papua.

Two families of *rodents* contain flying forms: the Anomaluridæ, including the genus *Anomalurus,* and the Sciuridæ, of which three genera, *Pteromys, Sciuropterus,* and *Eupetaurus,* are volant. *Anomalurus* has a well developed patagium extending from wrist to ankle but narrowing in front of the leg from the knee down. As a compensation, however, there is an interfemoral membrane from the heel to slightly beyond the base of the tail. The genus includes six flying species, all found in Africa.

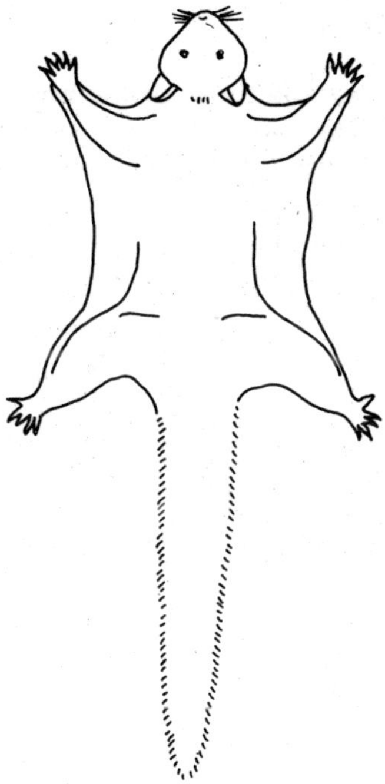

Fig. 18.13: Flying phalanger (marsupial), *Petaurs sciureus.*

Peromys has a highly developed patagium extending as far as the digits. There are also prepatagia and an interfemoral membrane, and the tail is large. The genus is found in the wooded districts of tropical southeastern Asia, Japan, and some of the Malasian Islands, and is said to soar through a distance of nearly 80 yards.

Sciuropterus has no interfemoral membrane, but has a much better developed tail than *Pteromys* as a compensation. This wide hairy tail is further supplemented by the hairy fringe on the patagium and along the rear of the thighs, and they give collectively a broad supporting area. While members of this genus are smaller than those of *Pteromys,* their geographical range is much greater, as it includes the northern part of the North American and Eurasian continents, and India.

Eupetaurus is of especial interest in that it is not arboreal but rock- and precipice-climbing. It inhabits the high elevations of northwestern Kashmir.

Among the *Insectivora, Galeopithecus,* the sole representative of the suborder Dermoptera, stands alone in its adaptation, for it exhibits the highest degree of aviation of any of the Mammalia except the bats, and while not of course ancestral to the latter, it is evidently derived from a common stock and gives a very clear idea of the manner in which the evolution of the bats was accomplished. In this genus the patagium reaches its highest development, as it extends from the rear of the head along the front of the arm (prepatagium), between the fingers to the base of the claws, between the fore and hind limb, webbing the toes as well as the finger, and between the hind limbs and the tail (interfemoral membrane), including the entire length of the latter organ as in the insectivorous bats (Microcheiroptera). The musculature and innervation of the patagium resemble those of the bats and differ decidedly from those of all other volant mammals. The hand is much larger than the foot, but the fingers show no trace of elongation. If they did the entire creature would be still more bat-like. As it is, the brain is midway in its development between that of a typical insectivore and that of a bat, and the alimentary canal is also bat-like except for an elongated colon or large intestine, which in the bats and birds is very short.

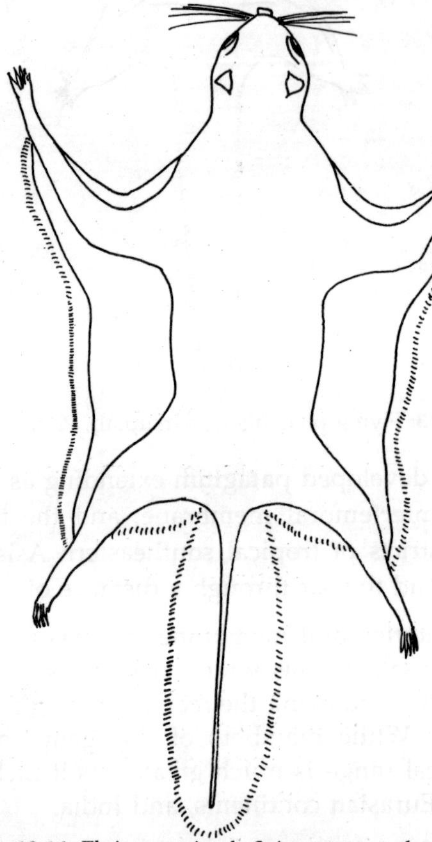

Fig. 18.14: Flying squirrel, *Sciuropterus volucella.*

Galeomithecus is nocturnal, as are most volant mammals, resting suspended, head down, from a branch. Its soaring powers are very great, for Wallace tells us of a record of 70 yards with a descent of not more than 35 or 40 feet or less than one in five. The genus includes two species: *Galeopithecus volans,* from the Malay Peninsula, Sumatra, and Borneo, and *G. philippinensis,* which inhabits the Philippine Islands.

The modifications which the bats have undergone to fit them for their aërial life have already been discussed.

Among *primates,* but one genus, *Propithecus* of the Lemuridæ, exhibits volant adaptation and even here it is not very marked, for the power of flight are quite limited. Lydekker gives 10 yards as the length of the leap, attributing the soaring to

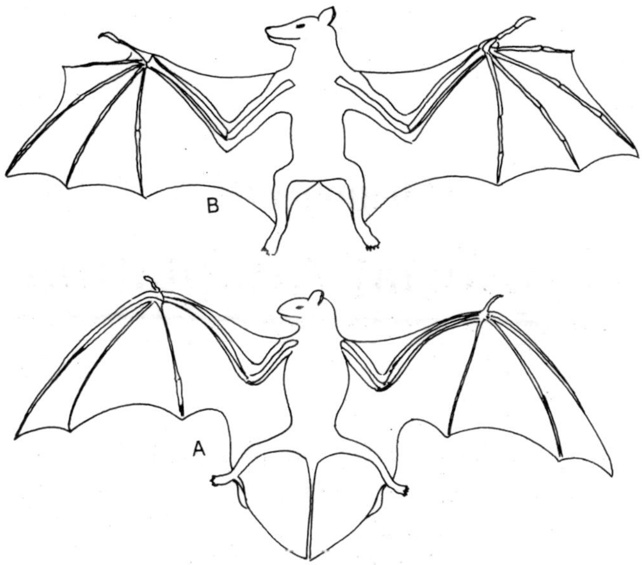

Fig. 18.15: Bats. A, insectivorous, *Vespertilio noctula;* B, frugivorous, *Pterapus* sp.

the powerful hind limbs, without even mentioning the existence of a patagium. Beddard, on the other hand, speaks of the "parachute-like fold of skin between the arms and the body, which suggests a commencement of the more complete parachute of flying foxes, etc." *Propithecus* includes three species, all from the island of Madagascar, and they are diurnal rather than nocturnal as is usual with volant mammals, although they are more active in the morning and evening than at other times.

Chapter—19

Scansorial Adaptation

Climbing on the part of arboreal animals is not necessarily a manifestation of ambition, but quite the reverse, in that relatively feeble creatures may take to the trees for safety and retreat and for abundant and easily procured food. Historically, arboreal life is of more than passing interest, for it is probable that paratically all flying vertebrates, except fishes, were derived from scansorial types and that, on the part of the mammals at least, during the long Age of Reptiles arboreal life was the one factor more than any other that safeguarded the race and rendered its subsequent evolution possible.

The list of climbing animals is very great, but of those which show very marked adaptation to arboreal life the numbers are relatively few. A partial list is as follows:

CLASS PISCES

Climbing perch (*Anabas scandens*).

Mud skipper (*Periophthalinus barbarus*).

CLASS AMPHIBIA

Stegocephalians from the Coal Measures tree trunks.

Tree frogs. Very large group with convergences. Cosmopolitan except for Africa.

CLASS REPTILLA

Lizards, especially geckoes and chameleon.

Tree snakes.

"Proavian"?

CLASS AVES

Passerine birds.

Hoatzin.

Parrots, woodpeckers, wood-hewers, and several instances among the typically terrestrial roders.

Galliformes; cruassows, guans, chachalacas.
Gruiformes: trumpeters.
Anseriformes: tree-ducks, muscovy ducks.
Pelecaniformes: snake birds, cormorants.

CLASS MAMMALIA

Order Marsupilia.
Didelphyidae, opossums; all but *Chironectes*, the water opossum.
Phalangeridae, phalangers.
Macropodidae: *Dendrologus,* the tree-kangaroo.
Dasyuridae: *Dasyurus, Phascologale.*

Order Edentata.
Bradypodidae, tree-sloths.
Myrmecophagidae, anteaters: *Tamandua, Cycloturus.*

Cohort Ungulata.
Hyracoidea: *Dendrohyrax,* the tree-hyrax.
Agriochaerus, an extinct oredont.

Order Carnivora
Felidae, cats: many partially, jaguar only, wholly.
Viverridae, civets, etc.: *Cryptoprocta, Viverra, Arctictis.*
Procyonidae: *Procyon,* the racoon; *Cercoleptes,* the kinkajou; *Nasua,* the coati; *Bassariscus, Bassaricyon.*
Mustelidae: the martens, and *Helictis.*
Ursidae: the brown bears.

Order Rodentia.
Anomaluridae, flying squirrels.
Sciuridae, squirrels.
Lophiomyidae.
Myoxidae, dormice.
Cercolabidae: only the American tree-porcupines (*Erethizon*).

Order Insectivora.
Tupaiidae, pen-tailed shrews.
Erinaceidae: *Gymnura* only.
Galeopithecidae, "flying lemurs."

Order Cheiroptera, bats.

Order Primates.
All but man, the baboons and a few ground monkeys.
Only expected orders: Monotremata. Cetacea, Sirenia.

BIONOMIC CLASSIFICATION

Wall and Rock Climbers

The classification of scansorial animals from the standpoint of their adaptation groups them into three subdivisions, of which the first are the wall and rock climbers. These are not necessarily tree-inhabiting at all, but are, like the gecko lizards, well suited for climbing on the walls of buildings as well as on similar surfaces in nature. The geckoes are, however, a very old and widely distributed group, and the range of their individual adaptation is great, hence it may well be that their scansorial adaptation is after all a response to arboreal life, and that the peculiar structure of their climbing organs rendered subsequent rock-climbing possible. Among mammals there is a genus of flying squirrels limited to high altitudes at Gilgit and perhaps in Thibet, and thought to live on rocks, perhaps among precipices (Beddard). Here, again, we have a form whose ancestry may have been arboreal, but if not, it would afford an interesting instance of volant adaptation without an intermediate arboreal habitat.

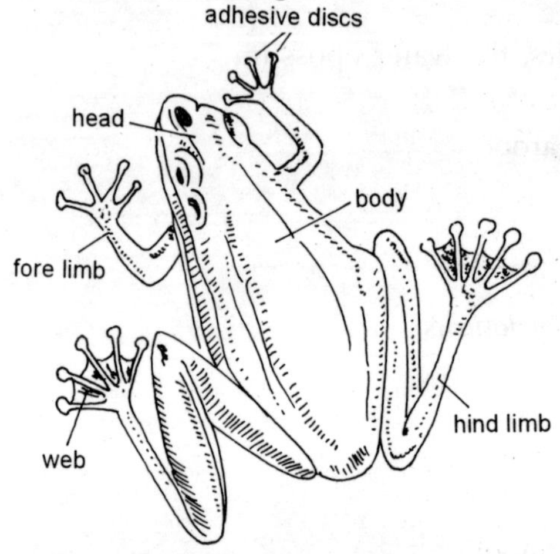

Fig. 19.1: Tree frog *Hyla*.

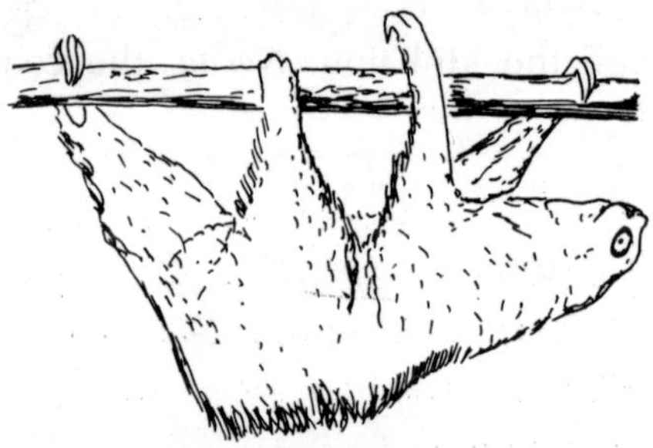

Fig. 19.2: Sloth, *Choloepus*, walking suspended from a branch. (After H. Allen).

Terrestrio-arboreal Forms

The second category, the terrestrio-arboreal, embraces a number of carnivores, rodents and insectivores which while capable of climbing, nevertheless are still perfectly at home upon the ground beneath the trees. They may nest in the trees with more or less extensive terrestrial excursions during the daytime, or they may climb for food and live on the earth unless impelled by hunger. Their climbing adaptations are not very marked.

Arboreal Forms

Still a third group embraces the wholly arboreal types, creatures which make the trees their home, and while some occasionally descend to the ground as in certain primates (gibbon), their terrestrial a progression may be slow and laborious compared with that in their true habitat. Arboreal forms, according to their mode of locomotion, may be grouped in the following subdivisions:

1. *Branch runners,* like the squirrels, marsupials, lemurs, and chameleons, which live and progress on all fours on the upper surface of the branches. The

group embraces, nevertheless, some instances of very perfect arboreal adaptation, as the great majority of tree-dwellers are here included.

2. *Forms suspended beneath branches.* The sloths (Fig. 19.2), for instance, are so constituted that they cannot walk upon the branches but rest and move suspended from them by the powerful recurved claws of all four limbs. Sometimes when quiescent, if a convenient branch lie sufficiently near, the sloth may rest his back on it and relax the hold of one or more of his feet, but the inverted position is rarely reversed. On the ground the animal progresses with the utmost difficulty. The bats should perhaps also be included under this head, as they rest suspended by the hind limbs, head down. The same position of rest is assumed by the so-called flying lemur, *Galeopithecus,* really not a lemur at all but an insectivore.

3. *Forms swinging by the fore limbs* (brachiators, Gr. *βραχίων,* arm). These forms show a very remarkable method of progression by means of the fore limbs, swinging with great speed and accuracy from limb to limb and from tree to tree. The hind limbs are comparable to those of the tree-dwelling marsupials and the creatures rest and progress on the tops of the branches at times, although the fore limbs are almost the sole organs of more rapid locomotion. Here belong many of the primates, more especially the great or manlike apes, and our pre-human ancestors.

MODIFICATIONS

Body

Climbing adaptation, as in the other lines of adaptive radiation, implies, certain body modifications as well as those of limbs. Body contour is of little moment in climbing, but strengthening of chest and ribs and of shoulder and hip girdles is of importance. Nevertheless, in thoroughly arboreal types the section of the thorax anteriorly is subcircular, and the ribs are much curved, in contrast with the compressed thorax and flat anterior ribs of quadrupedal running types (Anthony). The ribs, especially in the sloths, are numerous and afford ampel support to the contained viscera in their invested position. The dorso-lumbar series of vertebrae is often elongated, especially in the tree-sloths of the genus *choloepus,* where the number has apparently been increased from about nineteen (normal for the order) to from twenty-five to twenty-seven as a response to arboreal need. The same is true of other forms, *Capromys,* an arboreal rodent, has twenty-three as compared with the normal nineteen, and *Dendrohyrax,* the only arboreal ungulate now alive, has six more than its terrestrial, hoofed relatives.

Limb Girdles

The shoulder girdle especially is strong in that both elements, the calvicles and scapulae, are well developed, although in terrestrial types the clavicles tend to diminish, even in closely related forms, and may entirely disappear as in cursorial quadrupedal forms. The fore-and-aft swing of the limb of a deer or horse would be distinctly limited by a calvicle, but in a climbing type whose arms are subjected to much more varied and violent strains the calvicle is very essential, as it withstands the compression of the powerful breast muscles. The scapula is also well developed, but not exceptionally so.

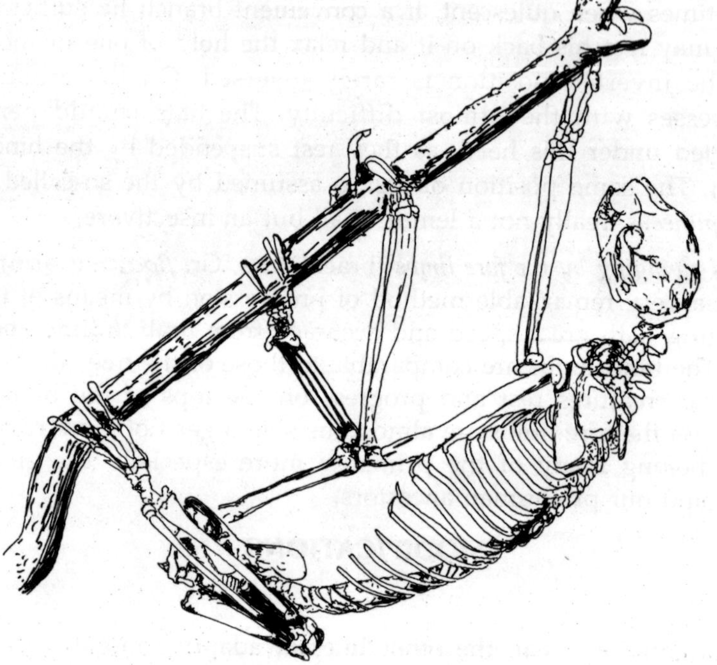

Fig. 19.3: Skeleton of sloth, *Choloepus didactylus.*

Pelvic Girdle

The ilium or hip-bone especially shows modifications in such types as the sloths and primates, as it is broadened out as a support for the visceaa. This is markedly true of the sloths, whose inverted posture necessitates additional support, since the mesenteries or membranes which sling the intestine to the dorsal wall lose much of their efficiency when the body is erect or inverted.

Limbs

In contrast with the cursorial types, it is the proximal limb segments which now elongate, especially in the suspended and brachiating forms, those of the sloths again being very long, while in the great apes the relative length bears a direct ratio to the creature's climbing powers, reaching the extreme in the gibbons (*Hylobates*) whose arms are so long that the knuckels of the hand touch the ground when the animal stands erect. The progress of the gibbon from branch to branch and tree to tree is remarkable. Climbing forms are generally plantigrade, some of the raccoons secondarily so. In certain lemurs (*Tarsius,* see; *Galago*) the tarsus may be elongate, but this is probably due to the fact that the creatures leap as well as climb, and the elongation of this segment is a response to the former need rather than to the latter.

Feet

The feet of arboreal animals may be either prehensile, that is, grasping, with more or less opposable digits, or non-prehensile. In the *non-prehensile type* the claws may be well developed as in the squirrels or the cats, giving a fairly tenacious hold. In the Canada tree-porcupine (*Erethizon*) the plantigrade feet are armed with long curved claws, in addition to which the soles bear spines and tubercles which aid in climbing.

Adhesive pads either on the tips of the digits or on the soles of the feet occur in several isolated instances, such as the tree-frogs, geckoes, and *Dendrophyrax* among mammals. The frogs are aided by a sticky secretion of their pads. "Tree-frogs, when hopping on to a vertical plane of clean glass, slide down a little, probably until the secretion stiffens, or dreis into greater consistency. . . . Wet leaves or moist glass-walls afford no hold. The adhesion of these frogs is assisted in most cases by their soft and moist bellies, just as a dead frog will stick to a pane of glass" (Gadow).

The geckoes, by means of their adhesive digits, climb up absolutely smooth and vertical surfaces, or, back downward, along a whitewashed ceiling. The apparatus, Gadow says, is complicated in its minute detail, but very simple in principle. The adhesion is effected not by sticky matter, but by small and numerous vacua.

Dendrophyrax, the tree-hyrax, is allied to the coney of Scripture. The tree-hyraxes frequent the trunk and larger branches of trees, sleeping in holes high up in the big trees, especially, according to Roosevelt's observations, the cedars. The adhesive organs have been described by G. E. Dobson, who says that these animals are enabled to climb perpendicular walls and trees without the use of claws. The thickly padded tuberculated soles are drawn up by certain flexor muscles, thus leaving a partial vacuum by means of which the animal retains its hold.

The primitive type of prehensile foot has been developed in the two great mammalian groups, that of the marsupial being represented by the opossum (*Marmosa*), and that of the early placentals by the credonts, the archiac flesh-eating mammals the foot of which has been shown to be a terrestrial modification of a grasping type.

Feet of the *prehensile type* are found to-day in the marsupials and primates. In the former group it is the hallux or great toe which is offset so as to oppose the fourth digit, the second and third being bound together in a common integument (syndactyly,) and so slender that their combined strength about equals that of the outermost or fifth digit. In marsupials which have become terrestrial the offset great toe has become vestigial or may entirely have disappeared, as in the kangaroos. In the primates, while the foot is perhaps most apt to show this opposable first digit, it also exists in the hand, although it is nowhere developed to the degree shown in mankind, wherein the final perfection of the hand as an organ of prehension has developed since its release from the necessity of arboreal locomotion.

Syndactyly (Gr. συν, together, and *δάκτυλος*, digit) has already been referred to as occurring in the marsupials, and even such as are no longer tree-inhabiting, like the kangaroo, still exhibit this feature in unreduced condition. It doubtless arose primarily, however, as an arboreal adaptation. The koala shows a rather remarkable modification for climbing, for the foot has a long, widely offset, clawless great toe, syndactylous second and third toes, which are clawed, and powerful clawed fourth and fifth toes, the former being the longer. The hand, on the contrary, has five equal digits, all of which bear sharp claws; but two digits, numbers 1 and 2, oppose the other three. Its clinging powers are so great that even death will not dislodge the creature from the tree in which it is shot.

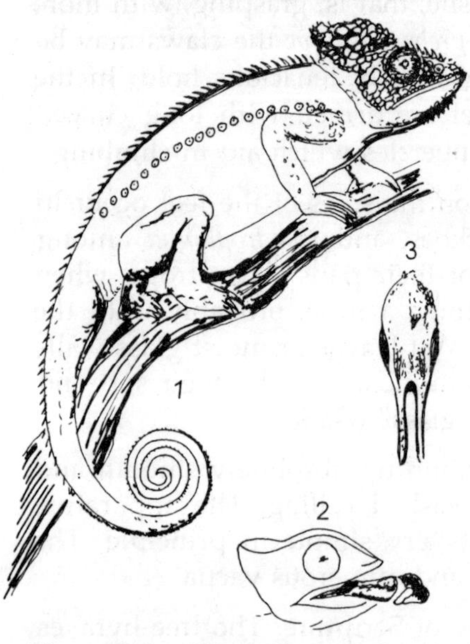

Fig. 19.4: Chameleon, showing syndactylous hands and feet, and prehensile tail. 2, head with partly protruded tongue; 3, head of horned chameleon, from above.

Among reptiles, the true African chameleons (Chamaelon) exhibit remarkable syndactyly, as it extends to both fore and hind feet. On the hand the first three fingers form the inner bundle and are opposed to the outer two which are likewise syndactylously bound. The foot is similar but reversed, in that the inner bundle contains two, the outer one three digits. These very admirable grasping organs are supplemented by a prehensile tail, so that the creature is very firmly anchored in position, which is rendered necessary perhaps in part by it is method of securing insect prey by the unerring aim of the enormously extensive tongue.

In the so-called scansorial birds such as the parrots, woodpeckers, and the like, the outermost toes has been rotated backward in such a way that it and the hallux oppose the second and third toes, the fifth, as in all birds, being absent. This gives a very firm grasp for the actual grip of a branch as in the parrots or, reinforced by strong claws, enables the animal to cling to the roughened bark of a tree trunk. In the parrots, woodpeckers, and cukoos the rotation of the outer toe is permanent and the foot is called zygodactylous (Gr. *ζυγόν*, yoke); certain others, owls, etc., may turn it backward or not at will.

While aboreal forms usually have need of all of their digits, occasionally one sees *digital reduction*. The foot of the koala, with syndactylous second and third toes, functions as four-toed, even though consisting structurally of five; certain of the primates (lemurs), on the other hand, some of which resemble the koala superficially very much, have actually lost the second digit so that the opposability of the first in grasping a limb is unimpeded. In the lemur (potto, etc.), the fourth digit is the largest

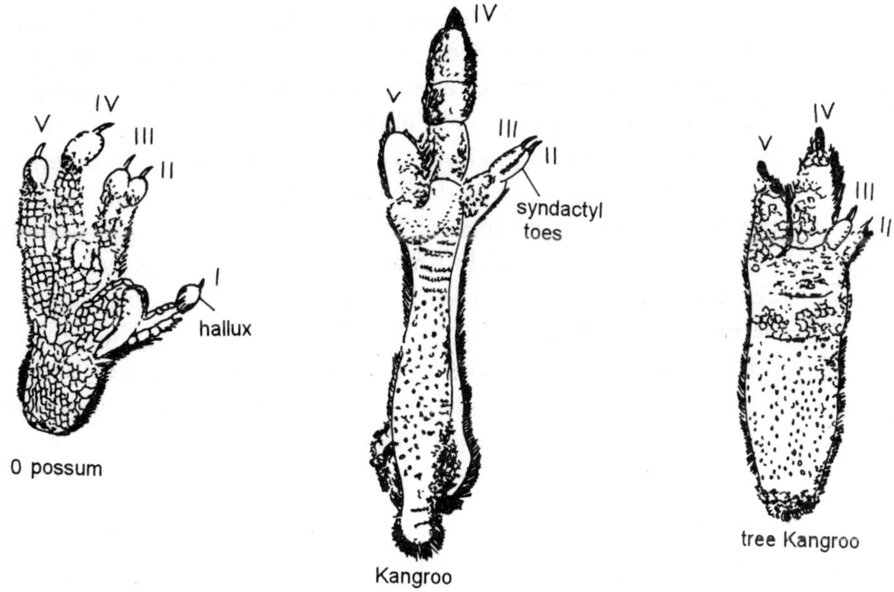

Fig. 14.5: Hind feet of marsupials.

as in the koala. Digital reduction is also seen in the tree-sloths, the two-toed sloth *Choloepus* having but two in the hand and three in the foot, while in the three-toed sloth *Bradypus* there are three in each, and the hand and foot are both somewhat elongated, especially in the powerful hook-like claws which, like the feet of the koala, retain their grip on the bough even after the animal has been shot.

Tail

The tail may be prehensile or not as in the case of the feet. If non-prehensile, there are ectodermal spines or scales on the under side, as in the flying squirrel *Anomalurus,* which prevent the animal from slipping down. The same effect is produced in the woodpecker by stiff spiny feathers which are braced against the tree trunk to which the creature clings. The posture is familiar, and enables the bird to drill into the wood for the grubs upon which it feeds, or to excavate cavities for its nest of for the storage of food.

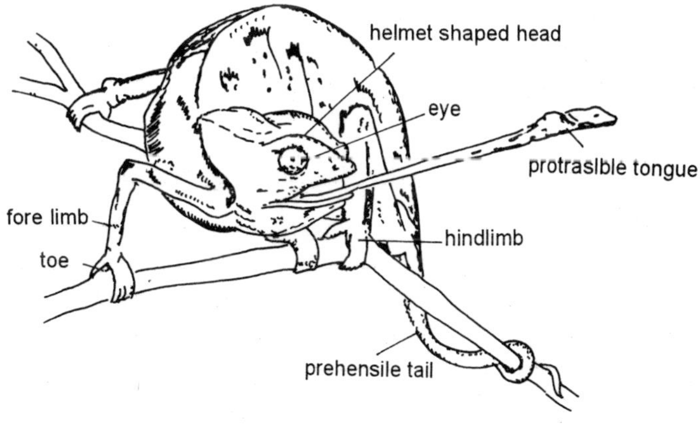

Fig. 14.6: *Chameleon.*

Fig. 19.7: Foot of a woodpecker, *Picus, viridicanus* showing fourth toe reversed for grasping.

Prehensile tails are found in a number of unrelated instances, as, for example, the chameleon lizards which have been mentioned, the opossums, the tamandua which is one of the anteaters, and certain of the New World monkeys (Cebidae) such as the spider monkey, the howlers, and the capuchins. Where the prehensile powers are well developed the tail is naked on the under surface near the tip. One of the most perfectly adapted of these forms is the spider monkey, *Ateles,* in which the tail is highly prehensile and functions as a "fifth hand." Perhaps as a correlation with this excellent grasping organ the real hands have lost the thumb, but the four long digits which remain form a splendid hooklike device for suspending the body. Not all South American monkeys have a prehensile tail; on the other hand, none of the Old World forms do, so that its presence is diagnostic of a New World ape.

Other Accessory Organs

Other accessory climbing organs might be mentioned, such as the parrot's beak and the spines and tubercles which are sometimes developed on the fore arm in certain lemurs (*Hapalemur griseus* male and upon the lowered of the ankle in *Galago garnetti*). In *Lemur catta* there is a patch of hardened skin on the fore arm which projects to a large extent and has been called a climbing organ, although it lacks the recurved spines; both this and the spiny patches of *Hapalemur* and *Galago* have glands connected with them, the function of which is doubtful.

Chapter—20

Geographic Distribution

The distribution of life over the earth's surface is world-wide, as no place is so forbidding as to be entirely without its flora or fauna; the dens tropical forest, the bleakest mountain, the scorching heat and drought of the desert, the devastating cold of the polar regions: each has its quota of inhabitants, living out their lives as best they may.

Closer study, however, reveals the fact that the distribution of life is by no means a uniform one, and, aside from differences in faunas, due to climatic or other causes, there are peculiar instances of isolated distribution. Thus, for example, Africa has elephants, antelopes, and great apes such as the gorilla and chimpanzee in its fama, while Brazil, with very similar environmental conditions, has none of these, but instead possesses the tapir, sloths, and prehensile-tailed monkeys. The tapirs themselves are found in Central and South American and again in the Malay Peninsula and Sumatra, a curious instance of discontinuous distribution, explicable only by assuming that the American and Malay tapirs are the last survivors of a widespread race, whose intervening representatives have been blotted out.

Again, it is seen that the faunæ of Great Britain and Japan, which are separated by thousands of miles, are very similar, whereas in Bali and Lombok, two small islands in the Malay Archipelago, separated by a strait only 15 miles wide, the animals are much more unlike.

Thus it is evident that the dispersal and distribution of animals is governed by laws which are far from simple.

Life occurs in almost all the habitats ranging from high mountains to the deepest sea bottom. One would find a series of contrasting conditions which of necessity profoundly affect the organism.

1. Distribution is Space

It is of two types:

(i) *Geographical Distribution.* It deals with horizontal or surfacial distribution of animals on the land and fresh water in different continents and on islands. The study of geographical distribution is called *Zoogeography*.

(ii) *Bathymetric Distribution*. It deals with the vertical or altitudinal distribution of animals on land and in water. The bathymetric distribution is further divided into:

(a) *Geobiotic or terrestrial.*

(b) *Limnobiotic or fresh water inhibiting.*

(c) *Holobiotic or marine water inhibting.*

2. Distribution Time

It is *Geological distribution* or the *durational distribution* of animals from the very beginning of the life on the earth upto recent time.

Pattern of Distribution

The distribution of life over the surface of earth is world wide. These are no such place where the living organisms are not found. Inspite of that the distribution of animals and plants is not uniform for example a particular species of animals is present in one area but entirely absent form others. Three patterns have been observed regarding distribution of animals.

1. Cosmopolitan Distribution
2. Bipolar Distribution
3. Discontinuous Distribution

1. Cosmopolitan Distribution

The animals which are found all over the world are said to have cosmopolitan distribution. *Mytilus, Artemia* (shrimp), bats, rats, hawks, cockroaches etc. are world wide in distribution. These animals have a wide range of adaptations to a wide variety of environmental conditions.

There are two types of cosmopolitan animals (a) *Eurytopic*. When they have wide range and uniform distribution and (b) *Stenotopic,* when they have restricted range of distribution.

2. Bipolar Distribution

In bipolar distribution the animals are only confined to the polar regions i.e. only in the waters of arctica and antarctica. A few examples are *Lampta, Grammaria, Sardina* etc. (Phylum coelenterata); *Limacina* (Mollusca); *Oncorhynchus,* (Pacific Salmon) and *Lamma cornubica* (shark).

3. Discontinuous Distribution

Some of the present day animals which are closely related are found distributed in widely separated areas. Some examples will illustrate the discontinuous distribution.

Marsupials are found in Australia, and South America. In Australia there are several kinds of marsupials like Kangaroos, Wombets, Koalas and Bandicoots. Only the oppossum is found in the eastern states of America and a few marsupials in the South America. Another example is that of Tigers. The Tigers are found only in India and neighbouring regions like Nepal. Lions are only found in India and Africa. Two species of Alligators are found one in South-Eastern United States and the other in China. The lung-fishes (Dipnoi) are represented by only three living genera, *Neoceratodus* in Australia, *Protopterus* in Africa and *Lepidosiren* in South America.

Another example of discontinuous distribution is the camel. They are present in parts of Africa, Asia and South America. The Llama found in Andes. Different species of Tapirs are found in Malaya, Central and South America and Islands of Jawa and Sumatra.

Many more examples are there for the discontinuous distribution. It can be explained if we understand the geological data and animal evolution.

From the Palaeozoic to the Mesozoic era, there were only two major land masses Gondwana and Laurasia and these were in contact at times. Gondawana centered around the South Pole while Laurasia overlapped the equator and extended well into the northern hemisphere. During the Cretaceous, these masses fragmented to form the present continents and these have since drifted apart very slowly toward their present positions. Arabia and India. Laurasia broke up into North America and Eurasia.

A species originates in a definite more or less restricted area and then tends to spread in all directions occupying suitable habitats. When an impassable barrier in encountered, the migration of the species stops unless it is removed, by geologic or climatic changes. Sometimes *bridges* are established and the animals are able to cross

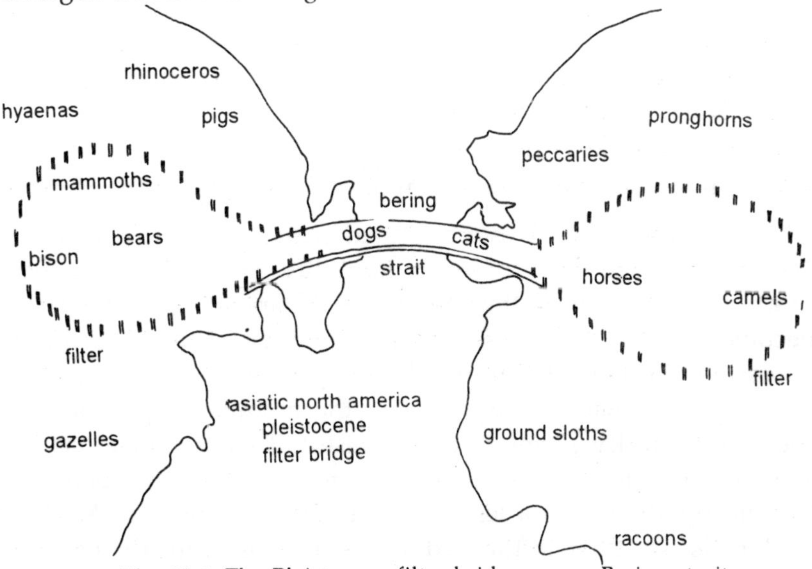

Fig. 20.1: The Pleistocene filter bridge across Bering strait.

the barriers. Three type of bridges have been recognised. (1) A *corridor bridge* is a broadly continuous connection as exists between Europe and Asia (2) A *filter bridge* is more temporary in duration as is the Bering strait which connects Asia with North America (3) A *Sweepstakes bridge* is the one which depends upon accidental transportation on floating ice, log of wood clinging mud of the birds claw or even on board the ships.

It is in connection with this type of history of the earth and animals that the problem of widely discontinuous distribution must be understood. Marsupials, Ratitae birds and lung fishes had a world wide distribution to begin with. Later on by substance of land and continental drift, they became isolated. Where the competition with superior animals was severe, they have been eliminated, in other places they survive till the present day. The barrier of deep sea between Bali and Lombock has existed since Palæozoic times and it has never been bridged. This is the reason for the distinctness of fauna in the two islands.

BARRIERS TO DISPERSAL

Topographic Barriers

Such barriers as high and extensive *mountain ranges* limit the distribution of many terrestrial forms, and in general are more effective if the mountains are more or less parallel with the equator, as in Europe and Asia. Here we find a marked difference between the species occupying the northern and those occupying the southern slopes. This is notably true of the great Himalaya Range in northern India, which rears its mighty summits far beyond the limits of perpetual snow. On the south we have the hot, moist plains of India, with a very distinct tropical fauna which in many respects resembles that of Africa. North of the barrier, conditions of climate, both in temperature and degree of moisture, are entirely changed, and with them appear animals, with some notable exceptions, of a totally different sort, more nearly comparable to those of Europe. In the New World where the mountain chains in general run north and south, their influence upon animal distribution is vastly less, and in no case do they form striking boundaries between zoögeographical realms, as do the Himalayas.

The mountain ranges in the western United States do exert a certain influence, however, principally through their control of humidity. The winds from the Pacific are often laden with moisture, but as they reach the mountain uplift they are deflected higher and become cooler, which causes the precipitation of the moisture in the form of rain. On crossing the barrier, the winds are practically moistureless and, as a consequence, aridity of climate prevails east of the mountains, producing conditions ranging all the way from dry plains to actual desert. This of course has its profound influence upon the character of the vegetation and consequently upon plant-feeding animals and indirectly, though to a less extent, upon carnivores. In North and South America the animals are thus influenced in their range directly by conditions of temperature and rainfall.

One important topographic barrier in North America is that which limits the Mexican plateau and again forms a line of demarcation between faunal realms. The plateau itself, though extending well into the tropics, has a temperate climate, and the fauna is similar to that of the region to the north and northwest. From the edge of the plateau the land drops away abruptly and conditions rapidly change to those of a hot, moist, tropical region, with a corresponding change in the plant and animal life, which is now that of tropical Central and South America.

Climatic Barriers

While *degree of heat* is of prime importance in limiting animal dispersal, Heilprin thinks that its influence has been largely overestimated. Notable instances in proof of his contention are the tiger and elephants. The former, while its normal home is in the host districts of India and the Indian archipelago, is not restricted to those regions nor to such as have similar climates, for it is found "in the elevated regions of the Caucasus and the Altai chain, and in the Himalaya range its footprints are not infrequently found impressed in the fields of snow. It is a permanent inhabitant of the cold plains of Manchuria and the Amoor region... prowling about even in winter along the icy margins of the Aral sea." The northern range of the tiger extends to about 53° north latitude, as far north as Lake Baikal in Siberia or Lake Winnipeg in Canada. Elephants likewise do not seem to suffer from cold, provided a sufficient amount of water be obtainable. Hannibal, who invaded Italy in 218 B.C., took a nerd of war-elephants, presumably African, over the Little St. Bernard Pass, which has an altitude of 7176 feet. Elephants also range in altitude in the Kilimanjaro region of eastern Africa from sea-level up to 13,000 feet, which must give a marked temperature range; the Indian elephant appears to be equally at home among the cool mountain heights and the hot and jungly lowlands.

The influence of temperature is much more marked in limiting the distribution of cold-blooded animals or those which depend upon external temperatures for the maintenance of their bodily heat, than in the warm-blooded forms. Thus the amphibians and reptiles are tropical and temperate in their distribution, rapidly diminishing in numbers toward either pole. Salamanders are limited to about the sixty-third parallel in Europe, while frogs and toads are absent only from the higher latitudes north and south.

Reptiles, on the other hand, are more restricted; the crocodiles are tropical or subtropical in their distribution, the northern limit of the turtles is 50°, while of the lizards there are few beyond 40°, and at 60° the order entirely disappears. Snakes have the widest range, but only three species in Europe are found beyond 55° and but one, the common viper, extends to the Arctic Circle. The terrestrial vertebrates of the frigid zones are therefore entirely birds and mammals, whose constant temperature and efficient bodily clothing enable them to maintain an active existence where the reptiles and amphibians would be reduced to a condition of torpor. This torpor can be endured for a while and the period of hibernation or winter sleep on the part of temperate cold-blooded animals is a normal manifestation, but it has its limits and can not extend over too great a sum total of the animal's life. The presence of fossil reptiles or amphibia in the rocks of a given geologic period is, therefore, highly indicative of past climatic conditions.

Lock of moisture controls the distribution of animals and plants principally where it is of sufficient extent to produce desert conditions, for while many forms are marvellously adapted to withstand the harshest drought, for others it forms an insuperable barrier.

The most notable desert barrier is that of the Sahara, which forms the area of demarcation between the Ethiopian or African zoögeographical realm and the Eurasian. So efficient is it that such mobile forms as the deer, which are found in the Americas from the Straits of Magellan to the Arctic and over the entire Eurasian continent except the Arabian peninsula and the adjacent area to the east, tare totally absent from Africa except in its extreme northern part, from the Straits of Gibraltar to Tripoli. Of course they are absent from the long isolated Australian region except where they have been introduced by human agency. The Sahara in Africa and the Arabian desert are impassable to such forms. Animals like the Amphibia, which require moisture for their larval life and have no great migratory powers, find even a small arid area an insuperable obstacle.

The *increase of moisture* renders a region unsuitable for certain forms of life, in part due to possible swampy conditions which may make the area impassable to creatures not adapted to them. Swamp-dwelling generally places its mark upon the animal, as in the splay feet of certain Permian reptiles. It is doubtful, however, whether even extended morasses act as a very great barrier to dispersal. Increased humidity does have a marked effect secondarily in affecting the vegetation, the spread of insect life, and the like, to be described below.

Vegetative Barriers

The profusion of vegetation depends very largely upon the climatic conditions mentioned above—temperature and degree of moisture—and its prevalence is favourable to the dissemination of certain types and unfavourable to others. The influence of vegetation is both direct and indirect—direct in the impossibility of forest-living, especially arboreal animals crossing regions where forestation does not occur; and in forming, especially in the tropics, so dense a growth of jungle that larger terrestrial animals are incapable of penetrating it. Thus, in Pleistocene times mastodons of the genus *Dibelodon* migrated from North into South America over the newly established isthmian land bridge, but during the Pleistocene, while there were several species of magnificent elephants (*Elephas* and *Mastodon*) in North America, none of them apparently succeeded in penetrating south of the southern limits of the Mexican plateau. The one possible recorded exception is that apparently of a true elephant

(*Elephas columbi*), a tooth of which has been reported from the Lower Pleistocene of Cayenne, French Guiana. What the conditions were which prohibited the passage of the later elephants into South America, while the earlier ones effected a crossing, we do not surely know, but he presence of a tropical jungle too dense to be penetrated by such huge beasts as these is highly suggestive, for at the present day the vegetation is said to be impenetrable to an animal as large as a man, without mechanical aid.

The lack of vegetation very effectively limits the distribution of certain forms, notably the primates, which with few exceptions are dwellers in tropical forests. Not only do they need trees for their safety and usual mode of progression, but also as a source of food, for the primates as an order live upon fruit, nuts, and blossoms, as well as upon insects and small birds, all of which are products of these forests. Primates were very prevalent during the Eocene in western North America. At the close of the Eocene period, however, they became utterly extinct in North America, nor is there the slightest evidence of their reappearance until the coming of man. The Eocene, especially during the middle part, when the primates were most abundant, was a time of subtropical or warm-temperate conditions, as the rocks contain palms and bananas mingled with many similar mild temperate trees. With the ushering in of the next period, however, that of the Oligocene, and more especially in the Miocene, the tropical forests gave way to those consisting largely of deciduous trees which could not support a primate population all the year, hence their local extinction.

This serves to illustrate not only the means whereby the absence of vegetation may limit dispersal, but also the way in which *food supply limits distribution*. Other illustrations would be the necessity for trees and shrubbery for animals with short-crowned browsing teeth, on the one hand, and of extensive pasturage of harsh grasses for those with long-crowned grazing teeth, on the other, each type of vegetation being unadapted to the other type of animal. Vegetation is important also in the case of many insects; caterpillars, for instance, may be addicted to a certain species of food plant and be coextensive with it in distribution and unable to exist where the plant does not occur. Conversely, some plants are absolutely dependent upon certain species of insects for pollination, without which they can not propagate their seeds. Parasitic insects which depend in turn upon the plant-feeders can not of course exist where the latter do not occur, for many parasites are limited exclusively to a certain genus and species of host, and this again may be restricted to a very small group of plants. Insectivorous birds may be in a measure dependent on certain specific insects, the limits of whose distribution determine theirs, while frugivorous birds, or those which, like the humming-birds, sip nectar from the flowers, are directly limited by the existence of certain types of vegetation. "Thus the whole fauna and flora of a district is bound together by a complicated network of particular conditions, and the slightest alteration in any detail may upset the balance of the whole and lead to far-reaching and unforeseen results" (Bartholomew).

Large Bodies of Water

Extensive bodies of water, when not frozen, form perhaps the most insuperable barriers of all, especially to terrestrial vertebrates like the amphibians, reptiles, and mammals, but not to those which have powers of sustained flight, such as birds and bats. Fresh-water fishes are also prevented from migration by large bodies of salt water, although certain fishes such as the shad, salmon, sturgeon, and smelt—the anadromous (Gr. *ἀνὰ*, up, and *ὄραμεῖν*, to run) fishes—pass from salt to fresh water annually, while in the catadromous (Gr. *κατὰ*, down) eels the reverse migration occurs. The barrier is effective, therefore, only in the case of exclusively fresh-water types, such as the carps, garpikes, catfishes, bowfin, and brook trout, some of which are quite locally restricted. To the modern Amphibia salt water constitutes a most effective barrier as common salt is poisonous to the amphibia, even a solution of 1 per cent preventing the development of their larvæ. Undoubtedly, however, many individuals have made long voyages across the seas on floating timber. Solutions of lime are likewise detrimental to many species, and as a rule limestone-terrain is poor in amphibian life; *Salamandra maculosa,* for instance, is absent in Central Germany on the Muschelkalk, but occurs in abundance in neighboring districts of sandstone or granite (Gadow).

Amphibia are, as one would infer, almost completely absent from oceanic islands, the Seychelles, New Caledonia, and the Fiji and Solomon islands forming island groups exceptional to a general rule. Another remarkable fact is the very nearly universal limitation of the tailed forms—sirens, newts, salamanders, etc.—to the northern hemishpere, for the southern land masses, Australia, Africa, and South America, are almost if not entirely isolated by the oceans, and the small migratory roads left open, which have proved sufficient for such creatures as the mammals, are inadequate for these inactive amphibians. The burrowing cæcilians, on the other hand, are confined to South America and Africa, with the exception of the Seychelles, southern India, and the East Indian islands of Sumatra, Java, and Borneo, all of which are supposed to be relics of an ancient southern continent, for whose existence prior to the Tertiary period there is much evidence. The Anura (frogs and toads) enjoy a much broader distribution than the tailed forms, being barred in their over-seas migration only from the most remote of oceanic islands. Their methods of dispersal will be discussed later.

For some reptiles, notably the Crocodilia and marine turtles, the seas of course afford no obstacle, the only practical limit to their distribution being that imposed by temperature. Among the most interesting of living reptiles are the giant tortoises, some of which have been in captivity for upwards of a hundred and fifty years. They are confined to-day entirely to certain oceanic islands—the Galapagos Islands off the coast of Ecuador (*galapago* being one of the Spanish terms for tortoise), and the islands of the western Indian Ocean, namely, the Mascarenes (Réunion, Mauritius, and Rodriguez), the Comoros, the Aldabras, the Amirantes, and the Seychelles. On the other hand, these tortoises are totally extinct on the mainland of South America, Africa, and Eurasia. Their shells are common as fossils, however, in India, Europe, and North America, in rocks of Miocene to Pleistocene age. As land tortoises are drowned within a few hours if they attempt to swim, their distribution could not have been accomplished by any overseas journey, but must indicate again former land connections over which they could travel. Their present distribution is also within the limits of the ancient Gondwana land mentioned above.

The serpents, many of which are fairly good swimmers, are incapable of passing large bodies of salt water of their own volition, with the exception of the specially adapted sea-snakes, which are found in the tropical seas, sometimes several hundreds of miles from land. Lizards in their adult condition are as incapable of passing an oceanic barrier as are the snakes, but, as Heilprin remarks, "it would appear that in some special way—whether as effected by the oceanic currents themselves or through the agency of birds—their eggs may be transported to very considerable distances out to sea, since this order of animals is sufficiently represented in remote islands where neither snakes nor amphibians have as yet been encountered."

With the birds, it is only the flightless forms, such as the ostrich, rhea, and cassowary, or the apteryx of New Zealand, which are debarred from trans-oceanic migrations, for even small birds are carried far on favouring gales.

With mammals, except those which, like the whales and seals, are especially adapted to marine life, bodies of water of a greater expanse than 20 to 50 miles are impassable when not frozen. Many mammals are, however, excellent swimmers, the jaguar being known to cross the broadest of the South American rivers, while the tiger and elephant, as well as the deer, will all take to the water freely; but it is doubtful whether any of them will venture out of sight of land. Heilprin summarizes as follows: although "it may safely be conceded, from our present knowledge on the subject, that while many of the land Mammalia can effect with safety and even readiness, such water passages as are most generally to be met with on continental areas, none, probably, would be prompted to undertake a journey across an arm of the sea whose width measured 50 or more miles, or even one much exceeding half that extent. To these difficulties or impossibilities in the way of dispersion must be attributed the circumstance that the vast number of oceanic islands are deficient, except where man has effected an introduction, in representatives of this particular class of animals." The finding of allied or nearly identical forms of mammals upon land masses now isolated generally points to former land connections where none now exist.

Just as bodies of water prevent the passage of land animals, so *land masses* form barriers to the spread of sea life. Such, for instance, is Cape Cod, which separates the relatively cold waters of Massachusetts Bay from the ocean to the south with a very marked distinction in the contained faunas. The Isthmus of Panama is another case in point, while certain relic seas like the Caspian contain seals, porpoises, and other marine vertebrates whose forefathers entered it when it was an arm of the sea; their descendants being in turn cut off from the ocean with the severance of communication. A species of shark in Lake Nicaragua is another such zoölogical exile.

Impurity and Lack of Salinity of Sea-Water

These afford an effective barrier to the dispersal of certain kinds of marine animals, such as the brachiopods, echinoids, crinoids, starfishes, squids, and forminifers, but notably the corals and sponges. While there are fresh-water sponges, they form a very small proportion of the entire number which are extant, and none have skeletal structures. On the other hand, the corals are exclusively marine although a few are known in brackish and even in tolerably fresh water. These exceptions, however, are notable, for as a rule all such forms require a water of maximum purity and salinity. Hence it is that extensive coral reefs or sponge-growing areas are never found near the mouths of large rivers such as the Amazon, Orinoco, or Mississippi, each of which bears a great load of sediment seaward. The Great Barrier Reef of Australia extends along the eastern shore of that continent for a distance of 1500 miles, and along the entire adjacent coast the map fails to reveal a single river of importance, although there are a few relatively small streams.

DISPERSAL OF ANIMALS

Necessity

The struggle for existence brought abut by the rapid rate of multiplication of all animals and the consequent scarcity of food within a given area renders the dispersal of animals imperative. Not only is this struggle between members of the same species, but between allied species or any sorts of animals whose needs are sufficiently similar to induce competition. Again, gradually changing climatic conditions, which render the old home no longer suitable, impel migration where migratory roads exist, otherwise gradual extinction is often the result. A study of the distribution of fossil animals such as the elephants or horses gives evidence of repeated and world-wide migrations which probably occurred as often as opportunity arose. The extension of geographicai range seems, therefore, to be a prime necessity in the great majority of organisms.

MEANS OF DISPERSAL

Animals and plants are so widespread over the surface of the globe that migratory routes must be many, despite the effective barriers which have been mentioned. The means of dispersal are:

Land Bridges

Land bridges, such as those at Suez and Panama, both of which are of comparatively recent origin, geologically speaking, have nevertheless afforded the means whereby continents received their faunas. The Panamanian bridge is of particular interest because we know the history of its existence so well. There was evidently land connection between North and South America at least up to late Eocene time, although it has not always have been where the Isthmus of Panama now lies. While that connection existed there was free intermigration between the continents, but during a long period of severance—until Middle Miocene time—no interchange of species was possible, and each fauna underwent a remarkable evolution, without the admixture of forms from the other. The re-formation of the land bridge, however, opened up the avenues of migration, and we find recorded the incursion of the more virile North American types into South America, such as the mastodons, horses, deer, wolves, and cats, and the immigration of South American forms, the ground-sloths, armadillos, and other species in exchange. Bering Strait, although now impassable except over the ice in winter, has been at various times in the geologic past a migratory route of the utmost significance, so important in fact that the land animals of North America and those of the Eurasian continent show the closest blood relationship, and this was especially true during Tertiary time. In fact, as Matthew says:

"The Alaskan bridge is in existence to-day, only a few yards of its planking removed, if one my so speak, the substructure intact, and the marks of the missing planks still showing on the undamaged portion." Actually the intervening water between America and Asia has a width of only 36 miles and a depth of from 23 to 30 fathoms, so that the comparatively slight elevation of 200 feet would afford dry and safe transit for the passing and re-passing hosts.

Natural Rafts and Driftwood

Terrestrial animals, either accidentally or with intent, occasionally take passage upon drifting material which enables them to accomplish over-water journeys which are sometimes of considerable extent. Many Arctic animals, such as the reindeer, will go out upon the ice to effect a crossing like that over Bering Strait, the Nova Zembla reindeer occasionally going to Spitzbergen, a distance of 240 miles; or such forms as the polar bear may venture out upon the shore ice for the seals or fish which constitute their prey. These bears are splendid swimmers, for one was observed by Captain Parry in Barrows Strait 20 miles from shore, with no ice in sight. Even such natatorial powers would not, however, take them to Iceland, and yet living bears are stranded on the Icelandic shores every year, no fewer than twelve having arrived in one season. In such instances, the journey was probably very largely accomplished by means of floating ice floes which had broken away from their moorings.

Natural rafts of vegetation are a very potent factor in dispersal. These are masses of driftwood and leaves held together by a tangle of vines, creepers, and other vegetation, and sometimes having a covering of soil sufficient to maintain living plants or even trees. Such rafts are formed by the natural accumulations of timber along the rivers, caused by the caving of the banks on the outside of the river bends, and as larger masses are formed they are swept out to sea by the action of the stream. Such natural rafts have been reported several times more than a hundred miles beyond the mouths of some of the world's great rivers and in one recorded instance the distance travelled was over a thousand miles! Heilprin tells us that floating masses of wood, with upright trees growing over them, were mistaken by Admiral Smyth in the

Philippine seas for true islands, until their motion made their real nature apparent. Such rafts often have quite an assemblage of animal life, such as monkeys, tiger-cats, squirrels, and arboreal mammals in general, together with reptiles and molluscs.

A boa, a huge species of snake, was transported 200 miles on a cedar tree to St. Vincent, and Heilprin records the landing of four pumas in a single night in the town of Montevideo. He goes on to say: "To what distance such a floating raft with its living cargo may ultimately be carried in safety, and without detriment to its inhabitants, over the oceanic surface there are as yet no data for determining. But there would appear to be no reason for assuming that they could not be transported to a distance of several hundreds of miles, seeing that the upright vegetation found on many of them would serve with powerful effect in the face of a wind. And while the majority of the animal inhabitants might be exterminated before the end of the voyage, the safe arrival on an island or distant shore of a very limited number of individuals, embracing both males and females, would serve in a short period, under favourable conditions, to stock the new land with the species. That an absolute limit is set, however, to migration as effected in this manner is proved conclusively by the utter absence in most of the oceanic islands of indigenous mammals, excepting bats."

Favouring Gales

The forms aided by this means are the aërial nekton—insects, birds, and bats—and while all three can progress without the aid of favouring winds their flight is very largely increased thereby. This aid is perhaps largely accidental, nevertheless migratory birds are known to take deliberate advantage of it.

Insects present numerous instances of great powers of sustained flight when aided by air currents, of which the most conclusive are naturally those recorded at sea. The ship *Pleione,* on reaching a point some 960 miles south-west of the Cape Verde Islands, came upon hundreds of moths belonging to a species which is common in the Eastern Tropics, but is not found in South America, the nearest land. From the direction of the wind during the previous four days, it appeared to be beyond doubt that the insects had come from the Cape Verdes, and therefore must have crossed nearly a thousand miles of ocean! The nuptial flights of many insects often aid, either incidentally or designedly, in their dispersal, when taken upon windy days.

Bird flight is marvellous, and some recorded instances are credible only in these days of perfection of the airplane. Many birds are known to pass several hundreds of miles on the wing without rest and it is highly probable that their unaided flight may extend over one or more thousand. The wild goose is supposed to fly at 60 to 75 miles an hour and the swallow at 90 or more, and many other birds may be nearly if not quite as speedy. A sustained flight of ten to twelve hours is not beyond belief, otherwise the enormous distances which certain migratory birds cover would not be possible. Land birds have been met with at all points in the transatlantic passage, but it is quite probable that passing craft afford temporary refuge to the weary ones. The prevailing westerly winds between 30° and 58° north latitude carry many birds all the way across, for the landing of American birds during or after heavy storms upon the coast of England and the island of Heligoland is no unusual thing. North of 58° the prevailing winds are easterly, so that the European birds are transported to America by way of Iceland and Greenland. The west to east passage is much easier, however, as the transatlantic airplane flights have repeatedly demonstrated. Bats are found the world over, even on the most remote sea isles, where they may be the only indigenous mammals. This is sufficient evidence for their migratory powers.

Chapter—21

Bathymetric Distribution

Bathymetric distribution concerns itself with the vertical range of organisms in space, and means much more than mere altitude, for in passing from the highest alpine peak to the abysmal depths of the sea one would find a series of contrasting conditions which of necessity profoundly affect the organism. Of these the primary conditions are three. (1) The air or water medium, which determines the method of breathing, is, with few exceptions, in each case prohibitive to the inhabitants of the other. (2) The presence or absence of light not only modifies the animal directly but indirectly through its effect upon the food supply, for assimilative plants, which form the ultimate nourishment of all animate nature, cannot exist where light is wholly absent. (3) The third primary condition is the presence or absence of a substratum, without which the organisms must be self-supporting, either buoyant or able to swim. This condition therefore determines the bionomic group to which the animal belongs, whether plankton or nekton on the one hand, or benthos on the other.

The secondary conditions limiting bathymetric distribution are whether the water be fresh or salt, and the increase of pressure with the depth—very slight for air-dwelling forms but relatively enormous for those dwelling in the sea.

Three Organic Realms

Hence from the bathymetric as well as from the geographic standpoint three organic realms may be recognized. The bathymetric divisions are:

1. *Geobiotic* (Gr. γῆ, land, and βιωτικός, pertaining to life) or terrestrial.
2. *Limnobiotic* (Gr. λίμνη, lake) or fresh-water inhabiting (lakes or rivers).
3. *Halobiotic* (Gr. ἅλς, sea) or marine or salt-water inhabiting (the sea).

These three realms stand in marked contrast to one another, although, as we shall see, the inhabitants do intermingle to a certain extent at their lines of contact, which also give opportunity for permanent intermigrations.

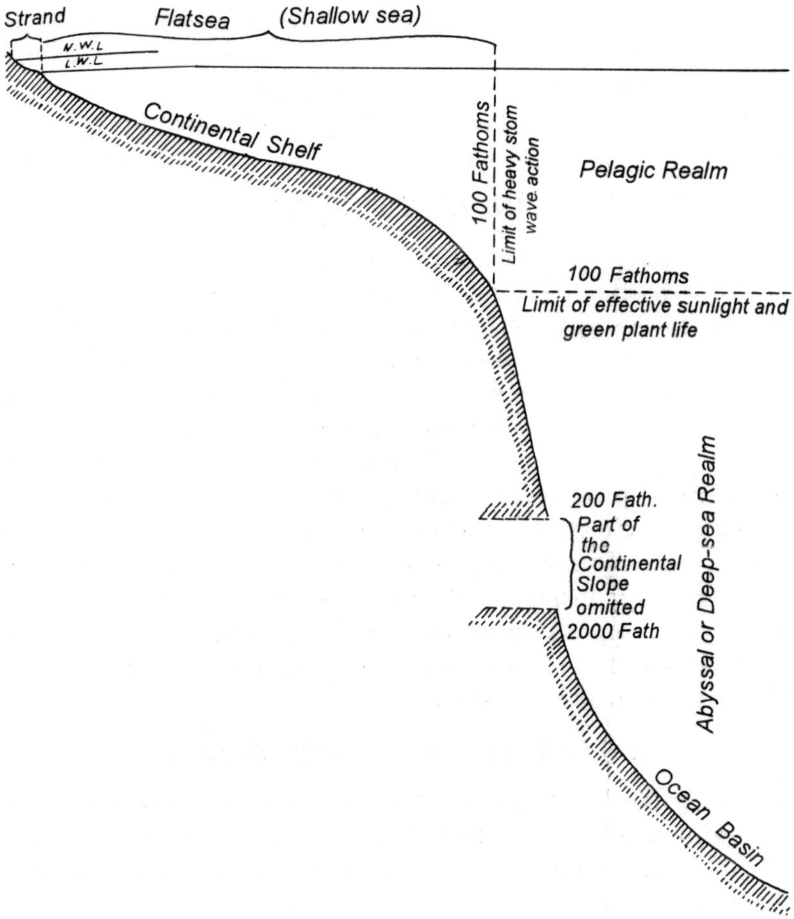

Fig. 21.1: Diagram showing marine bathymetric realms. Vertical scale highly exaggerated. Note: while 100 fathoms of depth is conventionally used to express the limit of the continental shelf, the average curve of the shelf profile is concave upwards to about 50 fathoms, when it becomes convex.

GEOBIOTICOR TERRESTRIAL REALM

The terrestrial ream extends from high-tide mark along the shores of all continents and islands to the summit of the highest elevation. It ranges in altitude, therefore, from lowlands to the Alpine sub-region, and each division—lowland, upland, prairie, high plain, or mountain range—has its own distinctive fauna and flora, governed by many influences, but in part at least by altitude. This last limits quite effectively the character of the flora, as for example the contrasting vegetation of the tropical Mexican coastal lowlands on the one hand and of the temperate high plateau on the other. The so-called timber-line, the limit of tree growth, is also governed by altitude, although varying in different regions, due to latitude and climate.

One additional terrestrial subrealm is the Cryptozoic (Gr. κρυπτός, to hide, and ζωή, life) or the subterranean caves, the only place in the geobiotic realm where light is absent. As we shall see in a later chapter this forms a limited biotic environment which profoundly affects its denizens. Internal parasites and the wood-boring larvæ of insect also dwell in a lightless environment and are consequently modified.

LIMNOBIOTIC REALM

The terrestrial waters contain a rather limited fauna, as comparatively few invertebrates have ever attained a foothold therein. This possibly is due in part to the freshness of the water, but also to the flowing character of the terrestrial waters, a condition to which the great majority of invertebrates with their mero-planktonic larvæ cannot adapt themselves. Certain lakes and relic seas are the only bodies of fresh water of sufficient depth to have the deep-sea characteristics of absence of light and increase of pressure, but we do not find such profound modifications in lacustrine forms as in those which people the abyss. This is probably due to large measure to the evanescent character of all lakes from the geologic point of view. They are individually so short-lived that there has not been time for any very marked adaptive characteristics to develop in their inhabitants. Thus, for instance, while luminescent or light-producing organs are so characteristic of deep-sea and nocturnal marine animals, none are found either in the deep lakes or in the equally evanescent caves, although in each instance they might be very useful in the struggle for existence.

HALOBIOTIC OR MARINE REALM

Biologically speaking, the most important bathymetric realm is the marine, for here we find all the contrasting characteristics abundantly developed, except, of course, the atmospheric medium, and in addition the ages during which the ocean has existed have afforded sufficient time for evolution to ruin its course.

The marine realm is divided into four subrealms, whose characteristics are as follows:

Strand,	light,	substratum	present	Littoral
Flatsea or shallow sea,	"	"	"	Littoral
Pelagic,	"	"	absent	Littoral
Abyssal,	dark,	"	present or absent	

Strand

The strand or tidal zone is the transitional area between the marine and terrestrial realms, for here the inhabitants are left bare twice daily by the receding tide and have to endure drying, either by means of closable shells or other devices, or burrow down into the moist sands, or must be able to breathe both the air and the water. The tidal zone is of course of very variable extent, owing to the differing height of tides and the declivity of the strand itself, and ranges from a width of a few feet to several and in rare cases to many miles; in the Bay of Fundy, the tides, running into a constantly narrowing area, grow proportionately higher until they attain the greatest altitude recorded in the world, about 50 feet at the time of the highest tides. Thus in Minas Basin, full-rigged deep-water ships were seen lying in the red mud with no water in sight, and yet a few hours later they were borne on the bosom of the incoming tide.

Flatsea

The term flatsea or shallow sea is applied to the waters overlying the continental shelf below low-water mark. This continental shelf is formed by the action of storm waves, which are continually cutting back the shore-line and depositing the débris, together with other land waste, upon the area, especially at its outer edge. This margin, which marks the extreme limit of wave action, is in round numbers about 600 feet below the surface of the sea, or at the 100-fathom line.

The continental shelf varies in extent, as does the strand, increasing in width at both the inner and outer margins, and is narrowest along the newer shores of continents and islands and widest where the coastal area is old or possibly subsiding. The slope of the continental shelf is, in general, long and smooth, the bottom near the outer edge descending more or less rapidly to the more profound depths of the sea.

The flatsea is of the utmost importance biologically, for it has both light with its attendant vegetation, and a substratum whereupon benthonic organisms may dwell. Furthermore, its relative shallowness makes the presence of comparatively low physical barriers effective for the isolation which is so essential in evolution. All of these factors, light, plant-food, movements of the waters, warmth, and isolation, make the flatsea a veritable hot-bed of evolution. Its importance as such has been abundantly recognized, so that it has been called the "cradle of evolution."

"The shore-fauna is certainly the most representative of all faunas. What pictures rise in the mind! Swiftly moving infusorians lashing their way through the water; Foraminifera with beautiful shells of lime slowly gliding on the fronds of sea-weed calcareous sponges like little vases and more irregular flinty- and horny-sponges, sometimes coating the rocks like the common crumb-of-bread sponge, sometimes growing in beds like the plants they were once supposed to be; hydroid zoöphytes like miniature trees on rock or sea-weed; sea-anemones and corals often like beds of flowers, living an easy-going life, waiting for food to drop into their mouths, or stinging small passers-by; unsegmented worms such as the 'living films' which glide on the sea-weeds or stones like mysteriously moving leaves, and the nemertines or ribbon-worms, also covered with cilia, but provided with a remarkable protrusible proboscis, sometimes ejected so violently as a weapon that it breaks off altogether and wriggles like a worm itself; the higher ringed worms or annelids in extraordinary number, like *Nereis*, *Phyllodoce*, and *Aphrodite* itself, so beautiful in themselves and in their names that we can understand the enthusiasm of the expert who is said to have named his seven daughters after seven favourite Polychæts; the starfish creeping up the rocks with their strange hydraulic locomotor system, the brittle-stars using their lithe arms like gymnasts, the sea-urchins tumbling along on the tips of their teeth, and the sluggish sea-cucumbers plunging their tentacles into the mud and then into their mouths; the beautiful colonies of 'moss-animals' or Bryozoa, crusting stone and weed as if with lace, or forming leaf-like fronds like the seamat (*Flustra*), which was one of the first animals Charles Darwin worked at, or growing into calcareous tufts as if in mimicry of corals; myriads of crustaceans, such as water-fleas, acorn-shells, beach-fleas, sandhoppers, nobody crabs, sea-slater, shrimps, hermit-crabs, and shore-crabs proper; strange sea-spiders, neither crustaceans nor spiders, like *Pycnogonum littorale*, clambering among the sea-weeds and hydroids;. . . bivalves innumerable, such as cockles and mussels, oysters and razor fish; herbivorous gasteropods like periwinkles, and voracious carnivores like the dog-whelks and buckies; sedentary limpets with a slight range of movement and a slight memory for locality, since beyond a narrow radius they fail to find their way home; an occasional cuttle-fish caught in a shore-pool and many more further out; a large representation of ascidians or sea-squirts, both simple and compound, which lie at the base of the vertebrate series; the lancets (*Amphioxus*) buried all but their mouth in the fine sand; true shore-fishes like sand-eels and gunnels and shannies; an occasional reptile like the lizard *Amblyrhynchus* which swims out among the rocks, or a poisonous sea-snake, or a turtle coming ashore to lay her egg; numerous shore-birds like oyster-catcher and rock pipit, gull and cormorant; and an occasional mammal like otter and seal—on the whole *a more representative fauna than in any other life-area.*"

Keen Struggle for Existence—"It is evident that the shore-area must be characterized by a keen struggle for existence. In the open sea there is practically no limit to the floating room and swimming room, but the shore is narrow and crowded. In a rock pool there is often no vacant niche. There is competition even for *foodhold*. It is important, for instance, that the limpet which makes little journeys in search of sea-weed to nibble should not go too far, else it will not find its way back, and will have lost the spot which its shell has grown to fit. It is curious, too, to see the American slipper-limpet—one growing on the top of another to the number of four or five—suggestive of the root-idea of a skyscraper" (Thomson).

The term littoral is variously used but may include both the strand and flatsea.

Pelagic Realm

The pelagic realm embrace all of the superficial waters of the ocean down to the depth to which effective sunlight penetrates. It is characterized physically by the presence of light and the absence of a substratum. In the upper portion there is variable temperature and frequent and violent wave action, while in its lower strata the movement of the waters and the temperature are greatly reduced.

The distance to which sunlight penetrates varies, being manifestly greater in the tropics where the sun's rays may be perpendicular to the surface of the sea, and less toward either pole where the rays become more and more oblique. By exposing photographic plates at various depths it has been ascertained that light penetrates in the tropics to a depth of 3250 feet, though this is true only of the ultraviolet and blue rays. Red and green rays fail to penetrate to 1625 feet, but at 325 feet all of the rays of ordinary bright sunshine are present, thought red light is the least strong.

Assimilating plant life, which forms the ultimate food supply of all animals, is dependent upon the presence of these red, orange, and yellow rays, which virtually restricts it to the upper 500 feet of oceanic waters. The importance of this is such that the vertical limit of the pelagic zone may be placed at about 600 feet or 100 fathoms; hence it becomes a seaward extension of the flatsea.

Owing to the absence of a substratum, however, no benthonic forms can exist in the pelagic realm but all must be either nektonic or planktonic. The pelagic realm constitutes therefore the high seas, and is not only the meeting place of the commerce of all nations but the great means of dispersal for countless forms of marine life, including the mero-planktonic larvæ of shallow- and deep-water benthos.

Abyssal Realm

The abyssal realm is that portion beyond the limits of the continental shelf and beneath the pelagic, and includes and waters below a depth of 100 fathoms. This realm may be divided into the abysso-pelagic zone, wherein there is no substratum, and all organisms, like those of the pelagic, must swim or float; and the abysso-benthonic zone, in which a substratum is present.

The main characteristics of the abyssal realm are:

1. *Absence of light*. Light of course must exist in the upper transitional strata but it lacks the rays essential to assimilating plant life, hence none exist. The animals therefore are all either carnivorous or feed upon dead organic matter, some of which is of plant origin (oozes). Below the transitional realm the darkness, except for organic luminescence, is profound.

2. *Quiescence*. There is an utter absence of movement aside from the sluggish ocean currents of the greater depths, the progress of which is immeasurably slow.

3. *Cold*. Below a certain depth, the waters in all oceans the world over have become permanently chilled nearly to the freezing point of fresh water, and all diurnal and seasonal variations of temperature and those due to the climatic zones cease to exist.

The mean temperatures of the Atlantic at the surface is 68° F., at 500 fathoms 37°, at 1000 fathoms 35.6.° The Mediterranean, on the other hand, is warmer, owing to the barrier at the Strait of Gibraltar, which prevents the circulation of the deeper and colder waters. At equivalent depths the Mediterranean temperatures are: surface, 75° F., 500 fathoms, 55° F., and 1000 fathoms, 55°.

4. *Pressure*. The pressure of the abyssal waters is enormous, increasing directly with the depth, the ratio of increase being about one ton to the square inch of surface for every thousand fathoms of depth, as compared with the normal air pressure of approximately 15 pounds to the square inch at sea-level.

Thus the abyssal realm constitutes a simple biotic environment of vast extent but of comparatively uniform and changeless character and hence one not conducive to rapid evolutionary change. None of the deep-sea creatures is old geologically speaking, for while from 25 to 35 Paleozoic genera are known in the present *shallow* seas, none of the animals which people the *deep* sea is older than the Mesozoic. They seem to be all migrants from shallow water which have become adapted to the deep-sea conditions, but there is in no instance the evolution of a new race of animals exclusively restricted to the abyssal realm.

INTERMIGRATIONS

Intermigrations between the various benthonic realms do occur where opportunity arises. They are, as in the geographic migrations, both permanent and temporary, the latter being divided in turn into seasonal migrations for breeding and occasional journeys for food or retreat.

Permanent migrations shoreward have not been many. Of the plants, there have been only the numerous fuci or sea-weeds which extend above the limits of the strand from just below high-water mark to the limits of salt spray. Thus while strictly terrestrial, they still feel the influence of their ancestral home. Beyond this point the migration apparently cannot extend.

Of the animals, certain molluscs are also shoreward migrants but these are gastropods only, for they alone seem able to combine air-living with locomotion; the bivalves, for instance, surviving exposure to the air by keeping the shell tightly closed, which makes any activity manifestly impossible. Examples of shoreward migrating snails are *Littorina* of Brazil, which climbs the mangrove trees, and *Ampullaria,* the connecting link between land and water snails. Of the Crustaca, there are several species of land crabs, of which one, *Birgus,* an ordinary crab-like form, is found far from the shore and every climbs mountains and trees, but returns to the sea-shore every year to breed, the marine habitat of the young indicating to us the ancestral home of the species. Another land crab, *Cenobita diogenes,* is one of the hermit crabs which utilize the cast-off shell of some gastropod for the protection of their otherwise defenseless body. A specimens from the Dry Tortugas group of the Florida Keys actually had a fossil shell (*Livona pica*) for its habitation. It is possible that the terrestrial isopods, woodlice and pill-bugs, were also derived from a littoral stock. Probably, however, most of the transitions have taken place through accident in the relic seas or lakes.

The transition from aquatic to terrestrial life on the part of the vertebrates, although doubtless occurring long ago, even previous to Devonian time, was another instance of permanent landward migration, but whether from the marine littoral, the land-locked seas (relic seas freshened into relic lakes), or from the originally fresh or terrestrial waters is not quite clear. Evidence, however, seems to point to the last supposition. The annual return of most amphibia to the limnobiotic rather than the halobiotic realm for the purpose of bringing forth their young points to the former as the ancestral habitat of the race.

Temporary shoreward migrations are seen in several fishes, notably the common eel, which may travel from one pond to another through moist meadow grass. *Periophthalmus,* a curious fish found along the shores of tropical seas, inhabiting the mud-flats at the mouths of the various tropical rivers of the Old World, is popularly known as the walking fish or mud-skipper, as it emerges freely from the water for hours at a time, progressing by means of its curiously modified pectoral fins. The lung-fishes (Dipnoi), whose swim-bladder is modified to serve as a lung, can also live a long time out of water. While there is evidence that the Dipnoi are the survivors of a widespread group of fishes, they are all confined to the fresh waters of Australia, Africa, and South America to-day. The climbing perch, *Anabs,* which inhabits fresh waters and estuaries in Africa and the oriental region, not only comes ashore but actually climbs trees to the height of several feet by means of the strong spines on its pelvic fins and gill-covers.

Permanent Seaward Migrations

Of plants there are eight species of sea-grasses which have a become adapted to salt-water life. The mangrove trees make their homes along tropical shores within the reach of the tides, but while the roots and portions of the trunk may be entirely submerged, the crown of leaves is always in the air; the grasses of which mention has just been made may, however, be wholly below the surface of the water.

Of arthropods, there are a number which have become entirely marine, although of terrestrial origin, notably the sargassum insects which live among the floating fronds of the gulf weed and a spider (*Desis*) found in rocky crevices along the shore. Of the vertebrates, the marine migrants belong only to the three higher classes, land-to-sea migrations on the part of fishes being *a priori* impossible and on the part of amphibia rare, as salt water is distasteful if not actually fatal to them Doctor Gadow says that common salt is poison to the amphibia, nevertheless we have reports of little frogs of the genus *Rana* hopping about on the flats of a tidal creek opening into Manila Bay, and two holes made by a crab were seen to be full of wriggling tadpoles newly hatched. The tadpoles were developing in only slightly diluted sea-water. However, even though occasional temporary migrations may occur, there is no instance at present or in the geologic past where amphibia have become permanently adapted to marine life.

Of marine reptiles, however, the tale is a very different one, for not only are there several different sorts—marine turtles, crocodiles, and sea-snakes—living to-day, but the geologic record is crowded with remains of sea-going Reptilia—plesiosaurs, the wonderfully adapted, dolphin-like ichthyosaurs, the Cretaceous sea-serpents or mosasaurs, and many others, all of which have become extinct.

Of birds, many, like the gulls, terns, frigate-bird, auks, and petrels, make their home on the bosom of the sea, but none perhaps is so thoroughly adapted as the great wandering albatross which follows a ship for miles without resting, and has almost entirely forsaken the land as an abiding place, except for breeding.

Among ancient forms, *Hesperornis* from the marine Cretaceous strata of our great West as an admirable instance of a sea-adapted bird. It had lost completely all powers of flight, while the retention of reptile-like grasping teeth in its jaws and its inclusion in marine sediments in association with plesiosaurs, mosasaurs, and other marine reptile and fishes give indubitable evidence of its habitat.

Of mammals, the seals, whales, and Sirenia or sea-cows are instances of permanent seaward migration, possibly by way of the terrestrial rivers or estuaries. The resultant modifications of these as well as of certain of the marine reptiles, are so profound as to render their return to the ancestral habitat either a relatively rare temporary migration, as in the sea-turtles and seals, or an impossibility, as among the ichthyosaurs and whales.

Temporary Seaward Migrants

Of temporary seaward migrants, the most noteworthy is the curious iguana-like lizard *Amblyrhynchus cristatus* which is found in the Galapagos Islands. Darwin says of it: "It is extremely common on all the islands throughout the group, and lives exclusively on the rocky sea-beaches, being never found, at least I never saw one, even ten yards in-shore. . . . Their tails are flattened sideways, and all four feet partially webbed. They are occasionally seen some hundred yards from the shore, swimming about.... When in the water this lizard swims with perfect ease and quickness, by a serpentine movement of its body and flattened tail—the legs being motionless and closely collapsed on its sides. . . . I opened the stomachs of several, and found them largely distended with minced sea-weed (Ulvæ), which grows in thin foliaceous expansions of a bright green or a dull red colour. . . . I have reason to believe it grows at the bottom of the sea, at some little distance from the coast. If such be the case, the object of these animals occasionally going out to sea is explained. . . . The nature of this lizard's food, as well as the structure of its tail and feet, and the fact of its having been seen voluntarily swimming out at sea, absolutely prove its aquatic habits; yet there is in this respect one strange anomaly, namely, that when frightened it will not enter the water. Hence it is easy to drive these lizards down to any little point overhanging the sea, where they will sooner allow a person to catch hold of their tails than jump into the water.... Perhaps this singular piece of apparent stupidity may be accounted for by the circumstance, that this reptile has no enemy whatever on shore, whereas at sea it must often fall a prey to the numerous sharks. Hence, probably, urged by a fixed and hereditary instinct that the shore is its place of safety, whatever the emergency may be, it there takes refuge."

Among mammals, temporary seaward migrants include the polar bear but especially the sea-otter, the latter being of particular interest in illustrating the course of evolution which the seals must have undergone in their adaptation to permanent marine life. This otter (*Enhydris*) was still comparatively plentiful all along the northern Pacific coast in the middle of the last century, but because of its valuable fur has been almost entirely extirpated. It is well adapted for aquatic life, with hind feet suited only for swimming, back teeth with smooth rounded crowns for crunching "shell-fish", and in the care which the mother shows for the pup,—dandling it and dividing with it.

Another very remarkable seaward migrant is the hippopotamus, which while really a terrestrial or river-inhabiting form, occasionally takes to sea in its passage from one river-mouth to another along shores which would be otherwise difficult to traverse.

Permanent Migrations from Sea to Fluviatile Realm

In all probability the molluscs (fresh-water clams, mussels, and snails), the crustaceans, and the crayfish represent permanent or landlocked migrants from salt to fresh waters; but the great host of marine invertebrates have never succeeded in gaining a permanent foothold in the fluviatile realm. There are some curious instances of creatures, vertebrates and invertebrates, however, which have made what might be called involuntary migrations into fresh water. Such animals are found in what are known as relic seas, formerly in direct open communication with the oceans but now cut off from them. The best-known examples of these are the Black Sea and the Caspian Sea, both of which once widely connected with the Mediterranean, but have been freshened gradually through the inflow of rivers. The animals which inhabit these seas are also relic faunas or relic species which upon being cut off from their ocean-living brethren have adapted themselves to the gradually freshening waters. In the Caspian Sea, for instance, while there are comparatively few different *kinds* of animals, the fishes are so abundant in *individuals* that the fisheries are equal to those of the northern Atlantic Ocean. The fauna includes sturgeon, salmon, herring, porpoises, and seals. American examples of relic seas are possibly Lake Ontario but more especially Lake Champlain. The elevated beach deposits of Lake Champlain contain an abundance of marine shells and the bones of seals and whales.

Temporary Migrations from Sea to Fluviatile Realm

There are a number of fishes, such as the shad, alewife, sturgeon, and salmon, which, while the major portion of their life is spent in the sea, ascend rivers periodically to spawn. Of these the most notable are the salmon, which ascend the Sacramento River to its extreme source, a distance of abut 400 miles. In the Columbia they ascend as far as the Bitter Root and Sawtooth mountains of Idaho, a distance of nearly a thousand miles; but their extreme limit is not known. In the Yukon a few ascend to Caribou Crossing and Lake Bennett, 2250 miles. "At these great distances, when the fish have reached the spawning grounds, besides the usual changes of the breeding season, their bodies are covered with bruises, on which patches of white fungus (*Saprolegnia*) develop. The fins become mutilated, their eyes are often injured or destroyed, parasitic worms gather in their gills, they become extremely emaciated . . . and as soon as the spawning act is accomplished, and sometimes before, *all* of them die. The ascent of the Cascades and the Dalles of the Columbia causes the injury or death of a great many salmon" (Jordan). This wonderful instinct again points to the salmon and other fishes having originally been fresh-water forms which have made a permanent migration seaward in some bygone period, but the homing instinct still impels them to return to their ancestral waters to bring forth their young, even though the act be suicidal so far as the individual is concerned.

Permanent Migrations from Fluviatile Realm to Sea

If, as certain authorities claim, the land waters are the ancestral home of all vertebrates, the marine fishes must have all come from one or more ancient seaward migrants from the fluviatile realm. It has also been argued that the Cetacea, some of which, like the blind fresh-water dolphins (Platanistidæ), inhabit the rivers of India, made the limnobiotic realm their transitional habitat in their ancestral migration from land to sea. This, however, has not been proved, although the supposition that they were at least estuary-inhabiting forms before taking to the high seas is certainly plausible.

Temporary Migrations from Fluviatile Realm to Sea

Of transitory migrants perhaps the most notable is the ordinary fresh-water eel. The eels of Europe spawn in a restricted area of the western Atlantic, in the vicinity of the West Indies, and the feeding grounds of the American forms are in the same vicinity. Thomson thus descries their life history: "The early chapters of the life-history remain obscure, but the young larva rises to the upper sunlit waters as a transparent, sideways-flattened, knife-blade-like creature, about three inches in length, with no spot of colour save in its eyes. It lives for many months in this state—known as a Leptocephalus—expending energy in gentle swimming, but taking no food. It subsists on itself, and becomes shorter and lighter, and cylindrical instead of blade-like. It is transformed into a glass-eel, about two and a half inches long, like a knitting needle in girth. It begins to move toward the distant shores and rivers. In some cases it may take more than a year to reach the feeding ground—those that ascend the rivers of the eastern Baltic having journeyed over three thousand miles. Their ranks are thinned, but large numbers succeed in finding the estuaries, and the passage of millions of elvers up our rivers is one of the most remarkable sights of Spring. There is a long period of feeding and growing in the slow-flowing reaches of the rivers and in the fish-stocked ponds. But there is never any breeding in fresh water, and after some years a restlessness seizes the adults as it sized the larvæ—a restlessness due, however, to a reproductive, not to a nutritive motive or impulse. There is an excited return journey to the sea—they don wedding garments of silver as they go and become large of eye. They appear to migrate hundreds of miles, often at least out into the Atlantic to the verge of the deep sea, where, as far as we know, the individual life ends in giving rise to new lives. In no case is there any return." This is an instance of former salt-water fish which having made a permanent migration into the fresh waters in some remote time, still seek not only their own birthplace but that of the race for the purpose of bringing forth their own young.

Migrations from Sea to Air and Air to Sea

The intermigrations between the sea and the air are relatively few. Under migrations from sea to air would come as temporary intermigrants the flying fishes, of which several genera, representing a number of separate evolutions, are known. Their aërial existence is very transitory, as the flights, if such they are, rarely exceed a hundred yards and are generally far shorter.

Of air to sea migrants the penguins are perhaps the best example, they as have lost entirely the power of aërial progression, but their wings, through the degeneracy of the feathers and a compensating broadening of the entire structure, have become admirable swimming devices for what may be called submarine flight.

Chapter—22

The Synthetic Theory of Evolution

Modern biology has two unifying concepts. One is the concept of organization. The tells us that at every level, from the molecule through the supra-molecular organelle, the cell, the tissue, the organism, the individual, and up to the population or the society, the properties of life depend only to a small degree upon the substances of which living matter is composed. To a much greater degree living things owe their nature to the way in which the components are organized into orderly patterns, which are far more permanent than the substances themselves. The other unifying concept of biology is that of the continuity of life through heredity and evolution. This tells us that organisms resemble each other because they have received from some common ancestor hereditary elements, chiefly the chromosomes of their nuclei, which are alike both in respect to the substances which they contain and the way in which these substances are organized. When related kinds of organisms differ from each other, this means that in the separate lines of descent from their common ancestor changes in the hereditary elements have taken place, and these changes have become established in whole populations.

THE FACT OF EVOLUTION

At the outset we should realize that the great majority of biologists accept as demonstrated the fact that organisms have evolved. To be sure, no biologist has actually seen the origin by evolution of a major group of organisms. Nevertheless, races and species have been produced by duplicating in the laboratory and garden some of the evolutionary processes known to take place in nature. The reason that major steps in evolution have never been observed is that they require millions of years to be completed. The evolutionary processes which gave rise to major groups of organisms, such as genera and families, took place in the remote past, long before there were people to observe them. Nevertheless, the facts which we know about these origins, some of which will be discussed provide very strong circumstantial evidence to indicate that the processes which brought them about were very similar to those found in modern groups of animals and plants which are evolving all around us today.

The state of our knowledge of the major steps of evolution which took place in the past is much like that of our knowledge of ancient history. We do not have any reliable eye-witness accounts of the events which produced the rise and fall of the civilizations of ancient Egypt, Sumeria, Babylon, and Crete, and the contemporary written records of these ancient times are fragmentary and unreliable. Nevertheless, the indirect evidence that these civilizations existed is so strong that it is accepted by scholars and laymen alike. Moreover, we teach as historical fact many of the probable events connected with their rise and fall. The evidence which biologists now have about the rise and extinction of major groups of animals in past geological eras is of a very similar nature, and carries with it about the same degree of high probability.

Since Darwin developed his theory of evolution more than a century ago, biologists have studied this subject in two different ways. Some have been interested in the course of evolution. By comparing a multitude of different kinds of animals or plants, they have tried to work out the evolutionary family tree of some particular group, such as the horses, the cone bearing trees, or mankind. Other biologists have asked themselves the question: "What makes evolution go?" By means of observations and experiments of various sorts on populations of living organisms, they are learning about the processes of evolution and the mechanisms responsible for them. The present book will consider only this second approach to the study of evolution. The study of the course of evolution, or phylogeny as it is often called, must be carried out in a somewhat different way in each separate group of organisms. Such studies have yielded few general principles which apply equally well to all kinds of living things. On the other hand the processes of evolution are in many ways similar even when we compare such different forms as bacteria, flies, grasses, and mammals. Consequently, an understanding of these processes is of far greater importance to the general biologist than is knowledge of the particular ancestry of any group.

THE MODERN SYNTHETIC THEORY OF EVOLUTION

The modern, synthetic theory of evolution recognizes five basic types of processes: GENE MUTATION, CHANGES IN CHROMOSOME STRUCTURE AND NUMBER, GENETIC RECOMBINATION, NATURAL SELECTION, AND REPRODUCTIVE ISOLATION. The first three provide the genetic variability without which change cannot take place; natural selection and reproductive isolation guide populations of organisms into adaptive channels (Figure 22.1). In addition, three accessory processes affect the working of these five basic processes. MIGRATION of individuals from one population to another, as well as HYBRIDIZATION between races or closely related species both increase the amount of genetic variability available to population. The effects of CHANGE, acting on small populations, may alter the way in which natural selection guides the course of evolution. The propose of the present book is to review our knowledge of each of these processes, and to show how they are interrelated with each other. The more we know about the five basic processes, the less reason we have for believing that any other basic processes remain to be discovered. We do not need to search any more for hidden causes of evolution. Nevertheless, we do need to understand much more abut the way in which known processes interact with each other.

At the outset we must recognize that at least in higher organisms, and perhaps in microorganisms as well, the three processes, mutation, genetic recombination, and natural selection, are equally indispensable for evolutionary change to take place. Speculations as to which of the three is the most important are completely pointless. Their interrelationships can be well illustrated by the following analogy. An evolutionary line of organisms which is changing through eons of time can be likened to an automobile being driven along the highway. Mutation then corresponds to the gasoline in the tank. Since it is the only possible source of new genetic variation, it is essential for continued progress, but it is not the immediate source of motive power. This source is genetic recombination, through the shuffling of genes and chromosomes which goes on during the sexual cycle. Since this process provides the immediate source of variability upon which selection exerts its primary action it can be compared to the engine of the automobile. Natural selection, which directs genetic variability toward adaptation to the environment, can be compare to the driver of the vehicle. Various lines of evidence, which will be explained in a later chapter, indicate that structural changes in the chromosomes, which change the order of arrangement of the genes along them, can have profound effects upon the interrelationships between genetic recombination and natural selection, and so can be compared to the transmission and accelerator of the automobile. Finally, reproductive isolation, which includes all of the barriers to gene exchange between populations, has a canalizing effect similar to that which the highway, with its limits and directive signs, exerts upon the driver of the automobile, thus permitting several vehicles to drive in the same direction at the same time.

The synthetic theory of evolution can be stated as follows:

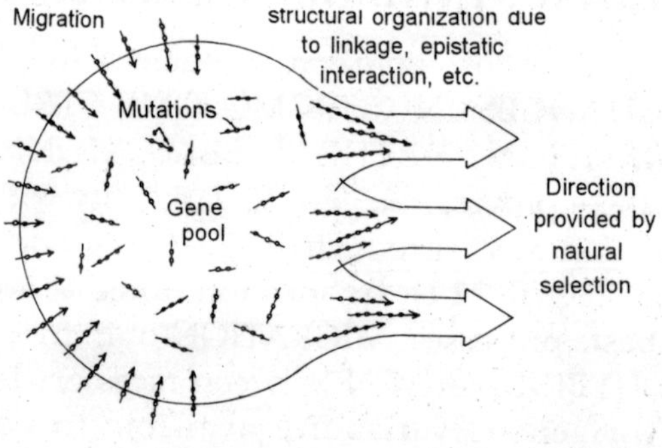

Fig. 22.1: Diagram to illustrate how the four basic processes—(1) mutation; (2) genetic recombination, which results from intercrossing betwen individuals of the population as well as between them and occasional new genotypes which enter by migration; (3) Structural chages in the chromosomes, which offect linkage and epistatic interaction of genes; and (4) Natural selection—interact to produce a progressive change in the population which keeps it adapted to the changing environment.

1. Biotic potential

It has been defined as the innate capacity of population to increase under optimal conditions with suitable age and sex ratios present. Under ideal conditions the biotic potential of most animals is extremely great. It has been calculated that a pair of housefly (*Musca domestica*) at the end of a season will multiply to occupy 2.5´10^9 cubic yards of space. A carp fish lays about 28 million eggs at a time. Even so a slow breeder as an elephant produces 6 youngs during its life span of 100 years and if all of them survive and breed, in 750 years they will become.

It is, however, known that most individuals of necessity do not survive in order that the energy cycle be maintained.

2. Variations

Darwin recognized two kinds of variations: (a) *Non-hereditary* and (b) *Hereditary*. The non-hereditary variations appear during the life of an animal and are not passed on to the offsprings. Herditary variations are important for evolution as they are passed on to the offsprings. Genetically controlled variations are produced by any of the three methods: (a) mutatious (b) chromosmal changes and (c) recombination. Besides some other phenomenon are also known such as polygeny, polymorphism and reproduction which accelerate the frequency of variations.

Since more individuals are produced than can survive and since variations of all grades do exist, it is only logical that an individual possessing some useful variations (for survival) will survive. Thus it cannot be asserted that all the genes of an organism are useful or that all of the characters of an organism have survival value.

3. Population Genetics

Evolution always takes place in a population. A population is a group of organisms of the same species, which interbreed with one another in a particular geographical area.

(i) *Genetic Equalibrium. G. H. Hardy,* an English mathematician and *W. Weinberg,* a german physicist in 1908, independently answered the question, "what happens within a population when random mating takes place in successive generation?". It is known as *Hardy-Weinberg's Law.* The law is the foundation of population genetics and of modern evolutionary theory.

This law state, "If alternate forms of genes (alleles) are present in a population in a given proportion and if random mating and equal survival of offspring exist, then the original proportion will be retained in all subsequent generations unless it is upset by some other factor, such as mutation or selection pressures".

The genetic equilibrium has been expressed mathematically as:

$p+q=100\%$

were p=frequency of one allele in a population (T).

q=frequency of second allele (t).

Substituting F_1 generation (*Mendel*) in terms of p and q the phenotypic equation will be:

$p^2 + 2pq + q^2 = 100\%$ or 1.

(TT) (2Tt) (tt)

If frequencies of p and q are equal, than

$(.5\times.5) + 2\ (.5\times.5) + (.5\times.5) = 1.$

$.25 + .50 + .25 = 1$

$1 = 1$

If frequencies of p is 60% and that of q 40%, than

$(.6\times.6) + (.6\times.4) + (.4+.4) = 1.$

$.36 + .48 + .16 = 1.$

$1 = 1.$

Salient features of Hardy - Weinber's Law

In the absence of all evolutionary forces:

1. The gene and genotype frequencies of each allele in a population remain at an equilibrium generation after generation.
2. The mating is completely a random phenomenon in a population.
3. The equilibrium in gene and genotype frequencies occurs in a large population.
4. All the genotypes in a population reproduce equally successfully.

(ii) *Selection Pressure.* Majority of the mutations have a harmful effect. This effect varies from lethel to very slight. Therefore, the individual possessing them are eliminated from the population. Moreover, all the mutations are harmful and some of them and their combinations are found to be beneficial to the individual possessing them and they can utilize the environment more successfully. Natural selection favours such genes. A classic case is the peppered moth (*Biston betularia*) studied by *R.A. Fischer, E.B. Ford and K. Well.* Until 1845 only white coloured moths were known in England. It had a great selective advantage in pre-industrial England as it metched the background colours. But in 1845 a single black moth appeared in the industrial centre of England. Upto 1895 their population increased more than 99%. While the population of white moths is represented by few individuals. The change from white to dark colour is due to sooty and coal-dust covered area of England.

Many other examples of selective pressure can be given such as resistance to insectisides in houseflies, mosquitoes, bedbugs etc.

3. Genetic Drift

The concept of genetic drift was developed by *Sewell Wright*. It is mainly concerned with changes in gene frequencies in small population while Hardy-Weinberg's law is effective in large populations. According to Hardy-Weinberg's law, in a large randomly mating population, without selection and mutation, the gene frequency remains constant. But in small populations, a gene may be fixed simply by chance and not under the control of natural selection. It is called genetic drift by *S. Wright*.

It is evident from the study of genetics that one gets actual genetic ratios, which correspond with the theoretically expected ratios. It is dependent upon having a large number of offspring. This is due to the reason that assortment of genes into gametes and the combination of gametes to form zygotes are random processes. In smaller breeding populations, where only a few offspring are formed, great fluctuations in expected ratios may take place by chance. Thus, in a very small population mode of two mating pairs, if one individual has a new mutation *a*, its genotype would be *Aa* and that of the other would be *AA*. It is completely possible, simply on the basis of chance that, if this individual with the new mutation and its mate give rise to six offsprings which live to maturity and the other pair produces four that live to maturity, five of the offsprings of the first pair will have the mutant gene *a*. Thus, out of the ten individuals to form the next generation, half of them would have the mutant gene. Suppose that there is no selection against this gene, it would have very fair possibilities of spreading further in the following generations. On the other hand, reverse may take place in the same situation. None of the offspring of the first pair, mainly by chance, might receive the gene *a* and it would be lost immediately.

Thus in small population, genetic drift may supersede the action of selection pressure. It plays an important role in evolution.

Among organisms having short life span (viz. insects) population, fluctuate drastically and two large populations of a species at different times are thus connected by an intermediate very low population. This type of low population of a species has been referred by *Stebbens* as a *"bottle neck."* Such bottle necks assume great importance because only the reproductively active survivors can pass on genetic material. Drift during the time of a bottle neck may result in the wide distribution in a large population of one or more non-adaptive phenotypes. It is a supplement of the *Sewell Wright* effect.

4. Speciation

The speciation has been mentioned in details in chapter. 17

5. Level of Evolution

Goldschmidt (1940) has devided the evolution into two: (1) *"microevoltion"* that of subspecies of origin of smaller evolutionary differences (2) *"Macroevolution"*, i.e., of species and genera or differentiation of a group of animals by adaptive radiation. *Simpson* (1953) has proposed additional term *megaevolution* i.e. large scale evolution such as that of families, orders, classes and phyla or evolution of a new type of animal from its predecessors.

It is now realized that evolution at all levels has employed the same mechanism as in speciation. When characters first appear they use distinctive of species and genera. As they diverge by the extinction of intermediate forms, the same characters defines higher and higher categories and this is determined by the continuation of known evolutionary process through long periods of tie. Thus, according to *Simpson* and other modern biologists, there is no reason to find other special processes to account for the evolution of higher categories.

Macroevolution is also called *adaptive radiation*. Characteristics of macroevolution are as follows: (1) subdivision of the group into many new groups, (2) an invasion of several new environmental situations and (3) diversification of structure and biology.

In microevolution many genes and gene combinations are influenced by hundreds of environmental factors. Thus, microevolution through the interplay of *mutation, speciation* and *genetic drift* probably produce change in all the populations from one generation to the next.

When a population successfully shifts from are major adaptive zone to another, it results in *megaevolution*. In such a shift attempts of most of the individuals is negativated

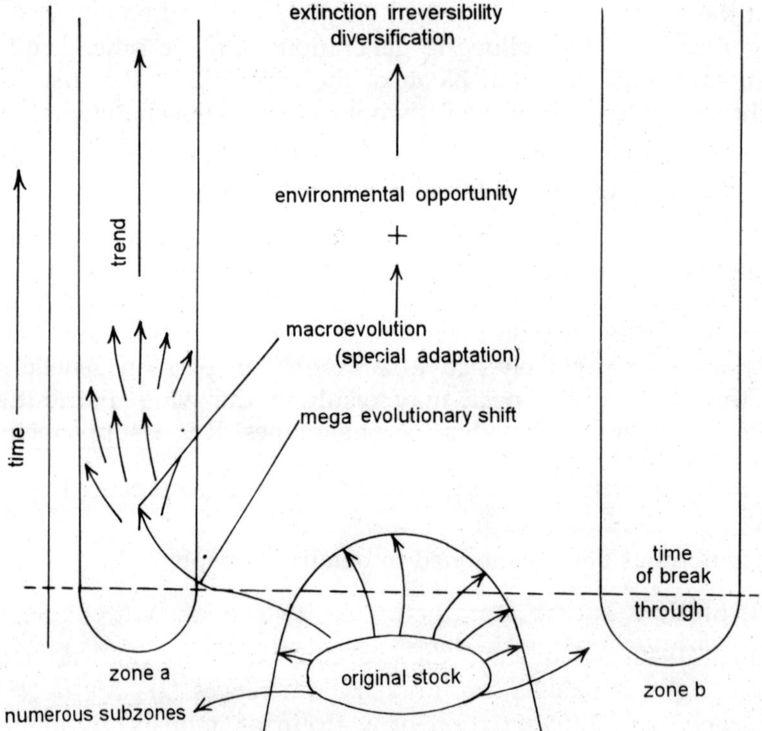

Fig. 22.2: The Process of Macro and Mega Evolution.

and only a few succeed in occupying new habitat. Evolution of amphibian from fishes and, birds and mammals from reptiles are such examples.

Simpson (1965) mentions the important conditions for megaevolution as follows;

1. Existence of a widespread group, which evolves progressively but has not become genetically immobilized for radical changes in relationship to environment.
2. A varied ecological terrain, responsible for diversification and fragmentation.
3. Shifting, unstable environmental conditions.
4. Shifting, severe environmental conditions responsible for extinction of specifically adapted populations and lessening or removing of competition within various ecological zones (adaptive zones) available to newly arising types.

Thus, megaevolution is a continuous series of changes, which can be divided taxonomically into horizontal, phyletic subdivisions of any size including subspecies.

Recently *J.M. Savage* (1965) of the University of Southern California divided the basis patterns of evolution into two (a) sequential evolution and (b) divergent evolution.

(a) *Sequential evolution* is the modification from generation to generation of one or more alleles in a population. (b) *Divergent evolution* takes place when two divergent population lines result form a single ancestral population.

Savage revised the classification of levels of evolutionary change. This includes *microevolution* or *speciation, macroevolution* and *megaevolution*.

SOME GENERALIZED TRENDS IN EVOLUTION

A number of laws or rules have been suggested for the general application to evolution.

Dollo's Law

It was proposed by *Dollo,* according to him the evolution is usually irreversible process.

Cope's Law

This was proposed by *E.D. Cope* according to him, organisms have a tendency, towards increase in size during their evolution.

Gause's Rule

It was proposed by *Gause.* According to him, two species with identical ecological needs cannot indefinitely occupy the same ecological niche.

Jordan's Rule

It was suggested by *K. Jordan.* According to him, closely related species or subspecies are usually, not found in identical regions but in adjacent regions separated by some barrier.

Bergman's Rule

Bergman pointed out that closely related species and subspecies of warm blooded (endothermic) birds and mammals are larger in the colder parts of their range and smaller in warmer areas.

Willistone's Rule

S.M. Williston, on there basis of fossil record, stated that, during the evolution of a linage, serially homologous parts tend to reduce in number but get more and more differentiated.

Evolution Process

Genetic stability can be altered by mutation and selection pressure as stated above because these factors influence the gene frequency in a population.

As mutation rates and selection pressures operate simultaneously both can be represented in a single formula:

$$q = p - sq^{2}(p)$$

Where q is the rate of change in frequency of gene q in the population.

up is the measure of the rate of mutation; p—q and sq^2 is the gain or loss of q alleles through selection pressures.

Chapter—23

Zoogeographical Distribution

Zoogeography is the scienc whichdeals with the distribution of all the animals over world. First serious attempt to map out the geographical regious was made by *Sclater* (1875) who based his conclusions upon the distribution of birds. There regious are: 1. Palaertic region, 2. Ethiopian region. 3. Indian region, 4. Australian region, 5. Neotropical region, 6. Neartic region. There are, however, very apparent objections to utilizing vagrant and barrier defying creatures like birds for this purpose. Accordingly, *Muray* in 1866 and more in detail *Alfred Russel Wallace* in 1876, divided the surface of earth into zoological regious based chiefly on the distribution of mammals.

Many other regional classifications have been made based an either discontinuous distribution of one particular class or upon the distribution of temperature variations and similar climatic factors. None of these is satisfactory. Therefore, *Wallace* combined Scalter's system. *Blanford* (1896), *Lydekk* (1896), *Heilprin* (1887), *Gudow* (1913), *Schmidt* (1954) ect. Also made their contribution for the study of zoogeogrophy.

Karl P. Schmidt (1954) divided the earth into three zoogeographical realms:

1. Arctogean,
2. Neogaean, and
3. Notogaean.

Arctogaean includes two regions the (a) Holarctic with sub-region: Arctic, Nearctic, Caribbean and Palearctic, (b) Paleotropical with sub regions: Oriental, Ethiopian and Malagasy.

Neogaean includes a single sub-region the Neotropical. ***Notogaean*** consists of regions: the Australian having the Australian and Papuan sub-regions and the Oceanian, which is formed of the New Zealandian, oceanic and Antarctic sub-regions.

Wallace's classification is generally considered to be accepted system.

The Indian region of *Scalter* is termed as Oriental region in the classification of *Wallace*.

Thus, six zoogeographical regions have been distinguished as follows:

1. *Palaearctic Region*. It includes Europe, temperate Asia, North Africa and Arabia.
2. *Nearctic Region*. It includes whole of north America and Greenland.
3. *Ethiopian Region*. It includes whole of Africa, Arabia, South of the tropic of Cancer, and Madagascar.
4. *Neotropical Region*. It includes whole of South and Central America and West Indies.
5. *Australian Region*. It New Guinea and neighbouring islands.
6. *Oriental Region*. It includes India, Sri Lanka, Indochina and, Malaya.

The exact boundaries of these regions are very difficult to recognize. The fauna of these regions are described only with reference to the vertebrates, although they form only 3-4% of animal kingdom.

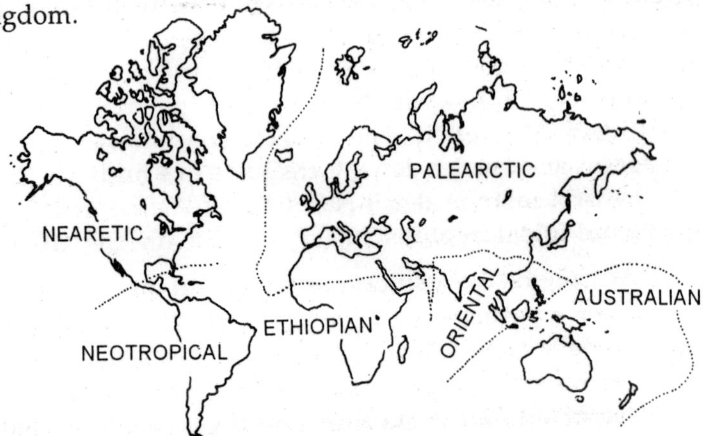

Fig. 23.1: Zoo geographical region of world.

PALAERCTIC REGION

Extent

This, the largest of the six regions, is estimated to cover an approximate area of 14,00,000 sq. miles. It includes the whole of Europe, Iceland, the Azors, Madeira, Canary Island, Cape Verde Islands, all the portion of Africa and Arabia which lies to, the north of the tropic of Cancer, Asia Minor, Persia, Afghanistan and Baluchistan and the whole of Asia north of a line which runs up to valley of the Indus, along the greater Himalayan range, thence eastwards to the Nanling Mountains, south of the Yang-tsk-kiang and out to sea just south of the Japanese islands.

Climate

The climate is temperate. This regions includes wet forest lands as well as dry open steppe lands. It also includes coniferous forests and a fringe of Tundra.

In the western portion there is a great inland seas which have an equalising effect on the climate. The Gulf stream materially helps to raise the temperature of weastern Europe. The eastern region is much cooler. The extreme of cold are felt in North Eastern Siberia. However, the highest summer temperature of the area is experienced in North-Western India, Arabia and Afghanistan.

The Northern part of this region is mostly low and flat, Scandinavian and the Ural mountain are the only elevated lands. In South a series of mountains ranges run across the region from west to east, including in Europe the Alps, Balkans, Caucasus etc., in Asia-the Tian shan, Altai Himalayan range etc.

Zoological Characteristic

The fauna of this region as a whole is very similar to the Nearctic region, that is why *Heilprin* proposed a great region, *Holarctic Realm* for both palaearctic and Nearctic regions.

The region possesses representatives of 135 families of terrestrial vertebrates, namely 33 of mammals, 68 of birds, 24 of reptiles and 10 of Amphibia. None of these, however, is peculiar to this regions but 9 are common and confined to the Palaearctic and Nearctic Regions. The nine families are the Talpidae (Moles) Ochotonidae (Picas) and Castordae (Beavers) among mammals, the Regulidae (Gold-crests), Colymbidae (Divers) and Tetraomide among birds; the Proteidae Protous, Salamanders and Amphiuma among Amphibia the fresh water fishes also support the union of the two regions, for five families namely the perches, stickle backs, Pikes, sturgeons and Polyodontidae are distinctly Holoarctic.

The palaearcitc region is subdivided into four sub-regions;

1. European sub-region,
2. Mediterranean sub-region
3. Siberian sub-region
4. Manchurian sub-region

1. **European sub-region** comprises Northern and Central Europe. In this sub-region only 85 families of terrestrial vertebrates are represented and of these the reptiles and Amphibia number only 6 each. Only one from of mammals are peculiar namely Myogale but such animals as the wolf, Hedgehog, shrew, mole and Darmouse are very characteristic as are also the wag tails, Pipits, Tits, Thrushes and many other birds.

2. **The Mediterranean sub-region** includes the rest of Europe, all the African and Arabian portion, Asia Minor, Persia, Afghanistan and Baluchistan. This is the richest region possessing representatives of 120 families of terrestrial vertebrates. The Fallow-Deer, Elephant Shrew, Civent, Ichneumon, Haena, Hyrax, Porcupine are all characteristic mammals.

3. **Siberian sub-region** embraces all Northern Asia, southwards to the Himalayas. It possess representatives of 94 families of the terrestrial Himalayas. It possess representatives of 94 families of the terrestrial vertebrates. Four genera of mammals i.e. the Yak, Musk deer and a mole are almost confined to this sub-region but also ranges into the oriental region. One of the most important members of the Siberian fauna is the fresh water seals, *Phoca ibirica* which inhabits the great fresh water lake- Baikal.
4. **The Manchurian Sub-region** includes China, Mongolia, Manchuria and Corea together with the whole of Japan. It has a rich and varies fauna with representatives of 102 families of terrestrial vertebrates. At least a dozen genera of mammals including Tibetan langur, Great Panda etc. are peculiar.

NEARCTIC REGION

Extent

It includes the whole of North America and extends south as for as the middle of Mexico. It includes Greenland in the east and Aleution islands in the west.

Climate

It has a great range of temperature. The chief physical features of the region are (1) Large lakes and inland sea in the north eastern portion and the important ranges of high mountains in the west. (2) In the east are smaller ranges, constituting the so-called Appalachian high land. (3) In the centre of this great continent is a vast extent of plain which in the north is frozen and hence barren, between altitudes 50° and 60° covered with forest, and in the south dry treeless desert.

Zoological Characteristics

The number of families of terrestrial vertebrates represented in the Nearctic Region is 120 viz. 26 of Mammals, 59 of birds, 21 of Reptiles and 140 of Amphibians. The Nearctic Region possesses five peculiar familiar; viz. the Haplodontidae and Antilocapridae among Mammals; the Chamaeidae among Birds; the Aniellidae among Reptiles and Sirenidae among Amphibians.

The Nearctic differs from the Palerctic region, in the possession of severals characteristic Mammals, such as oppossums (Didelphyidae) and Racoon etc., many birds such as the Blue-jays, and Turkey-buzzards etc., Reptiles such as Rattle snakes and Igunanas: Amphibia including Axolotl, Nectures, Siren and other large Urodeles.

There are, several peculiar genera of importance of which Scalops (web-footed Moles), Toxidae (American Badger), *Haplocerus* (Rocky mountain Goat) and Musk may be taken as examples.

Wallace has divided nearctic Region into four sub-regions:

(1) California sub-region

(2) Rocky Mountain sub-region

(3) Alleghany sub-region

(4) Canadian sub-region.

1. **California sub-region** embraces a narrow strip of countary between Sierra Nevada and the Pacific. In the North it includes island of Vancouver and the Southern part of British Columbia and in the South it extends up to the head of the Gulf of California. It has 86 families of terrestrial vertebrates and of which 21 are mammals, 49 of birds, 8 of reptiles and 8 of amphibians. These families are peculiar to this region, namely Haplodontidae, Chamaeidoe and Aniellidae. Vampires and free tailed Bats are characteristic of this region.

2. **Rocky Mountain sub-region.** It lies immediately to the east of California and includes the whole of the dry and elevated area covered by the mountains. It is the richest portion of nearctic region. It has 107 families of terrestrial vertebrates out of which 25 are mammals, 55 of birds, 18 of reptiles and 9 of amphibians. Although there are no peculiar families but are several characteristic genera like prong-buck (*Antilocapra*), Rocky mountain goat (*Haplocerus*), Prairie dog (*Cynomys*), American Bison and poisonous lizard (*Heloderma*).

3. **Alleghany sub-region** comprises the United States east of the Rocky mountain sub-region and South of the Great Lakes and includes Novo Scotia. Some 99 families of terrestrial vertebrates including, mammals 8, birds 53, reptiles 16 and amphibia 12 are represented in this sub-region. The peculiar animals are: opposum, star nosed moles, vampire bats etc. Turkeys and Sirenids (mud eel) are the other two families.

4. **Canadian sub-region**. All the remaining portion of North America and Greenland constitute the great Canadian sub-region. In this sub-region are to the found the representative of only 75 families including 20 of mammals, 44 of birds, 3 of reptiles and 8 of amphibian, thus it is the poorest region. But there are characteristic families including Deer, Rein deer, Elk, Bison, Sheep, Gluttons, Lemmings, the polar deer etc.

ETHIOPIAN REGION

Extent

The Ethiopian region consists of the whole of Africa and Arabia, south of the tropic of Cancer, together with Madagascar and the small adjacent islands. *Darlington* (1957 has not included Medagascar but has considered it as separate region on account of its distinctive fauna.

Climate

This region is mainly tropical. It is a large block of rain forests and isolated mountains. The eastern part has wide grassy plains. Its southern part is warm temperate with mixed vegetations.

Zoological Characteristics

The fauna of Ethiopian region is rich, varied and well marked. There are about 161 families of terrestrial vertebrates. Out of these 30 are peculiar.

Among mammals 12 families are peculiar which are; Chiromyidae (Aye Aye), Chrysochloridae (Golden moles, Centetidae (Tenress), Potamogalidae (Potamogale), Potelidae (Earth wolf), Battryergidal (African mole rats), Lophimyidae (crestedrat), Pedetidae (African Jamping Hares), Anomaluridae (African flying squirrels), Giraffidae (Giraffes), Hippopotamidae (Hippopotamus) and Orycteropodidae (Aard-Varks).

Among birds 13 families are peculiar; a Promeropidae (Promerops), Acerocharidae (Helmet Birds), Vangidae (Vangidal), Philepittidae, Musophagidae (Plankton eators), Coliidae (Colies), Serpentariidae (Secretary birds), Scopidae (Hammer-head bird) Mesoenatidae, Numididae (Guinea fowl) and Balaenicipitidae (Whale head).

Among reptiles following families are peculiar: Rhachiodontidae (Egg-eating snakes), Gerrhosauridae, Zonuridae (Girdled lizard) and Uroplatidae.

Among amphibian family Dactylethridae (clawed toads) is peculiar.

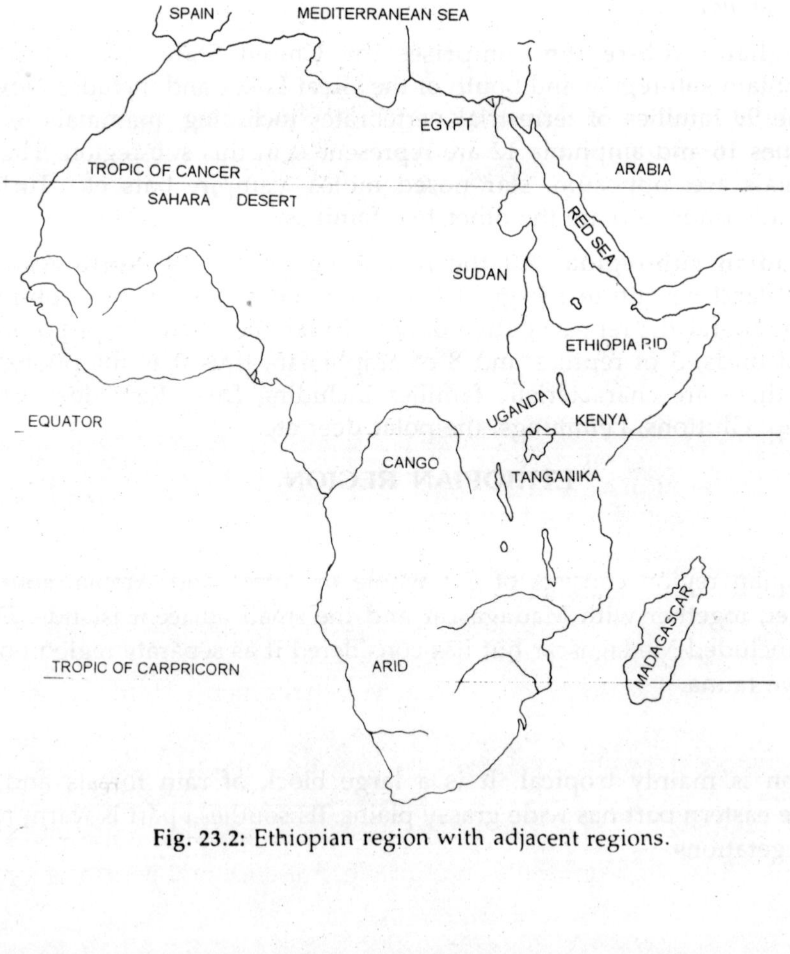

Fig. 23.2: Ethiopian region with adjacent regions.

In addition to these families, a large number of important genera are confined to this region, some of them are; Macroscelidae (Elephant-shrews), Hyaenidae (Hyaena), Elephantidae (Elephants) Rhinocerotidae (Rhinoceroses), Procavidae (Hyraces), Equidae (Horse etc.) Manidae (Pangolins); and among Birds the Ploceidae, Nectariniidae (Sundbirds), Zosteropidae (White Eyes); Indicatoridae (Honey Guides, Capitonidae (Babets), Bucerotidae (Hornbills) and Structhionidae (Ostriches).

The Ethiopian Region is divided into four sub-regions.

(1) South African sub-region

(2) East African sub-region

(3) West African sub-region

(4) Malagasy sub-region.

1. **South African Sub-region** comprises the southern portion of the continent from the Cape northwards. It is represented by 133 families of terrestrial vertebrates of which only one i.e. Prom ropidae is peculiar. The Golden Moles, Elephand Shrews, Earth Wolf, African Mole Rats, African Jumping Hares and Aard Varks are all characteristic mammals. Among birds—Secretary birds, ostriches and, in reptiles Egg-eating snakes are characteristic to this part.

2. **The east African sub-region** includes the rest of the tropical Africa and tropical Arabia. It has 145 families of terrestrial vertebrate of which only two namely Crested-Rats among mammals and Whale-Head among birds are peculiar. Certain other familiar mammals, though not absolutely confined but are highly characteristic, are Giraffes, Zebras and the Rhinoceroses.

3. **The West African sub-region** includes most of the African forest region from the river Gambia Eastwards to beyond Lake Chan and Southwards to embrace the water shed of the Congo. This area has no peculiar families although 134 occur in this subregion. The most characteristic mammals are the Gorilla, Chimpanzee, the Monkeys, Potamogal, the African flying squirrel and Okapi.

4. **Malagasy Sub-region.** It includes Madagascar, Mauritius, Secychelles and neighbouring islands. The fauna of this sub-region is one of most interesting in the whole world. 86 families of terrestrial vertebrates found here, no fewer than 8 are absolutely confined to it. These includes two of mammals Chiromyidae (Aye Aye) and Centetidae (Trenecs); 5 of birds i.e. Aerocharidae (Helmet Bird), Vangidae, Philepittidae, Mesoenatidae and Leptosomatidae,; and one of reptiles i.e. uroplatidae.

The most characteristic feature is that all the 35 known species of sub-family lemurinae of Family-Lemuridae are confined to this sub-region. Of birds 55 families are represented, 15 families of reptiles occur here, presence of 2 genera of Iguanidac is of special interest, since this is an essentially New World group. Chamaelions are very characteristic of Madagascar because they are more abundant than in any other part of the world.

NEOTROPICAL REGION

Extent

It embraces South America, most of Mexico and West Indies. It is joined to the Nearctic Region by Central American isthumus and Separated from other regions by sea.

Climate

Neotropical region is mainly tropical but southern south America continues into south temperate zone. From the west to the east runs the river Amazon with its hundred of square miles of evergreen forests. In the west the long range of the Andes has high mountain forests, plateau land and gentle slopes.

Zoological Characteristics

155 families of terrestrial vertebrates found here and not less than 39 families are absolutely confined to it.

Along mammals 10 families are peculiar: Cebidae (American Monkeys), Callitrichidae (Marmosets), Solenodontidae (Solenodonts), Dasyproctidae (Agouties), Chinchillidae (Chinchaillas), Caviidae Cavies), Dinomydidae (Dinomys), Bradypodidae (Sloths), Calnolestidae (Selvas) and Myremecophagidae (Ant-eaters).

Among birds 23 families are peculiar viz. Coerebidae (Honey Creepers), Phytotomidae (plant Cutters), Pipridae (Manakins), Oxyrhamphida, Dendrocolaptidae, Cononpophagidae, Formicarridae, Pteroptochidae, Galbulidae, Bucconidae, Rhamphastidae, Tinamidae, Steatornithidae, Momotidae, Todidae, Palamedeidae, Psophiidae, Aramidae, Cariamidae, Eurypygiade, Thimocorythidae, Opisthoconidae and Rheidae.

Among reptiles two families are peculiar viz. Xenosauridae and Dermatemydidae.

Among amphibians, 4 families are peculiar—Dendrobratidae, Dendrophryniscidae, Hemiphractidae and Amphignathodontidae.

Neotropical Region is divided into four sub-regions.

1. Chillian sub-region
2. Brazilian sub-region
3. Mexican sub-region
4. Antilean sub-region.

1. **Chillian sub-region.** It includes Western coast of South America and embraces the summits of Andes of Peru and Bolivia. This region does not contain any peculiar family but characteristic forms are Chinchillas, Llamas, Rheas etc.

Fig. 23.3: Typical South America placental mammals.

2. **Brazilian sub-region** includes all the rest of South America teminating Northwards at the Isthmus of Panama. This is the richest of the Neotropical sub-regions, containing representatives of 133 familiar of which 27 are of mammals, 71 birds, 23 reptiles and 12 amphibians. This sub-region is the great home of rearly all the arboreal vertebrates of South America such as American monkeys, Vampire bats, American porcupines, sloths and opossums.

There are in addition many other animals as Spiny-mice, Cavies, American Tapirs etc. found in the forest. Among reptiles and amphibians, the tree snakes, coral snakes, iguanas, alligators, solid-chested tree frogs, typical tree frogs etc. are common.

3. **Mexican sub-region** constitutes all the Neotropical country North of the Isthmus of Panama. It possesses representatives of 127 families of terrestrial vertebrates which include 24 of mammals, 67 of birds, 26 of reptiles and 10 of amphibians.

4. **The Artillian sub-region** includes all the west Indies except Tobago and Trinidad. Since the area is wholly made up of islands, most of which are small, it is hardly surprising to find that the number of families of terrestrial vertebrates is much less. Some 76 families, including 7 mammals, 47 birds, 16 reptiles and 6 amphibians are represented. There is a remarkable absence of mammals in this regions as in other insular sub-region for there are no Primates, Carnivores, Ungulates of Edentata. The only ordents are the spiny mice. The birds of the following important families are quite absent i.e. Plant cutters, Manakins, American creepers, Ant thrushes etc.

The negative characteristics of this region are also very remarkable. Except on Central America and West Indies, there are no insectivores, civets, oxen, sheep and antelopes.

AUSTRALIAN REGION

Extent

The Australian Region includes the whole of Australia New Zealand, New Guinea, the Molluca and other neighbouring islands, and practically the whole of islands in the Pacific ocean. Its boundary on the west is a line drawn between the islands of Bali and Lombok (Wallace's line), thence to the east of Celebes or the Philippine Islands. The line then due eastward along the tropic of Cancer to include the Sandwick Islands, thence curves round to the south of New Zealand and Auckland Islands, Tasmasnia and Australia.

Climate

The northern part of the Region, north Australia and New Guinea, lies with in the tropics with high summer temperature and much of the area is covered by rain forest. The interior of the Australia continent is also hot, but dry, while further south the climate becomes mainly temperate.

Zoological Characteristics

There are 134 families of terrestrial vertebrates about 30 families are peculiar.

Among mammals 8 families are peculiar viz. Macropodiae (kangaroos), Phalangeridae (Phalangers), Phascolomyidae (Wombats), Peramelidae (Bandicoots), Notoryctidae (Marsupial), Dasyuridae (Dasyurus), Echidnidae (Echidna) and Ornithorhynchidae (Ornithorhynchus).

Among birds 17 families are peculier namely: Paradiscidae (Birds of Paradise) Ptilonorhynchidae (Bower Birds), Meliphagidae (Honey Eaters), Drepanididae (Drepanis), Artichornithidae (Scrub Birds), Cyclopsittacidae, Stringopidae (Own-Parrots), Phinochaetidae (Kagu), Gonuidae (crowned pigeons), Didunculidae (Tooth-billed pigeon), Apterygiade (Kiwis), Dromaeidae (Emus) and Casuariidae (cassowaries).

Among reptiles 3 familiar are peculiar: Pygopodidae (scale footed Lizards), Hatteriidae (Tuatra) and Carettchelydidae (fly-River turtle).

Among amphibians 2 families are peculiar: Cerato-batrachidae and Genyophrynidae.

The Australian Region is also characterized by the absence of certain important groups. Thus, only few mammals are there except marsupials and monotremes. The other orders being represented by some bats and small rodents. Apes and monkeys, insectivores, carnivores, ungulates and edentates are entirely absent. Among birds the finches, buntings and wood packers are absent.

The Australian Region has been divided into four sub-regions.

(1) Australo-Malayan sub-region.

(2) Australian sub-region.

(3) Polynesian sub-region.

(4) New Zealand sub-region.

1. **Australo-Malayan Sub-region** comprises all the islands of the Malaya Archipelago not included in the Oriental Region together with New guinea, the Moluccs and the Solomon Islands, in this sub-region, there are 113 families of terrestrial vertebrates of which 4 are peculiar. These are crowned-region and Fly-River Turtle, both confined to New Guinea and two little known Amphibian families Ceratobatrachidae and Genyophyrynidae from the Solomon Islands and Sudest Island respectively. Since most of this area consists of small islands, a large number of peculiar genera and species exists. In New Guinea, three marsupial genera are peculiar. The Babirussa is another remarkable and peculiar member of the Austro-Malayan fauna occurring in

Celebas, Buru and the Sulu Islands. Among birds, Birds of Paradise, Bower Birds, Honey-Eaters, White eyes, Frog mouth, Lories and Megapodes attain their highest development in this area; while Honey-peckers, Cuckoo-Shrinks, Fly-catchers, Pittas and Fruit-pigeons are very numerously represented. Lastly in New Guinea and the neighbouring islands, the maphibian families Narrow—mouthed Toads and the Tree frogs are characteristic.

2. **Australian sub-region** consists of the whole of Australia and Tasmania. In the sub-region 98 families of terrestrial vertebrates are represented, namely 15 mammals 67, birds, 13 of reptiles and 3 of amphibians. Of these about half a dozen, which include wombats, marsupial mole Duck-Bills, scrub-Birds, Lyre-Birds and Emus are confined to this region. This area is notable as being the great home of marsupials, for out of 41 known genera, 34 are represented and 24 confined absolutely to it. The avifauna too is highly peculiar; though only 3 families are confined to this sub-region yet the proportion and peculiar species is larger than in any other sub-region in any part of the world.

The characteristic but not peculiar animals are Kangaroos, Bandicoots, Thylacine, Bower-birds, Honey-eaters, Creepers, Swallow-shrinks, Cockatoos, Bustard-Quails, Cobras, Scale-footed lizards, Varanus, Side-necked Tortoise. The cobras form about 2/3 of all the snakes found in Australia and all of them are poisonous so that the absence of vipers and Rattle snakes is more than compensated. Lastly, the entire absence of all tailed amphibians is noteworthy.

3. **Polynesian Sub-region.** The rest of the Islands as far north as the Tropic of Cancer and including the Sandwich Islands are embraced in the Polynesian sub-region. This is made p entirely of small islands and as such the absence of certain forms is to be regarded as more characteristic.

The families of terrestrial vertebrates is only 53, of these 3 are of bats and 37 birds, whose occurrence in these remote islands is probably due to their superior powers of dispersal. Of reptiles, 9 families occur and of amphibians only 2. The other two families of mammals include the Muridae (Mice) and the Cervidae (Deer).

The families peculiar to the polynesian are the Drepanididae confined to Sandwich islands; Kngu only found in New Caledonia and tooth-billed pigeons confined to Samoa.

4. **New Zealand subregion.** It includes New Zealand, Norfolk island, Auckland, Campbeli and Macquarie islands. These are 34 families of terrestrial vertebrates. Among mammals there are 3 families, family of Noctiliomidae (free-tailed bats) vespertilimoidae (Typical bats) and Muradiae (Micerats).

Among birds 5 families namely the Xeicides, Nestorids (nestor parrots), Stringopids (owl-parrots), Apterygids (Kiwis) and Hatterids are confined to this subregion.

Fig. 23.4: Typical marsupials of Australian Region

Among reptiles 13 families are peculiar. Snakes are absent. The amphibian is only represented by one frog, *Liopelma*.

The peculiar animal known as Tuatra (*Sphenodon punctatus*) belonging to family Rhynchocephalidae (class-Reptilia). It is commonly known as *living fossil*.

ORIENTAL REGION

Extent

This region includes those portions of continental Asia which are not comprised in the palaearctic and Ethiopian Region together with the Malaya Archipelago as for east including Bali, Borneo, the Pilippine islands and Formosa.

Climate

The Indian subregion is in its northern portion, chiefly composed of plain and desert, more particularly in the water-sheds of great rivers Indus and Ganges. Its fauna as a whole shows a desert affinity to that of the Ethiopian Region; while the desert in the north-west is debatable ground, and may be regarded as a transitional tract between the Oriental and Palaearctic Regions. The southern portion of India is more luxurient than the northern one, and is largely covered with tropical forest, with a series of elevated tracts culminating in Western and Eastern Ghats. Ceylong, the Indo-chinese subregion and most of the Malayan islands are almost entirely covered with tropical forests of the most luxurient character and possess varied and extremely rich fauna.

Zoological Characteristics

The terrestrial vertebrates are represented with 153 families of which 10 are peculiar.

Among mammals 4 families are represented by Hylobatiade (Gibbons), Tarsiidae (Tarsiers), Galeopithecidae (Flying lemures), Galeopithcidae (Flying lemures) and Tupaiidae (Tree-shrews).

Birds are with only one family—i.e. Eurylaemidae (Broad Bills).

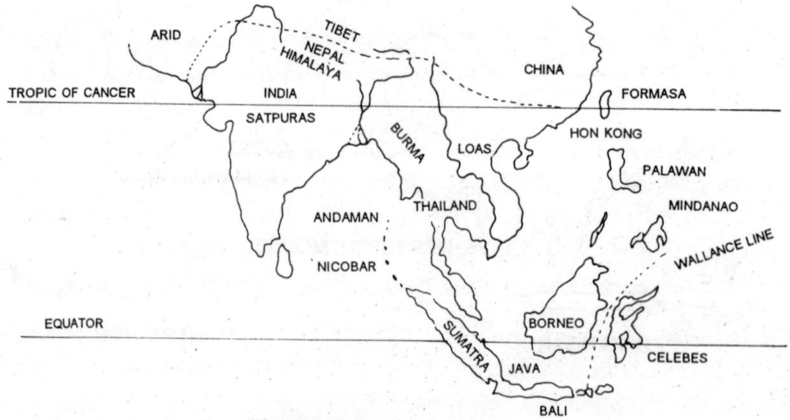

Fig. 23.5: Oriental region.

Among reptiles 5 families are peculiar they are Elachistodontidae, Uropeltidae (Shield Tails), Lanthanotidae, Gavialidae (Gavialis) and Platysternide (Big Head Tortoise).

In addition to the above, several well known species and genera are quite characteristic. Among Mammals the Orangutan (*Simia satyrus*) the Macaque monkeys (*Macacus*), the Tiger (*Felistigris*) and Indian elephant (*Elphas maximus*), Malayan Tapir (*Tapirs indicus*), and three out of the five known species of Rhinoceros are nearly or quite confined to the Region, while families Tragulidae (Chevrotains) and Manidae (Pangolin) are very characteristic. Although only one family of Birds is given above as peculiar, yet many other have their monopolies in the Oriental Region. They are Starlings drongos, Orioles, Honey Peckers, Bulbuls, Pittas and' many others. Reptiles and Amphibians are very well represented in this region for besides the five peculiar families above enumerated, the following are characteristic Illysiidae (Cylinder Snakes), Xenopeltidae, Acrochordidae (Wart snakes) and pitvipers etc.

Wallace divided Oriental region into four sub-regions:

(1) Indian sub-region.

(2) Ceylonese sub-region.

(3) Indo-Chinese sub-region.

(4) Indo-Malayan sub-region.

1. **Indian sub-region** consists of Central and Northern India, from the river Indus and the foot of the Himalayan southwards to Goa and the River Krishna, the line of demarcation taking a southward bend nearly as far as Mysore. It is inhabited by 123 families of terrestrial vertebrates of which only one is peculiar i.e. Elachistodontidae which include a single species of Colubrine Snake. There is no other peculiar feature. Indeed by *Sclater* and other, the first two of the Wallace's sub-regions which together include the whole of Indian peninsula are regarded as one.

2. **Ceylonese sub-region.** It includes the remaining portion of Indian Peninsula, not included in the Indian sub-region and the island of Ceylon. The main peculiarities of this area are the exclusive possession of the curious family of snakes knows as Shield-Tails and the presence of Loris which does not occur in Northern India, though recorded from Eastern Burma. The genera Platacanthomys (Spiny-Rat) is also characteristic, only occurring elsewhere in Cochin China. On the whole, 122 families of terrestrial vertebrates are represented and are indentical to the fauna of Indian sub-region.

3. **Indo-Chinese Sub-region.** It includes China, south to the Palaeartic region, Burma, Thailand and the islands of Hainan, Formosa and Andamans. It contains as many as 138 families.

Among mammals three genera i.e. Takin, Hapalomys and Panda are confined to this sub-region. It also contains Gibbons, Flying lemurs, Java Rhinoceros, Malyan Tapir etc.

Among birds wrens are common. One family of tortoise i.e. Platysternidae is peculiar. Among amphibians salamander and disc-tongued frogs occur.

4. **Indo-Malayan sub-region** includes the Malayan Peninsula and all those islands of the Malaya Archipelogo which fall with Oriental Region. In this region occur 132 families of terrestrial vertebrates of which one, the Lanthonotidae is peculiar and another i.e. Tarsiidae is practically so. In addition, several genera of mammals are peculiar. Among these may be mentioned Orang-Utan, Proboscis -monkey, Heringale, Cynogale, Malayan Badger and some small Rodents. The green Tupaia (Tree-shrews) is well represented and characteristic while the Gibbons, Flying lemurs, two species of Rhinoceros, the Malayan Tapirs and the Broad-bill are peculiarly Oriental forms which this region shows with the Indo-Chinese. The typical Australian Cockatoos are represented by a single species in the Philippines while the Magapodes another essentially Australian group occur in the Philippines, Borneo and the Nicobar Islands.

ZOOGEOGRAPHY OF INDIAN SUBREGION

Blanford (1901) in *Phil. Trans. Royal Soc. London* discussed the geographical distribution of animals on the earth in British India. In his discussion he included India, Pakistan, Kashmir, Gilgit, Ladak, Nepal, Bhutan, Sikkim, other Cis-Himalayan states, Assam, countries between Assam and Burma such as Garo, Khasi and Naga Hills, Manipur, Burma with Karenni, Tenasserim and the Mergui Archipelago, Andaman and Nicobar Islands. *B. Prasad* (1942) added Laccadive and Malolive islands.

Sub-regions of India

Owing to the heterogeny nature of the faunas of the different parts further divisions of Indian sub-region is not an easy task. Its divisions has been attempted by a number of scientists. A brief history of such attempts is as follows.

Jorden (1862) divided it on the basis or birds; *Gunther* (1864) on the basis of reptiles; *Blanford* (1876) on the molluscs; *Wallace* (1876) considered the distribution of all the groups of animals. *Blanford* (1888) on mammals, *M. Smith* (1931) based his findings of reptiles and amphibian; *B. Prasad* (1921) sub-divided Indian sub-region into 5 division. *Mahendra* (1942) divided Indian sub-region into 10 sub-divisions on the basis of distribution of reptiles and amphibians these are:

1. The arid and semiarid province of North India.
2. The western Himalayans.

3. Southern Burmese Province.
4. Trans-gangatic province.
5. Gangatic plain and adjacent part as for far south as 20° latitude.
6. South India below 20° latitude excluding Travancore.
7. Travancore province.
8. Ceylon.
9. Andaman Islands
10. Nicobar Islands.

1. **The arid and semiarid province of North India.** It consists of W. Pakistan, Punjab, Western Rajasthan and Cutch. The fauna is allied to that of the countries to the west of present W. Pakistan and also almost all genera and species are characteristic.

Among lizards are *Teratoscincus, Stenodactylus, Alsophylax, Agamura, Pristurus eremias* etc. Among snakes are *Leptotyphlops, Contia, Lytorhynchus* etc. are present. Many species are endemic. Very few turtles occur but none are endemic or peculiar to it.

Amphibians are rare the only genera found being the cosmopolitan *Rana* and *Bufo.*

2. **Western Himalayas.** It embraces Kashmir Simla, Kumaun and Garhwal Districts, adjacent mountainous regions of Western Tibet Alapine and Punjab etc. This zone is poor in amphibians and reptiles. There are certain endemic species eg. *Gymnodactilus fasciolatus, G. lawderanus, Japalura major, J. kumaonesis* and *Phynocephalus theobaldi* and *Leiolopisma ladacense.*

3. **The Trans-gangetic Province.** It consists of the northern part of Bihar and undivided Bengal, the whole of Assam, Burma north of 20° latitude Nepal, Himalayas to the east of Nepal, Sikkim and Bhutan. The herpetological fauna is both rich and characteristic. Amongst the endemic genera are *Ptyctoaemus, Mictopholis, Oriocalotes, Stolizteaia, Tylotriton etc.*

4. **Southern Burmese Province** comprises southern Burmese province including Burma approximately south of 20° latitude. It includes certain characteristic snake species and some endemic amphibians.

Among snakes are *Gymnodactylus oldhami, G. Consobinoides, Riopa anguina, Bungarus flaviceps, Doliophis intestinalis, Coluber melanurus.* Among turtles are *Platysternum megracephalum, Triony cartilagineus, Geoemyda grandis etc.*

Among amphibians *Kaloula macrodactyla, Glyphoglossues molossus, Callulla guttulata* are characteristic of this region.

5. **Gangatic plain and adjacent Region as far South as 20° lat.** The southern boundary is very vague and approximate. This sub-division is characterised

by the extreme poverty of fauna and has no endemic species. There is complete absence of house gecko. The Indian gharial, *Gavialis* ranges as much as in the Indus system as in Ganges, Brahmputra and Mahanadi.

6. **South India below 29° lat. excluding the Travancore Province.** This province is faunistically characteristic and has *Riopa lineala, Barkudia insularis, Gomnodactylus dekkanenois, Hemidactylus subtriedrus* and *Hemidactylus reticulatus.*

Only a few wide spread species of fresh water turtles are found.

7. **The Travancore.** Consisting of the hilly country south of Lat. 12° or 13° on the west, and south of the Coleroon river in the east. The distinctness of this division is well recognised, the fauna being extremely rich in endemic forms. The number of characteristic genera and spp. is to large.

8. **Ceylon.** It forms a distinct zogeographic province, with clear affinities to the Travancore sub-region.

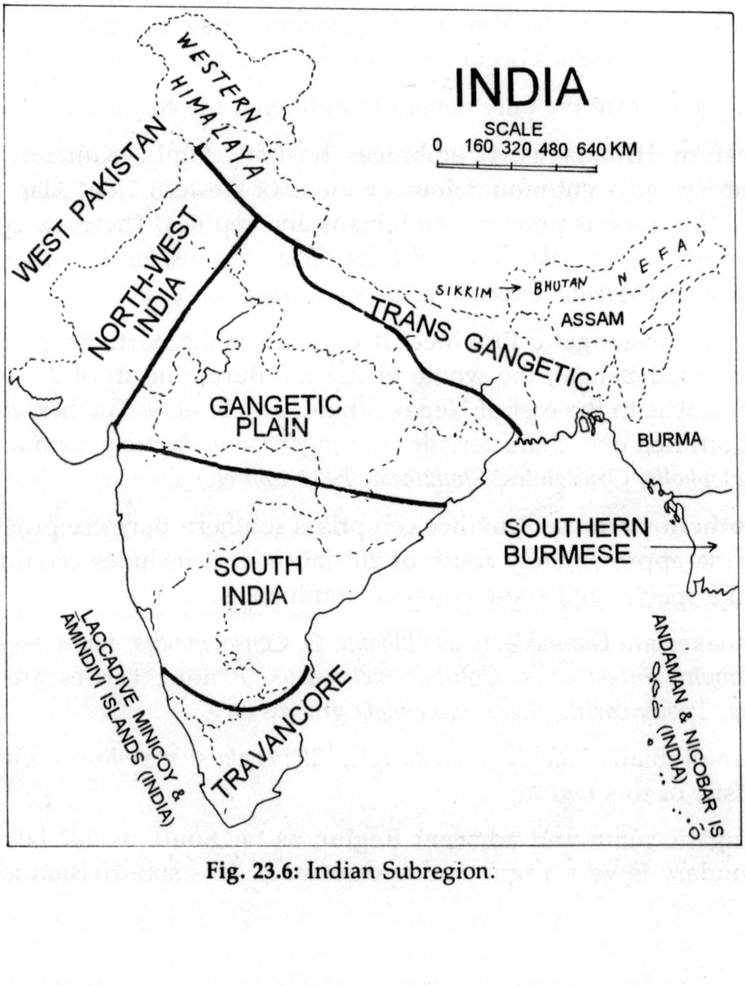

Fig. 23.6: Indian Subregion.

9. **Andaman Islands.** Besides having a few edemic forms, which differentiate the fauna of this region as separate from others, the Andaman are faunistically allied to the Burmese sub-region on one hand, and to Nicobars on the other. There are hardly any endemic genera owing evidently to the fauna being imported from the continent in comparatively recent times.

The endemic species are *Gymnodactylus rubidus, Phelsuma andamanesis, Cslotes andamansis, Typhlope andamansis, Boiga andamanesis* etc. There are no amphibians (except 2 spp. of *Rana*) and no fresh water turtles.

10. **Nicobar Islands.** Its fauna has affinities with Andaman Islands. It has fauna like *Calotes mystaceus, Mabuya andamanesis, Goniocephalus suberistatus.* The endemic species are few in number like *Mabuya rugitera, Calotes jubatuo, Leiolopisma macrotis* and some other.

WALLACE'S LINE

The Wallace's line is the supposed boundary between the Oriental and Australian faunas. The faunas of the Oriental and Australian Regions are extra-ordinarily different. What happens where they meet in the Indo-Australian Archipelago has long fascinated Zoogeographers. As first drawn by *Wallace* (1860), the line ran between Bali and Lombok, between Borneo and Celebes and between the Philippines and Sangi and Taland islands. Later authors debated the position of the line especially the northern part of it. *Huxley,* for example, who named Wallace's line, thought it should split the Philippine putting Palawan in the Orient and rest of the Philippine in the Australian region. Still later even the existence of the line was debated.

Later zoogeographers doubted the validity of Wallace's line. Majority of the scientists thought it be imaginary, some regarded that such as sharp boundary could not be drawn between any two faunal regions. Now Wallace's line is no longer accepted as the boundary between the two faunal regions. Now Wallace's line is no longer accepted as the boundary between the two regions, as there is great difference between the fauna of Bali and Lombok, that are separated by only about 32 km. The rich fauna of Borneo and comparatively poor faunas of Celebes and the Philippines cannot be denied. The difference is regarded due to impoverished young fauna in one area and an older richer fauna in the other.

WEBER'S LINE

In 1904, *Weber* draw another line between the Oriental and Australian regious. It demarcates fauna to the west as predominentally oriental and to the East Australian. However, one problem is that Australian elements for one group of animals, it may not be the same for another group. Majority of the zoogeographers do not recognize Weber's line and prefer a broad belt as a separate zone. This is due to the difference in motility of different kinds of animals. A further proof of these differences is that the flora of New Guinea is mainly Oriental while its fauna is strictly Australian.

Dickenson and others (1923 gave the term 'Wallacea' to separate eastern boundary of the Oriental region at or near Wallace's line and western boundary of the Australian

Region just west of New Guinea. In that the intervening area may be regarded as a separate subtraction transition area. *Darlington* (1957) also recognized that this is an appropriate and useful term. Wallacea bounds the Oriental and Australian regions.

INSULAR FAUNA

The distribution of animals on the islands is of special interest to the student of zoogeography because the islands, which are even closely placed, have their own distinctive fauna. This fauna is found to be different from main continent, although some island fauna is identical to the continental fauna. *Wallace* arranged all the islands in two categories viz. *Continental* and *Oceanic*. The *Continental island* is that which lying comparatively near one of the great continental masses and as a rule containing in its geological formation stratified rocks of various ages. The island has at one time or other been connected with the adjacent continent and got separated from the mainland by a stretch of sea either by sinking of land or by a rise of sea level. As examples of such islands we may take, Great Britain, Borneo, Java, Japan, Philippine and Formosa.

An *Oceanic island*, on the other hand, is one which has never been attached to any continent, but has been formed independently in mid ocean, either by volcanic agency or through the building up of coral reefs or again by a combination of both the methods. Many well known islands may be taken as examples as Galapagos, Sandwich, St. Helena, Azores, Bermuda and Karguelen.

THE OCEANIC ISLANDS

Galapagos Islands

It lies in the Pacific sea about 600 miles west of S. America across the equator It is a group of 15 volcanic islands with a total land area of about 2870 sq. miles. Out of 15 islands 5 are large and 10 are small. There are indigenous Mammals and no Amphibians. Reptiles, however are abundant and the gigantic land tortoises, found in these Islands are both peculiar and of remarkable interest. Five species of lizards occur, one of the family Geckonidae and four of Iguanidae while there are also two species of snakes.

The Birds are similarly interesting, in as much 72 species out of a total 100 occurring in the Islands, are peculiar 66 out of this number are land birds and of these only 3 are fond elsewhere.

The avian fauna is dominated by Darwin's finches. It constitute a natural group of several varied forms all derived from a single ancestor from South American mainland and radiated in different directions in accordance to their feeding habits. These are (i) ground feeders (ii) Cactus feeders (iii) insectivorous free finches (iv) wood pecker-like finches (v) vegetarian tree finches and some species of Penguins.

Of beetles about 40 species are recorded which are mostly peculiar.

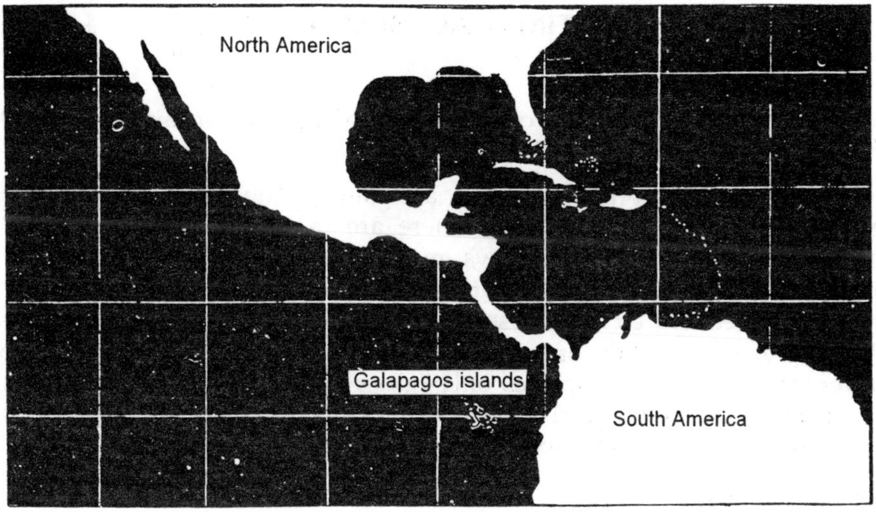

Fig. 23.7: Diagram representing Galapagos islands.

St. Helena

St. Helena is very small volcanic oceanic island about 10 miles long and 8 miles wide. It lies about 1000 miles of Africa in South Atlantic. In this Island there are no indigenous Mammals, land Birds, reptiles or fresh water fishes. One Wader is present namely a peculiar species of Plover (*Aegialities sanctae-helenae*). The beetles number 203 species of which 74 have probably been introduced. Strangely enough about 2/3 of the remainder are Rhynchephora (weevels)—a proportion approached in no other part of the globe.

Bermuda

No terrestrial Mammals occur in the islands but four species of bat occur rarely. Frogs and snakes are absent, but there is one lizard, which is a peculiar species. Of Birds 180 species are recorded, 85 of which are land birds. Of these, about a quarter are regular visitors, while only 11 are permanent residents. Eight of the latter are land and the other three aquatic birds, and all are common the North American mainland. Insects are comparatively scarce. Of Coleoptera 19 species are recorded. Of Hymenoptea 11, of Lepidoptera 26, Diptera 9, Hemiptera and Orthoptera, 9 each. All are common North American or West Indian species. The land molluscs comprise 28 species, six of which, are peculiar.

The Sandwich Islands

Indegious mammals are entirely absent, the Birds are numerous and present many pecularities. 58 species of aquatic and wading birds are recorded of which six are peculiar while four birds of prey have occurred in these islands. The perching birds are represented by 55 species, 41 of which belong to a peculiar family, the Drepanididae while parrots, pigeons and king fishers are absent. Two lizards also present.

CONTINENTAL ISLANDS

British Isles

It consists of two main islands off the coast of Europe Britain about 89000 sq. miles and Ireland about 32000 sq. miles.

Of terrestrial mammals only 40 species are found. Only one is peculiar, namely, the Orkey sole (*Microfus oreadenisis*), but there are in addition-several racial forms confined to three islands. These are shrews, moles, hedgehog, rabbits, hares, squirrels, ottors, marten, cat, reindeer, red deer, bats etc. while the hamsters, lemmings, bears, wolves, wild boar, beavers etc. are absent. Among birds the sea-birds form a large proportion of avian fauna. Only one species namely the Red Grouse is peculiar. The reptiles are represented by three species only (grass snake, adder and smooth snakes); two species of lizards (brown lizard and green sand lizard) *Anguis,* the limbless lizard is also present. Six species of amphibians present.

The Philippine Islands

In these very interesting islands there are peculiar 44 species of mammals (excluding cats), and of these no fewer than 33 are peculiar. Of the 22 Rodents found in the islands and peculiar to them 10 belong to quite peculiar genera. The other mammals confined to the Philippine are a *Galeopithecus* (Flying Lemur), a *Tupaia* (Tree Shrew), and three other Insectivores: a *Tragulus,* and 4 species of *Cervees* (Deer) and lastly, Bufellus mindorensis, a small Buffalos. Of land birds there are 542 species, with no less than 3/2 peculiar.

Formosa

In this island, whose southern portion enters the tropics 36 species of Mammals and 128 species of land Birds are recorded, and of these 10 and 35 are peculiar respectively. The proportion of over a quarter of peculiar land Birds is a very high one, considering the proximity of the island to the mainland. Nearly half the peculiar species of birds are most closely allied to forms occurring in the Himalayas, Southern India, the Malaya Islands and Japan, but absent from the nearest portion of the mainland; while there are eight species common to Formosa and India or the Malaya Islands, but not found in China. All these facts appear to indicate that a great change must have taken place in the distribution of the Asiatic fauna since the island now under consideration was separated from the adjacent continent.

Japan

Although occupying a very similar position with regard to adjacent continent of Asia to that occupied by the British Isles in relation to Europe the Japanese Islands possess a far large number of peculiar forms. Thus of 42 species of mammals no fewer than 27 are peculiar, including a Monkey, two Bats, six Insectivores, six carnivores, nine Rodents, two Deer and a Pig. The Erinaceidae (Hedgehogs) represented by a single species in Britain, are absent but the Ursidae (Bears) Bovidae (oxen) and Suidae (Pigs) are all found in Japan. About 200 species of Birds occur, 17 of which are peculiar. These are principally allied to North Chinese and Siberian forms, but three are of a tropical character, and include a Treron or fruit pigeon.

Chapter—24

Adaptation and Selection

It has been for some years the fashion among certain schools of biological thought to decry the study or even to deny the fact of adaptation. Its allege teleological flavour is supposed to debar it from orthodox scientific consideration, and its study is assumed to prevent the biologist from paying attention to his proper business of mechanistic analysis. Both these strictures are unjustified. It was one of the great merits of Darwin himself to show that the purposiveness of organic structure and function was apparent only. The teleology of adaptation is a pseudo-teleology, capable of being accounted for on good mechanistic principles, without the intervention of purpose, conscious or subconscious, either on the part of the organism or of any outside power. And to the second objection, the answer is that since adaptations are facts, it is the business of biologists to study them. If a biologist thinks that he has exhausted the study of a structure or a function merely by showing its adaptive advantage, he is a bad biologist; but so is he who thinks he has done so merely by giving a mechanistic account of its present condition and its embryological development. The truth is of course that every biological problem has its evolutionary as well as its immediate aspect, its functional meaning as well as its mechanistic basis; and both need to be studied.

Adaptation, in point of fact, is omnipresent. The field worker rightly laughs at the disbelievers in the adaptive significance of mimetic or protective coloration or of threat behaviour. I have been deceived in Africa by the resemblance of a mimetic spider to the ants with which it associates; have spent vain hours on a Surrey common searching for a nightjar's nest, so perfect was the bird's cryptic coloration, before stumbling accidentally upon it; have nearly fallen out of a tree when a wryneck on its eggs simulated a hissing snake. That the examples of protective coloration, afforded by the leaf-insect, the woodcock, the dab, or the twig-like larvae of geometrid moths, should be hackneyed is no argument against their biological validity. Nor does the disbelief of certain laboratory mechanists in warning coloration and other aposematic characters prevent chicks from associating the black and yellow of cinnabar caterpillars with nauseousness, or hinder human beings from paying attention to the rattle of a rattlesnake. The biologist who discovers by comparative study that the metabolism and respiratory pigments of animals are closely adjusted to their mode of life is not likely to imagine that the correspondence is fortuities. The physologist who unravels the postrural reflexes of a bird or investigates the chemical regulation of respiration-rate is not likely to dismiss organic function as non-adaptive; the naturalist who notes the constant correspondence between structure and inborn behaviour on the one hand and environment and way of life on the other-one has only to think of sloth and owl, anteater and flamingo, angler fish and whalebone whale—must believe either in purposive creation or in adaptive evolution; the evolutionary biologist who finds that the rise of each new dominant group in turn is associated with some basic improvement in organic mechanism, be it in the shelled egg, or warm blood, or placental reproduction, will have to admit that adaptation has been all-important in evolutionary progress.

It is perhaps unfortunate that the study of adaptations has been so closely associated with highly specialize and striking cases of the "wonders of nature" type, such as the resemblance of a butterfly to a dead leaf complete with mould-spots and imitation holes, or the almost fantastic contrivances of certain orchids which secure insect-pollination. For this tends to distract attention from the bedrock fact that some degrees of adaptation is omnipresent in life, and that this fact demands an evolutionary explanation.

However, in his recent very striking book Cott (1940) has shown that concealing and revealing coloration, when properly investigated, remain the paradigm of adaptive studies, and has thoroughly turned the tables on captious objectors. Such critics of the theories of protective coloration and mimicry have been in the habit of dismissing them as pure fantasies or armchair speculations. A. F. Shull (1936), for instance, goes so far as to state that the theories of aggressive and alluring resemblance "must probably be set down as products of fancy belonging to uncritical times" and concludes that "if the doctrine can emerge minus its sexual selection, its warming colours, its mimicry, and its signal colours, the reaction over the end of the century will have been a distinct advantage"! The array of facts presented by Cott shows that it is these objections which deserve the designation of "armchair": it is the field naturalist and the experimental biologist who provide the facts from which the theories are educed. Cott also summarizes the numerous experiments and observations which demonstrate the reality of selection operating in nature in favour of cryptic or aposematic coloration. He also points out the irrelevance of the criticism of McAtee (1932).

In addition, Cot analyses the features of pattern by which illusions of various sorts, whether for decrease or increase of conspicuousness, can be created, and then demonstrates their existence in nature. The particular method employed will be related to the type of habitat occupied. Thus inconspicuousness of the flat wing of a butterfly in low rough herbage is generally obtained by a false illusion of relief; the obliteration of sharp outline in a tangle of vegetation tends to be achieved by countershading together with ruptive markings, whereas with forms which must expose themselves on bark it involves arrangements for preventing marginal shadows, often coupled with an actual irregularity of the outline itself, achieved by irregular outgrowths. Most convincing are special correlations of pattern with unusual positions: an excellent example is the reversed countershading of sphingid caterpillars which feed at night, but rest in an inverted position by day, and of the peculiar Nile catfish *Synodontis batensoda*, which swims upside-down.

It is interesting that Suffer (1932, 1935), as the result of intensive studies pursued without knowledge of Cott's work, arrived at similar pursued without knowledge of Cott's work, arrived at similar conclusions. Three recent independent observers may also be quoted. Cornes (1937) cites the moth *Venusia veniculata*, which lives on a particular tree-lily. Its wings are marked with lines running at right angles to the body; and at night it invariably orientates itself across a dead leaf, so that its markings coincide with the conspicuous longitudinal lines on the leaf. Its antennae, which would destroy the resemblance if visible, are tucked out of sight under the fore-wings. When disturbed it settles down again, "after a few compass-like vacillations", in a similar position.

Again, W. W. A. Phillips (1940) describes the nest of *Hempus picatus leggei*, a shrike from Ceylon. The bird nests on bare limbs of trees. The nest is not only camouflaged with lichens and bark flakes, but its sides are built down flush with the branch so as to resemble a knot. Finally, the fledgling young, so long as the parents are away, sit motionless facing each other with eyes half-closed and beaks pointing upwards and nearly touching in the center. Their coloration is a mottled drab and blackish grey, so that they are almost invisible, even when nearly fledged. From a distance of little more than 12 feet the nest with the young bears a most remarkable likeness to a snag left on the upper side of a branch through the breaking off short of a smaller branch just beyond its junction with the major stem. The upward-pointing beaks help to heighten this similarity; they represent the sharp-angled fracture left at the top of the stump. This example may be compared with the protective coloration-*cum*- attitude of the brooding nightjar *Nyctibius griseus*, but is almost more remarkable as involving a co-operative attitude on the part of several birds.

Finally, Holmes (1940) describes the unique case of the common cuttlefish, *Sepia officinalis*, which can change its colour and pattern within the space of a second. By this means, it can draw on an amazingly varied repertoire of protective devices, including concealment by means of obliterative shading, close environmental resemblance, striking ruptive patterns, and flash patterns which bewilder an enemy by their extremely rapid sequence and great difference from each other, and also the scaring of enemies away by conspicuous threat patterns. Any particular one of these will be adopted according to circumstances. In addition, it employs special epigamic stimulative patterns in courtship. Related cephalopods do not show this multiform adaptation, which can be related to the particular habits of the species.

In regard to mimicry, the detailed following by the mimic of the pattern of the model, as the latter changes geographically from subspecies to subspecies, constitutes a beautiful case of detailed adaptation. This phenomenon is not due to any direct or indirect effect of climate.

As Cott rightly says, physiologists and anatomists do not dispute as to whether a wing is or is not adapted to flying: they set themselves to discover the extent to which, and the precise method by which, it is adapted to the function. Colour and pattern in this respect fall into line with any other functional attribute of organisms.

Actually, in view of the remarkable studies of particular kinds of adaptations made in the latter half of the nineteenth century, the incredulity shown by a certain school of modern biologists appears very remarkable. Thus, to take only one example—the various adaptations concerned with cross- and self-pollination in higher plants—we have the intensive work of Darwin (1877) on orchids, and the exhaustive survey, largely original, by Kener. After reading Kerner's account of the devices for securing cross-pollination, and those equally remarkable ones for securing self-pollination, the two often co-existent in the same flower as what the Germans call *doppelte Sicherung,* three would seem to be no room left for scepticism on this point. And if on one point, why on others? However, Cott's book deserves special attention, since it takes account of all the objections, theoretical, factual, and methodological, raised by the sceptics of the early twentieth century.

T. H. Morgan (1932,), in reviewing the subject, makes the following pertinent remarks. "A fact of some interest becomes apparent at once, namely, that what are usually cited as adaptations are instances in which a species shows some unusual type of structure, i.e. one in which it departs from most of the other species in the group. In other words, it is the exceptional that is often referred to as a typical case of adaptation. The reason for this is apparent. The exceptions stand out conspicuously as specialties for some particular situation. Nevertheless, a moment's thought should show that the general problem of adaptation is not to be found so much in these particular occasional departures as in the totality of the relations of the organism to its environment, which makes the perpetuation of the individual and of the species possible. The extreme cases catch our attention, and their special relation is sometimes more easily seen, or guessed at, than the more subtle physiological processes that make all life possible."

ADAPTATION AND FUNCTION; TYPES AND EXAMPLES OF ADAPTATION

Adaptation and function are two aspects of one problem. We may amplify this statement by reminding ourselves that the problem of adaptation is merely the problem of functional efficiency seen from a slightly different angle. There are certain basic functions, such as assimilation, reproduction, and reactivity, which are inherent in the nature of living matter, and can thus hardly be called adaptations. But any of them can be specialized or improved in various ways during evolution to meet the needs of the organism. The fact, for instance, that our gastric glands begin to secrete when our nose or eyes are stimulated by the smell or sight of food, is an adaptation concerned with assimilation, just as is the elaborate structural ruminating mechanism of the oxen and their allies.

The distinction between basic property and superposed adaptation may be well brought out by a historical example. Weismann considers the property of regeneration to be a special adaptation, acquired during the course of evolution by such animals as were especially exposed to los of limbs or other damage. Experiment, however, failed to confirm this conclusion: for instance, Morgan found that the abdominal appendages of hermit-crabs, though normally protected by the hard molluscan shell inhabited by the animal, regenerate just as readily as the exposed big claws or walking legs. Further, on general grounds it became more and more obvious that regeneration depended essentially on the basic capacity of living matter of or reproduction and growth. Regeneration is to-day universally looked upon as one aspect of an inherent quality of life, and the chief problem set by it to biology is not how to account for its presence in lower forms, but how to explain its restriction and absence in higher types.

Frequently associated with regeneration, however, is the faculty of autonomy or self-mutilation, whereby an animal detaches a limb, like a lobster, or a tail, like a lizard, sacrificing a part rather than risk the whole. In most cases autotomy take place at definite spots. The higher crustacea have special breaking-joints which enable them to throw off their claws and legs easily and with hardly any loss of blood; similar but less rigidly-predetermined breaking-joints occur in lizards' tails. It appears quite clear that whereas the regeneration of a lobster's claw is a survival of a basic property of life, its autotomy mechanisms is a more special adaptation—to the risk of the animal being unable to escape if it is seized by the claw, and to the dangers of loss of blood if the exposed claw is damaged.

In addition to the basic functions, others may arise in the course of evolution to meet the needs of the particular type. Thus active locomotion is absent in most plants and colour and pattern can only play an adaptive role in relation to higher animals with their elaborate sense-organs.

From the point of view of selection, adaptations fall into two categories—those of preadaptations fitting an organisms for a different environment or mode of life from the outset and adaptations in the ordinary sense, gradually evolved within the normal environment, whether stable or changing.

A biological classification shows that adaptations fall into a few main groups. In the first place there are adaptations to the inorganic environment. Some of these, like the temperature-adaptation of local races in *Drosophila,* or in frogs as described by Witschi, or of tropical as against arctic organisms, may be of a general physiological nature, unrevealed in any structural peculiarity. Others, like the climbing and parachuting habits of animals in tropical forests, or the black or red colour and the luminosity of deep-sea animals, are more specialized. Hesse and others (1937) in their

Ecological Animal Geography have produced an imposing array of the general types or regularities of adaptation imposed upon various types of fauna by the peculiarities of their inorganic environment. Frequently we can deduce an animal's mode of life and habitat from the structural adaptation which it possesses. Occasionally we may be puzzled, but find that fuller knowledge solves the puzzle. Thus the association of prehensile tails, indicating arboreal life, and fossorial forefeet indicating burrowing habits, in some of the South American anteaters, appears a paradox, until we remember that the fossorial claws are needed to open up the nests of tree-termites.

Next we may take adaptations concerning the organic environment—covering the functions of protection against enemies, the pursuit of prey, reaction against infectious disease and parasites, and the like. These are essentially interspecific. We also find intraspecific adaptations, concerned with competition or co-operation between individuals of the same species, e.g. the rapid growth of many plant seedligns, and the recognition marks of gregarious mammals and birds.

Finally there are adaptations of a more internal nature, concerned with improvement in functions such as digestion or excretion; or with general co-ordination, whether by nervous or endocrine means; or with the regulation of the internal environment. Reproduction may also be considered in this category. As examples of these various internal adaptations we may take the adaptation of the form of the digestive tract and the kinds and quantities of enzymes produced by it to the type of food normally eaten; in nervous co-ordination, we need only think of the inborn mechanism whereby everytime a limb-muscle is stimulated to action, it is normal antagonist is inhibited and relaxed, enabling the contraction of the other to be more effective; in internal regulation we may take the astonishingly delicate mechanism here by the acidity of the blood is kept constant in higher mammals; in reproduction we need go no further than the human species and reflect on the mutual reaction between early embryo and uterus by which the elaborately-organized placenta is produced.

These various classes of adaptations of course overlap an integrade. None the less, an enumeration of them is useful in reminding ourselves that adaptations are nothing else than arrangements subserving specialized functions, adjusted to the needs and the mode of life of the species or type. Most adaptations belonging to our first two categories subserve functions usually called ecological, while the functions of most of those in the last group are physiological. The concept of function has for so long been the preserve of physiology in the restricted sense that we are apt to forget that ecological function is of equal importance to the species.

Our enumeration will also serve as a reminder of the omnipresence of adaptation. Adaptation cannot by the universal among organisms, and every organisms cannot be other than a bundle of adaptations, more or less detailed and efficient, coordinated in greater or lesser degree.

On the other hand, adaptations subserving different functions may be mutually destructive, e.g. high specialization for sexual display is antagonistic to cryptic resemblance. In such cases, the balance between the opposing tendencies will vary in a very instructive way according to the ecology of the species. Artificial selection, as so often, provides valuable parallels. Thus some breeds of dogs, such as bulldogs and St. Bernards, owe their appearance to genes which are on the verge of inducing lethality, and can only be retained by selection of compensatory modifiers. Again, very high milk-producing capacity, rapidity of growth, or extreme conformation for meat purposes in cattle, pigs, etc., may close to the limit of physiological possibilities; in inferior environments (backward tropical regions) animals of this type cannot maintain themselves, so strong is counter-selection.

From the inexhaustible array of possible examples, we may select a few which have been subjected to quantitative analysis, which are unfamiliar or striking, or are of particular importance for evolutionary theory.

A. H. Miller (1937) has analysed in detail the structural peculiarities of the Hawaiian goose (*Nesochen sandvicensis*). This is an endemic of the Sandwich Islands, and exhibits specialization towards a non-aquatic running and climbing habit, with restriction of flying power and absence of migration. It habitat is arid, and not only does it appear never to enter water, in the wild state, but never to drink it except in the form of dew. In correlation with its specialized habits, the webs of the feet are reduced, the legs increased markedly in relative size; a number of muscle and tendon characters (quite different from those prominent in forms specialized for swimming) promote walking and running ability, while the long and flexible toes, with the large plantar pads, help it to climb among the steep irregular lava-flows; the wings and sternum, on the other hand, are definitely reduced. This example is of course much less striking than many classical cases, such as that of the giraffe or the mole, but is illustrates the general adaptive correlation of structure with habit, so clearly set forth by Boker (1924). Similarly the thrashers (*Toxo-stoma*) are adapted to digging.

From a rather different point of view the exhaustive work of Sick (1937) is worth mentioning. His detailed analysis of feather-structure in flying birds demonstrates that feathers exhibit adaptations for efficiency in flight down to the smallest and most unexpected details of structure and intercorrelation.

Desert animals show interesting behavioural adaptations against high winds. Thus various desert butterflies spend most of their active life flying about inside quite small bushes, in order to avoid being blown away; and various desert birds, like Clot-bey's lark (*Rhamphocorys clot-bey*), fortify the rim of their nest with ramparts of pebbles.

Our next set of examples concerns adaptations for the performance of a function overlooked by most biologists, that of toilet in mammals, on which Wood-Jones (1939b) has just published a valuable essay.

The most interesting cases are concerned with the care of the coat. Ungulates lack special structural adaptations for this function, and substitute the crude method of the rubbing-post, combined with a very restricted application of the tongue, and in some instances with the almost equally crude use of horns or antlers. In Equidae the subcutaneous muscle-sheet is highly developed so as to be capable of strong twitching; this, while mainly directed against flies, has a subsidiary toilet function.

In various mammals the tongue is the chief toilet organ. Its greatest specialization for this function is seen in the Feldiae, among which it is much rougher than in other mammals. Wood-Jones seems to be correct in maintaining that this roughness has been evolved primarily as a brush-and-comb. In regard to behavioural toilet adaptations also the cats are specialized: they are the only animals to lick their paw and use them to reach arts of the head not accessible to the tongue. Other organs that may show special toilet adaptations are the teeth and the feet. The most remarkable of these are the procumbent lower incisors and canines of the lemurs. These have all become strangely modified both in shape and position, so that they constitute a most efficient six-toothed comb, the downward strokes of which are well suited for dealing with the animals' thick woolly fur. Further, just as combs need cleaning, so do these teeth: this secondary toilet function is carried out by an abnormally developed sublingua. It should be noted that the development of teeth as toilet adaptations in lemurs is correlated with the almost complete substation of nails for claws on their digits.

The unrelated "flying lemur", *Galeopithecus,* and the bats are also precluded, though in another way, from the full use of their feet as toilet organs; and Wood-Jones points out that they, too, have lower front teeth which appear to be adapted as combs, though in a different way from the lemurs'. In *Galeopithecus* there is also a secondary toilet organ, in the shape of the serrated front edge of the tongue, which acts as a tooth-brush for the pectinated teeth. The toilet function of the special teeth has not been observed here as it has in the lemurs, but may with reasonable certainty be deduced.

In marsupials, Wood-Jones has observed that the polyprotodonts use their incisors as combs, so that the small size, large number, structure (and in some cases position) of these may be regarded as toilet adaptations, though the hind feet are also employed (as we employ both brush and comb). The few large incisors of the diprotodonts, on the other hand, are ill-suited for this purpose, and not employed in the toilet, Here, the united but much reduced syndactylous digits of the hind feet appear to be of use solely as toilet instruments. As with the teeth of the polyprotodonts, their size and shape pare correlated with the length and type of fur with which they have to deal. The bandicoots (Peramelidae) appear to be an exception, since they are polyprotodont but syndactylous. But observations shows that the shape of their teeth is not adapted to acting as a comb, so that the exception proves the rule.

Dusting instincts are among the important toilet adaptations, and may restrict habitat (e.g. in *Dipodomys:*).

Among the other cases cited by Wood-Jones we may mention the special bristly brushes on the feet of certain bats. But enough has been said to show the common characteristic of a particular type of adaptation. It is concerned with a function: the function may be carried out by different organs or combinations of organs in different forms: and the organs concerned how different degrees of structural modification correlated with efficiency in carrying out the function.

A word may here be devoted to the nest sanitation of birds, as this is a good illustration of an adaptation with two point of special interest—it is transitory, but unlike other transitory adaptations such as the foctal membranes of amniotes, the egg-tooth of birds, or the larval structural of echinoderms, it is wholly or mainly a matter of behaviour. In almost all birds, the nest-cup shows a degree of cleanliness which is astonishing until one reflects on the impossibility of rearing a brood of nestlings in their own filth. This cleanliness is secured in various ways. In some forms, such as birds of prey in the later nestling stage, the young defecate only after backing upto the nest-rim; in these, specially developed muscles ensure that the faces are projected well clear of the nest. In most passerines (and some other forms), the droppings are encased in a gelatinous sac secreted by the nestling's intestine. This makes it easy for the parent bird to handle the droppings, which are either eaten or carried away to a distance. In some cases they are eaten while the nestlings are small, but removed wen they grow larger, and in still other cases (e.g. starling, wren, swallow) a third stage is added in which the young evacuate backwards, clear of the nest. In some woodpeckers, the parents mix the nestlings' excreta with sawdust to faciliate handling. Young kingfishers appear to use the innermost part of the nest-tunnel as a latrine. In various species with domed nests, such as the willow warbler (*Phylloscopus trochilus*), the nestlings eject their faecal sacs on to the outer rim of the nest, outside the entrance hole and to the side of it, whence they are removed by the parents. But perhaps the most interesting fact is that in many species the nestlings will not defecate until the parent taps the cloaca with its-beak, often awaiting relief for long period with upturned; posterior! All these adaptations cease to operate, whether in parents or nestlings, as soon as the young birds leave the nest.

The delicacy of transitory adaptation is shown by the larval was of the parasitoid *Glypta haesitator*. These are feebly developed in the second and third instars, when only fluid food is taken, but are powerful in the first, when they are needed for eclosion, and the fourth, when they are required for feeding on solid food and for eating a way out of the host.

In conclusion, we may mention some cases of adaptation for display among birds, Stonor (1936, 1938, 1940), gives a detailed analysis for the birds of paradise (Paradiseidae) and shows conclusively that the remarkable variety of display structures and the equally remarkable variety of display attitudes found in the family are invariably combined to produce the maximum of visual effect. Two examples must suffice. The rifle-bird, *Epimachus (Ptilorhis) paradisea,* has a display quite unlike that of any other member of the family, in which the wings are spread in butterfly fashion; and the effect is enhanced by the broadening of some of the wing-feathers, resulting in a broader and more conspicuous wing. Again, the lesser superb bird of paradise (*Lophorina superba*), has two small patches of specially iridescent feathers on the head. For display, these are erected in such a way as to catch the light and appear as brilliant false eyes.

Stonor (1940) gives an equally illuminating functional analysis of the displays of the pheasants and their allies (Phasianide). We may cite one little-known example. In Bulwer's pheasant (*Lobiophasis bulweri*), the hinder feathers of the compressed tail are stiff and project downwards. In display, they are rapidly drawn through the dead leaves of the forest floor, and enhance the striking visual effect by means of sound.

Among the herons, the display of the Louisiana heron (*Hydranassa tricolor*) and the lesser egret (*egretta thule*) has been studied. Both have a crest, somewhat lengthened neck-feathers and special feathery aigrette plumes on the back. However, the latter are much more highly developed in the egret, the crest and neck-feathers in the heron. And in correlation with this, the egret in display bends down so as to render the fan of filmy aigrettes conspicuous, while the heron erects its head and neck, and the visual effect of the display depends mainly on the crest and much-bristled neck-plumes (Huxley, 1923b).

Conspicuousness is an essential of display: but this function runs counter to the need for concealment. The reconciliation of these opposing selective tendencies is effected in various interesting ways. Where the need for visual concealment is least, as in dense forest, selection for conspicuousness can have full play. It is certainly no coincidence that the most brilliant secondary sexual characters are found in forest forms such as birds of paradise, peacock, most pheasants, trogons, many humming-birds, etc. Where the need for more concealment is greatest, as in defenceless birds of open or relatively open country, display-coloration may be wholly absent, as in the skylark (*Alauda arvensis*), and visual stimulation must be effected solely by striking behaviour. In other cases, as with the prairie chicken (*Tympanuchus cupido*) of the American prairies, a compromise is effected by which the display characters are normally invisible and the bird is markedly cryptic, but become strikingly conspicuous (in this case by expanding of concealed patches of bare yellow skin on either side of the neck, until they look like half-oranges) during the display itself. The great bustard (*Otis tarda*) of the European plains is another striking example, which, by inflating an enormous throat-pouch and everything the wings to show normally concealed white feathers, transforms itself from an inconspicuous to a highly conspicuous object during its display.

Finally, the difference in reproductive habits in birds makes it possible to calculate the differing selective advantage that accrues from success in mating. We may distinguish fractional, unitary, and multiple reproductive advantage. Fractional reproductive advantage is provided by stimulative characters whose effect is merely to raise the reproductive efficiency of a single mate. Unitary reproductive advantage accrues to monogamous forms from characters adapted to securing a mate in the first instance: the male bird either secures a mate and reproduces, or does not do so and fails to reproduce. And multiple reproductive advantage accrues in polygamous forms from characters adapted to securing mates and in promiscuous forms from those adapted to securing coition: success here means transmitting successful characters to the offspring of many mates instead of only one. In correlation with these differences in selective value, characters with a fractional advantage, like display-characters in monogamous territorial passerine birds, in which display occurs only after a mate has been secured, are never very strongly developed. But in such forms a number of males regularly fail to secure a territory and a mate; and the characters concern with securing this reproductive advantage, such as song, are striking and may appear exaggerated or "hypertelic". Finally, where multiple reproductive advantage exists, display characters and display behaviour normally reach an extraordinary pitch of exaggeration, as in ruff (*Machetes,*), peacock (*Pavo cristatus*), various pheasants and grouse, birds of paradise, etc., and the display-characters may even be clearly disadvantageous to the individual in all aspects of existence other than the reproductive, as in the train of the peacock, the wings of the argues pheasant, or the plumes of some birds of paradise.

The giant panda (*Ailuropoda melanoleuca*) has recently been shown to possess an unexpected structural adaptation to its special feeding habits. As is well known, this aberrant carnivore lives almost exclusively on bamboo-shoots. In order to hold these properly while feeding, the sesamoid bone on the radial side of the hand has been much enlarged and furnished with a regular articulation with the scaphoid bone, and a muscle which normally run to the base of the pollex has become diverted to it. The sesamoid with its overlying horny pad has thus become modified into an organ functioning as an opposable thumb. The actual pollex was apparently too specialized to be modified in this direction. Through this remarkable adaptation the giant panda has become endowed with delicate grasping capacity far beyond that of any other member of the order, though the common panda (*Ailurus fulgens*) shows some modification in this direction.

As one more example of this type of adaptation we may take the external ears of the nocturnal bush-babies (*Galago*). These are very much enlarged, to catch and concentrate sound-vibrations. They are also mobile, like the enlarged pinnae of many other mammals, thus ensuring a considerable degree of directional hearing. Finally they are unique in having transverse discontinuities in the cartilages which enable them to be rapidly folded up, thus obviating damage to the delicate pinnae from contact with branches, etc. Here we have three sets of modifications all subserving one adaptive function.

A recent stud by Thorpe (1936) on the life-history of the chalcid *Encyrts infelix*, parasitic on a scale insect, will serve as an example of an unusual adaptation. On reaching its fourth instar, the parasitic larva changes its position and becomes invested with a membranous sheath produced by the host. The sheath then becomes attached in an extraordinary manner to the main lateral tracheal trunks of the host, in four (or six) separate places close above the larval spiracles, in such a way that air can pass through and the parasite from then on respires at the expense of its host. "The conclusion that the whole structure is an adaptation for the respiration of the parasite seems inescapable."

Such "induced adaptation", utilizing the tissues of a host organism, is of course also found in gall-producing animals; the galls they produce may be highly elaborate structures, clearly adaptive in protecting and sheltering the parasite. To quote from Went (1940), "The complexity of the structures induced by the gall insects is often astounding. The central part of the gall with the insect in it may become detached after it is full grown. Then the insect will be released from this box through opening of a pre-formed lid... The inside of the larval chamber is often lined with cells very rich in proteins."

Adaptation is as normal in instinct as in structure. The host-selective instincts of parasitoids hardly ever miscarry, the specificity of such instincts is secured by utilizing a distinctive combination of a few sensory clues. The curious roosting instincts of the hornbill *Lophoceros melanoleucos* are adaptations to secure its nocturnal safety.

Adaptation is just as often manifest internally as externally, in improvement of some physiological function as in better adjustment of some obvious external character like colour or pattern to the environment. Thus, to take a recently investigated example, the giant nerve-fibres of various cephalopods constitute an adaptation for quick and simultaneous contraction of the mantle to expel a jet of water. In *Loligo,* the size of the fibres is graded, larger fibres being found in longer nerves; "this is apparently a further device for securing more nearly simultaneous contraction".

The adaptation of parasites to their hosts comprises a wide range of physiological features, among which the degree of virulence may be singled out here. As is well known, many parasites are only mildly or not at all pathogenic to their natural hosts, though extremely virulent when given the opportunity of attacking "virgin" hosts, e.g. the trypanosomes of wild game when they obtain footing in domestic cattle. While this is in part due to an adaptive increase of resistance on the part of the hosts (cf. the resistance to measles, etc., of human populations which have been long exposed to the disease, while unadapted populations exhibit a high mortality), it may be in part due to the parasite developing an adaptive lower degree of virulence. For it is obviously a disadvantage, from a survival standpoint, for a parasite to kill its host, so that strains of too high virulence will tend to eliminate themselves.

Adaptations to symbiosis are sometimes very striking. Thus numerous animals are enabled to exist in wood by utilizing fungi which break down the wood and probably also act as a source of food for the animal. Special pockets are often produced by the animal, in which a supply of the fungus is carried. This occurs for instance in the larvae (probably only the females) of the wood-wasp *Sirex*. That the special organs are definite adaptations for ensuring a constant supply of the symbionts cannot be questioned. *Tridacna* has remarkable adaptations for exploiting algae, including lenses for increasing photosynthesis.

Again, the comparative study of respiratory pigments and respiratory behaviour in animals has revealed a series of respiratory adaptations to way of life. Recent work on animals with ciliary feeding has similarly revealed the existence of diverse and elaborate adaptations adjusting the ciliary mechanism to different modes of life. The total range of these functional devices is very large, and (once the hypothesis of special creation is ruled out) can only be ascribed to accurate selective adaptation.

REGULARITIES OF ADAPTATION

The perusal of such a work as Hesse, Allee, and Schmidt's *Ecological Animal Geography* (1937) shows that the study of faunas and floras confined to particular habitats will invariably reveal certain recurrent peculiarities. Sometimes these recurrent characters are obviously, or at least *prima facie*, adaptive, like the coloration of desert or pelagic forms, the prevalence of special touch-organs and of luminescence in the deep sea, webbed feet in aquatic birds and mammals, or prehensile tails in forest-living vertebrates. In other cases they are correlated characters in Darwin's sense: this applies, for instance, to some (though not all) of the reduction in relative size of exposed parts like ears or limbs, in subspecies or closely related species of mammals from high latitudes. In still other cases, their significance is doubtful, but even then the fact of their correlation with a particular habitat must be of some significance, and points the way to further analysis.

We have already mentioned certain regularities of variation in discussing clines, and given reasons for believing that most of them were genetic and adaptive, though the visible characters concerned might often by only correlates of the invisible physiological adaptations.

In other cases, we cannot be sure whether the regularities are genetic or purely modificational. Among these we may mention the parallel variation seen in many related species of fish with decreasing salinity, the tendency of fresh-water mussels to be more globes in larger waters, or the increase in thickness and spinosity of shell in the river snails of the genus *Io* as one proceeds down-stream. One must therefore suspend judgment as to the adaptive nature of such regularities pending experimental analysis.

As illustrating genetic regularities, we may take those of desert grasshoppers (Acrididae), as described by Uvarov (1938). This example is perhaps specially pertinent, since Uvarov is an opponent of all adaptational interpretations. He distinguishes four main faunas within the major climatic habitat afforded by desert—the deserticolous proper, inhabiting bare open ground; the saxicolous, inhabiting the rougher habitat provided by the rocky slopes of low eroded desert mountains; the arbusticolous, inhabiting the xerophilous shrubs of many deserts; and the graminicolous, inhabiting the perennial grasses of certain desert plains. These four types differ markedly in body-shape. The deserticolous forms have a depressed body (measured on the metathorax) with width-height ratio from 1.0 to 2.0. In saxicolous forms the ratio is from 0.7 to 1.0, much of the height being due to a prominent narrow dorsal crest. Arbusticolous forms have a similar ratio, though without the crest; and the graminicolous forms have the most compressed bodies of all.

In addition, deserticolous forms tend to be hairy, with punctured, wrinkled, or otherwise sculptured surface, and close resemblance in colour to the soil, often coupled with flash coloration in the hind wings and legs. Most of them are good fliers. In saxicolous forms, the sculpturing is much coarser (the above-mentioned dorsal crest being itself an example of this), and there is a considerably higher percentage of flightless forms. Coloration is similar to that of the first group. Arbusticolous forms possess "climbing legs", which differ in their proportions from the jumping legs of the first two types; they also exhibit concealing coloration, which here, however, tends to be greyish green. Finally, in the graminicolous forms hairiness and surface sculpture are usually negligible, flash coloration is absent, and general coloration is that of green or dry grass, frequently with the sharply-defined light longitudinal stripes that Cott (1940) has shown to be obliterative in grassy habitats.

In spite of Uvarov's anti-adaptional bias, it would seem clear that in these various faunas coloration, both general and flash, body form, and body sculpture are all adaptive. If the form of the legs in the arbusticolous forms can be designated by the functional term of "climbing", it would seem natural to designate the coloration as "concealing". The high pilosity of the open desert forms merits further study.

Another excellent recent example, the result of careful field study of a fauna inhabiting a region with well-marked ecological characteristics, is the work of Linsdale (1938) on the avifauna of the Great Basinin the western U.S.A. The region is arid, the climate severe, with prevalence of strong winds and somewhat scanty and usually low vegetation; the distribution of most birds therefore tends to be more scattered than in more luxuriant surroundings. The preponderating characters of the passerine birds correlated with those environmental features are as follows: a great development of flight-songs, in relation to the scarcity of high perches; a high percentage of protectively-coloured adults; a tendency for both netsling pulmage and nest-lining to be pale-coloured, in order to reflect excessive light; strong powers of flight, to cope with the wind; a high proportion of species are migratory, in relation to the severity of the climate; songs and calls are usually loud, to compensate for the scattered distribution of individuals; long-range vision is unusually acute, partly for the same reason, partly in correlation with the lack of obstruction by vegetation; a high proportion of forms nest on or close to the ground.

Dice (1940b) calls attention to adaptive regularities among the subspecies of the single genus *Peromyscus,* and we mention elsewhere the similar regularities in the Australian bird *Acanthiza.*

We have already referred to the frequent correlation of general tint with climate. Meinertzhagen (1934) gives a good example of this, in the darker plumage of a number of bird species in the Outer Hebrides. He concludes that reduced sunshine and increased atmospheric humidity, rather than higher rainfall, are the meteorological factors responsible.

Meinertzhagen (1919) also points out that in migratory species these regularities are correlated only with the climate of the breeding-quarters, not at all with that of the winter-quarters. This may be due to the greater intensity of selection during the breeding-season.

Buxton (1923) gives a valuable summary of the coloration of desert animals; but his rejection of their cryptic selective vale is much to sweeping. Though doubtless many instances of sandy pallor in deserts are examples of Gloger's rule, and correlated primarily with climate, many others are certainly cryptic. His objection that normally invisible areas, such as the soles of the feet in mice, are of the same colour as the visible parts may be accounted for by "correlated variation", the entire colour being affected except where selective counter-reasons exist. In general, pigmented chitin is tougher, more heat-absorptive, and less permeable to water-vapour. This accounts for various regularities of insect distribution, e.g. the frequency of black desert species, and the increase of pigmentation with altitude and latitude.

It should be mentioned here that some bird species have been experimentally darkened by exposure to humid conditions. The most interesting case for our purpose is *Munia flaviprymna,* a desert form of weaver from Australia. The dark experimental modification of this form, though rather variable, is somewhat similar to a dark form though rather variable, is somewhat similar to a dark form found in nature in a more humid region of Australia. It was at first concluded that the dark colour of this latter form (which was treated by Seth-Smith as a distinct species, but is to-day regarded as subspecific) was itself only modificational. It as much more likely, however, especially in view of its greater variability, that the experimentally darkened desert from was what Goldschmdit (e.g. 1940) calls a *phenocopy* of the genetic darkening of the subspecies from the humid region. If so, the two contrasted forms may have arisen by organic selection genetic adaptation having replaced an original adaptive modification. In any case it is worth noting that other climatic colour-forms are not modifiable in this way. Thus Sumner (1932) could obtain no darkening of pale forms of desert *Peromyscus* in more humid conditions, or lightening of dark humid forms in drier conditions.

When adaptive regularities exist, any exceptions to them immediately attract attention and call for analysis. For instance, the correlation of some sort of webbing on the feet with markedly aquatic habits is all but universal in birds. Ducks, geese, swans, gulls, terns, ptrels, frigate birds, pelicans, cormorants, gannets, and the like have either three or all four toes joined by a web; coots, moorhens, grebes, and phalaropes have lateral lobes on each toe. But the dippers (*Cinclus*) exhibit not a trace of webbing or any other aquatic adaptation, although they are restricted to steams, obtain much of their food below the surface of the water, and can swim on the surface. Structurally, they appear as terrestrial as a thrush or a wren. Can there be a reason for this exception to the rule, or are they still in the early stages of adaptation to a new habitat? Their wide distribution seems to negative the latter explanation. The suggestion may be made that they have adopted a unique type of aquatic food-seeking. Many birds that frequent stream-edges walk some way into the water in order to find food: the dippers have extended this habit and walk on until they are wholly submerged. They search for food by subaqukeous walking, and in this they not only do not require webbing but can get a better grip of submerged water-plants and rough surfaces if their toes are free. The exceptions is a clue to exceptional habits.

Determination of metabolism, temperature-resistance, etc., wen combined with accurate anatomico-physiological study of respiration and directed by ecological knowledge, often reveal regularities in the close adaptation of forms to their habitat.

As an example we may take the work of Wingfield (1939a, b) on mayfly nymphs belonging to various genera. Thus in *Baetis* from swift streams, the tracheal gills do not aid oxygen-consumption; in the pond-dwelling *Chloeon dipternum* they act as an accessory respiratory mechanism by promoting ventilation, but at low oxygen concentrations only while in the burrowing *Ephemera vulgata* they aid oxygen-consumption in all circumstances, apparently as true respiratory organs as well as by providing ventilation. Similarly, forms from swift streams have a lower thermal tolerance than those from slow streams, while those from ponds are most resistant. This is in accordance with the temperature extremes expected in nature.

Fox and his co-workers have studied the activity and metabolism of poikilothermal animals of very various kinds from different latitudes. Among closely-related species, the one living in higher latitudes is generally, but by no means always, adapted in some way to the lower temperatures of its normal habitat: at a given temperature, its heart-beat, respiratory movements, or other activity, is greater than that of its relative from warmer regions. The same phenomenon may also be found as between high-latitude and low-latitude populations of the same species. Differential heat-resistance also exists in many cases. As Fox points out, it is difficult to be sure whether the undoubted adaptation thus shown, enabling cold-water types to carry on the business of living at a reasonable rate, is modificational, genetic, or a mixture of the two. We are probably safe in assuming that, when the difference is one between different species and is of considerable extent, it is mainly genetic, although the recent work of Mellanby (1940) shows howrapi and extensive modificational adaptation may be. A critical analysis of the problem highly desirable. J. A. Moore (1941) has demonstrated a similar and undoubtedly genetic adaptation in different species of frogs, those from colder breeding habitats having a lower temperature-tolerance, and faster-developing eggs. Even the jelly-membranes and the form of the egg-mass are climatically adapted. Again, the field-mouse *Apodemus flavicollis,* in correlation with its distribution, prefers rather lower temperatures than the closely related *A. sylvaticus*; furthermore, within the species, individuals from higher latitudes preferred lower temperatures than those from warmer regions.

That the adaptation between geographical variety or subspecies of a single species may also the mainly genetic is shown by various researches, such as Timofeeff-

Resovsky's previously cited work on local variation of temperature-resistance in *Drosophila funebris*, but most exhaustively by the studies of Goldschmidt (1934, 1938b) on the gipsy moth *Lymantria dispar.* Here the genetic peculiarities of the geographic race, to use Goldschmidt's own words, "harmonize the life-cycle of the animal, especially the feeding season and the diapause, with the seasonal cycle of the inhabited region".

In many cases, notably in Japan, the lines of genetic demarcation between major groups of races are quite sharp. Originally it had been found impossible to correlate these lines with corresponding sharp changes in any single meteorological factor. Recently, however, as Goldschmidt (1938b) points out, it has been shown that they correspond with extreme accuracy with changes in soil type, and that the soil types in their turn depend upon the interrelation of several meteorological factors. This is a reminder of the fact that climate cannot be properly measured by variations in single meteorological phenomena, such as temperature or day-length, since it inevitably represents a complex summation of numerous factors; and further, that physical factors like soil or biological features such as geographical distribution may often prove the best indicators of such summations. In this instance, the discovery of physiological adaptations between subspecies of moths proved to be the first (and very accurate) indication of climatic regional differences. It should, be however, be noted that though the subspeciation of the gipsy moth is thus delicately adjusted to climate, adaptation to food plants may act as a limiting factor. For instance, in the U.S.S.R., the area of periodic mass outbreaks of rapid reproduction of *L. dispar* coincides with the distribution of its optimal food, the oak plant.

Plants, too, many show delicate climatic adaptation of geographical (ecoclimatic) subspecies. Thus according to Clausen, Keck, and Hiesey (1937) the coastal subspecies (which they call ecotypes) of most Californian plants have a constitution genetically harmonized with a climate providing mild winters and a long growing period. Transplanted to an alpine station, although their development is hastened through dwarfing, they can seldom or never mature any seed, and are often unable to flower on resuming growth in the summer. The alpine subspecies ("ecotypes") of the same plant have a cycle related to a climate of long cold winters and a short growing period. Transplanted to a coastal station, they flower poorly or not at all, and show a generally weak appearance in spring. "The adaptive a capacity (modificational plasticity) of coastal and alpine ecotypes is therefore insufficient to allow either to live and to compete n the habitat of the other. It is the difference in inheritance that enable them to succeed in their respective regions."

The exhaustive experimental studies of Turesson (see summary in Barton-Wright, 1932) have independently led him to similar conclusions. In different regions, adaptations arise which are jointly related to climate and life-cycle. He investigated both summer-flowering (festival) and spring-flowering types. In aestival forms, the more southerly populations showed a considerable (genetic) increase in height combined with lateness, while alpine populations showed earliness and reduced height as compared with lowland ones. In spring-flowering forms, on the other hand, it is the more northerly populations that show lateness, up to the latitude of southern Sweden; further north than this, earliness is again favoured. The low-latitude earliness appears to be related to the general earliness of trees in the region, for it is advantageous for the spring herbs to produce their leaves and flowers before the leafy canopy cuts off the sunlight; in very high-latitude spring forms, earliness is doubtless correlated with the shortness of the vegetation period. All the regional peculiarities of the plants investigated are thus adaptive.

Similar though less exact conclusions are reached by O.E. White (1926). For instance, black walnuts (*Juglans nigra*) from Minnesota are much winter-hardier than those from Alabam or Texas, though morphologically indistinguishable. Again a high mountain ecotype of the Cedar of Lebanon (*Cedrus libani*) is perfectly hardy in Massachusetts, where the normal form of this species shows poor cold-resistance. In support of adaptive climatic differentiation within the species, White cites the common practice of gardeners and foresters to use seed from the northern limit of a species' range when winter-hardness is desired.

ADAPTATION AS A RELATIVE CONCEPT

In cases like these, the physiological characters of the local groups must clearly have been adjusted during evolution to the climatic characters of their environments, and are thus in the strictest sense adaptive. But there are many examples where the evolutionary relation between physiology and climate is not so obvious. As illustration we may take some of the cases of plant distribution in Britain so interestingly discussed by Salisbury (1939). In the Scots pine (*Pinus sylvestris*), pollination occurs normally in May, but fertilization not for another thirteen months. Unless the temperature in both summers reaches a certain minimum combination the pollen-tubes will not reach the ovules. This provides quite a different set of meteorological conditions for fulfillment than does the attainment of a minimum level of temperature during one season, as would be the case for the fertilization of most species, and there is some evidence to show that it is a limiting factor for the northern distribution of the species.

In many cases is the temperature obtaining during the time of fruit-formation, not flower-formation, which is decisive. This is so, for instance, with many species of the southern element in the British flora, such as the common milkwort, *Polygala calcarea,* or the fluellin, *Linaria squira*. The form of the life-history may be of importance in various ways. For instance, the time at which flower-buds are laid down varies in different plants. In daffodils it is May, and the optimum is about 9ºC., while in hyacinths it is August and the optimum about 25º. Thus in daffodils either too high or too low May temperatures would inhibit flower production, while they would have no influence on the process in hyacinths. Again, the two British species of *Arum* differ in their winter habits, the common cuckoo-pint, *A. maculatum,* over-wintering as a deep-situated corm, immune from most frosts, while *A. neglectum* produces its new foligae in December. The latter is thus readily killed by frost, but where it can survive, its winter photosynthesis gives it an early start in spring before the trees above it have developed their leaves. The British range of *A. maculatum* extends far into Scotland, whereas that of *A. neglectum* is restricted to our southern and south-western coasts. We may say that *A. neglectum* shows an adaptation to woodland life—but only in mild temperate climates. A somewhat similar difference, with similar results on distribution, is seen between *Scilla verna* and *S. autumnalis.*

Numerous similar cases may be found in textbooks of plant ecology. We may add a recent example from animals. Nash (1937) has been able to study the ecology of the tsetse-fly *Glossina morsitans* both in East and West Africa. Both races appear to demand the same or very similar optimum conditions—a temperature of about 23°C, and a saturation deficiency of about 6 millibars. These conditions are much more nearly reproduced in the *Berlinia-Brachystegia* wood-lands of Tanganyika than in the rather different habitat provided by the small residual forest islands of North Nigeria. In both regions these forest areas constitute the "true habitat" of the species. In the dry season, as evaporation rises the flies become restricted to this true habitat; but in the wet season they show a much wider dispersal. Distribution is definitely controlled by climatic factors, not by abundance of game for feeding.

In Tanganyika, the species breeds mainly under fallen trees; in the rainy season dispersal is very extensive, and the comparative mildness of the dry seasons may allow it to consolidate some of its wet season advances and to form new fly-belts. In North Nigeria, on the other hand, the species breeds promiscuously on the forest floor (so that the logtraps so valuable in East Africa are useless); the west season dispersal is much less in extent, and the severity of the dry season is such that no new colonization can occur; the concentration of fly during the dry season is much more pronounced; the heat and aridity of the dry season is so great that certain habitats are never available, and mid-day inactivity occurs.

Nash considers that the West African from has remained essentially similar to the East African in its physiological requirements. "Having failed to adapt its constitution to the climate, it has perforce adapted its habits; had it evolved a constitution, which preferred a higher degree of evaporation and temperature, the greater frequency of optimum conditions would have enabled it to become as widespread a pest as its East African representative."

In a later paper Nash (1940) has applied these theoretical considerations in practice. Dealing with the three species of *Glossina* found in Nigeria, *G. tachinoides, G. palpalis,* and *G. morsitans,* he first established their basic ecological relations, and then introduced experimental clearing designed to accentuate the severity of pessimum conditions. With *G. tachinoides,* partial clearing on a small scale leads to local extermination, but this is followed by recolonization. With large-scale clearing, however, total extermination is obtained. With *G. palpalis,* this method appears to be of value only in the drier parts of the species' range. Finally, with *G. morsitans,* which has rather different ecological requirements, very extensive and ruthless total clearing is needed to effect extermination, and is not recommended "unless warranted by a large [human] population and abundant funds". The case is interesting as illustrating the practical applicability of an ecological viewpoint which thinks in terms of adaptation to environment.

The examples from tsetse-flies are illuminating in various ways. They illustrate, like the plant examples previously adduced, the importance of inherent physiological requirements, but also well demonstrate the role of modificational plasticity in ensuring adaptation.

As would be expected, plasticity in this sense is more widespread among plants than among animals. Following up the pioneer work of Bonnier (1895), its extent has been investigated by various authors. Many forms have an astonishing degree of plasticity. Thus Clements (1929) was able to demonstrate marked alpine dwarfing in lowland types of many species transplanted into alpine conditions. He was at first inclined to minimize the existence of genetic differences between types. Later, however, he admits that species differ in their plasticity. Thus in the genus *Mertensia, M. sibirica* has no plasticity, while *M. pratensis* and *M. lanceolata* can be made to resemble each other very closely.

Clausen and his associates, have shown how complex is the interrelation of genetic and modificational factors in such cases. For instance, four major ecotypes (ecoclimatic subspecies) are differentiated in the majority of plants in the U.S. Pacific slope—a coast range, a lower mountain, a subalpine, and an alpine form. Yet corresponding ecotypes of different species may react quite differently when transplanted. Thus the alpine race of *Potentilla diversifolia* is relatively stunted when transplanted to a mid-altitude station but the alpine races of *P. glandulosa* and *P. gracilis* become largest at the mid-altitude station and are most dwarfed in their natural habitat: the alpine races of *Achillea millefolium* and *Aster occidentalis,* on the other hand, while tallest at the mid-altitude station, are more dwarfed in lowland than in alpine conditions.

Meanwhile Marsden-Jones and Turrill (1938, etc.), though failing to corroborate some of the more sweeping claims of Bonnier and of Clements, have demonstrated how different is the range of modificational plasticity in different species. Thus the knapweed *Centaurea nemoralis* and the kidney-vetch *anthyllis vulneraria* are little modifiable by different soil conditions, while the plantain *Plantago major* is extremely plastic. In higher animals, behavioural adaptation seems to take the place of modificational plasticity in plants.

In some of these cases, the modification can hardly be regarded as adaptive. This applies, for instance, to the stunting of organisms by unfavourable conditions as the limit of the range of the species is approached. This is of course common in plants, but may also occur in higher animals. Thus the American frehswater fish *Xenotis megalotis* is markedly smaller in the northern part of its range, in correlation with the mean temperature and the length of the warm season. Fully-grown forms from northern

Michigan are 20 per cent smaller than those from the south of the state. Though such modification appears to be wholly non-genetic, it must alter the ecological relations of the species.

These examples of correlation between organic constitution and climate or habitat begin to shed light on the problem of adaptation as a whole. Some climatic adaptations show high specialization—for instance, the run-off mechanisms of plants exposed to constant moisture, or contrariwise the water-storage mechanisms of certain desert plants; some of these latter from the Arizona desert can store enough water to last for more than one rainless year—in certain cases (e.g. *Ibervillea sonorae*) up to ten or more!. Other correlations with climate the more general, though clearly adaptive in the narrow sense of having been accumulated by selection over a long period. Here we may reckon the various adaptations of plants to cold winters—deciduousness in broad-leaved trees; restriction of transpiration in needle-leaved trees; over-wintering as bulbs, corms, or seeds, etc., in herbs; general resistance to low temperature. Adaptations of mammals in cold climates to hibernation, to the reduction of heat-loss, or to adjust the breeding season to the needs of the growing young, fall under the same head. For example, in the roedeer fertilization occurs in autumn as with other north tempeate Cervidae, but the embryo does not develop beyond the early segmentation stage until spring, thus ensuring birth in the favourable period of early summer. Similar definite climatic adaptations, but of a much more delicate nature, are to be found between closely related species, or, as we have seen in *Lymantria,* between races of a single species.

At the other extreme there are organisms with ranges limited by climatic factors, rather than closely adjusted to them. We have just seen excellent examples in *Glossina morsitans* and in various plants. Such forms of course show some climatic adaptation—no tsetse-fly could exist in the arctic, for instance, or in a full desert—but it is of a very general nature. The correlation between the organisms and its environment is in this respect neither delicate nor exact: there is an absence of the lock-and-key correspondence to be seen, for instance, in some colour-adaptations, or in various devices for securing cross-pollination in plants—and apparently in the climatic relations of *Lymantria.* Similarly, many higher animals are found in a number of different habitats. Adaptation is then to a range of habitat-types, not to a single habitat. Certain features in the environment (here often in the plants rather than in physical characters) act as limits to the distribution of the species, but adaptation is not close or detailed.

The common heron (*Ardea cinerea*) shows a marked ecobiotic adaptation to securing food from shallow waters. In addition, it is restricted, and presumably adapted, to a certain climatic range. But the environment also acts selectively in yet another way. During exceptionally severe winters, herons may starve through the freezing of the waters which they frequent. The careful records compiled annually by the British Trust for Ornithology show that herons from colonies within easy range of salt water were least affected by the very severe winter of 1939-40. In 1940, the heron population of the British Isles showed a general decrease in number of occupied nests, compared with the average for the previous three years. But whereas for inland heronrieis (more than 25 miles from tidal waters) the decrease was 31 per cent, for coastal heronries (less than 2 miles from tidal waters) it was only 13 per cent. Thus low winter temperatures and distance from the sea, sometimes separately, sometimes jointly, are bound to be factors limiting the northern distribution of the species. This is borne out by the facts. The heron breeds up to 70° N. in Norway but only to 66° in Sweden and to about 60° in the U.S.S.R.

We find further that organism may be adapted to climatic (and other environmental) factors either narrowly or broadly. Stenothermic species, for instance, have a narrow range of temperature-tolerance, eurythermic forms a wide one: Moore points out that eurythermy is very rare in aquatic animals, and is itself an adaptation to the fluctuating temperature of land life. Stenohaline and euryhaline forms may be similarly distinguished. We may extend the concept to individual plasticity by distinguishing "stenoplastic" and "euryplastic" forms. Euryplasty may grade over into general high viability, which is itself an adaptation, though internal or intrinsic rather than related to particular external conditions. Range of ecological habitat may also be broad or narrow.

We must also remember that adaptations may be very close and detailed, and yet, like mimicry in Lepidoptera, of no or negligible value to the species as a whole, since they have arisen entirely by intra-specific selection, and are thus biologically subsidiary to adaptations affecting general viability, resistance to parasitoids, etc.

This is perhaps the place to mention some interesting cases which suggest the evolutionary adaptation to recent climatic change may now be active. Timofeeff-Ressovsky (1940) cites several cases, of which we may mention the serin finch (*Serinus canarius serinus*) and the yellow-breasted bunting (*Emberiza aureola*). The former has since 1800 extended its range northwards from southern France almost to the English Channel, and from the eastern Alps almost to the Baltic. The latter since 1825 has extended its range westwards from the Urals to west Leningrad. Similarly the roller (*Coracias garrulus*) has shown a northern range-extension in eastern Europe, and the warbler *Acanthopneuste viridana* a westward extension very similar to that of *Emberiza aureola;* while the black redstart (*Phoenicurus ochrurus*) has more or less paralleled the serin, and the greater spotted woodpecker (*Dryobates major*) has within fifty years extended its British range from south of the Tweed to the northernmost woodlands in southern Caithness. The last two cases are not cited by Timofeeff-Ressovsky.

Meinertzhagen (1919) cites other cases, such as the shorelark, *Eremophila flava,* which has not only expanded its breeding-range westwards, but about 1847 established a new migration route, in this differing from other species which have extended their range in a similar way. He also mentions the crested lark, *Galerida cristata,* as an example of the same phenomenon on a more extended time-scale, and accompanied by subspeciation. The fulmar petrel (*Fulmarus g. glacialis*) has shown a marked southern extension of range within the last sixty years along the coasts of Britain. The old supposition that the spread was initiated by a reduction of human depredations seems to be erroneous.

Among Lepidopetra, the moth *Plusia moneta*, first recorded in S. England in 1890, is now common there, and has reached Scotland. The comma butterfly, *Polygonia c-album*, hardly known in Britain outside Gloucestershire, Monmouthshire, and Herefordshire before 1920, has since markedly extended its range E., S., and N. (numerous reports in *The Entomologist*).

It may be suggested that, whenever the effects of human interference can be shown not to have been operative, such range-changes will generally be the result of a changed ecological adaptation. The ranges of forms like the fulmar would be much restricted by the amelioration of climate since the last ice-age, and any genetic changes in temperature-tolerance or nest-site selection which enabled the species to regain some of the lostground would be subject to positive selection. The other species mentioned are extending into milder climates: here presumably a climatic preadaptation was already present, which changes in habitat-preference or nest-site selection have finally enabled the species to utilize. The matter is a complex one, however, and needs thorough investigation before we can conclude that the range-change are the result of adaptive change.

It must not be forgotten that, in the long perspective, dynamic evolutionary trend are as important as are static interrelations at any given moment. The worsening of the climate at the end of the Mesozoic reduced the general adaptiveness of the dinosaurs, pterosaurs, and other reptilian groups, while increasing that of the early mammals and birds. The recent glacial period enabled the cold-climate preadaptation of many tetraploid plants to become dominant over the other adaptive features of the corresponding diploids in higher latitudes, leading to extensive spread of the former. The spread of man favoured that of organisms preadapted to be commensal or semi-parasitic on him or his crops, like house-sparrow, rat, house-martin, or "weds" in general.

Adaptation is thus seen, not as a hard-and-fast category, but as something relative. It is not an all-or-nothing phenomenon, but takes many forms and exists in all degrees. Like other biological categories, it can only be properly understood by detailed and where possible quantitative analysis. Furthermore, the mistake must never be made of thinking of adaptational adjustment solely or primarily in relation to the physical environment: the biological environment is just as important. In some cases plants are restricted to special habitats not because of special climatic adaptations but because they possess a wide range of tolerance towards climatic conditions, with a low degree of what we may call biological or competitive vigour. Thus competition presents their establishment in most habitats; only where their extra margin of tolerance removes them from the swamping effect of their biological competitors can they flourish. Salisbury (1929) cites various cases of this phenomenon. Thus *Ranunculus parviflorus* is in Britain restricted to very unfavourable habitats, e.g. dry shallow soil overlying rock. In cultivated ground (unmanured) it not only grows well, but much better than in nature, and produces ten to twenty times as many fruits. Again, the sorrel *Rumex acetosa* is notorious as a plant of acid soils. In cultivation, however, it shows an increased growth on limed soils, proving that its restricted distribution in nature is due to the competition of plants which are less tolerant of acid conditions.

Such examples "sufficiently illustrate the fact that plants grow not where they would, but rather where they must". The same sort of thing is probably true of many plants characteristic of marginal conditions, e.g. alpine habitats.

In a letter Professor Salisbury has kindly furnished me with some further striking examples. One concerns the rosette plants of open grazing land: for instance, *Senecio campestris* and *Filipendula hexapetala* "are confined, as wild plants, to our chalk astures, but their vigorous growth in other types of soil, when they are protected from competition by cultivation, indicates that their restriction in nature is due to the competition factor". The continual grazing prevents other plants from growing high enough to affect the rosette plants' growth, whereas their peculiar growth-habit flattens their own leaves down in such a way that they cannot readily be eaten.

Another example, involving quite different factors, is that of the hellebore (*Helleborus foetidus*), which in Britain is almost entirely confined to ash woods on calcareous soils. Here again in cultivation this species grows and reproduces well in non-woodland and non-calcareous situations. Its peculiar restriction appears to be due to a combination of two factors. In the first place, it seems susceptible to competition, and any woodland habitat suppresses competitors which are vegetatively active only in summer, whereas it, being evergreen, can assimilate also in winter. On the other hand, most woodlands are too shady in summer; but the unshaded phase of ashwoods, which lasts for seven months out of the twelve, is sufficient for the hellebore. As Salisbury (1929) well says, "dominance may be the consequence of unfavourable conditions acting by *Selective depression*, or to favourable conditions acting as a *selective stimulus*, but in either case the dominance is determined by the *relative* vigour of the species and its competitors."

The perfection of adaptation is also correlated with the degree of competition and other forms of selection-pressure. We discuss this phenomenon more at length elsewhere of this chapter. Here we will merely recall the well-known fact that the intensity of life in the tropics is correlated with a greater prevalence and a greater perfection of various adaptations, of which mimicry is perhaps the best studied. Similarly arboreal adaptations such as prehensile tails are most fully developed where the arboreal habitat is developed in most extreme fashion — in South America. Conversely, where selection-pressure is lower, adaptation tends to be less perfect. We have seen a small-scale example of this in the cichlid fish of African lakes. A large-scale example is provided by the marsupials of the Australian region. Tree-kangaroos, for instance, show an adaptation to arboreal life so incomplete that one cannot imagine their survival in the tropical forests of Malaya or the Amazon. And in general, the Australian marsupials seem unable to compete successfully with introduced forms from other regions, whether predators or direct competitors.

The brings us back to what already been said about adaptation and function. Adaptation, we there said, "is merely the problem of efficient function seen from a slightly different angle". But it is a commonplace that all grades of efficiency of every function coexist in nature. The function of vision ranges from mere response to high light-intensities up to binocular colour-vision. Aquatic locomotion is at a low level in *Amoeba*, at a high pitch in a dolphin or mackerel. Thus we find in nature, not merely every possible type of adaptation, but every grade within each type. Efficiency of function at its most general consists in all-round viability, and this is largely a matter of harmonious adjustment of parts and part-functions. Thus whereas specialized adaptation may push its possessors close to the limits of biological possibility, extremes of all sorts will be discouraged in what we may call generalized adaptation. This is illustrated by the classical work of Bumpus (1899) who picked up a number of sparrows (*Passer domesticus*) found helpless in a storm, and compared those which died with those which revived. The survivors were in general more uniform, while those which died showed greater variability. What is possibly the result of a similar selection of a central type was found by Weldon (1901) and by Cesnola (1907) in two types of land-snail (but not in a second species of *Clausilia*). The inner whorls of adults shells were found to be considerably less variable than young shells of the same size, though the mean was the same. This appears to indicate selective elimination of extreme types.

This is what we would expect of a selection its view. Organisms are selected, not on the basis of conformity to an ideal plan, not in relation to complete functional efficiency, but on the basis of survival. The forms that exist are those that have managed to survive; and survival may be and often is achieved by means of curiously makeshift devices. Not only that, but a high degree of adaptation in one character or function may be a measure of low efficiency in some other respect. It seems, for instance, to be no chance that the most elaborate devices for cross-pollination occur in somewhat rare species of orchids; and Batesian mimicry can only develop in types which are much rarer than their models. Again, specialization which brings success in one set of conditions may involve a loss of plasticity, and so be a real disadvantage if conditions change.

Thus the study of adaptation seems destined to take a new turn. The first stage concerned itself with the fact of adaptation —is such-and-such character an adaptation, or is it not? In the next stage biologists were interested in the mechanisms of adaptation—do adaptations arise through natural selection, by Lamarckian means, or in what other way? To-day the emphasis is on the analysis of adaptation itself, and the bearing of that analysis on other branches of biology—how well-developed are the different types of adaptations shown by a particular organism, and what light does its particular adaptive complex throw on its ecology and on the direction and the strength of the selection to which it has been exposed? The significance of adaptation can only be understood in relation to the total biology of the species.

PREADAPTATION

The subject of preadaptation demands a section to itself. By preadaptation (sometimes styled passive adaptation) we mean either than an existing species (or subspecies) is by its peculiarities predispose to take advantage of a certain type of environment, or that a particular mutant or natural variety is from the outset adapted to particular conditions, whether those in which it originates, or others into which it might be thrown by chance. We may distinguish the two as constitutional and mutational preadaptation respectively. Let us take this latter category first. Lamoreux and Hutt (1939) find that White Leghorn follows are markedly more resistant to vitamin B deficiency than other breeds, such as Rhode Island Reds or, still more, Barred Plymouth Rocks. On normal diet, this characteristic is without any effect on survival, but on a somewhat deficient diet it could be decisive.

A somewhat similar type of variation in a physiological (and therefore potentially adaptive) character is seen in the response of the crop-sac of pigeons to the pituitary hormone known as prolactin. Some breeds proved not less than eight times as responsive as others. Similarly, among plants different strains of the same species may differ markedly, e.g. in water-requirements.

We have already drawn attention to the marked preadaptation of certain mutants in fowls to warm climates, a preadaptation which has been taken advantage of by man. Hutt (1938) has shown that other breeds show minor differences in genetic heat-resisting capacity, which could well serve as pre-adaptive features.

An interesting case was found by Strohl and Kohler (1934) in the meal-moth *Ephestia kuhneilla*. Here a mutation to brown colour, though accompanied by certain unfavourable properties —reduction in egg-number and length of life—also involved a markedly higher heat-tolerance. This differs from the thermal preadaptation of the cladoceran mutant previously described in the complex of characters involved, one of them a visible colour-mutation. What appears to be an example of mutational preadaptation is the replacement of normal by melanic forms in various warm-blood vertebrates in certain areas. As pointed out, the dark forms appear to be preadapted to a moister and cooler climate.

Polyploidy in plants is frequently a thermal preadaptation, but in this case usually towards cold-resistance. An excellent example of preadaptation in a hybrid is the ice-grass *Spartina townsendii,* which has proved better adapted than either of its parent species to their own habitat of salien marsh and mud-flats. In general, it is clear that any form arising by a sudden large change, as by autopolyploidy, or still more by hybridization and allopolyploidy, must be preadapted in some way if it is to survive.

Another possible case of preadaptation, here as regards coloration, is that of the lapwing *Lobipluvia malabarica,* which on a belt of brick-red soil along the Malabar coast lays highly cryptic red eggs in place of the "earthy-coloured" ones seen elsewhere. As suggested in *Nature* this may be due to local selection of types laying the erythrystic eggs found sporadically in so many species. However, the facts concerning egg-mimicry in cuckoos cannot be explained on the basis of preadaptation, and show that elaborate true adaptations may be brought about in egg-colour, so that further analysis of this case is required.

Kalmus (1941a and b) finds that various body-colour mutations in *Drosophila* are potential preadaptations to changes in humidity. Thus yellow flies are less resistant to desiccation than wild-type, but ebony and black flies are more so. This appears to depend on the fact that darkening of the cuticle is associated with a tanning process which renders it less pervious, a fact probably to be connected with the frequency of black insects in deserts.

If is of course true that many such preadapted and markedly distinct new forms are later modified by the selection of small gene-mutations; and it is equally obvious that even the most triflingly beneficial gene-mutation to be found in the constitution of a wild species, must in one sense have been preadapted at its first occurrence. But there is a real and important distinction between the two types of occurrence. For one thing, many (or most) gene-mutations appear to be of necessity carried on in the recessive state until such time as they can be made part of some especially favourable combination: when this is so, they were not preadapted as regards their original phenotypic effect. The truth, as so often in biology, is that a continuous series exists, but that the two ends of it are very distinct. In general, when the origin of a successful new form is due solely or mainly to a single large step (or at least to one that is readily perceived as large by the biologist) we speak of mutational preadaptation, but when a new form arises by a series of small and in general imperceptible stages, we speak of adaptation in the accepted Darwinian sense.

We next come to constitutional preadaptation, where the existing constitution of a stock predisposes it to certain modes of life rather than to others. Salisbury (1929) points out that annual plant species are preadapted to desert conditions. Another example is afforded by the adhesive digital pads or discs of various frogs, which are best developed in the arboreal tree-frogs, though also present in fully functional form in various non-arboreal types. Adaptation to tree-life here seems to have been secured by enlarging these pre-existing structures relatively to body-size. Engels (1940) shows that the digging habit of the thrashers (*Toxostoma*) depends on pre-adaptive peculiarities of musculature—an interesting case of structure preceding function in birds.

In other cases, the preadaptation is less immediate, an organ subserving one function being readily modified for another. The classical example is the evolution of the lungs of land vertebrates from the air-bladders of certain fish, but there are of course numerous other cases of *Funktionswechsel* which illustrate this long-range type of constitutional preadaptation. In all these, however, a great deal of adaptation in the ordinary sense is also necessary, so that it could be better to exclude them from the category of preadaptation proper, and style them predisposition.

General predisposition is shown in the ease with which secondary aquatic life is resumed by terrestrial types. Terrestrial life involves numerous progressive advances in general physiology: the possession of these predisposes such forms to be able to compete successfully with aquatic types in their own environment. Predisposition in the endocrine field is found where an organism which lacks a certain hormone, yet contains tissues capable of responding to the hormone. An example is the response of bird oviducts to progesterone, which appears to be produced only by mammals.

Returning to true constitutional preadaptation, we have the well-worn example of flightlessness in the insects of small oceanic islands. Although very numerous groups may be represented among them, a disproportionate number belong to groups which are not in general good fliers, or are characterized by reduced wings. Flightlessness is here thus the accentuation of a pre-existing tendency. We may here mention the interesting experimental results of L'Heritier, Neefs, and Teissier (1937) on *vestigial*, the wingless mutant of *Drosophila*. When a mixed population of winged (wild type) and functionally wingless (vestigial) individuals was reared in the open air in such a way that they were moderately exposed to the wind, the result after thirty-eight days was an increase in the percentage of homozygous vestigials from 12.5 to 67 per cent, through the wind carrying away more of the winged flies. When the culture was transferred to a large room, the wind could no longer act as a selective agent, and in fifteen further days the percentage of pure vestigials was halved.

A less familiar example is cited by Eigenmann (1909), who maintains that modern fresh-water fish must have been recruited from ancestors preadapted to fresh-water existence by possessing non-pelagic types of eggs. But the *locus classicus* of discussion concerning preadaptation is the blind cave fauna. The out-and-out Darwinians believe that their sightlessness is due to selection gradually ridding the stock of useless organs, while some out-and-out preadaptionists have gone so far as to maintain that mutational blandness came first, and that the sightless type then found a favourable environment ready-made in caves. The truth would appear to be between these two views. No proof has ever been given of full mutational preadaptation in this case, and it is in any event most unlikely. But it is a fact that the cave fauna is drawn preponderantly from types that normally shun light and therefore live in holes and corners. Such forms are constitutionally preadapted to enter caves, and will frequently be visually under-equipped. Their later evolution will consist in their further adaptation to a completely cavernicolous existence, accompanied by further reduction of eyes.

Thus Eigenmann (1909) points out that the fish fauna of the Kentucky caves must have been sifted out by this type of preadaptation from a normal riverine fish fauna when a certain stretch of river became subterranean. The types that were negatively heliotropic, nocturnal, or stereotropic remained subterranean, and then developed further adaptations to cave life; while other ecological types moved out into a connecting river which remained in the open. "The major adaptation to cave existence, the power of finding their food and mates without the use of light, they [the ancestors of the existing cave-fish] possessed before the formation of the caves, and it is responsible for their present habitat." The same general view, with certain modifications, is taken by more recent workers in this field. Hubbs (1938), for instance, after presenting an analysis of the characters and relationship of the thirty-five or so known cave-fish, concludes that this "confirms the theory that cave animals have arisen from species moderately preadapted to cave life". To take an example, the "weak-eyed, long-barbelled, nocturnal catfishes" have given rise to an undue percentage of cave-fish.

There is one notable exception to the general rule, namely the Mexican characid cave-fish, *Anoptichthys jordani.* Although this, as its name implies, is blind, it must have been derived from a form very similar to *Astyanax fasciatus,* which is a large-eyed open-water form, without any obvious preadaptation to caves. Hubbs suggests that the lack of competition, as evidenced by the absence of other cave-fish in this region, faciliated its colonization of caves, sightlessness evolving later.

Hubbs concludes by pointing out that preadaptation has constituted but the first step in the evolution of cave-fishes: later changes, such as further degeneration of skin pigmentation and eyes, and further specialization of sensory barbels and the like, must have been produced by progressive adaptation after the cave habit had been established.

It may be pointed out that, in general, the preponderance of degenerative (loss) mutation will result in degeneration of an organ when it becomes useless and selection is accordingly no longer acting on it to keep it up to the mark. In other cases, as in the hind limbs of Cetacca, degeneration may be actively promoted by selection, as the organ's presence externally is disadvantageous.

Thus, while normal Darwinian adaptation adjust a species to a constant or a changing environment *in situ,* constitutional preadaptation acts as a preliminary sifting device, restricting the inhabitants of specialized habitats in the main to forms with some definite predisposition to the peculiar mode of life involved. Mutational preadaptation is intermediate in the nature of its action, providing a preliminary sifting of lesser extent and shorter range.

Some writers, e.g. O.E. White (1926), consider that a constitutional preadaptation towards cold-resistance has led to certain natural orders of plants being able to survive in higher latitudes when the uniform warm conditions of the earlier cenozoic later give place to a sharply-zone climate, while other groups, not similarly preadapted, became restricted to the tropics. Among the former, he cites the willows (Salicaceae) and horsetails (Equisetaceae), among the latter the palms (Palmaceac) and the Artocarpaceae.

A somewhat different constitutional preadaptation to temperature is found in Crustacea. The osmoregulatory mechanism of various marien Crustacea is such that they are able to tolerate waters of low salinity much more readily at high than at low temperatures. This fact is reflected in the natural distribution of fresh-and brackish-water Crustacea, and very possibly of other invertebrate groups.

Goldschmidt in various of his writings has suggested that preadaptation may play a rather different role by means of large mutations giving what he styles "hopeful monsters", which then can serve as the starting-point for quite new evolutionary trends. As one example of where he thinks this must have occurred he cites the flatfishes, since he considers it impossible for their asymmetry to have arisen gradually. He then extends the principle to other less cogent cases. Mr. J.R. Norman, of the Britis Museum, however, tells me that there exist a few less extreme forms of flatfish, which must be similar to earlier evolutionary stages, and that there seems to be no reason against assuming a gradual evolution of the group from the beginning. Many fish occasionally rest on the bottom, some on their bellies, others on their sides. If benthic existence for any reason were advantageous, selection would set in to improve the type in this respect—with the belly-resting forms by dorso-ventral flattening and lateral extension (as has happened in sharks and rays), with the sideways-resting forms by behavioural and structural asymmetry of the eyes and head.

However, Goldschmidt goes further than this. In his latest book (1940) he maintains that there is a fundamental distinction between micro- and macro-evolution. The former, depending on gene mutation and recombination, may lead to subspecific and other diversification within the species, but cannot produce new species, or, *a fortiori*, higher categories. These come into being through macro-evolutionary change, which, according to him, demands a radical change in the primary chromosomal pattern or reaction-system. Such a change in reaction-system he calls a *systemic mutation*, though he states that it may have to be accomplished in several steps. Only after the repatterning has reached a certain threshold value does the new species-type emerge. He considers (1940) that in some cases at least the initial stages arise only in the absence of selection-pressure against the heterozygote and under certain conditions of inbreeding. But once a new stable pattern, viable as a homozygote, is produced, "selection acts only upon the new system as a whole". In other words, if it survives, it survives as a preadaptation in viability. In other cases he considers that the early steps, too, may be favoured by selection on account of viability effects on development, and that the change will be much quicker than any micro-evolutionary effect dependent on single genes; this process could be regarded as halfway between normal adaptation and preadaptation.

There is a great deal of evidence that gene-mutations *are* involved n specific differences, and that subspecies may evolve into full species. Many of Goldschmidt's analogies between "monstrous" forms found in nature and large mutational steps observed n the laboratory (e.. partial or total wing-rudimentation) are valueless until we know that the natural forms have arisen at a single bound; they may well be merely phenotypically similar to the mutants, but be due to the accumulation of small gene-mutations. Such accumulations may evolve into "gene-patterns" characteristic of species; and the association of gene-mutations with sectional rearrangements may produce relatively large effects: but these are not systemic changes. When he states that the evolution of the *Drepanididae* "by a series of micromutations controlled by selection is simply unimaginable", one can only reply that his imagination differs from that of many other biologists. He rightly insists on the importance for evolution of mutations with consequential developmental effects but these are presumably gene-mutations.

However, even if we dismiss Goldschmidt's views as unproven or unnecessary, preadaptation of various kinds has clearly played a not inconsiderable role in evolution.

THE ORIGIN OF ADAPTATIONS: THE INADEQUACY OF LAMMARCKISM

How has adaptation been brought about? Modern science must rule out special creation or divine guidance. It cannot well avoid frowning upon entelechies and purpositive vital urges. Bergson's *elan vital* can serve as a symbolic description of the thrust of life during its evolution, but not as a scientific explanation. To read *L'Evolution Creatice* is to realize that Bergson was a writer of great vision but with little biological understanding, a good poet but a bad scientist. To say that an adaptive trend towards a particular specialization or towards all-round biological efficiency is explained by an *e'lan vital* is like saying that the movement of a railway train is "explained" by an *e'lan locomotif* of the engine. Moliere poured ridicule on the similar pseudo-explanations in vogue in the official medical thought of his day.

Modern biology, taken by and large, also repudiates lamarckism. They are based not on scientific fact and method, but upon wish-fulfillment. Shaw, in his preface to *Back to Methuselah,* says in effect that he dislikes the idea of a blind mechanism such as Natural Selection underlying evolutionary change-*ergo,* such as a blind mechanism cannot be operative. *Pace* Mr. Shaw, this reasoning does not commend itself to scientists. One of the main achievements of science has been to reveal that the facts of nature frequently fail to accord either with the wishes or with the apparently logical preconceptions of human beings. *Per contra,* we may remind ourselves that, as was pointed out nearly half a century ago by Ray Lankester, lamarckism is self-contradictory, since it maintains that "a past of indefinite duration is powerless to control the present, while the brief history of the present can readily control the future".

An unfortunate suspicion rests on Kammerer's work, and his results on salamanders have not been confirmed by Herbst (1924). Heslop Harrison's adaptive induction of melanic mutations in moths could not be re-obtained by McKenny Hughes (1932) or by Thomsen and Lemcke (1933). Repetition of Guyer's work on induced inheritance of immunity by other investigators has yielded entirely negative results 1924). Pavlov himself withdrew his claim to have demonstrated the inheritance of experience in mice. Recently Crew (1936) has repeated the elaborate researches of McDougall on the hereditary transmission of the effects of training in rats: his results entirely contradict McDougall's lamarckian claims, and he is inclined to ascribe the discrepancy to an insufficiency of controls and an inadequate attention to genetic method on McDougall's part.

Other work, such as that of Heslop Harrison on the feeding habits of insects, is capable of alternative explanation, and is therefore not crucial. Indeed, the researches of Thorpe have made the alternative explanation the more likely, by demonstrating the role of larval conditioning to food in determining the egg-laying reactions of the adults. The researches of Durken (1923) on the colours of butterfly pupae are also capable of alternative explanations, here in terms of unconscious selection of predispositions, and/or of *Dauemodifikationen.*

There remain one or two results, such as that of Metalnikov (1924) on immunity in waxmoths, and of Sladden and Hewer (1938) on food-preferences in stick insects which seem *prima facie* to demand a lamarckian explanation. However, in view of the fate of other claims, and of the theoretical difficulties we shall discuss below, too much weight must not be attached to such isolated cases.

Nor need we pay attention to the view advanced by certain lamarckians, that the inherited effects of function or environmental modification are so slight that they cannot be detected experimentally but require cumulative action through thousands of generations to become obvious. Exceedingly minute differences can be detected by proper technique. The total failure of sixty-nine generations of disuse to affect the eyes or the phototropic responses of *Drosophila,* as shown by Payne (1911), is a good example of the failure of disuse to produce lamarckian effects.

To plead the impossible of detection is a counsel of despair. It is also unscientific: the only scientific procedure would be to refine technical methods until the postulated effects were capable of detection. The experiment has nothing impossible about it with pure-bred stocks and in a rapidly-breeding species.

It is, however, necessary to realize that important indirect objections can be made to any lamarckian view. In the first place there is the fact of mendelian recessivity. A recessive character can be rendered latent indefinitely by keeping the gene concerned in the heterozygous condition; yet when the recessive gene is allowed to unite with another like itself, the resultant character is identical with that of pure-bred recessives in which it has been manifested, and therefore exposed to environmental stimuli, throughout.

An equally fundamental difficulty concerns all those almost innumerable cases in which the two sexes differ in adaptive characters of structure or behaviour. For we know with certainty that the genetic constitution, in the shape of the chromosomes, is distributed irrespective of sex. The chromosomes of a sire will be distributed among his descendants of the second and later generations according to the laws of chance, in a purely random way, and equally among his male and female descendants. What lamarckian mechanism could ensure that the hereditary effects of functions confined to males are transmitted to male descendants only? The situation can just possibly be saved by subsidiary hypotheses, but only at the cost much superfluous complexity, as the geocentric hypothesis was formally saved by the doctrine of epicycles.

Apart from this, we find numerous cases where lamarckian inheritance, even if it exited elsewhere, must be either impossible or exceedingly restricted. Let us first take the case of the higher mammals. These have their internal environment regulated to an extraordinary degree of constancy. The temperature of the blood and to a still higher degree its salt composition and its acidity, are kept constant by elaborate special mechanisms. The reproductive cells, like all other cells in the body, are exposed to the internal environment supplied by the blood-stream. How then can changes in the external environment be transmitted to them? The regulation of the internal environment provides an effective shock-absorber for all the more obvious alterations which could occur in the external environment. Yet higher mammals have evolved as rapidly and in as obvious adaptive ways as any lower types in which this buffering does not exist.

The social Hymenoptera provide another natural experiment of great interest. In them, as is well known, the bulk of the work of the colonies is carried on by neuter females, while reproduction is entrusted to the much less abundant full females and the males. How is it possible on any lamarckian view to discover a mechanism by which the special instincts and structures of the workers have been evolved? They cannot transmit them in reproduction, for they do not reproduce; and the males and females do not practise the instincts nor possess the structures. Attempts have been made to obviate the difficulty by pointing to the fact that occasionally neuter females will lay unfertilized egg, so producing males, If, however, such occasional abnormalities of reproduction suffice to generate the elaborate special characters of neuter ants and bees, then lamarckian transmission operating through normal reproductive channels should have such strong effects as to be detectable by the crudest experiment; and this is certainly not the case.

Insects, indeed, provide a number of hard nuts for lamarckian cracking. All higher insects emerged from the pupa into an adult or imago stage, during which they never moult, and so are incapable either of total or local growth (save by mechanical stretching of membranous parts of the exoskeleton), or of alteration in the form of hard parts. Here again it seems all but impossible to imagine any mechanism by which any modification involving structural change in hard parts could be transmitted. Indeed, such modifications cannot very well be produced at all in the individual: thus the only lamarckian mechanism conceivable is one by which a tendency or an attempt to alter the structure of hard parts would have its first visible effects in the next or late generations! Yet adaptations of hard parts are striking in insects.

A very similar objection applies to mammalian teeth. These, as is well known, exhibit remarkable adaptations to the type of food on which they are normally used. Yet the only effect of use upon them is mechanical abrasion, tending to wear away the structure which has been built up in the plastic stage when the tooth is not used at all.

The origin of the general cryptic resemblance of animals to the prevailing colour of their habitats, whether desert, open sea, green foliage, tundra, or snow, has often been ascribed to a direct effect of the environment. If this rather vague statement means anything, it must imply that the characters in question are either pure modifications, or have been genetically induced by some from of lamarckism. Granted that there exist, notably in insects, some cases of modification, we can now safely assert that most of these characters are genetic. Some of these in their turn may be merely correlated with physiological adaptations; but there are some examples where we can show that selection in favour of cryptic resemblance must have been the agency at work.

In an earlier chapter we mentioned the case of dark subspecies of *Peromyscus* inhabiting local lava-flows. Precisely similar examples are known in birds, e.g. in the desert lark, *Ammomanes deserti,* of which the darkest and the palest subspecies live close together in North Arabia, the one on black lava-desert, the other on pale sand-deserts. This is cited by Meinertzhagen (1934), who makes the pertinent comment that such cases of protective resemblance to soil are largely confined to ground-loving birds. Thus in the black Ahaggar desert, the *Ammomanes* are very dark, while the local babbler (*Argya fulvus buchanani*) is even paler than on the sand of the Sahara—presumably in relation to climate. An even more striking case has recently been described for other mice of another genus, *Peroganthus.* In a valley of southern New Mexico a black lava area of between 100 and 200 squire miles exists quite close to an area of gleaming white gypsum. *P. intermedius* exists on the lava beds in an almost black form, while the representatives of *P. apache* on the gypsum area are nearly white. Both species have normal "mouse-coloured" forms on neighboring rocky areas. Other mammals, reptiles, and insects from the two special areas show corresponding but less extreme colour modifications. Here again no climatic or other influences capable of bringing about the colour-differences can be detected, and we are drive to conclude that the colour is a protective adaptation originated by selection. If selection can be effective in such cases, there is no reason to postulate any lamarckian effect for any examples of general cryptic resemblance.

Hovanitz (1940) cites a similar and very striking case from the butterfly *Oeneis chryxus,* which in the Sierra Nevada exists in two sharply-contrasted dark and pale subspecies or forms, restricted to dark volcanic and pale granitic rocks respectively, the two pure forms connected by clines extending over 10 to 40 miles, where the rock-types are intermingled. Dark rock outcrops in the granitic area below a certain size are inhabited by light forms, being apparently too small to support a dark population that can maintain itself against swamping by crossing. Hovanitz is forced to the view that the two forms owe their origin or at least their maintenance to selection, but rejects the idea that this is exercised via predators in relation to concealing coloration. His objections may be profitably analysed. In the first place, since the upper surface resembles the environmental background much more closely than the lower, he states that visual selection by predators could only occur when the upper surface is exposed, namely, in flight. However, his own photographs show a certain degree of difference in the lower surface. Secondly, he states that when not in flight, they rest in "relative darkness" between rocks, among herbage, etc., "where colour is of no value". This last statement is a mere assertion, as no evidence is given as to possible predators in such situations. (In other habitats, Lepidoptera are frequently captured when at rest.) Finally, he states that almost the only possible predators are two species of birds which only occasionally take insects, and therefore cannot act selectively. In the first place, because his search for predators has not been successful, that is no reason for concluding that they do not exist. Cases must indeed be rare where a small butterfly has no enemies. But further, he appears to disregard the quantitative findings of students of the mathematics of selection, such as Haldane (1932a) and R. A. Fisher (1930a). A 1 per cent advantage-i.e. the average survival of 101 members of one form as against 100 of another—would be almost impossible to detect, yet it would promote an evolutionary change of considerable rapidity, markedly modifying the stock within a few hundred generations.

Hovanitz also makes the general objection to the theory of protective coloration that "the animals getting along best in nature are those which are not `protected" - a fallacy so hoary that it hardly needs serious discussion (but see the general analysis, and on hypertely. Most naturalists will prefer to regard such a case as this as *prima facie* one of concealing coloration until definite evidence to the contrary is produced.

Finally we may mention various special examples of protective resemblance and mimicry. The resemblance of certain moths to birds' droppings or of a stock insect to a stick cannot very well be put down to the inheritance of environmental modifications or the effects of use! In mimicry, the resemblance of model to mimic is often achieved by way of a trick—a similar effect is produced by quite a different mechanism. The "painting in" of a waist on various beetles or bug mimics of ants is a good example: numerous others may be found in Carpenter's little book on mimicry, or in the more general work of Cott (1940).

These are some of the most striking cases in which a lamarckian explanation cannot, it seems, apply. We have already seen that, merely from the standpoint of logic and theory, most adaptations or functional evolutionary changes could be interpreted equally readily on the basis of indirect control by selection as on that of direct control by environment and use. We are therefore driven to ask why, when numerous adaptations like those just cited are shown to be incapable of lamarckian explanation, we should postulate lamarckism to account for the others, which are no different *qua* adaptations. To do so would be to sin against the economy of hypothesis and demand the application of William of Occam's razor.

Thus we are driven back on to direct experimental proof, and that, as we have already set forth, is meagre and confusing. It is for these reason that the majority of biologists, including the very great majority of those who have experience of actual genetic work, repudiate lamarckism, or, at best, assign to it a subsidiary and unimportant role in evolution. Even if lamarckism be operative at all, it seems clear that some other mechanism must be invoked to account for the major part of evolution.

Most biologists also look askance at orthogenesis, in its strict sense, as implying an inevitable grinding out of results predetermined by some internal germinal clockwork. This is too much akin to vitalism and mysticism for their liking: it removes evolution out of the field of analysable phenomena; and it, too, goes contrary to Occam's razor in introducing a new and unexplained mechanism when known agencies would suffice. Furthermore, as R. A. Fisher has cogently pointed out, the implications of orthogenesis, like those of lamarckism, run directly counter to the observed fact that the great majority of mutations are deleterious. In any event, as we shall see in a later chapter, the cases in which a true orthogenetic hypothesis is demanded in preference to a selectionist one are very few, and even in these few it may turn out that it is our ignorance which is responsible for the lack of alternative explanations. As set forth elsewhere, numerous cases exist where evolutionary potentiality is restricted; but these are quite distinct from orthogenesis in the strict sense of a primary directive agency in evolution.

Selection itself often produces an apparent orthogenetic effect. This was realized by H. W. Bates over three-quarters of a center ago in his classical paper on mimicry (1862), where he wrote "the operation of selective agents gradually and steadily bringing about the deceptive resemblance of a species to some other definite object, produces the impression of their being some innate principle in species which causes an advance of organization in a special direction., It seems as though the proper variation always arose in the species, and the mimicry were a predestined goal". However, these and the similar examples drawn from paleontology on analysis turn out to be much better explicable on selectionist principles. Just as the apparent purpose of adaptation is only a pseudo-teleology, so its apparent inner direction is only a psudo-orthogenesis.

THE ORIGIN OF ADAPTATIONS: NATURAL SELECTION

There remains natural selection. Before discussing some concrete examples of selection at work to produce adaptation and of adaptations illustrating the work of natural selection, a few general points deserve to be made. In the first place there is the aged yet apparently perennial fallacy that such-and-such an arrangement cannot be adaptive, since related organisms can and do exist without it. This is, quite frankly, nonsense. It is on a par with saying that electric refrigerators are not useful, because many people, even among those who can afford the expense manage to get on happily without them, or even that alphabets and wheeled vehicles are useless luxuries or accident because the negro and other human stocks never invented them.

There are in fact numerous possible explanations of such a state of affairs. It may be that mutations in that direction did not crop up, or were not available before mutations in some other direction set the stock specializing along other lines; it may be that there are differences in the genetic make-up or the environment of the two forms, as yet undetected by us, which make such an adaptation less advantageous to one than to the other.

All the natural selection can ensure is survival. It does not ensure progress, or maximum advantage, or any other ideal state of affairs. Its results, in point of fact, are closely akin to those of commercial business. In business, what gets across—i.e. is sold—is what can be sold at a profit, not by any means necessarily what is best fitted to meet the real needs of individuals or of the community. The reason for the failure of a commodity to be sold may be lack of purchasing power in the community as much as poor quality, or lack of persuasive (and not necessarily truthful) advertising as much as inefficient production methods.

In the same way a species or a type may survive by deceiving its enemies with a fraudulent imitation of a nauseous from just as well as by some improvement in digestion or reproduction, by degenerate and destructive parasitism as much as by increased intelligence. There still exist those who, even while rejecting the view of Paley and his school that adaptation is a proof of divine design, continue to approach evolution in a rather reverential attitude and to attach some sort of moral flavour to natural selection. They should be reminded of adaptations such as those by which the ant-parasite *Lomechusa* obtains its food, or the orchid *Cryptostilis* ensures its reproduction. *Lomechusa* produces a substance which the ants so dote upon that they not only feed the adult beetle in return, but allow its grub to devour their own larvae—a sacrifice to a gin-producing moloch *Cryptostilis* practices an ingenious variety of prositution: by resembling the females of a fly both in form and in odour, it induces the males to attempt copulation with its flowers, thus securing its own pollination.

We should finally remember that the incidence of selection is different for rare and for abundant species, and that an adaptation forcibly promoted by intraspecific selection in an abundant species might have little or no biological value when worked upon by interspecific selection in its rarer relatives.

It is another fallacy to imagine that because the major elimination of individuals occurs, say, in early life that therefore selection cannot act with any intensity on a phase of minimum numbers, say the adult stage. It has, for instance, been argued that because the main elimination of butterflies take place by parasitization or enemy attack during the larval stage, therefore elimination of the imagines by birds or other enemies can have no appreciable selective effect, and therefore any protective or warming or mimetic colouring which they exhibit cannot have any adaptive significance. But selection need not act with equal intensity at all stages of the life-cycle: even if it should be more intense in early life (and much early mortality appears to be accidental), it could still produce effects on adult characters.

The same argument applies to adaptive colouring shown in the larval stage. Even if this has no effect in protecting the larvae from parasitization, it will have selective value if it protects from attack by other enemies. Granted that on the average 90 per cent of larvae will in any case succumb to parasites, selection can clearly act in other ways on the remaining 10 per cent, just because they are the survivors. In general, selection may promote highly specialized adaptations not only in any particular organ or function, but at any particular phase of the life-history. The elaborate pelagic specializations of many invertebrate larvae at once come to mind, or the adaptations of seeds. Salisbury (1929) cites an interesting case of juvenile adaptation: most plant species, even if light-demanding forms, show greater shade-tolerance in early life, which militates against suppression by shading in the crowded conditions soon after germination.

However, as Professor Salisbury points out in a letter, since the adult phase follows the juvenile in time, many adult characters may well be non-selective *qua* adult characters, but merely consequential results of juvenile adaptations. Some cases of this sort are discussed later.

It is, after all, the adults which reproduce, and a 1 per cent advantage of one adult type over another will have precisely the same selective effect whether the adults represent ten, one, or one-tenth of 1 per cent of the number of fertilized eggs originally produced. The same applies to those plants in which the main elimination occurs during the seedling stage. Selection, in fact, can and does operate equally effectively at any stage of the life-cycle, though it will operate in entirely different ways at one time and another. Further, elimination is far from being the only tool with which selection operates. Differential fertility of the survivors is also important, and in man and many plants is probably the more influential.

Popham (1941) has made a careful investigation of the biological significance of the variation in colour (measured in terms of shade of grey) in various water-boatmen (Corixidae). The animals tend to resemble the backgrounds of the ponds where they are found. This is due partly to habitat-selection: animals confined in surroundings markedly different in background from their own shade become restless and leave to seek other waters. Secondly, it is due to developmental colour-adaptation, the nymphal and adult shade approximating to that of the surroundings in which they have lived in the previous instar (though during a given instar there is no power of colour-adjustment). And finally, it is due to selection, predators (rudd, *Scardinius erythrophthalmus*, were used) taking a heavier toll of "unproteced" animals, i.e. those which are least like their backgrounds.

In regard to selection, a number of intreseting results were obtained. For one thing, two fairly similar colour-varieties for which there was marked differential predation when one of them closely resembled the background, were equally attacked when the background was markedly different from both. It is thus, as would be expected, the relative difference of the two forms from the background, not the absolute difference between them, which acts selectively. Again as expected, a decrease in the population-density of the prey increased the advantage of the protected form. This of course implies a self-regulatory mechanism as regards predator-prey balance, protection conferring maximum benefit when most necessary to the species. A decrease in the predators' population-density also increased the selective advantage of protective colour, presumably on the purely mathematical basis of a reduction in the number of encounters.

Quantitatively, the selection in certain circumstances was very intense. E.g. in one experiment, in which there were employed equal numbers of insects of the same shade as the background ("protected") and differing from it by one colour-standard ("unprotected") the relevant results were as follows:

Insects eaten		Selective advantage
Unprotected percent	Protected percent	(Protected ÷ Unprotected)
75.5	24.5	3.08

In another set of experiments, three types of insects were used, differing from other by one colour-standard, and used against backgrounds of various standards. The results show the variation in intensity of selection with change in the relative difference of coloration between insects and background.

	Per cent eaten of insects, of colour-standards:—			Selective advantage		
	x	(x + 1)	(x + 2)	(x + 1) / x	(x + 2) / x	(x + 2) / (x + 2)
Differences from background (in colour-standard units)						
4, 5, 6	34	33	33	All approximately equal		
3, 4, 5	28	36	36	1.29	1.29	1.00
2, 3, 4	27	32	41	1.19	1.32	1.28
1, 2, 3	11	36	53	3.27	4.82	1.47

A further set of experiments was carried out with species of water-boatmen of different sizes. It was found that the predator used, the rudd, is almost entirely restricted to those of a certain intermediate size. Large forms were difficult to capture (14 per cent taken as against 86 per cent of a medium-sized species) while small species were apparently not noticed at all.

Selective advantage is here, in certain conditions, very large. But we must remember that an advantage which it would be extremely difficult to demonstrate experimentally, say of 1 per cent, would have an effect which, biologically speaking, would be rapid.

Various cases where a selective balance is involved show as forcibly as any laboratory experiment the strength of selection-pressure. The best of all is probably that of industrial melanism. Ford (1940b) has recently shown that in unfavourable conditions (feeding only on alternate days) the dominant melanic form of the moth *Boarmia repandata* has a selective advantage of nearly 2 to 1 (52 blacks: 31 normals surviving to the imago stage where equality was expected).

Even where optimum food-conditions were provided, the ratio was 101: 91. Yet in spite of this enormous constitutional advantage of the melanics, the selective advantage conferred by cryptic colouring on the non-melanics has prevented their replacement by melanics in all non-industrial areas. Mr. Ford informs me that in another case (not yet fully analysed), the melanic form is more cold-resistant. Yet it has not managed to oust the cryptic form, even in the extreme north of Scotland, far to the north of the industrial regions where it has become the type.

An elaborate and large-scale demonstration of selection in action has been given by the of Quayle (1938) on the gradual development, by various scale-insect pests of citrus fruits, of a high degree of genetic resistance to the hydrocyanic acid used to try to kill them. As long ago as 1914 Quayle's attention was drawn to the unsatisfactory results from tent fumigation of lemon trees against red scale (*Aonidiella aurantii*) in the Corona district of California. In most localities it was not then necessary to repeat fumigation for two, three, or even four years. At Corona, however, neither increased dosage nor repetition of fumigation every year or even every six months was effective.

Controlled experiments were later carried out in which the scales from different areas were grown on the same tree and exposed to different concentrations of gas in the same chamber. The results showed that whereas in insects from many localities the normal dosage was reasonably effective, and no scales survived a 50 per cent increase of dosage, in those from the resistant areas, about five times as many survived normal dosage and almost as many survived the increased dosage as survived the normal one in the case of non-resistant strains.

In 1915 evidence turned up of a resistant local strain of the black scale (*Saissetia oleae*), and since then the area of resistance has spread and the degree of resistance has been increased. In 1925 a resistant strain of the citricola scale, *Coccus pseudomagnoliarum*,, was first observed. Prior to this date, fumigators had guaranteed their work with this pest and offered a second fumigation free if the first proved unsatisfactory. In the next few years, the area of resistance spread rapidly, and the highest dosages compatible with the health of the trees failed to give satisfactory results, with the result that fumigation could not be guaranteed, and was eventually abandoned in favour of spraying. Controlled experiments showed that a dose four times the "danger dose" for trees was needed to kill all the resistant insects. In another experiment all insects of a non-resistant strain were killed by sixty minutes' exposure to 0.05 per cent HCN. But after sixty minutes' exposure to a sixfold increase of gas (0.3 per cent), many insects of a resistant strain were alive and a few survived ninety minutes. In general the resistant strain, in concentrations which killed 60 to 100 per cent of non-resistant strains, proved from two to four times more resistant.

Scale insects are not the only forms to show this phenomenon. Hough (e.g. 1934) experimentally proved not only that strains of codling moth (*Cydia pomonella*) from different areas differ markedly in the capacity of their larvae to enter apples sprayed with lead arsenate, but that, when the strain is raised on freshly sprayed fruit in the laboratory, the percentage of larvae capable of this increases from generation to generation.

Resistance in red scale is genetic and it remain unlettered after many generations in the laboratory. An interesting fact is that the resistant strain of red scale has shown itself more resistant to various other toxic substances, to which it has not been exposed in the orchards, e.g. to the fumingants methyl bromide and ethylene oxide, and to oil sprays. It is also probable that it shows greater ability to withstand desiccation. Thus its newly evolved resistance appears to be a general rather than a specific one: the same is true of the codling moth.

Quayle concludes that the resistant strains have developed locally, as a result of intense selection due to the fumigation methods in vogue. When, as appears usual, they have developed earlier in some localities than in others, this is presumably due to the availability of actual or potential variance of the right type, or of new mutations in the right direction. In all cases the area inhabited by resistant strains has rapidly increased. Quayle gives reasons for thinking that this is in the main due to the rapid spread of local resistant types, as soon as these are available through mutation or recombination, rather than to immigration of the resistant forms from the localities where they first appeared. Quayle further points out that the standard fumigation dosage in California, in non-resistant as well as resistant areas, is now much higher than originally. "The schedules have been revised several times and always upwards. It is interesting to note that in Australia, South Africa, and Palestine, countries much younger than California in fumigation practice, the dosage used against the same insect is much lower than in California."

The large-scale experiment with its laboratory controls is of great interest in showing that intense selection may be very effective in bringing about important changes, and in giving indications as to the rate at which the process can operate.

As regards the intensity of selection operating in nature, R. A. Fisher (1939) has been able to calculate the selection operating against (a) homozygosity as against heterozygosity of the various single dominants giving the numerous colour-patterns other than the normal or basic one, (b) combinations of two of these dominants, in the grasshopper *Paratettix texaus*.

The selection against homozygous single dominants varies from about a 7 per cent to a 14 per cent disadvantage, while the elimination of double dominants is estimated to be not less than 40 per cent in each generation. This Fisher considers points to "powerful and variable ecological causes of elimination", whereas the selection in favour of single-gene heterozygotes is probably to be accounted for solely in terms of viability differences.

In any case, if we repudiate creationism, divine or vitalistic guidance, and the extremer forms of orthogenesis, as originators of adaptation, we must (unless we confess total ignorance and abandon for the time any attempts at explanation) invoke natural selection—or at any rate must do so whenever an adaptive structure obviously involves a number of separate characters, and therefore demands a number of separate steps for its origin. A one-character, single-step adaptation might clearly be the result of mutation; once the mutation had taken place, it would be preserved by natural selection, but selection would have played no part in its origin. But when two or more step are necessary it becomes inconceivable that they shall have originated simultaneously. The first mutation must have been spread through the population by selection before the second could be combined with it, the combination of the first two in turn selected before the third could be added, and so on with each successive step. The improbability of an origin in which selection has not played a part becomes larger with each new step.

Most adaptations clearly involve many separate steps or characters: one need only think of the detailed resemblance of a close mimic to its model, the flying qualities of a bird's wing, the streamlining of secondary aquatics like ichthyosaurs or whales. When we can study actual adaptive evolution with the aid of fossils, as with the hooves of horses or the molar teeth of elephants, we find that it is steadily directional over tens of millions of years, and must therefore have involved a very large number of steps. The improbability is therefore enormous that such progressive adaptations can have arisen without the operation of some agency which can gradually accumulate and combine a number of contributory changes: and natural selection is the only such agency that we know. In such cases it is especially evident that what is selected is not a particular gene, but a whole complex of genes in regard to their combined interacting effect.

R. A. Fisher has aptly said that natural selection is a mechanism for generating a high degree of improbability. This is in a sense a paradox, since in nature adaptations are the rule, and therefore probable. But the phrase expresses epigrammatically the important fact that natural selection achieves its results by giving probability to otherwise highly improbable combinations—and "in the teeth of a storm of adverse mutations".

This is an important principle, not only for the conclusion that adaptations as seen in nature demand natural selection to explain their origin, but also for its bearing no the "argument from improbability", used by many anti-Darwinians against Darwinism in general. Bergson has employed this with regard to the origin of the eye. Haldane (1932a) and others, however, have pointed out that a gradual improvement of the visual mechanism from pigment-spot to fully-developed eye is to be expected, and that the parallel development in vertebrates and cephalopods of eyes with lenses is, on the basis of the laws of optics, not in the least unlikely. Indeed, on more general grounds, the properties of natural selection entirely nullify the argument from improbability in this and other cases.

Thus T. H. Morgan and Hogben have asserted that natural selection is seen, in the light of modern genetics, to be essentially destructive: in the absence of natural selection, all the known forms of life would exist, and in addition a vast assemblage of other types which have been destroyed by selection. Though both have now adopted a much more selectionist standpoint, these past views must be refuted as anti-selectionists still often cite them.

T. H. Morgan (1932) writes: "If all the new mutant types that have ever appeared had survived and left offspring like themselves, we should find all the kinds of animals and plants now present, and countless others." The catch here is in the *if;* and the answer, of course, is that every type immediately ancestral to a mutant has been brought into existence only with the aid of selection.

In point of fact the general thesis is entirely untrue. It is on a par with saying that we should expect the walls of a room to collapse on occasion owing to all the molecules of gas inside the room moving simultaneously in one direction. Both are of course only improbabilities—but they are improbabilities of such a fantastically high order as to be in fact entirely ruled out. Each single existing species is the product of a long series of selected mutations. To produce such adapted types by chance recombination in the absence of selection would require a total assemblage organisms that would more than fill the universe, and overum astronomical time.

It should further be remembered that the degree of adaptive specialization is correlated with intensity of selection-pressure.

Elsewhere we have noted how, in the balance between the opposed adaptive tendencies towards cryptic coloration and display coloration in birds, the degree of development of display (epigamic) adaptations is directly proportional to the reproductive advantage it confers upon an individual male. The greater abundance and development of cryptic and aposematic adaptations to be found in the tropics, where selection-pressure is highest, is also to be noted.

The converse of this positive correlation is the tendency of originally adaptive structures or functions to degenerate in the absence of further selection-pressure in their favour. We have spoken of this in relation to the eyes and pigmentation of cave animals but the fact is one of the commonplaces of evolutionary biology, e.g. in parasites. The vestigial wings of ratite birds provide an excellent example. These are in all cases degenerate as regards the adaptations needed for flight. Where, however, they are employed in epigamic display as in the ostriches (*Struthio*) (*Rhea*), they remain of considerable size; but where this further function is absent, as in the emus (*Dromaeus*) and cassowaries (*Casuarius*), they are reduced to vestiges. This tendency towards degeneration of useless structures—i.e. those on which selection-pressure is no longer maintained—is, as we have seen, automatic in most organisms, owing to the accumulation of small degenerative mutations that throw the delicate mechanism of adaptation out of gear. This may be further generalized in terms of gene-effects. Most genes have multiple effects. Organs under direct selection will be modified by a system of genes; but the genes of such a polygenic system will also have secondary effects tend to promote degeneration in size or function. Further, when two linked polygenic systems are lodged in the same chromosome or chromosomes, and selection is acting to alter the main character controlled by one system, while that controlled by the other is useless, the resultant recombination will "break up" the useless character; in virture of the tendency of random change to be towards decreased efficiency, this also will promote degeneration.

None of this reasoning, however, should apply in the case of organisms which do not practise outcrossing. Here, the recombination of "loss" mutations is impossible, and thus degeneration should be exceedingly slow. Furthermore, since many loss mutations need recombinational buffering to survive, they will be automatically eliminated where recombination is impossible. The result should be the persistence of originally adaptive but now functionles structures. The natural place to look for such "relict adaptations" is the floral mechanisms of plant species which have wholly abandoned outcrossing.

At first sight there would appear to be numerous examples of this. For instance, in various Compositae, such as dandelions (*Taraxacum*) and hawkweeds (*Hieracium*) there exist a number of forms which, in spite of producing all their seed by obligatory apomixis, continue to form showy flower-heads, obviously adapted to attract insects. However, the persistence of these erst-while adaptations may be due to the short time elapsed since the change to apomixis. On the other hand, in *Taraxacum* Dr. Turrill informs me that apomixis. On the other hand, in *Taraxacum* Dr. Turrill informs me that apomixis very probably dates back at least 10,000 years.

A more serious objection is the existence of numerous "correlated characters" of the captiulum which still have functional significance. Various parts of the mechanism provide the developmental scaffolding for the adaptive pappus; the ray florets still play a protective role during the night closure of the head, though this protection itself is perhaps a relict adaptation as it probably concerns the pollen. However, such considerations would not apply to obligate apomicts in grasses, where the relict floral mechanism was adapted to anemophilous cross-pollination, nor to the vegetatively reproducing coral-root, *Dentaria bulbifera,* which still makes the unnecessary gesture of producing obviously entomophilous flowers without any apparent subsidiary function.

Obligatory self-pollination should produce the same result. Here the difficulty is to find satisfactory examples, since in most cases some outcrossing still occurs. Thus the orchis *Epipactis leptochila* is normally self-pollinating, but cross-pollination can occur during a brief period. The closely allied *E. latifolia* is exclusively cross-pollinated. However, Dr. Mather informs me that, in Britain at least, the tomato (*Solanum lyosperiscum*) shows no cross-pollination(save in one amomalous variety); yet its obviously entomophilous flowers persist. In some cereal strains, the frequency of cross-pollination is so low (only 2 per cent) that it should enormously reduce the speed of degeneration.

There is thus a prima facie case for the persistence of "relict adaptations" whenever cross-breeding is absent but more investigation is required for full confirmation.

ADAPTATION AND SELECTION NOT NECESSARILY BENEFICIAL TO THE SPECIES

So far, we have been discussing adaptation more or *less in vacuo.* We must now draw attention to the important fact that it will have different effects according to the type of selection operating.

This is best illustrated by the distinction between interspecific and intraspecific selection. In one sense, almost all selection is intraspecific, in that it operates by favouring certain types within the specie at the expense of other types. The only exceptions would be when species spread or become extinct as wholes. The former occurs with such species as are produced abruptly, e.g. by allopolyploidy after hybridization. The latter occurs when no strains within a species are capable of adjusting themselves to a change of climate or to the arrival of new competitors or enemies. Selection in such cases no longer operates by any differential action between different strais, and the whole species spreads or disappears in competition with other species.

The term intraspecific selection can, however, properly be used in a more restricted sense, to denote selection concerned only with the relations of members of one species. One the same basis, interspecific selection is then selection which is ultimately concerned with the environment or with other species. Thus selection for speed in an ungulate will *operate* intraspecifically in the broad sense, btu is *directed* interspecifically in being concerned with escape from predators. Similarly selection for cold-resistance in a period of decreasing temperature is directed environmentally, and may favour the entire species in competition with others. But selection for striking epigamic plumage in male birds is directed intraspecifically, in being concerned with the advantage of one maleoer another in reproduction. It would thus be more correct to speak of selection concerned with intra-or interspecific adaptation; however, it is more convenient to use the terms in the sense I have just outlined.

We have already discussed intraspecific selection briefly in relation to the numerical abundance of species. In scarce species, competition will be more with other species and selection will be related more directly to the environment, while in abundant species there will be more competition between individuals of the species itself. Of course inter- and intraspecific selection will often overlap and be combined; but the intensity of one or the other component may vary very greatly.

An interesting type of selection which is in a certain sense intermediate between interspecific and intraspecific, may occur in forms which exist in numerous and relatively isolated local populations, particularly if the local populations are subject to large fluctuations in numbers. In such cases alpoptiariveations that turn out to have general, in steafmeylcal, value, and which thuayndt displace all other local strains by...exces migration". Wright calls this *intergroup selection.* When this operates, groups compete *qua* groups, on the basis of elaborate gene-combinations restricted to the separate groups. It is probable that this type of evolution has played a considerable role in some kinds of species: cf. Sumner (1932) for *Peromyscus.*

Intergroup selection, however, may operate between groups with a functional basis as well as between those with a regional basis (local populations). Intergroup selection of this sort we may perhaps call social selection, since it will encourage the gregarious instinct and social organization of all kinds. As Allee (1938) has recently stressed with the aid of a wealth of examples, the bases for social life in animals are deep and widespread. There exist numerous cases where it has been experimentally shown that aggregations of a certain size enjoy various physiological advantages over single individuals. Once that occurs, selection will encourage behaviour making for aggregation and the aggregation itself will become a target for selection.

In a late paper (1940) Allee develops this theme further. He shows that when degree of crowding is plotted against efficiency for a large number of functions, the resultant curves are of two sharply distinct types. In the first type the performance has optimum efficiency of lowest population density (e.g. a single pair for maximum fertility per pair in various insects). But in the second there is a phase of "undercrowding" during which the efficiency of the function increases with population densit, finally reaching a peak and then descending in a phase of overcrowding. A special case is the reproductive advantage conferred by size of colony in colonial-nesting birds.

Processes of this type will of course give curves differing in shape, slope, and so forth, and will have correspondingly different results. Wherever such a curve occurs, it means that an aggregation near the peak value will constitute "a supraindividual unity on which natural selection can act.... Such low or feeble social units may be poorly integrated, but still possess demonstrable survival value"; and out of such primitive groupings, intergroup social selection can evolve such specialized group-units as the ant or termite colony.

Finally, since processes giving curves of the multiselective type have been discovered in every major group of animals, it becomes clear that social selection will be widespread, and that "sociality is seen to be a phenomenon whose potentialities are as inherent in living protoplasm as are the potentialities of destructive competition".

In general, the intraspecific type of selection is much commoner than is generally supposed. Thus to think of natural selection as first and foremost a direct struggle with enemies or with the elusive qualities of prey is a fallacy. An equally important feature of the struggle for existence is the competition of members of the same species for the means of subsistence and for reproduction. Surprise has been expressed by some biologists at the fact that in New Zealand, domestic pigs which have become feral have, in spite of the absence of predatory enemies, reverted to something like the wild type; but incompetition for food and reproduction the leaner and more active wild type must clearly have a strong relative advantage over the fatter and more sluggish domestic forms, so that reverse mutations or reversionary recombinations will be favoured by selection.

Elsewhere we consider other examples of intraspecific selection. Sometimes the competition is restricted to individuals of one sex, as in intrasexual selection; sometimes to individuals of a single litter, as in the intrauterine selection of mammals. Again it may be especially intense at a certain period of life, as is the competition for light and space between the sedlings of many higher plants. Another example from plants concerns the competition between the haploid male plants produced by the pollen-grains. Genetic research has shown that these may be affected in various ways, including the rapidity of their growth down the style, by the genes they bear. As a result of this, *certation*, or a "struggle for fertilization" between genetically different types of pollen-grain, may and often does occur, and genes which induce rapid growth of pollen-tubes will often be at a premium. Nothing of the sort, however, appears to take place in higher animals. The only known exception is the gene described by Gershenson (1928) in *Drosophila,* with lethal effects only on Y-bearing sperms. There is also the alleged differential activity of the two types of sperms informs with male heterogamety; if this be a fact, it is probably due to some effect of differential size, the male-determining appearing to have in many cases a smaller head.

Even in most of the relations between species with its enemies, competition is intraspecific. Normally, a certain number of individuals are bound to be killed: when so, the main pressure of selection is directed to keeping an individual out of the category of the relatively unprotected, where it will be an almost certain victim, into that of the well-protected where at least it has an even chance of survival. Any improvement in the protection of some individuals will lead to the bulk of the population being placed at a disadvantage, so that they will once more come under selection-pressure. Such considerations will apply to speed in escape, cryptic and mimetic resemblance, and many other adaptations against predators.

In a different sphere, most competition within civilized human societies is between individuals. The difference of course is that success in this competition is not biological, measured by increased survival to later generations, but social, consisting of monetary and other satisfactions; in fact social and biological success are usually inversely correlated.

Artificial selection is clearly intragroup in its methods. Thus racehorses are selected for reproduction almost entirely on the basis of their individual performances. In most domestic forms, however, once marked breed characteristics have been established, intergroup (interbreed) competition may operate, and reduce or wholly eliminate certain types.

The dependence of the results of selection on the type of competition prevailing is well seen in the case of the social hymenopetra, such as honey-bees, wasps, and ants, where reproductive specialization prevails, and therefore the extinction of individual neuters can have no effect on the constitution of later generations, provided that the community survives.

Haldane (1932a) has demonstrated that only in such a society, which practises reproductive specialization, so that most of the individuals are neuters, can very pronounced altruistic instincts be evolved, of a type which "are valuable to society, but shorten the lives of their individual possessor". Thus unless we drastically alter the ordering of our own reproduction, there is no hope of making the human species much more innately altruistic than it is at present.

The existence of intraspecific selection, i.e. selection between genetically different types within a species, enables us to expose another widespread falacy—namely, that natural selection and the adaptations that it promotes must be for the good of the species as a whole, for the good of the evolving type pursuing a long-range trend, for the good of the group undergoing adaptive radiation, or even that it must promote constant evolutionary progress. In actual fact we find that intraspecific selection frequently leads to results which are mainly or wholly useless to the species or type as a whole. Thus the protection afforded by a cryptic or a mimetic resemblance of moderate accuracy might speedily approach the limit so far as its value to the species is concerned, if there were any way in which selection could be restricted to effects on the species as a species. But as a matter of fact selection acts via individuals, and this intraspecific competition between individuals will often lead to the process of adaptation being continued until almost incredibly detailed resemblances are reached. The perfection of the resemblance of *Kallima* to a dead leaf is one of the marvels of nature; not the least marvellous aspect of it is that it is of no value to the species as a whole.

A. J. Nicholson (1933) has pointed out how advantages operating at one stage of the life-history may be compensated for by increased mortality in other stages, so that the species does not benefit as a whole. Thus in most Lepidoptera a cryptic pattern favouring survival of adults will result in more larvae, which in turn will permit a disproportionate increase in parasitoid infection, thus bringing down the number of adults again. Wherever this balance of elimination as between stages is approximately self-regulation, factors affecting it will be over-riding as regards interspecific selection, while selection for other characters must be intraspecific.

In such examples, the adaptation is at least not deleterious. In other cases, however, it may lead to deleterious results. This is perhaps especially true of selection which is not only intraspecific —confined to competition between members of the same species —but also intrasexual—confined to competition between members of the same sex of the same species. When polygamy or promisiscuity prevails, the selective advantage conferred by characters promoting success in mating will be extremely high accordingly in such forms we meet with male epigamic characters of the most bizarre sort which, while advantaging their possessor in the struggle for reproduction, must be a real handicap in the struggle for individual existence. The train of the peacock, the tail of the argus pheasant, the plumes of certain birds of paradise, the horns and antlers of certain ungulates, are obvious examples. In such cases of course a balance will eventually be struck at which the favourable effects slightly outweigh the unfavourable; but here again extinction may be the fate of such precariously-balanced organisms if the conditions change too rapidly.

We may, however, go further and suggest with Haldane (1932a) that intraspecific selection is on the whole a biological evil. The effects of competition between adults of the same species probably, in his words, "render the species as whole less successful in coping with its environment. No doubt weaklings are weeded out, but so they would be in competition with the environment. And the special adaptations favoured by intraspecific competitions divert a certain amount of energy from other functions, just as armaments, subsidies and tariffs, the organs of international competition, absorb a proportion of the national wealth which many believe might be better employed".

Intraspecific competition among anemophilous plants has led, it seems, to a real overproduction of pollen; among male mammals to unwidely size as in sea-elephants, or to over-developed weapons and threat-organs as in deer and various horned groups; among parasites to their often monstrous exaggerations of fertility and complications of reproductive cycle.

There can be little doubt that the apparent orthogenesis which pushes groups ever further along their line of evolution until, as with size in some mesozoic reptiles and armour in others, they are balanced precariously upon the edge of extinction is due, especially in its later stages, to the hypertely induced by intraspecific competition.

This conclusion is of far-reaching importance. It disposes of the notion, so assiduously rationalized by the militarists in one way and by the *laisser-faire* economists in another, that all man need to do to achieve further progressive evolution is to adopt the most thoroughgoing competition: the more ruthless the competition, the more efficacious the selection, and accordingly the better the results.... But we now realize that the results of selection are by no means necessarily "good", from the point of view either of the species or of the progressive evolution of life. They may be neutral, they may be a dangerous balance of useful and harmful, or they may be definitely deleterious.

Natural selection, in fact, though like the mills of God in grinding slowly and grinding small, has few other attributes that a civilized religion would call Divine. It is efficient in its way—at the price of extreme slowness and extreme cruelty. But it is blind and mechanical; and accordingly its products are just as likely to be aesthetically, morally, or intellectually repulsive to us as they are to be attractive. We need only think of the ugliness of *Sacculina* or a bladder-worm, the stupidity of a rhinoceros or a stegosaur, the horror of a female mantis devouring its mate or a brood of ichneumon-flies slowly eating out a caterpillar.

Both specialized and progressive improvements are mere by-products of its action, and are the exceptions rather than the rule. For the statesman or the eugenist to copy its methods is both foolish and wicked. As well might the electrical engineer copy the methods of the lightning or the heating-engineer those of the volcano. It indubitably behoves us to study the methods of natural selection, but this will be to discover how to modify and control them in new ways and, very definitely, to see what to avoid. Not only is natural selection not the instrument of a God's sublime purpose; it is not even the best mechanism for achieving evolutionary progress. An important step towards a rational applied biology will be the full analysis of the various modes of operation of selection with a view to its eventual control and its intensification for our own purposes.